Methods in Cell Biology

VOLUME 86

Stem Cell Culture

Series Editors

Leslie Wilson
Department of Molecular, Cellular and Developmental Biology
University of California
Santa Barbara, California

Paul Matsudaira
Whitehead Institute for Biomedical Research
Department of Biology
Division of Biological Engineering
Massachusetts Institute of Technology
Cambridge, Massachusetts

Methods in Cell Biology

VOLUME 86

Stem Cell Culture

Edited by
Dr. Jennie P. Mather
Raven Biotechnologies, Inc.
One Corporate Drive
South San Francisco
California, USA

AMSTERDAM • BOSTON • HEIDELBERG • LONDON
NEW YORK • OXFORD • PARIS • SAN DIEGO
SAN FRANCISCO • SINGAPORE • SYDNEY • TOKYO
Academic Press is an imprint of Elsevier

Cover Photo Credit: Spermatogonial stem cells (SSCs) are unique among stem cells in the adult in that they are the only cells that undergo self-renewal and directly transmit genes to subsequent generations. The figure illustrates the testes of infertile mice that have been transplanted with SSCs from mice transgenic for reporter genes: LacZ (upper panel) or green fluorescent protein (GFP, below). Each blue or green stretch of cells represents a colony of germ cells (spermatocytes and sperms) that has arisen from a single SSC. For details see Kubota & Brinster, Chapter 4. Photo by Dr Hiroshi Kubota.

Academic Press is an imprint of Elsevier
84 Theobald's Road, London WC1X 8RR, UK
Radarweg 29, PO Box 211, 1000 AE Amsterdam, The Netherlands
30 Corporate Drive, Suite 400, Burlington, MA 01803, USA
525 B Street, Suite 1900, San Diego, CA 92101-4495, USA

First edition 2008

Copyright © 2008 Elsevier Inc. All rights reserved

No part of this publication may be reproduced, stored in a retrieval system or transmitted in any form or by any means electronic, mechanical, photocopying, recording or otherwise without the prior written permission of the publisher

Permissions may be sought directly from Elsevier's Science & Technology Rights Department in Oxford, UK: phone (+44) (0) 1865 843830; fax (+44) (0) 1865 853333; email: permissions@elsevier.com. Alternatively you can submit your request online by visiting the Elsevier web site at http://elsevier.com/locate/permissions, and selecting *Obtaining permission to use Elsevier material*

Notice
No responsibility is assumed by the publisher for any injury and/or damage to persons or property as a matter of products liability, negligence or otherwise, or from any use or operation of any methods, products, instructions or ideas contained in the material herein. Because of rapid advances in the medical sciences, in particular, independent verification of diagnoses and drug dosages should be made

ISBN: 978-0-12-373876-9
ISSN: 0091-679X

For information on all Academic Press publications
visit our website at books.elsevier.com

Printed and bound in USA
07 08 09 10   10 9 8 7 6 5 4 3 2 1

Working together to grow libraries in developing countries

www.elsevier.com | www.bookaid.org | www.sabre.org

ELSEVIER   BOOK AID International   Sabre Foundation

CONTENTS

Contributors ... xi

Preface ... xv

Biography ... xix

1. Derivation of Human Embryonic Stem Cells in Standard and Chemically Defined Conditions

Eric Chiao, Muriel Kmet, Barry Behr, and Julie Baker

I. Introduction	1
II. Methods	2
III. Discussion	12
IV. Materials	12
References	13

2. Autogeneic Feeders for the Culture of Undifferentiated Human Embryonic Stem Cells in Feeder and Feeder-Free Conditions

Andre Choo, Ang Sheu Ngo, Vanessa Ding, Steve Oh, and Lim Sai Kiang

I. Introduction	16
II. Methods and Materials	17
III. Results	21
IV. Discussion and Summary	25
References	28

3. Microporous Membrane Growth Substrates for Embryonic Stem Cell Culture and Differentiation

Steven D. Sheridan, Sonia Gil, Matthew Wilgo, and Aldo Pitt

I. Introduction	30
II. Rationale	32
III. Methods	32
IV. Summary	54
References	55

4. Culture of Rodent Spermatogonial Stem Cells, Male Germline Stem Cells of the Postnatal Animal

Hiroshi Kubota and Ralph L. Brinster

I. Introduction	60
II. Rationale	63
III. Methods	67
IV. Materials	78
V. Discussion	80
References	81

5. Characterization of Human Amniotic Fluid Stem Cells and Their Pluripotential Capability

Laura Perin, Sargis Sedrakyan, Stafano Da Sacco, and Roger De Filippo

I. Introduction	86
II. Amniotic Fluid Cell Composition and Stem Cells	87
III. *In Vitro* Potential of Amniotic Fluid Stem Cells	91
IV. *Ex Vivo* and *In Vivo* Application of Amniotic Fluid Stem Cells	94
V. Discussion	97
References	98

6. Method to Isolate Mesenchymal-Like Cells from Wharton's Jelly of Umbilical Cord

Kiran Seshareddy, Deryl Troyer, and Mark L. Weiss

I. Introduction	102
II. Rationale	103
III. Methods	103
IV. Materials	110
V. UCMSCs and Culture Characteristics	115
VI. Discussion	116
References	118

7. Isolation, Characterization, and Differentiation of Human Umbilical Cord Perivascular Cells (HUCPVCs)

J. Ennis, R. Sarugaser, A. Gomez, Dolores Baksh, and John E. Davies

I. Introduction	122
II. Methods	123
III. Concluding Remarks	133
References	134

8. Hepatic Stem Cells and Hepatoblasts: Identification, Isolation, and *Ex Vivo* Maintenance

Eliane Wauthier, Eva Schmelzer, William Turner, Lili Zhang, Ed LeCluyse, Joseph Ruiz, Rachael Turner, M. E. Furth, Hiroshi Kubota, Oswaldo Lozoya, Claire Barbier, Randall McClelland, Hsin-lei Yao, Nicholas Moss, Andrew Bruce, John Ludlow, and L. M. Reid

I. Introduction	139
II. Epithelial–Mesenchymal Relationship	139
III. Stem Cells and Lineage Biology	140
IV. Liver Processing	150
V. Fractionation of Liver Cell Subpopulations	162
VI. Feeder Cells	170
VII. Extracellular Matrix	174
VIII. Culture Media	176
IX. Monolayer Cultures of Cells	180
X. Three-Dimensional Systems	185
Appendix 1: Nomenclature, Glossary, and Abbreviations	195
Appendix 2: Details on the Preparation of Buffers/Reagents	207
Appendix 3: Commercial Sources of Reagents	211
Appendix 4: Sources of Primary and Secondary Antibodies for Immunoselection and for Characterization of the Cells	212
References	213

9. Culture of Pluripotent Neural Epithelial Progenitor Cells from E9 Rat Embryo

Ronghao Li and Jennie P. Mather

I. Introduction	228
II. Rationale	229
III. Methods	229
IV. Materials	237
V. Discussion	237
VI. Summary	238
References	239

10. Primary and Multipassage Culture of Human Fetal Kidney Epithelial Progenitor Cells

Deryk Loo, Claude Beltejar, Jeff Hooley, and Xiaolin Xu

I. Introduction	242
II. Rationale	243
III. Methods	245
IV. Materials	250
V. Discussion	253
VI. Summary	254
References	254

11. Isolation of Human Mesenchymal Stem Cells from Bone and Adipose Tissue

    Susan H. Bernacki, Michelle E. Wall, and Elizabeth G. Loboa

I.	Introduction	258
II.	Isolation of hMSCs from Trabecular Bone	259
III.	Isolation of hMSCs from Adipose Tissue	263
IV.	Growth, Subculture, and Cryopreservation of Undifferentiated hMSCs	265
V.	Surface Marker Characterization of hMSCs	267
VI.	hMSC Differentiation Assays for Osteogenic, Adipogenic, and Chondrogenic Pathways	270
VII.	Summary	276
	References	276

12. Culture of Mesenchymal Stem/Progenitor Cells in Adhesion-Independent Conditions

    Dolores Baksh and John E. Davies

I.	Introduction	280
II.	Suspension Culture of Bone Marrow-Derived MPCs	280
III.	Analysis of Suspension Culture	283
IV.	Mechanism of MPC Expansion in Suspension	288
V.	Discussion	291
	References	292

13. Purification and Long-Term Culture of Multipotent Progenitor Cells Affiliated with the Walls of Human Blood Vessels: Myoendothelial Cells and Pericytes

    Mihaela Crisan, Bridget Deasy, Manuela Gavina, Bo Zheng, Johnny Huard, Lorenza Lazzari, and Bruno Péault

I.	Introduction	297
II.	Materials and Methods: Cell Analysis and Purification by Flow Cytometry	298
III.	Materials and Methods: Long-Term Culture of Human Blood Vessel-Associated Progenitor Cells	301
IV.	Materials and Methods: Proliferation Kinetics of Human Myoendothelial Cells and Pericytes	303
V.	Materials and Methods: Genotypic and Phenotypic Analyses of Long-Term Cultured Cells	304
VI.	Conclusion	306
	References	308

14. Isolation and Culture of Colon Cancer Stem Cells

    Patrizia Cammareri, Ylenia Lombardo, Maria Giovanna Francipane, Sebastino Bonventre, Matilde Todaro, and Giorgio Stassi

I.	Introduction	312
II.	Identification of CSCs in Colon Cancer Tissue	314

Contents

    III. Isolation and Propagation of Colon CSCs   315
    IV. Discussion   322
        References   323

15. Isolation and Establishment of Human Tumor Stem Cells

Penelope E. Roberts

    I. Introduction   326
    II. Rationale   327
    III. Methods   327
    IV. Results   336
    V. Discussion   338
    VI. Materials   340
        References   341

16. Stem Cells from Cartilaginous and Bony Fish

David W. Barnes, Angela Parton, Mitsuru Tomana Jae-Ho Hwang, Anne Czechanski, Lanchun Fan, and Paul Collodi

    I. Introduction to Models and Uses   344
    II. Embryonal Stem Cells   348
    III. Tissue-Specific Stem Cells   353
    IV. Outlook and Future Contributions   361
        References   362

Index   369
Volumes in Series   379

CONTRIBUTORS

Numbers in parentheses indicate the pages on which the authors' contributions begin.

Julie Baker (1) Department of Genetics, Stanford University School of Medicine, Stanford, California 94035

Dolores Baksh (121, 279) Organogenesis Inc., Canton, Massachusetts 02021

Claire Barbier (137) Departments of Cell and Molecular Physiology, UNC School of Medicine, Chapel Hill, North Carolina 27599

David W. Barnes (343) Mount Desert Island Biological Laboratory, Salisbury Cove, Maine

Barry Behr (1) Department of Obstetrics and Gynecology, IVF/ART Laboratories, Stanford University, Stanford, California 94062

Claude Beltejar (241) Raven Biotechnologies Inc., South San Francisco, California 94080

Susan H. Bernacki (257) Joint Department of Biomedical Engineering at NC state University and UNC, Chapel Hill Burlington Laboratories, Raleigh, North Carolina 27695

Sebastiano Bonventre (311) Department of Surgical and Oncological Sciences, University of Palermo, 5 – 90127 Palermo, Italy.

Ralph L. Brinster (59) Department of Animal Biology, School of Veterinary Medicine, University of Pennsylvania, Philadelphia, Pennsylvania 19104

Andrew Bruce★ (137) Vesta Therapeutics, 801-8 Capitola Drive, Suite 801, Durham, North Carolina 27713

Patrizia Cammareri (311) Department of GENURTO, University of Palermo, 129–90127 Palermo, Italy

Eric Chiao (1) Department of Genetics, Stanford University School of Medicine, Stanford, California 94035

Andre Choo (15) Stem Cell Group, Bioprocessing Technology Institute, Agency for Science Technology and Research, Singapore 138668

Paul Collodi (343) Department of Animal Sciences, Purdue University, West Lafayette Indiana

Mihaela Crisan (295) Hillman Cancer Center, University of Pittsburgh Cancer Institute, Pittsburgh, Pennsylvania 15213, Stem Cell Research Center, Children's Hospital of Pittsburgh of UPMC, Pittsburgh, Pennsylvania 15213, and Department of Pediatrics, Children's Hospital of Pittsburgh of UPMC, Pittsburgh, Pennsylvania 15213

★ *Current address*: Tengion Laboratories, 3929 Westpoint Blvd., Suite G, Winston-Salem, NC 27103

Anne Czechanski (343) Mount Desert Island Biological Laboratory, Salisbury Cove, Maine

John E. Davies (121, 279) Faculty of Dentistry, Institute of Biomaterials and Biomedical Engineering, University of Toronto, Ontario, Canada M5S 3G9

Bridget Deasy (295) Stem Cell Research Center, Children's Hospital of Pittsburgh of UPMC, Pittsburgh, Pennsylvania 15213, and Departments of Orthopaedic Surgery and Bio-Engineering, University of Pittsburgh, Pennsylvania 15213

Vanessa Ding (15) Stem Cell Group, Bioprocessing Technology Institute, Agency for Science Technology and Research, Singapore 138668

Jane Ennis (121) Tissue Regeneration Therapeutics, Suite 512, Toronto, ON M5G 1N8

Lanchun Fan (343) Department of Animal Sciences, Purdue University, West Lafayette Indiana

Roger De Filippo (85) Childrens Hospital Los Angeles, Saban Research Institute, Developmental Biology Program, Keck School of Medicine, University of Southern California.

M. E. Furth (137) Institute for Regenerative Medicine, Wake Forest Baptist Medical Center, Winston Salem, North Carolina 27157

Manuela Gavina (295) Department of Orthopaedics, University of Pittsburgh School of Medicine, Pittsburgh, Pennsylvania 15213, and Stem Cell Research Center, Children's Hospital of Pittsburgh of UPMC, Pittsburgh, Pennsylvania 15213

Sonia Gil (29) Millipore Corporation, Bioscience Division, Danvers, Massachusetts 01923

Maria Giovanna Francipane (311) Department of Surgical and Oncological Sciences, University of Palermo, 5–90127 Palermo, Italy

A. Gomez (121) Institute of Biomaterials and Biomedical Engineering, University of Toronto, Toronto, ON M5S 3G9, Canada

Jeff Hooley (241) Raven Biotechnologies Inc., South San Francisco, California 94080

Jae-Ho Hwang (343) Mount Desert Island Biological Laboratory, Salisbury Cove, Maine

Johnny Huard (295) Department of Orthopaedics, and Stem Cell Research Center University of Pittsburgh School of Medicine, Pittsburgh, Pennsylvania 15213

Lim Sai Kiang (15) Institute of Medical Biology, Agency for Science Technology and Research, Singapore 138673

Muriel Kmet (1) Department of Genetics, Stanford University School of Medicine, Stanford, California 94035

Hiroshi Kubota[†] (59, 137) Departments of Cell and Molecular Physiology, UNC School of Medicine, Chapel Hill, North Carolina 27599.

Lorenza Lazzari (295) Fondazione Ospedale Maggiore Policlinico, Department of Regenerative Medicine, Milan, Italy

Ed LeCluyse (137) CellzDirect, Inc., Pittsboro, North Carolina 27312

Ronghao Li (227) Research and Development, Raven Biotechnologies, Inc., South San Francisco, California 94080

[†] *Current address*: Laboratory of Cell and Molecular Biology, Department of Animal Science, School of Veterinary Medicine, Kitasato University, Towada, Aomori 034-8628, Japan

Elizabeth G. Loboa (257) Joint Department of Biomedical Engineering at NC state University and UNC, Chapel Hill Burlington Laboratories, Raleigh, North Carolina 27695

Ylenia Lombardo (311) Department of Surgical and Oncological Sciences, University of Palermo, 5–90127 Palermo, Italy

Deryk Loo (241) Raven Biotechnologies Inc., South San Francisco, California 94080

Oswaldo Lozoya (137) Joint Department of Biomedical Engineering at NCSU and UNC School of Medicine, Chapel Hill, North Carolina 27599

John Ludlow[‡] (137) Vesta Therapeutics, 801-8 Capitola Drive, Suite 801, Durham, North Carolina 27713

Jennie P. Mather (227) Research and Development, Raven Biotechnologies, Inc., South San Francisco, California 94080

Randall McClelland[$] (137) Departments of Cell and Molecular Physiology, UNC School of Medicine, Chapel Hill, North Carolina 27599

Nicholas Moss (137) Departments of Cell and Molecular Physiology, UNC School of Medicine, Chapel Hill, North Carolina 27599

Ang Sheu Ngo (15) Stem Cell Group, Bioprocessing Technology Institute, Agency for Science Technology and Research, Singapore 138668

Steve Oh (15) Stem Cell Group, Bioprocessing Technology Institute, Agency for Science Technology and Research, Singapore 138668

Bruno Péault (295) McGowan Institute for Regenerative Medicine, Pittsburgh, Pennsylvania 15219, Stem Cell Research Center, Children's Hospital of Pittsburgh of UPMC, Pittsburgh, Pennsylvania 15213, and Department of Pediatrics, Children's Hospital of Pittsburgh of UPMC, Pittsburgh, Pennsylvania 15213

Angela Parton (343) Mount Desert Island Biological Laboratory, Salisbury Cove, Maine

Laura Perin (85) Childrens Hospital Los Angeles, Saban Research Institute, Developmental Biology Program, Keck School of Medicine, University of Southern California

Aldo Pitt (29) Millipore Corporation, Bioscience Division, Danvers, Massachusetts 01923

L. M. Reid (137) Departments of Cell and Molecular Physiology, Joint Department of Biomedical Engineering, Program in Molecular Biology and Biotechnology, UNC School of Medicine, Chapel Hill, North Carolina 27599, and Institute for Regenerative Medicine, Wake Forest Baptist Medical Center, Winston Salem, North Carolina 27157

Penelope E. Roberts (325) Raven Bio-technologies, Inc., South San Francisco, California 94080

Joseph Ruiz (137) Vesta Therapeutics, 801-8 Capitola Drive, Suite 801, Durham, North Carolina 27713

Stafano Da Sacco (85) Childrens Hospital Los Angeles, Saban Research Institute, Developmental Biology Program, Keck School of Medicine, University of Southern California.

[‡] *Current address*: Tengion Laboratories, 3929 Westpoint Blvd. Suite G, Winston-Salem, NC 27103
[$] *Current address*: Admet Technologies, 801-8 Capitola Drive, Suite 801, Durham, North Carolina 27713

R. Sarugaser (121) Institute of Biomaterials and Biomedical Engineering, University of Toronto, Toronto, ON M5S 3G9, Canada

Eva Schmelzer (137) Departments of Cell and Molecular Physiology, UNC School of Medicine, Chapel Hill, North Carolina 27599.

Sargis Sedrakyan (85) Childrens Hospital Los Angeles, Saban Research Institute, Developmental Biology Program, Keck School of Medicine, University of Southern California.

Kiran Seshareddy (101) Department of Anatomy and Physiology, College of Veterinary Medicine, Kansas State University, Manhattan, KS 66506

Steven D. Sheridan (29) Millipore Corporation, Bioscience Division, Danvers, Massachusetts 01923

Giorgio Stassi (311) Department of Surgical and Oncological Sciences, University of Palermo, 5–90127 Palermo, Italy

Matilde Todaro (311) Department of Surgical and Oncological Sciences, University of Palermo, 5–90127 Palermo, Italy

Mitsuru Tomana (343) Mount Desert Island Biological Laboratory, Salisbury Cove, Maine

Deryl Troyer (101) Department of Anatomy and Physiology, College of Veterinary Medicine, Kansas State University, Manhattan, KS 66506

Rachael Turner (137) Joint Department of Biomedical Engineering at NCSU and UNC School of Medicine, Chapel Hill, North Carolina 27599

William Turner[¶] (137) Joint Department of Biomedical Engineering at NCSU and UNC School of Medicine, Chapel Hill, North Carolina 27599

Michelle E. Wall (257) Joint Department of Biomedical Engineering at NC state University and UNC, Chapel Hill Burlington Laboratories, Raleigh, North Carolina 27695, and Flexcell International corporation, Hillsborough, NC 27278

Elaine Wauthier (137) Departments of Cell and Molecular Physiology, UNC School of Medicine, Chapel Hill, North Carolina 27599

Mark L. Weiss (101) Department of Anatomy and Physiology, College of Veterinary Medicine, Kansas State University, Manhattan, KS 66506

Matthew Wilgo (29) Millipore Corporation, Bioscience Division, Danvers, Massachusetts 01923

Xiaolin Xu (241) Raven Biotechnologies Inc., South San Francisco, California 94080

Hsin-lei Yao[∥] (137) Joint Department of Biomedical Engineering at NCSU and UNC School of Medicine, Chapel Hill, North Carolina 27599.

Lili Zhang[#] (137) Departments of Cell and Molecular Physiology, UNC School of Medicine, Chapel Hill, North Carolina 27599

Bo Zheng (295) Stem Cell Research Center, Children's Hospital of Pittsburgh of UPMC, Pittsburgh, Pennsylvania 15213

[¶] *Current address*: 29 Levant Street, San Francisco, Cal. 94114
[∥] *Current address*: Dr. Stephen Crews, Department of Biochemistry and Biophysics, Program in Molecular Biology and Biotechnology, 321 Fordham Hall, UNC School of Medicine, Chapel Hill, North Carolina 27599
[#] *Current address*: Department of Infectious Diseases, The First Affiliated Hospital of Nanjing Medical Center, Nanjing, PR of China 210029.

PREFACE

Studies of development, physiology, and cell biology have historically benefited from the ability to culture cells outside the body in *in vitro* systems. This has been true from the first tissue cultures done at the turn of the 20th century by Alexis Carrel and Ross Harrison through the establishment of functional hormone-secreting tumor lines mid-century by Gordon Sato and his colleagues. Further advances in culturing a wide variety of cell types were dependent on work done by Charity Waymouth, Hayden Coon, Renato Dulbecco, and Richard Ham on improving the nutrient portion of cell culture media and the discovery, purification, and availability of a large number of hormones, growth factors, and attachment factors so that the serum could be reduced, and eventually eliminated for many cell types, in the last decades of the 20th century. Thus, the very process of learning how to keep cells alive and functioning *in vitro* has taught us a great deal about cell biology and animal physiology.

Interestingly, stem cells have been among the last of the cell types to be routinely cultured. This seems to be due to the complex biological requirements of these cells, which interact closely with each other and their own microenvironment, or niche. In addition, keeping stem cells in culture requires not only providing positive signals for growth and survival, but also removing signals driving the cells toward differentiation. Finally, at least in the adult, tissue stem cells, or progenitor cells, are a minor component of the tissue and thus present a challenge in identifying and isolating them for culture. The chapters in this book reflect these challenges.

This book strives to provide methods for the isolation and culture of representative stem or progenitor cells from embryonic, fetal, adult, and cancer tissues. It is neither exhaustive nor definitive since this is an exciting and rapidly evolving field. Most of the examples are taken from the culture of human stem and progenitor cells, since this is a strong focus for the development of stem cell therapies, and also the most rapidly advancing part of this field. Two examples focus on rodent stem cell cultures, those of very early neural plate neural progenitor cells, which come from a developmental stage difficult to obtain in humans and the spermatogonial stem cell cultures derived from newborn testis, again a tissue difficult to obtain. We also present methods for culturing fish and invertebrate stem cells at the end of the book. However, some of the principles in these methods will also apply to the culture of other stem cells from other species, including humans. In fact, the aim of the book is to allow the reader to not only have at their command detailed methods for the isolation and culture of the cell types mention explicitly but also to extract general principles that will become a good starting point for the culture of other stem and progenitor cell types not yet published.

In recognition of the difficulty of isolating the minor stem cell component, there is a considerable amount of effort dedicated to a description for methods of isolating each of the stem cell types discussed. These vary from manual dissection of the stem cells as in embryonic stem (ES) cell lines to using the medium itself to select for stem cells and against the non-stem components of the tissues as demonstrated in the last three chapters. The initial three chapters deal with the isolation of new ES cells from mice and humans, the use of self-feeder layers generated from the human ES cells themselves to avoid xenoprotein contamination, and finally, the growth of ES cells on filters to aid in the separation of ES cells from feeders, or conditioned medium, and observation of the cells.

The second section deals with the culture of cells from amniotic and fetal tissue, frequently a richer source of tissue stem or progenitor cells. One chapter discusses the use of spermatogonial stem cells as a source of totipotent stem cells that allow direct re-introduction into the germ line. This is followed by chapters on isolation of stem-like cells from amniotic fluid and from Wharton's jelly and umbilical cord perivascular cells from the discarded placenta and umbilical cord. These sources clearly provide the advantage in that no destruction of an embryo is required for the isolation of stem cells.

These chapters are followed by a discussion of the isolation of tissue stem or progenitor cells from fetal and adult tissues. The fetal source tissues discussed are rodent and human liver, rodent neural stem cells (mentioned above), and human kidney. In these chapters, the role of calcium in selecting for the stem cell component is emphasized, as well as the use of conditioned medium from non-fibroblast fetal cell types. These chapters are followed by chapters detailing a series of methods that deal with adult mesenchymal stem cell isolation and growth from a number of sources including adipose tissue, muscle, and bone. Isolation of stem cells from adult pericytes is also described.

The last part of the book deals with the culture of solid tumor stem cells. This is a rapidly growing field where the majority of the current literature uses partially purified populations of tumor cells enriched for stem cells. This field could clearly benefit from the establishment of long-term, purified, and well-characterized cancer stem cell cultures.

Finally, the chapter on stem cells from invertebrates and fish covers several species, including zebrafish and fugu—two systems that are well characterized genetically. This should be of interest for all of those who are using, or might consider using, these systems as models for studying development. In addition, some of the culture techniques and approaches might be applicable to culturing other types of stem cells.

In general, several themes have emerged from this volume which should be considered by those investigators involved in all stem cell culture. First, stem cells do not like growing on plastic surfaces so a feeder layer, attachment to attachment factors or matrix proteins, or growth as spheres, where the cells in the sphere provide the attachment, is important. Second, stem cells like to remain in close communication with other cells so feeder layers, or conditioned medium, are frequently helpful and conditioning by other cell types than fibroblasts might prove useful. Keeping the cells at a high density and not completely dissociating them when plating or passage also seems to be a common theme. Finally, the nutrient needs and the growth factors and hormones required to keep stem cells viable, replicating, and in an undifferentiated state will be different from those required for more differentiated cells. The ability to grow cells without serum is adding to our understanding of these requirements, as well as removing serum, per se being a requirement for some stem cell types.

In summary, this volume should provide a good overview for anyone doing stem cell work, particularly with human tissues, and detailed protocols for culturing a range of stem cells from embryonic, fetal, adult, and cancerous tissues.

Jennie P. Mather, PhD

BIOGRAPHY

Dr. Mather established Raven on pioneering research performed in her laboratory at Genentech. With more than 30 years of experience in cell culture and cell biology research, Dr. Mather is a recognized leader in the application of cell biology to technology and pharmaceutical product development. She has unusually broad experience that spans basic research in cancer biology and reproductive endocrinology as an Assistant Professor at The Rockefeller University to applied research in development and product discovery at Genentech. Prior to founding Raven, Dr. Mather was a Staff Scientist for 15 years at Genentech, engaged in all phases of drug discovery and development, from project conception through scale-up and the development of potential new products. Dr. Mather led or participated in 12 project teams that produced a number of Genentech's marketed products, including Herceptin®, a monoclonal antibody for treatment of patients with metastatic breast cancer; Activase®, a biosynthetic form of the human tissue plasminogen activator (t-PA) for treatment of heart attack, acute ischemic stroke, and acute massive pulmonary embolism; Pulmozyme®, an inhalation solution for management of cystic fibrosis; and Genentech's anti-IgE antibody currently in late-stage clinical trials for asthma. Dr. Mather's work led to a number of breakthroughs in cell technology and several key patents for Genentech, including serum-free media for the commercial production of t-PA and other products, genetically engineered production cell lines, several tissue progenitor cell lines, and use of patents on several Genentech pipeline products. She also contributed to the design of the cell culture biomanufacturing processes used for commercial production of four of Genentech's marketed protein therapeutics. Dr. Mather is an inventor of 30 issued patents, the author of more than 150 publications, and the author or editor of five books on animal cell culture. She is on the board of directors of Healthcare Businesswomen's Association and serves on the scientific advisory board of Springboard Enterprises as well as two bioscience companies. Dr. Mather is the recipient of the First Innovator of the Year Award from the Healthcare Businesswomen's Association (2003) and was named as one of the Top 10 Innovators for scientific and business aptitude by Red Herring Magazine (2002). She received a PhD from the University of California, San Diego, and was an NIH-INSERM exchange scientist in Lyon, France.

Jennie P. Mather, PhD (Founder)
President and Chief Scientific Officer

CHAPTER 1

Derivation of Human Embryonic Stem Cells in Standard and Chemically Defined Conditions

Eric Chiao,[*] Muriel Kmet,[*] Barry Behr,[†] and Julie Baker[*]

[*]Department of Genetics
Stanford University School of Medicine
Stanford, California 94035

[†]Department of Obstetrics and Gynecology
IVF/ART Laboratories
Stanford University
Stanford, California 94062

I. Introduction
II. Methods
    A. Feeder Preparation
    B. Blastocyst Preparation
    C. Passage and Expansion
    D. Cryopreservation of hESCs
    E. Characterization
III. Discussion
    A. What to Expect
IV. Materials
    A. Media and Reagents
    B. Miscellaneous Equipment
    References

I. Introduction

Human embryonic stem cells (hESCs) are derived from the inner cell mass (ICM) of *in vitro* fertilized embryos, which are donated after consent from both maternal and paternal gametes. We, and others, have used cryopreserved and subclinical grade embryos to successfully derive hESC lines. Derivation and subsequent maintenance of hESCs has been described using numerous conditions, including a variety of growth substrates and several types of feeder and nonfeeder systems (Chen *et al.*, 2005; Cowan *et al.*, 2004; Fletcher *et al.*, 2006; Hovatta and Skottman, 2005; Kim *et al.*, 2005; Klimanskaya *et al.*, 2005, 2006; Lee *et al.*, 2005; Mateizel *et al.*, 2006; Oh *et al.*, 2005; Simon *et al.*, 2005; Thomson *et al.*, 1998; Zhang *et al.*, 2006). Recently, progress has been made toward deriving hESCs using chemically defined, xeno-free media, which may expedite further therapeutic approaches (Chen *et al.*, 2007; Ellerstrom *et al.*, 2006; Fletcher *et al.*, 2006; Genbacev *et al.*, 2005; Hong-mei and Gui-an, 2006; Hovatta, 2006; Inzunza *et al.*, 2005; Ludwig *et al.*, 2006). This chapter provides stepwise methodology using different conditions and techniques in order to promote the derivation and maintenance of hESCs, including conditions that support the culture of human blastocysts and the basic characterization of newly established lines. What is not provided is an in-depth analysis of the legal, moral, and ethical challenges that face the researcher when pursuing these endeavors. We exclude this discussion since it is greatly dependent upon local governmental restrictions. However, we emphasize that researchers need to be well acquainted with the legal and ethical standards governing their countries, states, and institutions and refer readers to the National Academy of Sciences (NAS) and California Institute of Regenerative Medicine (CIRM) guidelines for establishing patient consents and ethical standards. Furthermore, derivation of hESCs requires approval from an authorized Institutional Review Board (IRB).

II. Methods

A. Feeder Preparation

If feeders are used, they must be mitotically inactivated either by treating with mitomycin C or by irradiation. We have successfully derived hESC lines using both irradiated human foreskin fibroblasts (ATCC) and mitomycin C-treated e12.5 primary embryonic fibroblast from CF1 mice (Charles River).

1. Preparation of primary mouse embryonic fibroblasts

On the bench: Observe clean technique.

Sterilize all instruments needed for the surgery.

1. Sacrifice pregnant females 12 days after the morning the plugs were observed (e12.5).
2. Swab the abdomen with 70% ethanol.
3. Grasp the skin of the abdomen with forceps and make a large incision through the skin with scissors.

1. Derivation of Human Embryonic Stem Cells

4. Grasp the body wall with fine forceps and make an incision with scissors.
5. Remove the uterus with blunt forceps.
6. Embryos can be released from the uterus by inserting the tip of fine-tipped scissors into the anterior end of one of the uterine horns, then cutting along the length of the uterus.
7. Release each embryo and place them into a 10 cm dish with 1× PBS, phosphate-buffered saline.
8. Remove the yolk sac and placenta and place the embryo into a new 10 cm dish with PBS.
9. Remove the head and the intestines with forceps (all the red parts of the embryos) and place into a new 10 cm dish with PBS.
10. Transfer the dish to a laminar flow hood.

Under the Hood: Observe sterile technique.

11. Wash twice with PBS by transferring each embryo from the same litter from one 1× PBS dish to new dish with sterile 1× PBS.
12. Transfer the embryos to a new 10 cm dish with 5 ml of .5% trypsin–(ethylenediaminetetraacetic acid) EDTA.
13. Mince the embryos with sterilized scissors.
14. Pipette up and down to break down the embryos into single cells.
15. Leave the 5 ml trypsin/minced embryos dish tilted into the 37 °C incubator for a 5-min incubation.
16. Bring back the dish under the hood and pipette vigorously to disperse the cells.
17. Add fibroblast dissection media to neutralize the trypsin. Plate ~2 embryos per T175 flask. If desired, the largest clumps of tissue can be settled out in a 50 ml conical tube prior to plating.
18. Change the media the day after plating. Incubate flasks until confluent, usually 1–2 days.
19. When the T175 are confluent, freeze down 2 vials/T175. This is passage 1.

2. Mitotically inactivating feeders

Before mitotically inactivating the cells, we typically expand the cells to the largest number of flasks we can manage, typically around 36–48 T175 flasks per person performing the prep. In addition, we try to inactivate primary mouse embryonic fibroblasts by passage 4. Therefore, for a routine prep, we would thaw 2 vials of primary mouse embryonic fibroblasts into 2 T175 flasks (P1), pass 1:3 to 6 flasks (P2), pass 1:3 to 18 flasks (P3), then pass 1:2 into 36 flasks (P4). We grow the cells in media containing antibiotics for the first two passages and switch to antibiotic-free media upon the third passage.

A. Mitotically inactivating feeders with mitomycin C.
   1. Remove media from a confluent flask of cells.
   2. Add 10 ug/ml mitomycin C in media to the cells.
   3. Incubate for 2.5 h.
   4. Wash with 1× PBS
   5. Trypsinize the cells.
   6. Spin down.
   7. Wash the cell pellet with media. Spin down.
   8. Repeat step #6 twice.
   9. Resuspend in media and count the cells. Spin down.
   10. Resuspend the cells in freezing media at a convenient density (i.e., 10^6 cells/ml).
   11. Aliquot the cells into cryovials and freeze.

B. Mitotically inactivating feeders by irradiation
   1. Wash with 1× PBS.
   2. Trypsinize the cells.
   3. Irradiate the cells to 6000 rads.
   4. Count the cells. Spin down.
   5. Resuspend the cells in freezing media at a convenient density (i.e. 10^6 cells/ml).
   6. Aliquot the cells into cryovials and freeze.

3. Plating feeders to support hESCs growth
   1. Gelatinize tissue culture grade dishes with 0.25% gelatin for 30 min.
   2. Thaw a vial of mitotically inactivated feeders and resuspend in at least 5 volumes of feeder media.
   3. Pellet the cells in a clinical centrifuge at low speed (~1000 rpm) for 1 min.
   4. Remove supernatant. Resuspend the cells and plate at a density between 3×10^4 and 5×10^4 cells per cm^2.
   5. Incubate the feeder cells overnight before plating hESCs.
   6. Before plating hESCs, remove feeder media. Wash two times with 1× PBS then add hESC media.

B. Blastocyst Preparation

Basic formulations for human embryo culture media are composed of a balanced salt solution, carbohydrates, and a protein source. While a single culture medium can support the development of human embryos up to the cleavage stage, blastocyst formation and viability is poor. Increasing understanding of both the physiological requirements of the embryo as it develops from the zygote to the blastocyst stage and changes in metabolite concentrations (including glucose, pyruvate, and lactate), in the reproductive tract in the human female at different stages of the menstrual cycle, has led to the development of sequential culture media (Behr, 1999; Behr et al., 1999; Gardner et al., 1996). In accordance with the switch from pyruvate-based to glucose-based metabolism, blastocyst culture media contain a higher concentration of glucose and lower levels of pyruvate and lactate when compared to culture medium for cleavage stage embryos. In humans, the embryonic genome is activated between the 4-cell and 8-cell stages. Accordingly, cleavage stage embryos at 4–8-cell stage on day 3 after insemination are transferred to blastocyst medium for an additional 2–3 days for blastocyst formation.

The availability of sequential culture media also provides an opportunity to produce viable blastocysts for derivation of hESC lines. The commercially available sequential culture media includes G1/G2 (Scandinavian IVF Science, Gothenburg, Sweden), Sage Fertilization/Cleavage/Blastocyst Media (CooperSurgical, Trumbull, CT), P1/Blastocyst Medium (Irvine Scientific, Santa Ana, CA), Sydney IVF Cleavage/Blastocyst Media (Cook, Brisbane, Australia), and ISM1/ISM2 (Medicult, Denmark). Potential sources of human embryos for stem cell research include excess embryos donated by the patients typically after they have been cryopreserved for greater than 1 year; those embryos unsuitable for embryo transfer or cryopreservation which would otherwise be discarded; those that have been identified by preimplantation genetic diagnosis as carrying genetic defects.

We have derived hESC lines using both fresh, subclinical grade embryos and thawed, cryopreserved embryos. Although immunosurgery is not required to derive hESC lines, derivation rates may be improved if the ICM is first isolated by immunosurgery for blastocysts with prominent ICMs. Finally, in our experience the embryo quality is not an absolute predictor of the ability of the embryo to give rise to an hESC line. Embryos with perfectly formed ICMs often fail to give rise to new lines while embryos with no visible ICM have been used to establish lines.

1. Thawing Cryopreserved Embryos

   1. Embryo thawing solutions: The basal thawing medium is HEPES (4-(2-hydroxyethyl)-1-piperazineethanesulfonic acid)-buffered human tubal fluid medium (mHTF) or similar medium supplemented with 20% "protein source."
   2. Cryostraws with frozen day 3 or day 5 embryos are removed from liquid nitrogen and exposed in air for 30 s, followed by 30 °C water bath for 45 s.
   3. Expel embryos into 0.5 M sucrose solution. Leave for 10 min.
   4. Transfer embryos to the 0.2 M sucrose solution. Leave for 10 min.
   5. Wash embryos in mHTF-20% protein source, then transfer to blastocyst medium with 20% protein source and further culture at 37 °C with 5% CO_2, 5% O_2, and 90% N_2 until use.
   6. If the thawed embryos are frozen on day 3, they will be cultured in blastocyst medium with 10% protein source for additional 2–3 days, when the embryos will develop to the blastocyst stage.

2. Removal of the Zona Pellucida

Before immunosurgery to isolate the ICM or direct plating of the blastocyst, the blastocyst must be liberated from the transparent glycoprotein layer surrounding the blastocyst called the zona pellucida. Assisted hatching by making a hole on the zona pellucida of cleavage-stage embryos can help the subsequent blastocyst hatch out from the zona. However, assisted hatching should be performed under an inverted microscope. In general, assisted hatching can be performed in three ways: chemically (using acidic Tyrode's solution), mechanically (partial zona dissection), and by recently introduced laser technology.

A. Assisted hatching by acidic Tyrode's solution

1. Use a Falcon 1006 dish, pipette 10 µl of acidic Tyrode's solution, and make four 5–10 µl drops of mHTF or PBS + 10% protein source into the center of the dish. Cover the drops with warm mineral oil and place it on a warmed surface until ready to use.

2. Gently aspirate the embryo on the holding pipette (80–120 µm in diameter) at 9 o'clock in such a way that the acidic Tyrode's solution filled microneedle at 3 o'clock is opposed to an area of empty perivitelline space or to cellular fragments.

3. Lower the acidic Tyrode's solution filled needle (10–12 µm) and gently expel acidified solution over a small 35–50 µm area while holding the needle tip immediately next to the zona. The size of the hole should be about 1/2–2/3 the diameter of a blastomere of an 8-cell stage embryo. Use small circular motions to avoid excess acid in a single region.

4. Immediately cease expulsion of acidic medium when the inside of the zona is pierced and move the embryo away from the acidic medium to the opposite side of the drop.

5. At the completion of assisted hatching, the embryos are washed through 3 × 100 µl drops of cleavage medium and then placed back in their culture drops in the CO_2 incubator.

B. Assisted hatching by laser

1. Place embryos to be hatched in a separated drop in the culture dish. Mark with a circle on the bottom of the culture dish.

2. Set up infrared 1.48-µm diode laser (Hamilton Thorne Research, Beverly, MA) and then locate an unfertilized oocyte or abnormally fertilized embryo to use for laser calibration.

3. The embryo is positioned to place a region of the zona pellucida on the aiming spot. Fire laser once through the width of the zona in an area between blastomeres in order to reduce any heat damage to the cells.

4. It is important to perforate the zona completely without harming the embryonic cells with the laser shot. After embryos are hatched return the culture dish to the CO_2 incubator.

C. Assisted hatching by partial zona dissection (Cieslak *et al.*, 1999)

   1. Microtool requirements consist of one holding pipette and one microneedle. The embryo is held loosely by gentle suction from the holding pipette and the microneedle is passed through the zona pellucida at the largest perivitelline space and advanced tangentially.

   2. Release the embryo from the holding pipette and hold it by the microneedle. The microneedle is brought to the bottom of the holding pipette and pressed to it, pinching a portion of the zona pellucida.

   3. By gently rubbing the microneedle against the holding pipette with a sawing motion, the first cut is made and the embryo is released.

   4. Once again, the embryo is rotated with the microneedle, vertically, until the slit is clearly visible at the 12-o'clock position as a dark vertical line in the zona pellucida. The embryo is held firmly by suction from the holding pipette and a second cut is made in a similar manner creating a cross-shaped slit.

   5. The embryo is then transferred back to the culture medium for further culture.

D. Zona removal by acidic Tyrode's solution

   1. Expose blastocyst to acidic Tyrode's solution (Irvine Scientific, Santa Ana, CA; pH 2.1–2.5) for 10 s followed by immediate repeated rinsing in the culture medium.

   2. Remove the embryo immediately and rinse several times in the culture medium to get rid of any traces of the acidic Tyrode's solution.

   3. In order to avoid unnecessary exposure to the acidic solution, care must be taken not to rupture the zona completely during the manipulation.

E. Zona removal by pronase

   1. Another option for zona removal is enzymatic treatment (Fong *et al.*, 1998). Transfer blastocyst to 0.5% pronase (Sigma, St. Louis, MO) in the culture medium under oil for 1–2 min at 37 °C in a 5% CO_2 atmosphere.

   2. Monitor zona expansion and the increase in perivitelline space on the heated stage under an inverted microscope at 400×.

   3. Most of the embryos show stretching and softening of the zona by 1 min. If not, incubate the embryo for a further 30–60 s.

   4. Transfer the blastocyst to the culture medium and gently wash it several times just before the complete disappearance of the faint zona.

   5. Culture zona-free blastocyst in blastocyst medium with 10% SSS until use.

3. Immunosurgery

   1. Wash a hatched blastocyst 2 times in antihuman serum antibody 1:5 in blastocyst media.
   2. After the second wash, place the blastocyst in a new drop of antibody mix and incubate for 30 min.
   3. Wash 3 times in fresh, prewarmed blastocyst media.

4. Wash 3 times in guinea-pig complement serum.
5. Incubate for 10 min in guinea-pig complement serum.
6. Wash 3 times extensively in blastocyst media.
7. Plate the isolated ICM clump.

C. Passage and Expansion

Either intact blastocysts or ICMs dissected by immunosurgery can be plated. If intact blastocysts are plated, mechanically dissecting away trophoblast cells from the initial outgrowth or dissecting the primary outgrowth into multiple clumps on 3–4 days after plating may be beneficial (see Fig. 1 for outgrowth appearance).

Approximately once per week, the outgrowths are passed to a fresh plate of feeders. Colonies can be mechanically dissected using borosilicate Pasteur pipettes that have been pulled into fine needles and blunted over a flame. Mechanical passage is best performed within a laminar flow hood outfitted with an inverted phase contrast microscope. When deriving new lines, we pass the colonies mechanically until we have expanded the line to enough cells to populate a 35 mm or 60 mm dish, after which the lines may be passed either mechanically or enzymatically.

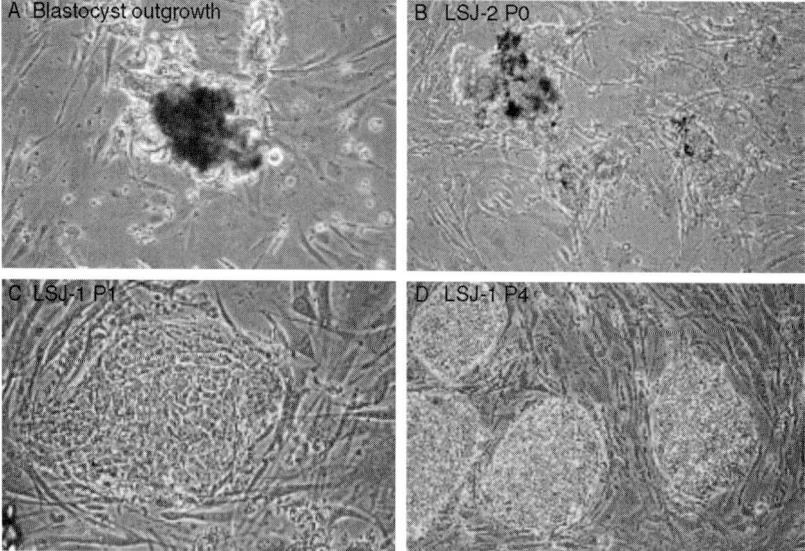

Fig. 1 Appearance of human embryonic stem cells (hESCs) under phase contrast microscope. (A) Initial outgrowths from blasted blastocysts take a couple of days to completely plate down and spread out. (B) Dissecting the initial outgrowth on the original plate before the first passage may be beneficial. hESC-like clumps can be seen on the right side of the picture. (C) Usually, hESC colonies are recognizable by the first passage. (D) hESC colonies are characterized by well-defined borders with small, tightly packed cells with a high nucleus to cytoplasm ratio and prominent nucleoli. (See Plate no. 1 in the Color Plate Section.)

Several hESC media variations based on Thomson's original recipe have been reported to successfully support the growth of hESCs (Genbacev et al., 2005; Hong-mei and Gui-an, 2006; Inzunza et al., 2005; Thomson et al., 1998). Most use Knockout Serum Replacer from Invitrogen and FGF2 (basic FGF, fibroblast growth factor). Our standard hESC media recipe is a modification of the media recommended by WiCell (http://www.wicell.org) with twice as much FGF2 (8 ng/ml vs 4 ng/ml). Since Knockout Serum Replacer contains bovine serum albumin, efforts have been made to develop completely xeno-free media for growing hESCs. Recently, xeno-free, chemically defined media has been used to derive hESCs (Ludwig et al., 2006). We have successfully derived two new hESC lines using a synthetic media (SM-1, described in Section IV) that is a further modification of Ludwig's 2006 recipe (Chiao and Baker, unpublished data).

1. Mechanical passage

    1. Heat the middle of a sterile borosilicate Pasteur pipette over a flame. Stretch the pipette so that it forms a narrow bore where heated.
    2. Bend the pipette, snapping the glass where it narrows so that the end creates a needle with a fine point. Polish the end of the needle over a flame.
    3. Using either a single needle or a pair of needles, cut the colonies into ~2–4 pieces.
    4. Transfer dissected colonies to a new plate with fresh feeders.

hESCs can be successfully passaged with collagenase, dispase, trypsin, and TrypLE. When we pass the cells enzymatically, we routinely use collagenase as described below.

2. Enzymatic passage

    1. Remove the old media.
    2. Add fresh media with 200 units/ml collagenase in hESC media.
    3. Incubate for 5–15 min at 37 °C. hESC colonies will not detach but feeders may begin to detach from the surface of the dish.
    4. Scrape the cells off the surface of the dish with a sterile cell scraper.
    5. Break up the colonies by pipetting up and down semivigorously in collagenase solution.
    6. Spin down and remove the collagenase solution.
    7. Wash 3 times in hESC media.
    8. Remove PMEF media from the new plate of feeders.
    9. Wash the new feeders 2 times with 1× PBS.
    10. Add hESC media.
    11. Plate the hESCs (typically pass at 1:3 dilution).
    12. Disperse cells sliding plate front to back then side to side several times.

D. Cryopreservation of hESCs

hESCs can be cryopreserved in liquid nitrogen using slow freezing. We routinely use 90% fetal bovine serum (FBS)/10% dimethyl sulfoxide (DMSO) as the freezing media. We have also successfully substituted human serum albumin for FBS. When hESCs are dissociated in preparation for cryopreservation, we try to maintain the average size of the colonies larger than when we passage the colonies. The recovery rate of hESCs following cryopreservation is low. Therefore, we freeze down cells at a high concentration and thaw at a high density. For master stocks, we freeze 1/3 of one 10 cm dish per vial and thaw each vial into a single 60 mm dish.

E. Characterization

1. General Criteria

Standard characterization of newly derived hESC lines should include measuring their identity, stability, pluripotency, and differentiation ability (Loring and Rao, 2006). Genetic identity of the new lines can be established by determining the HLA isotype, single tandem repeat genotyping, or SNP genotyping. Stability of the lines requires confirming the ability to recover the lines after cryopreservation and determining the karyotype of the hESC following long-term culture (i.e., following 3 months of culture or ~24 passages). Some of the commonly used markers used to identify undifferentiated hESCs are the following: SSEA3, SSEA4, Tra-1-60, Tra-1-80, Oct-4, Nanog, Sox2, alkaline phosphatase, and telomerase. The pluripotent state of hESCs can be demonstrated by *in vitro* and *in vivo* differentiation as described below.

2. *In Vitro* Differentiaton and Embryoid Body Formation

hESCs can be easily differentiated by growing them in media other than hESC media such as Dulbecco's Modified Eagle's Medium (DMEM) + 20% FBS. Differentiation can also be achieved by growing the cells in suspension as embryoid bodies. Specific markers of differentiated cells can be identified either by reverse transcriptase–polymerase chain reaction (RT–PCR) or immunohistochemistry.

1. Embryoid body formation

    1. Dissociate the hESC colonies either mechanically or enzymatically, taking care to maintain the colonies at a larger size than when passaging.
    2. Wash the hESC colonies 3 times with media in conical tubes, allowing the hESC colonies to settle at the bottom of the tube. This helps to remove the feeder cells from the suspension.

3. Plate the hESC colonies on ultralow adhesion dishes.
4. The following day, change the media by settling out the embryoid bodies (EBs) and removing the old media. Change the media approximately every other day.

3. Teratoma Formation

Assessing the *in vivo* ability of hESCs to differentiate into cells from all three germ layers is accomplished by generating benign teratomas in immunodeficient mice. After teratomas are generated, they are dissected out of the animal and analyzed by standard histological procedures. Three anatomical sites are commonly used to implant the hESCs, under the testes capsule, intramuscularly, and under the kidney capsule. This protocol describes the procedure for generating teratomas under the kidney capsule.

1. Plate the hESC cells onto Matrigel for a few days before the surgery.
2. Harvest the cells the morning of the procedure by scraping the cells and resuspending them with a small amount of 1× PBS so as to make a slurry of cells. An alternative way is to make embryonic bodies the night before, settle the EBs before the surgery, and resuspend them in a small amount of PBS.
3. Anesthetize the mouse according to approved animal protocol.
4. Make a single horizontal incision across the spinal region (back of the mouse) about 0.5 cm below the bottom rib.
5. To visually locate the kidney, slide the incision to one side of the body wall just to the corner of the rib cage and the spinal column. The bright red and bean shaped kidney is visible through the body wall, accessible from the side of the back.
6. Grab the back skin with sharp forceps and make a small cut with scissors through the body wall outside of the kidney, taking care to avoid blood vessels and nerves. Insert the tip of the scissors in the cut and open the scissors up to expand the hole.
7. Locate the kidney through the hole in the body cavity, using blunt forceps.
8. Pull the kidney out by its fat pad or express the kidney through the hole in the body cavity by pushing in the abdomen around the kidney.
9. With two sharp forceps tear a small hole in the capsule membrane on the extreme end of the kidney.
10. Collect the cells into the transfer pipette.
11. Insert the transfer pipette into the hole in the capsule and push the tip all the way to the end of the kidney capsule, along the surface of the kidney. Move the transfer pipette back and forth creating a cavity on the far end of the kidney capsule. Push gently the syringe so as to push the cell aggregates in the cavity as the tip of the pipette is gently removed from the kidney cavity.
12. Remove the transfer pipette.
13. Using blunt forceps, replace the kidney back inside the body. Be sure that the kidney goes back into the body cavity and not between the body cavity and the skin.

14. Close the skin incision with wound clips or adhesive sutures.

15. Place the mouse back into the cage and use a heating lamp for a short time to provide some heat while the mouse is waking up.

16. Monitor the mouse for signs of pain or distress.

17. Tumors will become visible as lumps under the skin between 2 and 3 months following implantation.

III. Discussion

A. What to Expect

While recovery rates of frozen embryos are highly variable, one could expect as many as 50% of the embryos to survive the thawing process. Depending on the stage of the frozen embryo, another 30–50% of those that survive thawing may develop to blastocyst stage.

In our hands, the majority of intact blastocysts form outgrowths when plated. For outgrowths destined to generate an hESC line, we could observe colonies with typical hESC morphology within the first two passages (see Fig. 1). The hESC colonies clearly proliferate with each passage while non-hESC clumps do not.

As the hESC lines begin to expand, spontaneous differentiation will be observed. Undifferentiated hESC colonies have clearly defined borders, with uniform, tightly packed cells that exhibit a shiny, refractory morphology under the phase contrast microscope. Under higher power, hESCs have a high nucleus to cytoplasm ratio with prominent nucleoli. If the number of differentiated colonies becomes greater than ~40% of the total colonies, they should be removed by aspiration. If the number of undifferentiated colonies becomes extremely low compared to differentiated colonies, the undifferentiated colonies can be individually dissected out and plated into a smaller dish. In general, hESCs are more difficult to grow in synthetic media when compared to standard hESC media, exhibiting a higher rate of spontaneous differentiation.

Recovery rates following the first cryopreservation can be extremely low. If only a few small colonies are visible one week after thawing, additional feeders can be plated onto the existing culture to allow further growth of the hESCs.

IV. Materials

A. Media and Reagents

Immunosurgery:

Antihuman serum antibody produced in goat; Sigma H9640-2 ML
Complement sera from guinea pig; Sigma S1639-1 ML

Feeder preps:

Human foreskin fibroblasts; ATCC SCRC-1041

Freezing media: 10% DMSO in 90% FBS or human serum albumin

Fibroblast dissection media: DMEM; 1 × Pen/Strep; 10% FBS

Mitomycin C; Sigma Cat # M4287-2 MG

Passage and expansion

Collagenase Type IV; Invitrogen Cat # 17104-015

1 × hESC media: DMEM:F12; 20% Knockout Serum Replacer (Invitrogen, Cat # 10828); 0.1 mM nonessential amino acids (Invitrogen, Cat # 11140); 2 mM L-glutamine; 0.1 mM 2-mercaptoethanol; 8 ng/ml FGF-basic (Peprotech, Cat # 100-18BX).

1 × Synthetic Media (SM-1): DMEM:F12; 5 mg/ml human serum albumin (Irvine Scientific, Cat # 9988); 1 × ITS-X supplement (Invitrogen, Cat # 51500-056); 0.1 mM nonessential amino acids (Invitrogen, Cat # 11140); 2 mM L-glutamine; 1 mM LiCl (Sigma Cat # L7026); 0.1 mg/ml GABA (Sigma Cat # A5835-25 g); 0.1 mM 2-mercaptoethanol; 80 ng/ml FGF-basic (Peprotech, Cat # 100-18BX); 10 ng/ml NT-4 (Peprotech, Cat # 450-04); 10 ng/ml Activin (Peprotech, Cat # 120-14); 400 ng/ml Noggin (R + D Systems, Cat # 719-NG).

B. Miscellaneous Equipment

Nalgene Cryo 1 °C Freezing Container; Cat # 5100-0001

References

Behr, B. (1999). Blastocyst culture and transfer. *Hum. Reprod.* **14**(1), 5–6.

Behr, B., Pool, T. B., Milki, A. A., Moore, D., Gebhardt, J., and Dasig, D. (1999). Preliminary clinical experience with human blastocyst development in vitro without co-culture. *Hum. Reprod.* **14**(2), 454–457.

Chen, H., Qian, K., Hu, J., Liu, D., Lu, W., Yang, Y., Wang, D., Yan, H., Zhang, S., and Zhu, G. (2005). The derivation of two additional human embryonic stem cell lines from day 3 embryos with low morphological scores. *Hum. Reprod.* **20**(8), 2201–2206.

Chen, H. F., Kuo, H. C., Chien, C. L., Shun, C. T., Yao, Y. L., Ip, P. L., Chuang, C. Y., Wang, C. C., Yang, Y. S., and Ho, H. N. (2007). Derivation, characterization and differentiation of human embryonic stem cells: Comparing serum-containing versus serum-free media and evidence of germ cell differentiation. *Hum. Reprod.* **22**(2), 567–577.

Cieslak, J., Ivakhnenko, V., Wolf, G., Sheleg, S., and Verlinsky, Y. (1999). Three-dimensional partial zona dissection for preimplantation genetic diagnosis and assisted hatching. *Fertil. Steril.* **71**(2), 308–313.

Cowan, C. A., Klimanskaya, I., McMahon, J., Atienza, J., Witmyer, J., Zucker, J. P., Wang, S., Morton, C. C., McMahon, A. P., Powers, D., and Melton, D. A. (2004). Derivation of embryonic stem-cell lines from human blastocysts. *N. Engl. J. Med.* **350**(13), 1353–1356.

Ellerstrom, C., Strehl, R., Moya, K., Andersson, K., Bergh, C., Lundin, K., Hyllner, J., and Semb, H. (2006). Derivation of a xeno-free human embryonic stem cell line. *Stem Cells* **24**(10), 2170–2276.

Fletcher, J. M., Ferrier, P. M., Gardner, J. O., Harkness, L., Dhanjal, S., Serhal, P., Harper, J., Delhanty, J., Brownstein, D. G., Prasad, Y. R., Lebkowski, J., Mandalam, R., *et al.* (2006).

Variations in humanized and defined culture conditions supporting derivation of new human embryonic stem cell lines. *Cloning Stem Cells* **8**(4), 319–334.

Fong, C. Y., Bongso, A., Ng, S. C., Kumar, J., Trounson, A., and Ratnam, S (1998). Blastocyst transfer after enzymatic treatment of the zona pellucida: Improving in-vitro fertilization and understanding implantation. *Hum. Reprod.* **13**(10), 2926–2932.

Gardner, D. K., *et al.* (1996). Environment of the preimplantation human embryo *in vivo*: Metabolite analysis of oviduct and uterine fluids and metabolism of cumulus cells. *Fertil. Steril.* **65**(2), 349–353.

Genbacev, O., Krtolica, A., Zdravkovic, T., Brunette, E., Powell, S., Nath, A., Caceres, E., McMaster, M., McDonagh, S., Li, Y., Mandalam, R., Lebkowski, J., *et al.* (2005). Serum-free derivation of human embryonic stem cell lines on human placental fibroblast feeders. *Fertil. Steril.* **83**(5), 1517–1529.

Hong-mei, P., and Gui-an, C. (2006). Serum-free medium cultivation to improve efficacy in establishment of human embryonic stem cell lines. *Hum. Reprod.* **21**(1), 217–222.

Hovatta, O. (2006). Derivation of human embryonic stem cell lines, towards clinical quality. *Reprod. Fertil. Dev.* **18**(8), 823–828.

Hovatta, O., and Skottman, H. (2005). Feeder-free derivation of human embryonic stem-cell lines. *Lancet* **365**(9471), 1601–1603.

Inzunza, J., Gertow, K., Strömberg, M. A., Matilainen, E., Blennow, E., Skottman, H., Wolbank, S., Ahrlund-Richter, L., and Hovatta, O. (2005). Derivation of human embryonic stem cell lines in serum replacement medium using postnatal human fibroblasts as feeder cells. *Stem Cells* **23**(4), 544–549.

Kim, S. J. (2005). Efficient derivation of new human embryonic stem cell lines. *Mol. Cells* **19**(1), 46–53.

Klimanskaya, I., Chung, Y., Becker, S., Lu, S. J., and Lanza, R. (2006). Human embryonic stem cell lines derived from single blastomeres. *Nature* **444**(7118), 481–485.

Klimanskaya, I., Chung, Y., Meisner, L., Johnson, J., West, M., and Lanza, R. (2005). Human embryonic stem cells derived without feeder cells. *Lancet* **365**(9471), 1636–1641.

Lee, J. B., Lee, J. E., Park, J. H., Kim, S. J., Kim, M. K., Roh, S. I., and Yoon, H. S. (2005). Establishment and maintenance of human embryonic stem cell lines on human feeder cells derived from uterine endometrium under serum-free condition. *Biol. Reprod.* **72**(1), 42–49.

Loring, J. F., and Rao, M. S. (2006). Establishing standards for the characterization of human embryonic stem cell lines. *Stem Cells* **24**(1), 145–150.

Ludwig, T. E., Levenstein, M. E., Jones, J. M., Berggren, W. T., Mitchen, E. R., Frane, J. L., Crandall, L. J., Daigh, C. A., Conard, K. R., Piekarczyk, M. S., Llanas, R. A., and Thomson, J. A. (2006). Derivation of human embryonic stem cells in defined conditions. *Nat. Biotechnol.* **24**(2), 185–187.

Mateizel, I., De Temmerman, N., Ullmann, U., Cauffman, G., Sermon, K., Van de Velde, H., De Rycke, M., Degreef, E., Devroey, P., Liebaers, I., and Van Steirteghem, A. (2006). Derivation of human embryonic stem cell lines from embryos obtained after IVF and after PGD for monogenic disorders. *Hum. Reprod.* **21**(2), 503–511.

Oh, S. K., Kim, H. S., Ahn, H. J., Seol, H. W., Kim, Y. Y., Park, Y. B., Yoon, C. J., Kim, D. W., Kim, S. H., and Moon, S. Y. (2005). Derivation and characterization of new human embryonic stem cell lines: SNUhES1, SNUhES2, and SNUhES3. *Stem Cells* **23**(2), 211–219.

Simon, C., Escobedo, C., Valbuena, D., Genbacev, O., Galan, A., Krtolica, A., Asensi, A., Sánchez, E., Esplugues, J., and Fisher, S. (2005). First derivation in Spain of human embryonic stem cell lines: Use of long-term cryopreserved embryos and animal-free conditions. *Fertil. Steril.* **83**(1), 246–249.

Thomson, J. A., Itskovitz-Eldor, J., Shapiro, S. S., Waknitz, M. A., Swiergiel, J. J., Marshall, V. S., and Jones, J. M. (1998). Embryonic stem cell lines derived from human blastocysts. *Science* **282**(5391), 1145–1147.

Zhang, X., Stojkovic, P., Przyborski, S., Cooke, M., Armstrong, L., Lako, M., and Stojkovic, M. (2006). Derivation of human embryonic stem cells from developing and arrested embryos. *Stem Cells* **24**(12), 2669–2676.

CHAPTER 2

Autogeneic Feeders for the Culture of Undifferentiated Human Embryonic Stem Cells in Feeder and Feeder-Free Conditions

Andre Choo,* Ang Sheu Ngo,* Vanessa Ding,* Steve Oh,* and Lim Sai Kiang[†]

*Stem Cell Group
Bioprocessing Technology Institute
Agency for Science Technology and Research,
Singapore 138668

[†]Institute of Medical Biology
Agency for Science Technology and Research,
Singapore 138673

Abstract
I. Introduction
II. Methods and Materials
   A. Culture of hESC
   B. Characterization of hESC
III. Results
   A. Growth and Morphology of hESC Cultured in Feeder and Feeder-Free Conditions Using HuES9.E1
   B. Characterization of Pluripotent Markers
IV. Discussion and Summary
   References

Abstract

Human embryonic stem cells (hESC) are pluripotent cells that proliferate indefinitely in culture while retaining their ability to differentiate to any cell type in the body. Conventionally, hESC are cultured either directly on feeders or on an extracellular matrix supplemented with conditioned medium (CM) from feeders. To minimize the risk of xenozootic infections, several sources of primary human feeders have been identified. However, this does not eliminate the risk of contaminating hESC with infectious agents from the donor human feeders. In this study, we evaluated the use of the CD105+/CD24 hESC-derived mesenchymal stem cell (MSC) line, HuES9.E1, for its ability to support the growth of undifferentiated hESC in feeder and feeder-free cultures. This line was previously reported to be karyotypically stable and phenotypically displayed MSC-like surface antigens and gene transcription profiles. In addition, like adult MSC, HuES9.E1 can be differentiated to adipocytes, osteocytes, and chondrocytes *in vitro*. When tested for its ability to support hESC growth, it was found that hESC maintained the undifferentiated morphology for >12 continuous passages in coculture with HuES9.E1 and >8 passages in feeder-free cultures supplemented with CM from HuES9.E1. Furthermore, the hESC cultures continued to express the pluripotent markers, Oct-4, SSEA-4, Tra-1-60, Tra-1-81, and retained a normal karyotype. When injected into severe combined immunodeficient (SCID) mice, hESC differentiated to form teratomas comprising of tissues representative of the three embryonic germ layers. Potentially, the ability to derive and use autogeneic feeders may provide a safe and accessible source of feeders for the expansion of hESC required in clinical applications.

I. Introduction

Human embryonic stem cells (hESC) lines were successfully isolated from the inner cell mass of blastocysts and cultured *in vitro* by Thomson *et al.* in 1998. These pluripotent cells proliferate indefinitely under specific culture conditions but still retain the ability to differentiate into cell types representative of the three embryonic germ layers. hESC can be cultured either directly on feeder layers (Feeder coculture) or on extracellular matrices supplemented with conditioned medium (CM) from feeder layers (Feeder-free culture) (Reubinoff *et al.*, 2000; Xu *et al.*, 2001). Conventionally, primary mouse embryonic fibroblast (MEF) has been used to support undifferentiated hESC growth; however, a variety of human cell lines as feeders have also been reported in the literature. These include adult marrow cells, newborn foreskin fibroblasts, fetal muscle, fetal skin, and adult fallopian tubal fibroblasts (Amit *et al.*, 2003; Cheng *et al.*, 2003; Choo *et al.*, 2004; Hovatta *et al.*, 2003; Reubinoff *et al.*, 2000; Richards *et al.*, 2002; Xu *et al.*, 2001). Despite the advantage of using feeders from human sources, there are still concerns that hESC can be contaminated by infectious agents from the donor (Stacey *et al.*, 2006). One approach to circumvent this problem is to derive autogeneic feeder cells from hESC itself, which in turn can be used to support undifferentiated hESC growth (Stojkovic *et al.*, 2005; Wang *et al.*, 2005).

In this study, we demonstrated that two hESC lines previously grown on the immortalized MEF line, ΔE-MEF (Choo *et al.*, 2006), readily adapted to the hESC-derived mesenchymal stem cell (MSC) line, HuES9.E1 (Lian et al., 2007), on both feeder coculture and feeder-free culture. Morphologically, the hESC retained the undifferentiated phenotype and pluripotency was confirmed by the positive detection of cell surface markers and intracellular transcription factors. Furthermore, the hESC cultures maintained a normal karyotype (46 X,X) and formed teratomas when injected into severe combined immunodeficient (SCID) mouse.

II. Methods and Materials

A. Culture of hESC

hESC lines are available from a variety of different sources. A comprehensive list is provided by the National Institutes of Health (NIH) hESC Registry (http://stemcells.nih.gov/research/registry/defaultpage.asp), which outlines information about the providers and the characteristics of the hESC lines available. In this study, hESC lines used, HES-2 (46 X,X) and HES-3 (46 X,X), were obtained from ES Cell International. The cells were cultured at 37 °C/5% CO_2 on mitomycin-C-inactivated feeders, HuES9.E1 (4×10^4 cells/cm^2) on gelatin-coated organ culture dishes (Feeder cocultures) or on Matrigel-coated organ culture dishes supplemented with CM from HuES9.E1 (Feeder-free cultures). Medium used for culturing hESC was KNOCKOUT (KO) medium, which contained 85% KO–Dulbecco's Modified Eagle's Medium (DMEM) supplemented with 15% KO serum replacer, 1 mM L-glutamine, 1% nonessential amino acids and 0.1 mM 2-mercaptoethanol, and 5 ng/ml of fibroblast growth factor-2, FGF-2. (All media components were obtained from Invitrogen.)

1. Passaging of hESC

To passage hESC from a confluent organ culture dish (Falcon, surface area~2.4 cm^2)

- Aspirate spent culture medium from the dish of hESC and rinse cells twice with phosphate-buffered saline (PBS+, Invitrogen) to remove residual culture medium.
- Add 1 ml collagenase IV (200 U/ml in PBS) and incubate at 37 °C for 5–7 min.
- Aspirate the collagenase and rinse cells gently twice with KO medium. Six to eight hundred microliters (600–800 µl) of KO-medium (feeder coculture) or HuES9.E1-CM (feeder-free culture) is added to the cells before dissociation.
- Cells are dissociated into small clumps (~100–1000 cells/clump) by repeated pipetting and seeded at 1:3–1:4 dilutions on new mitomycin-C-inactivated feeders or Matrigel-coated plates (~200 µl of cell suspension). It is important not to break the cells too small as this will affect the ability of the cells to recover following passaging.

- Allow the cells to settle briefly before carefully transferring the dish into the CO_2 incubator.
- Medium is changed daily and the cultures are passaged weekly upon confluency.

2. Preparation of Inactivated Feeders

The derivation of the hESC-derived MSC line, HuES9.E1, was previously described by Lian *et al.* (2007) and will not be covered in this chapter.

- For use of HuES9.E1 as feeders in hESC culture, seed cells into 175 cm^2 T-flasks and grow until confluent in HUES medium, which contains 88% DMEM high glucose, 10% fetal bovine serum (FBS), 1% L-glutamine and 1% nonessential amino acids (FBS from Hyclone, all other components obtained from Invitrogen).
- To inactivate feeders, aspirate spent culture medium and add 45 ml of 10 µg/ml mitomycin-C (Sigma-Aldrich, diluted in HUES-medium) to each flask and incubate at 37 °C/5% CO_2 for 2.5–3 h.
- After treatment, aspirate the mitomycin-C solution and wash the cells 3 times with PBS+.
- To detach the cells, add 3 ml of 0.25% trypsin–EDTA, ethylenediaminetetraacetic acid (Invitrogen) and incubate at 37 °C for 5 min.
- Add 7 ml of HUES-medium to neutralize the enzyme activity and dissociate cells.
- Collect the cell suspension and centrifuge the cells at 400 × g for 5 min.
- Decant supernatant, resuspend cells in freezing mix (90% FBS and 10% dimethyl sulfoxide (DMSO), Hyclone and Sigma-Aldrich, respectively) at a density of 1 × 10^6 cells/ml (Feeder cocultures) or 4 × 10^6 cells/ml (Feeder-free cultures) and aliquot 1 ml into each cryovial for freezing at −150 °C.

3. Plating of HuES9.E1 for Feeder Cocultures

- Thaw a vial of mitomycin-C-inactivated feeders (1 × 10^6 cells) and transfer to 9 ml HUES-medium.
- Centrifuge the cells at 400 × g for 5 min. Decant and resuspend the pellet in 10 ml HUES-medium.
- Aliquot 1 ml of cell suspension (1 × 10^5 cells) into each organ culture dish which is pre-coated with 0.1% gelatin (Sigma-Aldrich, diluted in water) for 1–2 h at room temperature.
- Allow the cells to adhere for 24 h at 37 °C/5% CO_2 before changing the medium from HUES-medium to KO-medium. Allow an additional 24 h to equilibrate before adding hESC.

4. Preparation of HuES9.E1 CM for Feeder-Free Cultures

- Thaw 2 vials of mitomycin-C-inactivated feeders (4 × 10⁶ cells/vial) and transfer to 18 ml HUES-medium.
- Centrifuge the cells at 400 × g for 5 min. Decant and resuspend the pellet in 30 ml HUES-medium.
- Inoculate the cell suspension into a 175 cm² T-flask and allow the cells to adhere for 24 h at 37 °C/5% CO_2 before changing the medium from HUES-medium to KO-medium. Allow an additional 24 h for equilibration.
- For the preparation of CM, aspirate the spent culture medium and replace with 30 ml of fresh KO-medium. Incubate for 72 h at 37 °C/5% CO_2 to allow conditioning of the medium by the inactivated feeders.
- Collect the CM and filter the medium through a 0.22 μm membrane filter (Nalgene). Aliquot the CM into tubes and store at −20 °C.
- When required for feeding of hESC cultures, thaw the CM in a 37 °C water bath and supplement with an additional 5 ng/ml before use.

5. Preparation of Matrigel Plates for Feeder-Free Cultures

- To minimize the formation of gel, thaw Matrigel (Becton-Dickinson) overnight at 4 °C. Aliquots (0.5–1 ml) of Matrigel can be made from the stock bottle (10 ml) and refrozen for future use. It is important to ensure that all pipettes and tubes used are prechilled.
- Dilute thawed Matrigel 1:30 fold with cold KO-medium and aliquot 1 ml of Matrigel solution into each organ culture dish. Incubate the dishes at 4 °C overnight.
- On the day of use, aspirate the Matrigel solution and rinse the culture dishes once with KO-medium. Add 800 μl of CM and the dishes are ready for hESC passaging.

B. Characterization of hESC

1. Flow Cytometry Analysis of Oct-4, SSEA-4, and Tra-1–60 Expression

- Aspirate spent culture medium from the organ culture dish of hESC and rinse cells once with 1 ml of PBS+ to remove residual culture medium.
- Add 200 μl of 0.25% trypsin-EDTA and incubate at 37 °C for 5–7 min.
- Dissociate hESC to a single-cell suspension by repeated pipetting and transfer cells to a microfuge tube containing 800 μl of HUES-medium.
- Centrifuge the cells at 15000 × g for 30 s. Decant and resuspend the pellet in 50 μl fluorescence-activated cell sorter (FACS) solution (1% Bovine serum albumin (BSA) in PBS).
- Aliquot 10 μl of cell suspension into microfuge tubes.

- Add 100 µl Reagent A (Fix and Perm Kit, Caltag) to the tubes, mix by pipetting, and incubate at room temperature for 15 min.
- Centrifuge the cells at 15000 × g for 30 s. Decant and wash the cells once with 200 µl of FACS solution.
- Resuspend the pellet in 100 µl Reagent B (Fix and Perm Kit) together with individual primary antibodies to Oct-4 (Santa Cruz), SSEA-4 (Developmental Studies Hybridoma Bank), and Tra-1-60 (Chemicon) at 10 µg/ml, 100 µl neat culture supernatant, and 20 µg/ml, respectively, and incubate at room temperature for 15 min.
- Centrifuge the cells at 15000 × g for 30 s. Decant and wash the cells once with 200 µl of FACS solution
- Resuspend the pellet in 100 µl fluorescein isothiocyanate (FITC)-conjugated goat anti-mouse antibody (DAKO) at 1:500 dilution with FACS solution and incubate at room temperature in the dark for 15 min.
- Centrifuge the cells at 15000 × g for 30 s. Decant and wash the cells once with 200 µl of FACS solution
- Resuspend pellet in 200 µl of FACS solution and analyze for antibody binding on a FACScan (FACSCalibur, Becton Dickinson).

2. Immunocytochemistry

- Aspirate spent culture medium from the organ culture dish of hESC and rinse cells once with 1 ml of PBS+.
- Fix cells at room temperature for 45 min in 4% paraformaldehyde.
- Wash the cells thrice with PBS+.
- Add 200 µl of primary antibody to Tra-1-60 (Chemicon) at 30 µg/ml (diluted with FACS solution) and incubate at room temperature for 1 h.
- Wash the cells thrice with PBS+.
- Add 200 µl of phycoerythrin (PE)-conjugated goat anti-mouse antibody (DAKO) at 1:500 dilution with FACS solution and incubate at room temperature in the dark for 30 min.
- Wash the cells thrice with PBS+ and visualize for cell surface staining on a fluorescent inverted phase contrast microscope (Olympus).

3. *In vivo* SCID Mouse Models

- Inject hESC (4×10^6 cells) after harvesting by enzymatic treatment (Section III.A.1) into the rear leg muscle of 4-week-old female SCID mice.
- Observe for the formation of teratomas 8–10 weeks after injection.
- Sacrifice the animal by CO_2 asphyxiation and dissect out the teratoma.
- Fix the teratoma in 10% formalin (Sigma-Aldrich), embed in paraffin, and then section and examine histologically after hematoxylin-eosin staining.

4. Karyotyping
- Culture actively growing hESC cultures in organ culture dishes with 25 μl of 0.6 μg/ml colcemid solution diluted in 1 ml KO-medium for 15–16 h at 37 °C/5% CO_2 to arrest the cells in metaphase.
- Cytogenetic analysis after metaphase arrest is routinely outsourced to the Cytogenetics Laboratory at the KK Women's and Children's Hospital.

III. Results

A. Growth and Morphology of hESC Cultured in Feeder and Feeder-Free Conditions Using HuES9.E1

We have previously shown that hESC lines, HES-2 and HES-3, can be routinely cultured using the immortalized MEF cell line, △E-MEF, in both feeder and feeder-free conditions (Choo et al., 2006). To evaluate if the hESC-derived MSC line, HuES9.E1, can also support the undifferentiated growth of hESC, cells were seeded directly into both these conditions using HuES9.E1 instead. Similar to the cultures on △E-MEF, hESC formed distinct colonies (Fig. 1) that were tightly clustered and retained a high nucleus to cytoplasmic ratio (not shown). The cells reached confluency after 5–7 days and morphologically, no cystic differentiating regions were observed. Under both feeder and feeder-free conditions, hESC continued to maintain the undifferentiated hESC morphology for greater than 12 and 8 passages, respectively (42 and 28 population doublings).

B. Characterization of Pluripotent Markers

Apart from observing the cellular morphology, hESC have to be further characterized to establish the supportive ability of HuES9.E1 in maintaining pluripotency. Several different assays are conventionally used, which includes the detection of cell surface markers (SSEA-3/4, Tra-1-60/81), intracellular transcription factors (Oct-4/Nanog), and the formation of teratomas in *in vivo* SCID mice models.

Using flow cytometry, the expression of Oct-4, SSEA-4, and Tra-1-60 in hESC was compared. Like HES-3 cultured on △E-MEF (cocultures), >89% of cells continue to stain positive for the three pluripotent markers after six continuous passages (Fig. 2). Similarly, for HES-3 maintained in feeder-free condition with CM from HuES9.E1 after seven continuous passages, >90% of cells still stain positive for the pluripotent markers (Fig. 3). Hence, the continuous culture of HES-3 cells using HuES9.E1 did not result in a decrease in the population of undifferentiated cells compared to hESC cultured using △E-MEF. Furthermore, this result was confirmed by the homogenous Tra-1-60 staining of HES-2 and HES-3 colonies cultured on △E-MEF (Fig. 4A), HuES9.E1 (Fig. 4B), or HuES9.E1 CM (Fig. 4 C and D).

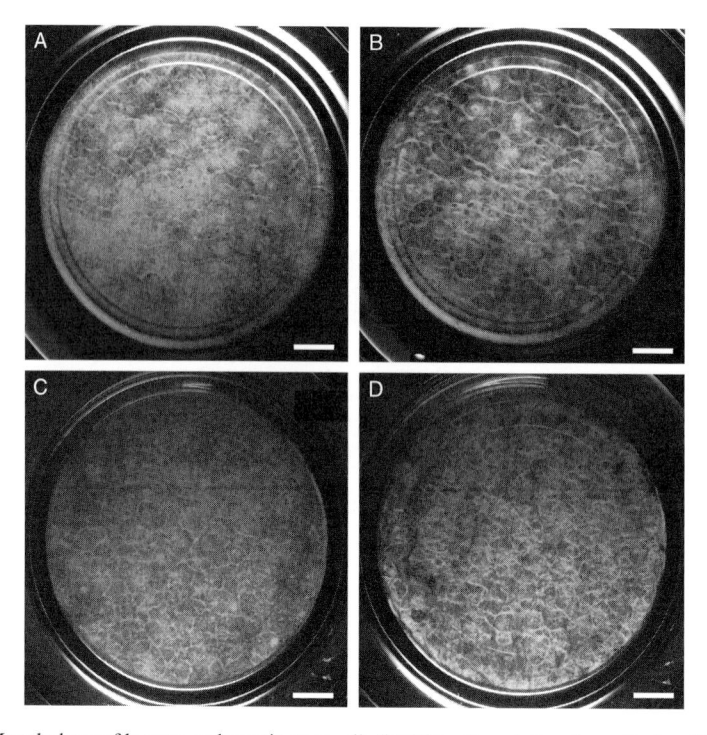

Fig. 1 Morphology of human embryonic stem cells (hESC) cultured directly on feeders (A and B) or on Matrigel supplemented with conditioned media (CM) from feeders (C and D, respectively). HES-3 cells were cocultured on △E-MEF (A) and HuES9.E1 (B); feeder-free cultures of HES-2 and HES-3 cells supplemented with CM from HuES9.E1 (C and D, respectively). Representative images taken using a stereomicroscope of hESC grown in organ culture dishes. Scale bar = 2 mm.

As a confirmation that the hESC expanded on HuES9.E1 are pluripotent, passage 12 HES-3 cells were harvested and injected into SCID mice. Teratoma formation was observed 10 weeks post injection. Histological analysis of the sections from the teratoma identified representative tissues from the three embryonic germ layers, including neural epithelium (ectoderm, Fig. 5A), gut epithelium (endoderm, Fig. 5B), and muscle (mesoderm, Fig. 5C). As a negative control, HuES9.E1 cells were injected into SCID mice; however, no teratoma formation was observed, indicating that HuES9.E1 is not tumourigenic (results not shown). Lastly, cytogenetic analysis performed on HES-3 cells from both feeder and feeder-free cell populations showed that the cells continue to retain a normal karyotype (46 X,X) under both culture conditions (Fig. 6A and B, respectively). Taken together, these results indicate that autogeneic feeders derived from hESC can be used to support undifferentiated growth of hESC in both feeder and feeder-free conditions retaining morphology, pluripotent marker expression, *in vivo* differentiation potential, and karyotypic stability.

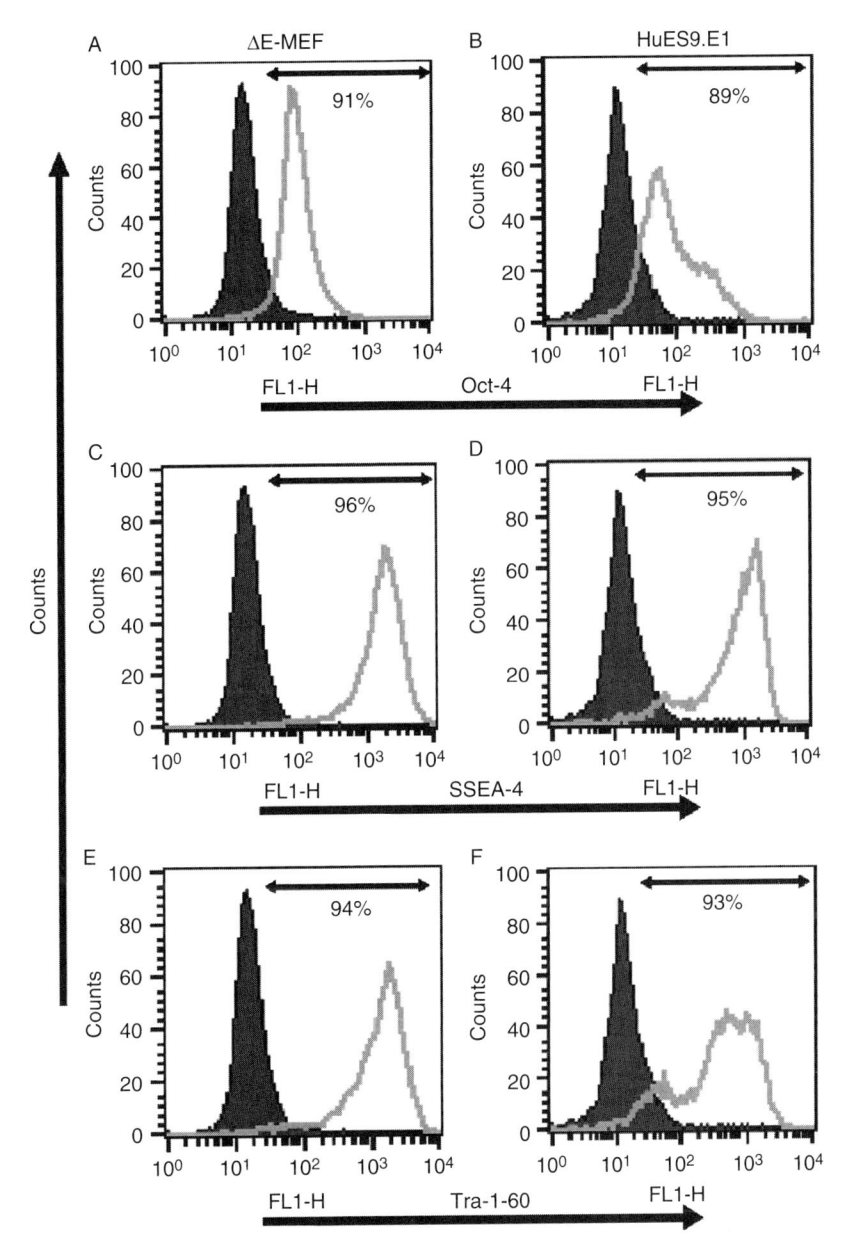

Fig. 2 Flow cytometry analysis of Oct-4, SSEA-4, and Tra-1-60 in HES-3 cells after six continuous passages on HuES9.E1 feeders compared with △E-MEF. Single-cell suspension of HES-3 cells was fixed, permeabilized, and labeled with antibodies specific for Oct-4 (A and B), SSEA-4 (C and D), and Tra-1-60 (E and F, respectively). The labeled cells were then incubated with fluorescein isothiocyanate (FITC)-conjugated α-mouse antibodies. The shaded histogram represents staining with the negative isotype control and open histograms represent staining with the corresponding antibodies.

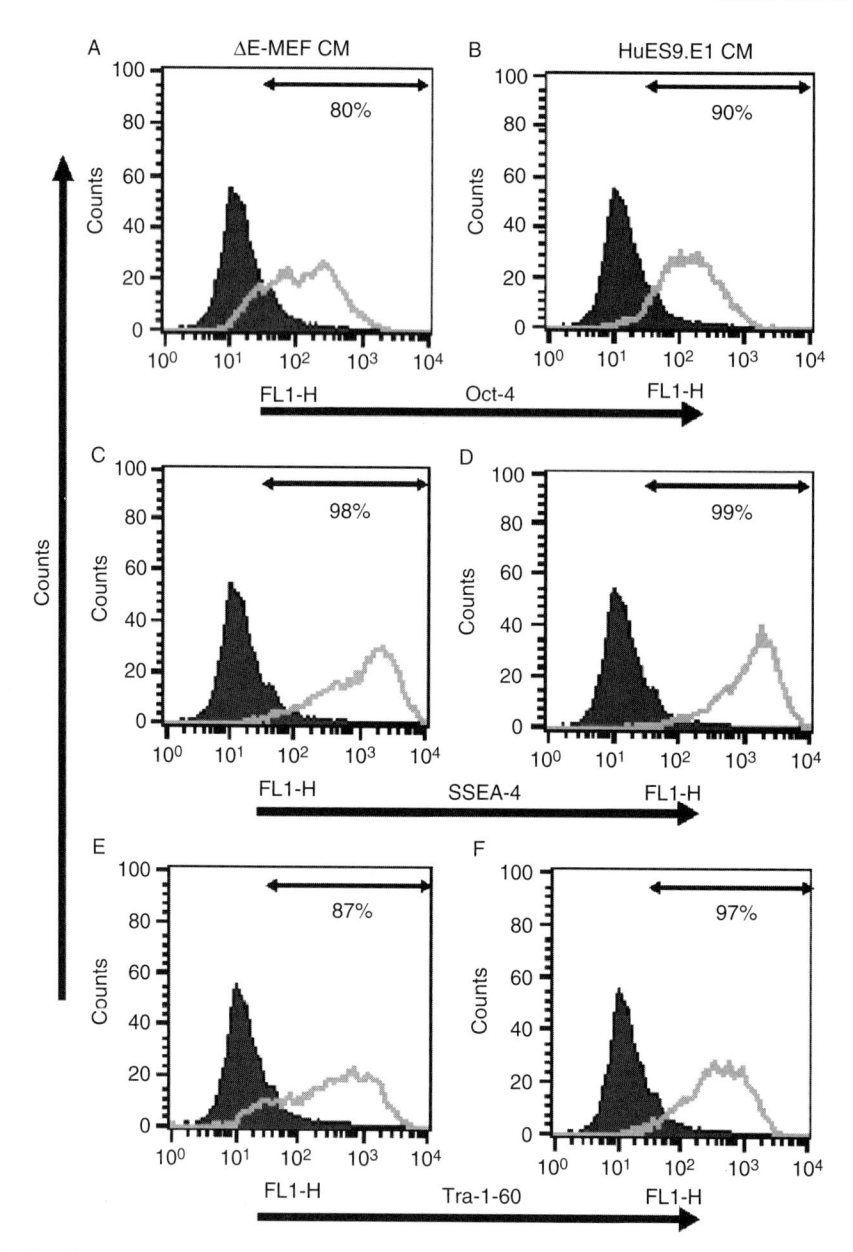

Fig. 3 Flow cytometry analysis of Oct-4, SSEA-4, and Tra-1-60 in HES-3 cells cultured on Matrigel supplemented with conditioned medium (CM) from HuES9.E1 (Passage 7) compared with ΔE-MEF. Single-cell suspension of HES-3 cells was fixed, permeabilized, and labeled with antibodies specific for Oct-4 (A and B), SSEA-4 (C and D), and Tra-1-60 (E and F, respectively). The labeled cells were detected with fluorescein isothiocyanate (FITC)-conjugated α-mouse antibodies. The shaded histogram represents staining with the negative isotype control and open histograms represent staining with the corresponding antibodies.

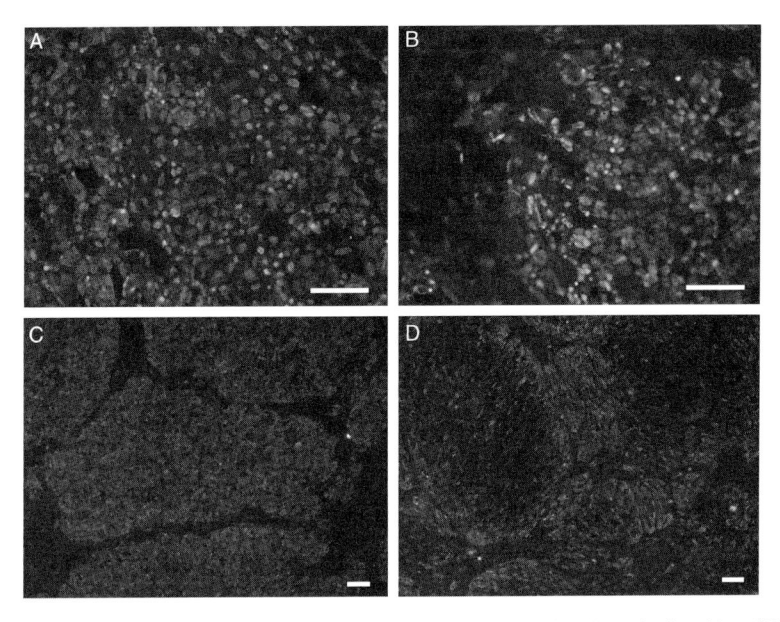

Fig. 4 Immunofluorescent staining for Tra-1-60 on hESC cultured directly on feeders (A and B) or on Matrigel supplemented with conditioned media (CM) from feeders (C and D, respectively). HES-3 cells cocultured on △E-MEF (A) and HuES9.E1 (B); feeder-free cultures of HES-2 and HES-3 cells supplemented with CM from HuES9.E1 (C and D, respectively). Colonies were stained with Tra-1-60 followed with phycoerythrin (PE)-conjugated α-mouse detection antibody. Scale bar = 100 μm.

IV. Discussion and Summary

Despite the intent to culture hESC in a fully defined environment, the use of feeders either in coculture with hESC or for the production of CM for feeder-free cultures still proves to be the most robust strategy of maintaining undifferentiated hESC. There have been numerous reports on the culturing of hESC with human feeders; however, there are still concerns of the risk of contamination by human infectious agents from the donor. These would potentially include viral contaminants that have oncogenic risk or viruses that are not routinely screened for its risk of transmission is considered low; nonviral contaminants as well as the emergence of new pathogens (Stacey *et al.*, 2006).

Several groups have recently shown that hESC can be differentiated into fibroblast-like or MSC-like cells, which, in turn, can be used to support the undifferentiated growth of hESC (Stojkovic *et al.*, 2005; Wang *et al.*, 2005; Xu *et al.*, 2004). The advantages of using these hESC-derived feeders are that they are autogeneic to the hESC, thus providing more homogeneity and consistency in the culture platform. Also, if the hESC are screened for the absence of viral contaminants, it eliminates the concern of contamination or screening of the feeders from a second donor source.

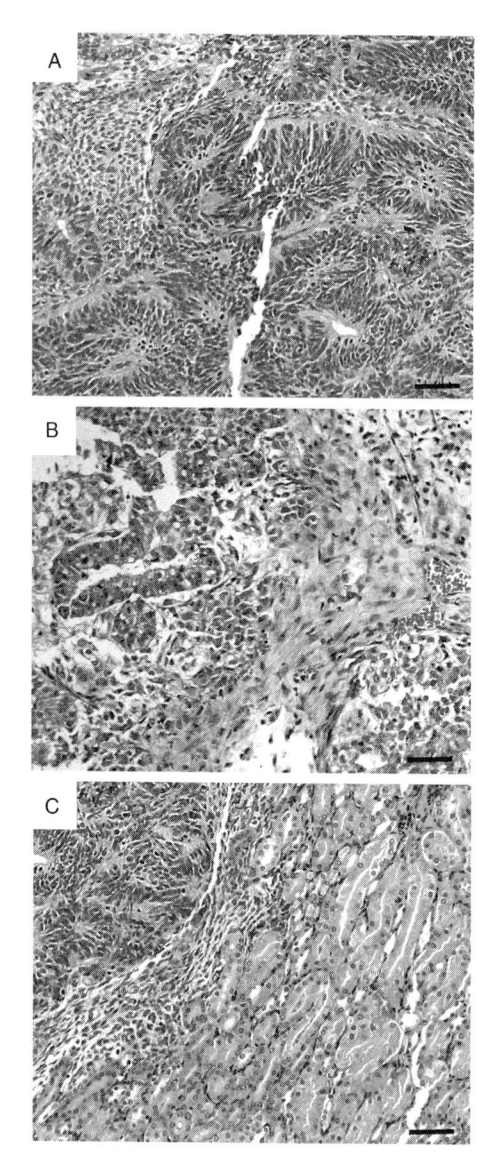

Fig. 5 Histological analysis of teratomas formed in severe combined immunodeficient (SCID) mice after intramuscular injection of HES-3 cells cocultured on HuES9.E1 feeders after 11 passages. (A) neural epithelium, (B) gut epithelium, and (C) muscles. Scale bar = 30 μm.

In this study, the hESC-derived MSC line, HuES9.E1, was shown to support the undifferentiated growth of hESC in both feeder and feeder-free conditions. This cell line was previously derived by Lian *et al.* (2007) and was shown to have similar morphology, phenotype, and function compared to adult MSC. The cells are also

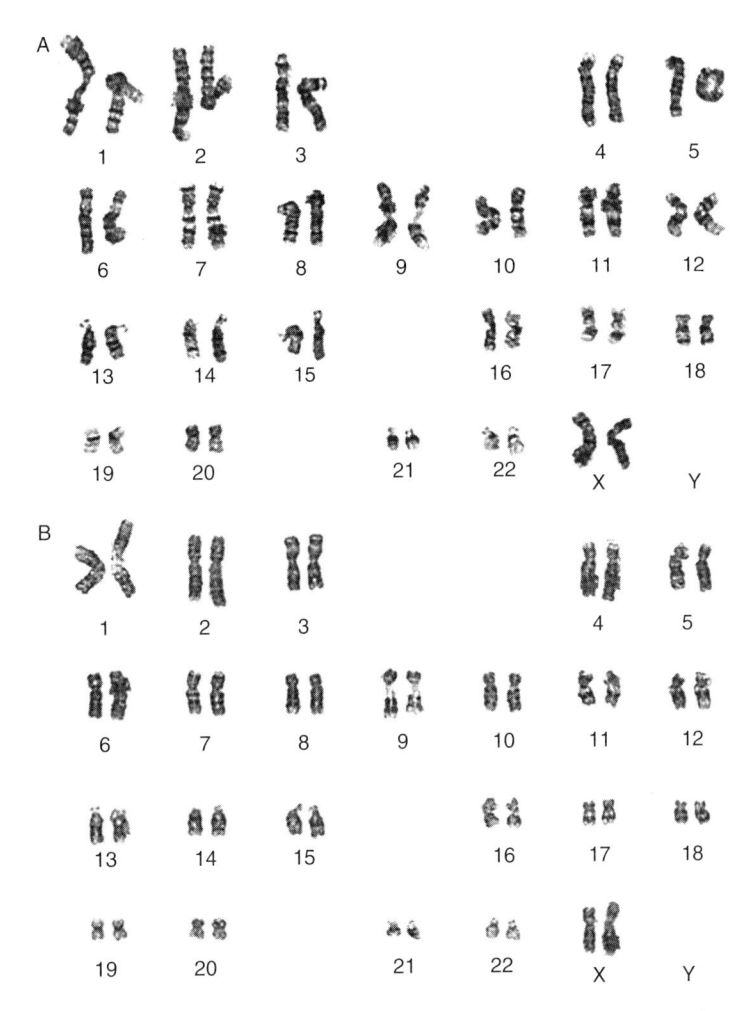

Fig. 6 Cytogenetic analysis of HES-3 cells cultured on HuES9.E1 (A) or supplemented with conditioned medium (CM) from HuES9.E1 (B). Normal karyotype (46 X,X) was observed for human embryonic stem cells (hESC) from both conditions after six passages.

karyotypically stable and have a higher proliferative capacity compared to adult MSC. A significant difference between HuES9.E1 compared to other hESC-derived feeder lines is that instead of using a heterogeneous population of fibroblast-like cells obtained directly after differentiation, a positive and negative cell sorting step for CD105 and CD24, respectively, was incorporated into the derivation protocol to ensure that a consistent and reproducible cell type is achieved. Furthermore, since CD24 is present on hESC and differentiated hESC-derived cell types (Cai et al., 2006), selecting for a CD24-negative cell type is advantageous in reducing contamination by these potentially teratoma-forming cells.

In summary, we have successfully adapted and cultured undifferentiated hESC on the hESC-derived MSC line, HuES9.E1. Characterization of the hESC showed that they retain the pluripotent phenotype, have the ability to differentiate to the embryonic germ layers, and maintain a stable karyotype after continuous passaging. Potentially, the ability to derive and use autogeneic feeders may provide a safe and accessible source of feeders for the expansion of hESC required in clinical applications.

References

Amit, M., Margulets, V., Segev, H., Shariki, K., Laevsky, I., Coleman, R., and Itskovitz-Eldor, J. (2003). Human feeder layers for human embryonic stem cells. *Biol. Reprod.* **68**, 2150–2156.

Cai, J, Chen, J, Liu, Y, Miura, T, Luo, Y, Loring, J. F., Freed, W. J., Rao, M. S., and Zeng, X (2006). Assessing self-renewal and differentiation in human embryonic stem cell lines. *Stem Cells* **24**, 516–530.

Cheng, L., Hammond, H., Ye, Z., Zhan, X., and Dravid, G. (2003). Human adult marrow cells support prolonged expansion of human embryonic stem cells in culture. *Stem Cells* **21**, 131–142.

Choo, A., Padmanabhan, J., Chin, A., Fong, W. J., and Oh, S. K. (2006). Immortalized feeders for the scale-up of human embryonic stem cells in feeder and feeder-free conditions. *J. Biotechnol.* **122**, 130–141.

Choo, A. B., Padmanabhan, J., Chin, A. C., and Oh, S. K. (2004). Expansion of pluripotent human embryonic stem cells on human feeders. *Biotechnol. Bioeng.* **88**, 321–331.

Hovatta, O., Mikkola, M., Gertow, K., Stromberg, A. M., Inzunza, J., Hreinsson, J., Rozell, B., Blennow, E., Andang, M., and Ahrlund-Richter, L. (2003). A culture system using human foreskin fibroblasts as feeder cells allows production of human embryonic stem cells. *Hum. Reprod.* **18**, 1404–1409.

Lian, Q., Lye, E., Suan, Y. K., Khia Way, T. E., Salto-Tellez, M., Liu, T. M., Palanisamy, N., El Oakley, R. M., Lee, E. H., Lim, B., and Lim, S. K. (2007). Derivation of clinically compliant MSCs from CD105 +. *Stem Cells* **25**, 425–436.

Reubinoff, B. E., Pera, M. F., Fong, C. Y., Trounson, A., and Bongso, A. (2000). Embryonic stem cell lines from human blastocysts: Somatic differentiation *in vitro*. *Nat. Biotechnol.* **18**, 399–404.

Richards, M., Fong, C. Y., Chan, W. K., Wong, P. C., and Bongso, A. (2002). Human feeders support prolonged undifferentiated growth of human inner cell masses and embryonic stem cells. *Nat. Biotechnol.* **20**, 933–936.

Stacey, G. N., Cobo, F., Nieto, A., Talavera, P., Healy, L., and Concha, A. (2006). The development of 'feeder' cells for the preparation of clinical grade hES cell lines: Challenges and solutions. *J. Biotechnol.* **125**, 583–588.

Stojkovic, P., Lako, M., Stewart, R., Przyborski, S., Armstrong, L., Evans, J., Murdoch, A., Strachan, T., and Stojkovic, M. (2005). An autogeneic feeder cell system that efficiently supports growth of undifferentiated human embryonic stem cells. *Stem Cells* **23**, 306–314.

Thomson, J. A., Itskovitz-Eldor, J., Shapiro, S. S., Waknitz, M. A., Swiergiel, J. J., Marshall, V. S., and Jones, J. M. (1998). Embryonic stem cell lines derived from human blastocysts. *Science* **282**, 1145–1147.

Wang, Q., Fang, Z. F., Jin, F., Lu, Y., Gai, H., and Sheng, H. Z. (2005). Derivation and growing human embryonic stem cells on feeders derived from themselves. *Stem Cells* **23**, 1221–1227.

Xu, C., Inokuma, M. S., Denham, J., Golds, K., Kundu, P., Gold, J. D., and Carpenter, M. K. (2001). Feeder-free growth of undifferentiated human embryonic stem cells. *Nat. Biotechnol.* **19**, 971–974.

Xu, C., Jiang, J., Sottile, V., McWhir, J., Lebkowski, J., and Carpenter, M. K. (2004). Immortalized fibroblast-like cells derived from human embryonic stem cells support undifferentiated cell growth. *Stem Cells* **22**, 972–980.

CHAPTER 3

Microporous Membrane Growth Substrates for Embryonic Stem Cell Culture and Differentiation

Steven D. Sheridan, Sonia Gil, Matthew Wilgo, and Aldo Pitt

Millipore Corporation
Bioscience Division
Danvers, Massachusetts 01923

Abstract
I. Introduction
    A. Membrane-Based Cell Culture Historical Perspective
II. Rationale
III. Methods
    A. General Considerations of Membrane-Based Cell Culture
    B. Preparation of Membranes
    C. Co-culture of ES Cells with MEFs for Pluripotent Expansion
    D. Membrane-Based Differentiation of ES Cells
    E. Microscopic and Immunochemical Analysis Using Membrane Inserts
IV. Summary
    References

Abstract

As the field of embryonic stem cell culture and differentiation advances, many diverse culturing techniques will ultimately be necessary in order to fully reproduce the various environments these cells normally encounter during development. Although most of the work to date has been performed on solid plastic supports, this growth support has several limitations in its representation of the *in vivo* environment.

Impermeable substrates force the cells to exchange their gas and nutrients exclusively through the top side of the cultured cells. In contrast, cells growing *in vivo* are exposed from several directions to factors from the blood, other cells, soluble factors, and liquid–air interfaces. Additionally, solid plastic presents a smooth two-dimensional surface that is not experienced *in vivo*. Therefore, the use of traditional plastic presents limitations upon normal cellular morphology, function, and differentiation.

An important alternative to growth on solid plastic is the growth of cells on microporous membranes. One of the many advantages to cell growth on porous membrane substrates is their ability to provide a surface that better mimics a three-dimensional *in vivo* setting. A porous membrane allows multidirectional exposure to nutrients and waste products. In addition, the membrane separation of dual chambers allows for the coculture of cells of different origin to study how cells interact through indirect signaling or through providing a conditioned niche for the proper growth and differentiation of cell types.

I. Introduction

A. Membrane-Based Cell Culture Historical Perspective

As an alternative to solid two-dimensional (2-D) plastic substrates, microporous membranes have been used for over five decades in a vastly diverse footprint of cell biology (Boyden, 1962; Cook *et al.*, 1989; Grobstein, 1953; Verfaillie, 1992). Cell culture on a microporous substrate allows cells to achieve multidirectional influence from factors and nutrients for proper function and differentiation while providing a three-dimensional (3-D) growth substrate for more *in-vivo* like attachment, growth, and differentiation.

Microporous membranes have been indispensable in the segregation of direct cell-to-cell contact from diffusible factors on cellular function and differentiation (Liu *et al.*, 2000; Verfaillie, 1992). Many cellular functions, including differentiation, can be influenced separately by direct cell-to-cell contact or by diffusible factors which may result in the conditioning of the environment from a distance (i.e., through the blood or semidirect contact during development). One of the very earliest references to use porous membranes for cell culture was to segregate the effects of direct cell-to-cell contact from diffusible factors. In 1953, Clifford Grobstein noted that mesenchymal tissue isolated from mouse embryos was able to induce the appropriate differentiation of nephric epithelium when separated by a microporous membrane (Grobstein, 1953). Since these early observations, microporous membranes have been the definitive method for separating the role of direct versus indirect soluble factor-mediated communication for cell-to-cell influences for a number of cellular and developmental processes.

Early experimentation with epithelial cells demonstrated that growth on membranes provided a more conducive environment for proper growth as based on cellular morphology (Cook *et al.*, 1989), differentiation (Widdicombe *et al.*, 2003), hormone response, and asymmetric expression of appropriate surface receptors

(Fuller and Simons, 1986; Mostov and Deitcher, 1986). In addition to benefiting the differentiation of epithelial cells, epithelial monolayers grown on porous membranes form an effectively separated dual chamber system for the study of molecular transport across the monolayer. The ability to sample from both separate chambers allows for measurement of the transport of nutrients and drug compounds from one chamber through the cell monolayer to the other chamber in a controlled, *in vitro* fashion. Membrane-based epithelial cell systems have been shown to be a standard method for evaluating the transport of potentially therapeutic compounds. This method effectively simulates *in vivo* transport processes since the appropriate differentiation and polarity of the cell is maintained by the structure and porosity of the membrane on which they are grown (Byers *et al.*, 1986). Microporous membrane-grown epithelial cells have been proven to be able to model the transport of these chemicals as an *in vitro* simulation of the *in vivo* characteristics of the appropriate cell asymmetry and function.

Microporous membranes also provide an excellent substrate for the study of cellular migration. The use of membranes of defined pore size allows the cells to selectively migrate through the pore to the other side of the membrane. Often referred to as Boyden Chambers (Boyden, 1962), membrane inserts with pore sizes ranging from 3 to 8 µm have been pivotal in the *in vitro* study of cellular migration (Chen, 2005), chemotaxis (Baum *et al.*, 1971), and invasion (Albini *et al.*, 1987). The membrane establishes a dual chamber system to study the migration of cells from one chamber to the other. The chamber toward which the cells migrate (the receiving chamber) typically contains factors and/or other cells that act as an attractant to induce the directional migration of cells. The ability to score the migration speed and frequency of cellular migration (Gildea *et al.*, 2000) has enabled cancer and immunology studies (von and Mackay, 2000). These assays have the ability to differentiate simple migration and/or chemotaxis from the active process of invasion as seen in processes such as immune response or cancer metastasis once a layer of extracellular matrix (ECM) is applied to the membrane that the migrating cells need to digest in order to pass through the membrane barrier (Albini *et al.*, 1987).

Porous substrates also have an important advantage over typical 2-D plastic culture in that they provide an extra dimension of access to growth factors, media components, and gaseous environments (Rutten *et al.*, 1990). Typical cell culture has been monolayer driven, using immortalized lines for the specific purpose of studying specific cellular events. Stem cell culture has an important attribute apart from studies with immortalized cell lines in that these cells are more profoundly responsive to their physical and chemical environment. This is particularly evident in the differentiation of particular tissue types that normally occurs in the 3-D organism where cells are exposed, in many directions, to several influences. In addition, many of the stem cell differentiation procedures involve structures that are not monolayer in nature, such as embryoid bodies (EBs) (Boheler *et al.*, 2004; Chen and Kosco, 1993; Hwang *et al.*, 2005; Ng *et al.*, 2005), which are limited by their exposure to media components and gas exchange. In this need, substrates, such as microporous membranes, that offer a more 3-D level of exposure to examine the roles of growth factors allow for more typical contact to these factors. Additionally, the ability to dissect the effect of direct versus indirect cellular communication facilitates the understanding of diffusible factors on expansion and differentiation, which will help improve media formulations for these directed purposes. This level of understanding in directed differentiation will be pivotal in the understanding of the role of stem cells in developmental biology and cellular therapy.

II. Rationale

Stem cell biology is a unique opportunity to redefine the paradigms in cell biology. A full understanding of the requirements to control and manipulate these stem cells into desired tissue types will necessitate a cross-functional expertise of cell biologists, material scientists, and engineers. Microporous membranes have been instrumental in the study of many biological functions during development and basic biological understanding of cell-to-cell influences on cell function and fate. This well-established technology will prove greatly beneficial in the understanding of stem cell isolation, expansion, and differentiation.

The methods in this chapter are presented as an introduction to the basic procedures for porous membrane-based stem cell culture. Although the experimental and protocol specifics can be diverse for different analyses, the basic protocols presented here for membrane-based stem cell culture are meant to facilitate these studies by providing initial procedural recommendations. The membrane-based methods presented include treatment for cell growth and attachment, coculture, expansion and passage of embryonic stem cells, ESCs (murine and human), basic differentiation, and microscopic sample preparation and analysis.

III. Methods

A. General Considerations of Membrane-Based Cell Culture

The earliest membranes used for cell culture were typically composed of natural product-derived materials such as mixed cellulose esters (Grobstein, 1953). These materials were ideal for their high porosity, strong cell adhesion, and structured surface for cellular function and differentiation. Further developments such as optically transparent membrane materials and track etching (a process resulting in a tight distribution of defined pore size formed through a combination of charged-particle bombardment, or irradiation, followed by chemical etching) have expanded the effectiveness, sensitivity, and reproducibility of membrane-based cellular assays (Goel *et al.*, 1992). These membrane materials include transparent hydrophilic polytetrafluoroethylene [PTFE—also referred to as Biopore (Pitt *et al.*, 1987)], and track-etched thin film polycarbonate (PCF) and polyethylene terephthalate (PET) membranes with defined pore sizes ranging

3. Membrane-Based Embryonic Stem Cell Culture

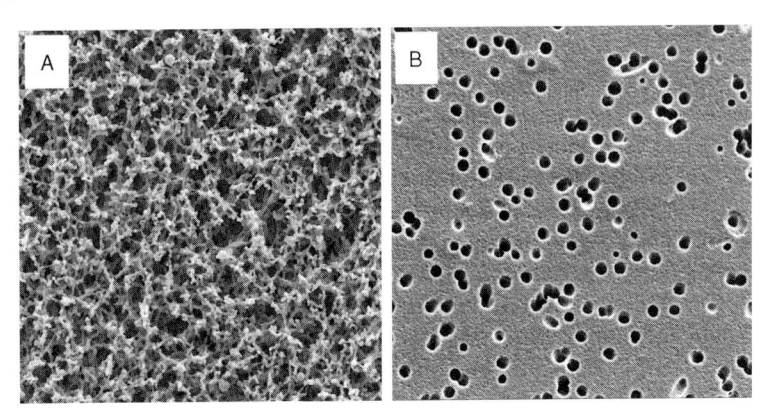

Fig. 1 Mixed Cellulose Ester and Track-Etched Thin Film Membranes. Scanning electron micrographs of the surfaces of (A) mixed cellulose ester (Millipore HA) and (B) track-etched 3-μm polyethylene terephthalate (PET) membranes show diverse surface structure, pore size, and distribution. Mixed cellulose ester membranes offer very high porosity for soluble factors and have a structured surface for cellular attachment. These membranes typically have a small functional pore size (0.45 μm) and are generally opaque. Track-etched PET membranes provide superior diffusion, homogeneous pore size (0.45–12 μm) as well as good optical properties for microscopy.

from 0.2 to 12 μm (Fig. 1). PET membranes are particularly well suited for stem cell studies because of their good optical quality, high diffusibility, and large range of pore size and density.

Although membrane-based cell culture techniques are very similar to those on solid tissue culture-treated polystyrene, there are some considerations that need to be taken when using membranes both in nomenclature and media exchange.

Since membrane-based cell culture provides asymmetric exposure to the cells, terminology has been developed in order to describe the two compartments. Because of their long history in epithelial cell studies, it is often common to refer to the underside of the membrane (the side opposite of the cultured cells on top of the membrane) as the basolateral side of the membrane based on the polarity observed when epithelial cells are grown on membranes. As per this convention, the literature often refers to the upper side of the membrane as the apical side. In addition, the support plates that hold the membrane inserts or plates are often referred to as receiver plates (the multiwell plates in which membrane inserts and plates are placed, in which each well is separated) or feeder trays when describing a large single-well companion plate in which all wells of a multiwell membrane plate are exposed to the same reservoir of media.

Procedurally, the most notable difference in membrane-based cell culture from plastic-based systems is the hydrostatic force of media from under that membrane that could cause a net upward flow of media through the membrane if there is no media on top of the membrane or the head height of the media inside the insert is lower than outside. This media flow could result in cell detachment and loss of cells that are poorly attached to the membrane. In order to prevent this possibility, it is important to aspirate and dispense media in the following order: Aspirate media from companion well (basolateral side) > aspirate from inside insert (apical side) > dispense fresh media back inside insert > dispense media outside insert in companion well. The volumes of the two chambers should be such that the heights of the media are the same inside and outside of the insert (see Table I) to prevent net media flow.

B. Preparation of Membranes

Modern membrane inserts for tissue culture typically come tissue culture treated to provide a charged surface conducive to cellular attachment. Although this treatment is the traditional method used on many different types of tissue cultureware to allow the attachment and proliferation of many cell types, the addition of surface coatings that provide a more *in vivo*-like extracellular environment, such as collagen, can be used depending on specific protocols and cellular needs to coat membranes in much the same ways, and with many similarities, to coating procedures for plastic tissue cultureware. Note: Biopore PTFE membranes are not tissue culture treated and require ECM coating for cell attachment and function.

In addition to coating the membrane of the inserts, companion plate wells or feeder trays often are not tissue culture treated and will require appropriate coating procedures in order to attach cells for the purpose of coculture experiments.

Below is a listing of typical membrane coating procedures for popular ECM proteins. These protocols describe coating with four common types of ECM used on membrane plates and inserts, collagen, fibronectin, laminin, and Matrigel®. These protocols are general in nature and should be modified for particular experimental or culture needs.

Table I
Typical Media Volumes Used for Millicell® (Millipore Corp.) Cell Culture Inserts and Plates

	Well diameter (mm)	Membrane surface area (cm^2)	Apical volume (ml)	Basolateral volume (ml) (24- or 96-well companion/receiver well)	Basolateral volume (ml) (single-well feeder tray)
6-well inserts	27	4.5	2	4.2[a]	N/A
12-well inserts	15	1.1	0.4	2.0[a]	N/A
24-well inserts	9	0.3	0.2	1.3[a]	N/A
Millicell-24 cell culture plate	11	0.7	0.3	0.6	28
Millicell-96 cell culture plate	5	0.1	0.075	0.25	32

[a]The basolateral volumes are measured in standard plastic culture plates with 18-mm well heights and are at equal head height with the apical well.

Values are given for Millipore Millicell inserts and plates. Since values will change by design, proper volumes should be empirically determined with other vendors.

1. Extracellular Matrix Coatings

   a. Coating with Collagen, Fibronectin, Laminin, and Matrigel
   1. Collagen Type 1 Coating (Table II)

   Materials
   - Millicell inserts (12 mm or 30 mm), 24- or 96-well membrane plates
   - 60% Ethanol
   - Rat tail collagen, Type 1, ~3 mg/ml

   Protocol
   1. Dilute the collagen 1:4 in 60% ethanol (1 part collagen and 3 parts 60% ethanol) and vortex until the collagen is solubilized.
   2. Place the applicable number of 12 mm Millicell inserts into the well(s) of a 24-well cell culture plate. For 30 mm inserts, use 6-well cell culture plate. Twenty-four- or 96-well plates should be placed in accompanying companion plate.
   3. Using a sterile pipette, add 50 µL of the collagen/ethanol mixture to each of the 12 mm Millicell insert(s). For 30 mm inserts, add 400 µL.
   4. Gently shake the cell culture plate until the collagen and 60% ethanol mixture evenly coats the inside of the Millicell insert or plate well.
   5. Air dry inserts in a laminar flow hood. Leave cell culture plate cover ajar to allow airflow and prevent condensation.

      Note: Although drying typically takes anywhere from 3 h to overnight, overnight drying is recommended.
   6. Insert or plates are ready for use.

   2. Fibronectin Coating

   Materials
   - Human fibronectin, 1 mg/ml reconstituted. Handle and store according to manufacturer's instructions.

   Protocol

Table II
Typical Extracellular Matrix (ECM) Coating Volumes for Millicell Inserts and Plates

	Volume (µl) per 12 mm (24-well) well	Volume (µl) per 30 mm (6-well) well
Collagen/Ethanol mix	50	400
Fibronectin/DMEM mix	100	700
Laminin/DMEM mix	100	700

DMEM; Dulbecco's modified Eagle's medium.

1. Dilute the fibronectin 1:10 in serum-free Dulbecco's modified Eagle's medium, DMEM (1 part fibronectin in 9 parts DMEM), and vortex until fibronectin is solubilized.
2. Place the applicable number of 12 mm Millicell inserts into the well(s) of a 24-well cell culture plate. For 30 mm inserts, use a 6-well cell culture plate.
3. Using a sterile pipette, add 100 µL fibronectin/DMEM coating mixture to each of the 12 mm Millicell insert(s). For 30 mm inserts, add 700 µL.
4. Gently shake the cell culture plate until the fibronectin/DMEM mixture evenly coats the Millicell insert(s). If using Biopore (Millipore CM) membrane, proceed to step 7.
5. Incubate for 45 min at room temperature.
6. Remove excess fibronectin. Proceed to step 8.
7. For Millipore CM membrane: air-dry inserts overnight in a laminar flow hood. Leave cell culture plate cover ajar to allow airflow and prevent condensation.

Note: Drying typically takes overnight (16 h) in a laminar flow tissue culture hood or clean equivalent

8. Seed Millicell insert with appropriate cell density

3. Laminin Coating

Materials
- Laminin, 1 mg/ml. Handle and store according to manufacturer's instructions.

Protocol
1. Dilute the laminin 1:10 in serum-free DMEM (1 part laminin in 9 parts DMEM) and vortex until the laminin is solubilized.
2. Place the applicable number of 12 mm Millicell inserts into the well(s) of a 24-well cell culture plate. For 30 mm inserts, use 6-well cell culture plates.
3. Using a sterile pipette, add 100 µL of laminin/DMEM mixture to each of the 12 mm Millicell insert(s). For 30 mm inserts, add 700 µL.
4. Gently shake the cell culture plate until the laminin/DMEM mixture evenly coats the Millicell inserts. If using Biopore (Millipore CM) membrane, proceed to step 7.
5. Incubate for 45 min at room temperature.
6. Remove excess laminin. Proceed to step 8.
7. For Millipore CM membrane: air-dry inserts overnight in a laminar flow hood. Leave cell culture plate cover ajar to allow airflow and prevent condensation.

Note: Drying typically takes overnight (16 h) in a laminar flow tissue culture hood or clean equivalent

8. Seed Millicell insert with appropriate cell density

4. Matrigel Coating Protocol

Matrigel (Becton Dickinson Biosciences) is a prepared basement membrane mixture rich in extracellualar matrix proteins produced from a mouse sarcoma. Its major components are laminin, collagen IV, heparan sulfate proteoglycans, and entactin (Kleinman *et al.*, 1982). This matrix is often used as both a thin layer for general attachment and invasion studies (Albini *et al.*, 1987) (Table III) or as a coating for human ESC attachment and expansion (see "Methods" section C-2 below for human ES cell procedure).

Materials
1. Matrigel—thawed and mixed according to manufacturer's instructions.
2. Chill (~4 °C) the hanging Millicell plates or inserts.
3. The pipette tips should be chilled.
4. Sterile-chilled Dulbecco's phosphate-buffered saline, DPBS (for pre-wetting the membrane).

Procedure for Thin Layer Coating
1. All steps must be performed using standard cell culture aseptic technique.
2. Pre-wet the membrane with sterile chilled DPBS (Table II). Dispense the following volumes of DPBS into each well.
3. Aspirate out the DPBS.
4. Gently dispense appropriate volume (Fig. 2) of the Matrigel with chilled pipette tips. Avoid generating bubbles during dispensing.
5. Cover plates and place in incubator and incubate for 2 h at 37 °C.
6. Once incubation is complete, the plates are ready to use.

Table III
Typical Pre-Wetting and Coating Volumes for Matrigel Thin Gel Coating of Millicell Inserts and Plates

	PBS Pre-wet volume (µl)	Thin gel coating minimum volume Matrigel (µl)
24-well hanging inserts	100	40
12-well hanging inserts	200	100
6-well hanging inserts	1000	300
Millicell 24	200	65

PBS; phosphate-buffered saline.

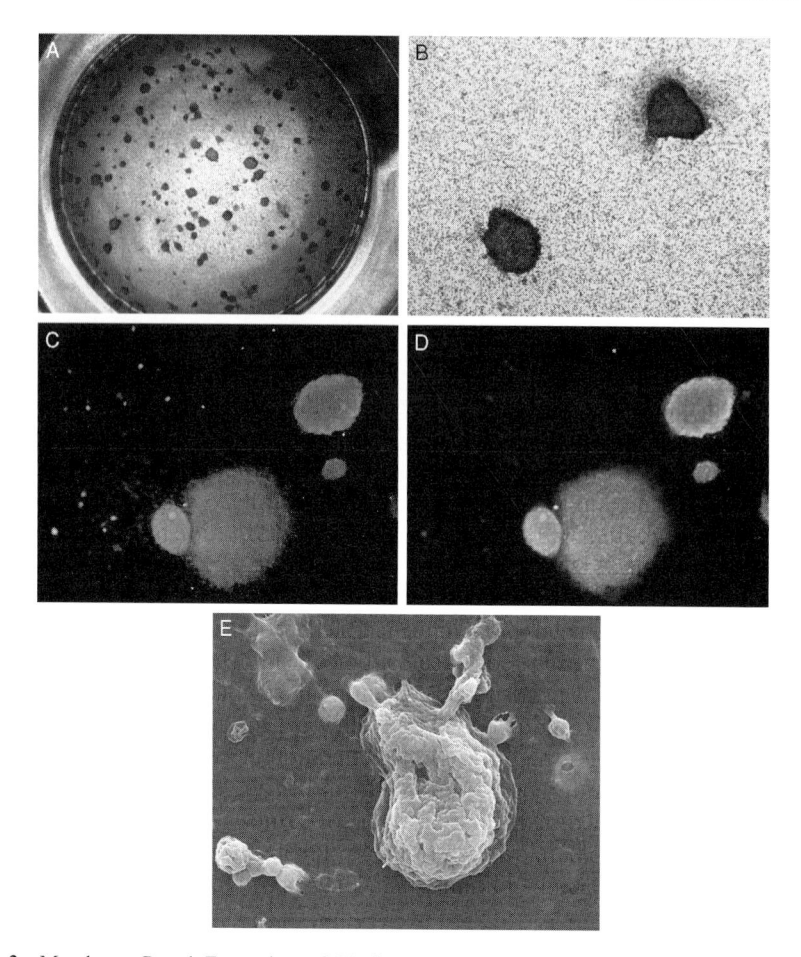

Fig. 2 Membrane-Based Expansion of Pluripotent Embryonic Stem Cell Colonies by Dilution Cloning. ES cells (strain 129/S6) were plated at ~200 cells per well of 96-well (A), 24-well Millicell (B–D) 1-μm polyethylene terephthalate (PET) tissue culture plates or 1-μm PET Millicell 6-well inserts (E). Indirect coculture was performed as described with ~2 × 10^4 mouse embryonic fibroblasts, MEFs (strain CF-1), per cm^2 in the feeder tray (plates) or companion wells (inserts). Pluripotency of murine embryonic stem cell (mESC) clonal colonies were determined by positive alkaline phosphatase (A, 20× and B, 100×), and immunoflourescence staining with anti-Oct 4 (C, 100×) and anti-Nanog (D, 100×). Images A–D were taken through the PET membrane with an inverted microscope. Alkaline phosphatase staining: Colonies were fixed with 3.7% formaldehyde and treated with Napthol/Fast Red Violet solution (Millipore Corp.) Immunocytochemistry analysis: Colonies were fixed with 90% methanol and permeablized with 0.1% saponin. Primary antibodies, mouse anti-Oct-4 (Millipore #MAB4401) and rabbit anti-Nanog (Millipore #AB5731) were followed by the secondary antibodies anti-mouse Cy3 and anti-rabbit Cy2, respectively. Scanning electron microscopy (E) samples were prepared by fixation in 4% gluteraldehyde followed by ethanol drying and OsO$_4$ treatment (see text). Images were acquired by coating with ~150Å gold for contrast enhancement and electrical continuity. Representative images were collected in the FEI Quanta FEG ESEM 200 under high vacuum at 15 keV. (See Plate no. 2 in the Color Plate Section.)

C. Co-culture of ES Cells with MEFs for Pluripotent Expansion

The requirement that the ESCs are maintained in the undifferentiated state is a general challenge during clone isolation and expansion. Carefully controlled culture techniques are required in order to have a homogeneous population of pluripotent cells for sensitive subsequent manipulations such as transgenic animal production by blastocyst injection or reproducible tissue-specific differentiation. Traditional methods with both human and mouse ESCs typically accomplish the expansion of pluripotent ESCs by a labor-intensive multistep direct coculture of the ESCs on a mitotically inactivated (typically mitomycin C or gamma irradiation treated) mouse embryonic fibroblast (MEF) feeder layer in tissue culture-treated plates or flasks. In more recent improvements to prevent cross-species contamination, several researchers have shown the use of feeder layers composed of fibroblasts from other origins such as human embryonic fibroblasts (Amit *et al.*, 2003), human ESC-derived fibroblasts (Xu *et al.*, 2004), or other cells types (Choo *et al.*, 2004; Hovatta *et al.*, 2003; Lee *et al.*, 2004; Miyamoto *et al.*, 2004) (also see Chapters 1 and 2 this volume) as suitable feeder layers for human ESCs. Additionally, the ability for the feeder layer to prevent the differentiation of the ESCs by indirect influence is shown by the ability of preconditioned media (exposed to feeder cells before use) to maintain pluripotent ESC culture (Smith and Hooper, 1987; Wang *et al.*, 2005; Xu *et al.*, 2001).

Microporous membranes offer the opportunity to allow coculture of ESCs and MEFs in an indirect manner whereby the MEFs continually condition the medium for ESC pluripotent expansion while facilitating their separation for subsequent downstream manipulations such as gene expression analysis, immunochemical analysis, and differentiation.

1. Multiwell Expansion of Murine Embryonic Stem Cells

This section illustrates the expansion of pluripotent colonies in a membrane-based cell culture plate. It has been demonstrated that the ability of MEFs to maintain ESCs in an undifferentiated state does not require direct contact between the two cell types (Wang *et al.*, 2005; Xu *et al.*, 2001). This has been shown with both mouse and human embryonic stem cells (hESCs). In this example, we illustrate the ability of MEFs to continually condition the media for the growth and undifferentiated clonal isolation of ESCs in a porous membrane-based indirect coculture system.

The configuration of this coculture setup involves growing the MEFs in a feeder tray below wells containing ESCs separated by a porous membrane filter. This arrangement allows for a physical separation between the two cell types eliminating the need for mitotically inactivating the MEFs while continuing to allow the MEFs to condition the media for maintenance of ESCs pluripotence (Fig. 2). Since the ESC-containing wells share a larger volume of media, frequency of media exchange can be reduced. In addition to improving the culturing and expansion of the ESC clones in 96-well format, the separation of the cell types in larger membrane insert wells using this filter-based coculture eliminates the requirement for removing the MEFs before downstream manipulations such as differentiation, embryoid body formation, or blastocyst injection.

Although this example uses the ability of MEFs to expand ESCs in an undifferentiated state, variations in this coculture system can easily be adopted for the study of indirect cell–cell interactions to direct the differentiation of stem cells by using cell types that may induce differentiation and for screening of factors that may interfere with these processes.

a. Materials
- MEF Media:

    DMEM (5.6 mg/ml glucose)
    10% fetal bovine serum (FBS)
    2 mM glutamine
    1 mM sodium pyruvate
    100 units/ml penicillin/streptomycin
    1% nonessential amino acids (from 100× stock)

- ESC Media:

    DMEM (5.6 mg/ml glucose)
    20% ESC qQualified FBS
    2 mM glutamine
    1 mM sodium pyruvate
    100 units/ml penicillin/streptomycin
    1% nonessential amino acids (from 100× stock)
    55 µM β-mercaptoethanol
    1000 units/ml LIF (ESGRO®—Millipore Corp)

- DPBS
- Fibronectin 0.1% solution
- Primary MEFs

b. Protocol

i. Fibronectin Coating and MEF Seeding of Single-Well Feeder Trays

1. Coat single-well feeder trays with fibronectin for 45 min at room temperature.

    a. Stock is 0.1% (1000 µg/ml)
    b. Working concentration is 25 µg/ml in sterile DPBS
    c. 5–10 ml of fibronectin solution per single-well tray

2. Remove excess fibronectin, trays are now ready for use

3. Thaw MEF from −80 °C freezer.
   a. Gently shake MEF vial in a 37 °C water bath.
   b. Transfer MEF to 15 ml tube containing 10 ml prewarmed MEF media.
   c. Pellet cells, 1000 rpm for 4 min.
   d. Remove supernatant and resuspend cells in fresh prewarmed MEF media.
4. Seed fibronectin-coated single-well feeder trays with MEF feeder cell suspension.
   a. 1.7×10^6 MEF cells per single-well tray ($\sim 2 \times 10^4$ cells per cm^2) will result in 95% confluence within 24 h.
5. Cover with lid and incubate single-well trays at 37 °C overnight.

ii. ESC/MEF Indirect Coculture on Membrane-Based Cell Culture Plate

1. Harvest ES cells from MEF–ESC direct cocultures [previously cultured as described (Robertson, 1987)].
   a. Wash cells 2× with prewarmed DPBS 10 ml per T75 flask (let them sit for 1–2 min per wash).
   b. Add 0.25% trypsin, 3 ml per T75 flask, should take about 3 min, as observed under microscope.
   c. Add ESC media to inactivate trypsin.
   d. Mix well and wash flask wall to remove all cells.
2. Enrichment of ESCs from MEF feeder layer cells by selective adhesion.
   a. Add cell suspension to new TC flask.
   b. Incubate at 37 °C for 30–45 min.
   c. Remove cell suspension and repeat steps a and b. Three incubations will typically remove most MEFs, leaving a majority of ES cells in the cell suspension.
3. Seed cell culture filter plate (or insert) wells with ESC suspension.
4. Remove MEF media from single-well feeder trays (or companion wells) and replace with ESC media.
5. Add ESC seeded cell culture filter plates to single-well trays (or companion wells).
6. Incubate assembly at 37 °C until appropriate confluence level for passage.

iii. Passage of ESCs from Membrane-Based Cell Culture Plate

1. Wash cells 2× with prewarmed DPBS (let them sit for 1–2 min per wash).
2. Add 0.25% trypsin, 100 μL per 12 mm insert; 500 μL per 30 mm insert should take about 3 min; observe under microscope when cells are balled up.

3. Add ESC media to each well to inactivate trypsin.
4. Gentle scraping may be necessary to detach cells from the membrane surface. Using a P200 pipette, scrape colonies off of insert with pipette tip, taking care not to puncture the membrane.
5. Mix well to remove all cells.
6. Seed membrane cell wells with ESC suspension.
7. Incubate at 37 °C overnight.
8. *Repeat as desired when colony morphology dictates need to passage.*

2. Membrane-Based Expansion of hESCs by Indirect Coculture

Since their first isolation from preimplantation blastocysts in 1998 (Thomson *et al.*, 1998), hESCs have been cultured by a variety of methods using both direct culture methods with mitotically inactivated feeder cells (Akutsu *et al.*, 2006) or by indirect feeder-free systems through the use of fibroblast preconditioned media (Lebkowski *et al.*, 2001; Rosler *et al.*, 2004; Xu *et al.*, 2001) on Matrigel, laminin (Xu 2006) or fibronectin (Amit and Itskovitz-Eldor, 2006)-coated tissue culture plastic.

Porous membranes are ideal for use in the indirect coculture of hESCs with MEFs in order to provide separation of the pluripotent hESCs for downstream manipulation while allowing a convenient system that continually conditions the media without the need for mitotically inactivating the feeder cells (Fig. 3). In addition, the ability to passage pluripotent hESCs on a membrane without the need for careful dissection of the colonies from the feeder layers allows for a more controlled and efficient passage methodology.

This method can be easily transferred to a system for the study of indirect cellular interactions and diffusible factors on differentiation by the coculture of cells that induce differentiation of hESCs.

a. Materials
- hESC media

    DMEM-F12 (5.6 mg/ml glucose)
    10% serum replacement
    2 mM glutamine
    1 mM sodium pyruvate
    100 units/ml penicillin/streptomycin
    1% nonessential amino acids (from 100× stock)

- 55 µM β-mercaptoethanol
- Basic fibroblast growth factor (bFGF) at final concentration of 20 ng/ml
- MEF media

    DMEM (5.6 mg/ml glucose)
    10% FBS

3. Membrane-Based Embryonic Stem Cell Culture

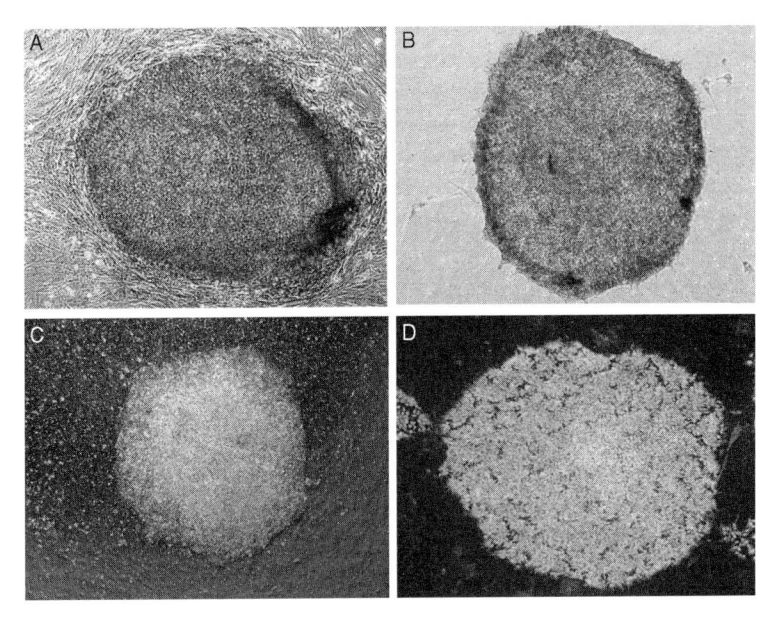

Fig. 3 Membrane-Based Expansion of Pluroptent Human ES Cell Colonies. H9 human embryonic stem cells were cultured in DMEM/F-12 media supplemented with basic fibroblast growth factor (bFGF) either by direct coculture on a mouse embryonic feeder layer in 6-well TC plate (A) or by indirect coculture in Matrigel-coated Millicell 6-well hanging 1-μm polyethylene terephthalate (PET) inserts (B–D). Indirect coculture was performed by plating mouse embryonic fibroblasts (MEFs) at ~40 K cell/cm^2 onto 6-well companion plates into which the inserts were placed. Images were taken at 50× of alkaline phosphatase stained (A and B), dark phase (C), and anti-Oct-4 stained (D) colonies. Alkaline phosphatase and anti-Oct-4 staining performed as described in Fig. 2. PET membrane samples (B–D) were imaged through the PET membrane with an inverted microscope. (See Plate no. 3 in the Color Plate Section.)

2 mM glutamine
1 mM sodium pyruvate
100 units/ml penicillin/streptomycin
1% nonessential amino acids (from 100× stock)

- Matrigel

b. Protocol

Technical note: The thawing of cryopreserved hESCs typically results in a low percentage of viable cells even under optimized, standard procedures (Fujioka *et al.*, 2004). For this reason, it is recommended to initially thaw frozen hESC stocks directly onto a MEF feeder layer until a critical mass of pluripotent colonies is obtained before the transfer, culture, and passage of hESC on membrane inserts.

Although membranes are tissue culture treated, ECM coatings, such as Matrigel, laminin, or fibronectin, are required for proper attachment and expansion of hESCs.

i. Gelatin Coating and MEF Seeding
   1. Dilute gelatin in DPBS for a 0.1% solution.
   2. Add enough 0.1% g solution to each well of a solid bottom companion plate, ~1 ml per well.
   3. Growth surface area of a 6-well companion plate is ~9.5 cm^2.
   4. Incubate plate at 37 °C for at least 30 min.
   5. Remove remaining gelatin and replace with MEF media.
   6. Thaw MEFs from −80 °C freezer.
      a. Gently shake MEF vial in a 37 °C water bath.
      b. Transfer MEF to 15 ml tube containing 10 ml prewarmed MEF media.
      c. Pellet cells at 1000 rpm for 4 min.
      d. Remove supernatant and resuspend cells in fresh prewarmed MEF media.
   7. Seed gelatin-coated 6-well companion plate with MEF feeder cell suspension.
      a. 5×10^5 MEF cells per well (~2×10^4 cells per cm^2) will result in 95% confluence within 24 h.
   8. Cover 6-well plate with lid and incubate at 37 °C overnight.

ii. Passage of hESCs from MEF Direct Cell Cultures to Membranes Inserts

This procedure is for the transfer of hESCs from direct MEF cocultures to membrane inserts for subsequent indirect coculture. See Akutsu *et al.* for details on establishing and maintaining direct hESC–MEF cocultures (Akutsu *et al.*, 2006).

   1. Prepare a 1 mg/ml solution of collagenase IV in DMEM/F12 and filter sterilize.
   2. Prepare Matrigel-coated inserts:
      a. Dilute Matrigel 1:28 in sterile DMEM/F12.
      b. Add enough Matrigel in DMEM/F12 to coat each insert, ~500 µL per 6-well insert (Note: Growth surface area of a Millicell 6-well hanging insert is ~4.5 cm^2.)
      c. Incubate at room temperature or 37° C for at least 2 h to overnight.
   3. Remove hESC culture (typically in 6-well tissue culture plate with MEF feeder layer) from incubator.
   4. Aspirate media.
   5. Add collagenase IV solution to each well.
      a. 1 ml per well for a 6-well solid bottom companion plate.

3. Membrane-Based Embryonic Stem Cell Culture

6. Incubate hESCs and collagenase IV solution at room temperature for 10 min.
7. Using a P200 pipette, carefully scrape colonies off of MEF feeder layer with pipette tip, taking care not to remove the feeder layer.
8. Collect colony suspension with 5 ml pipette along with collagenase IV solution to a 15 ml conical tube.
    a. Pipette gently to mix and break up colonies, careful not to break colonies into single cells.
9. Pellet cells at 1000 rpm for 4 min.
10. After spinning, remove supernatant and resuspend hES cells in fresh hESC media
    a. Gently mix cell suspension.
11. Carefully aspirate Matrigel solution from treated 6-well inserts, place inserts into MEF-seeded 6-well companion plates.
12. Dispense 1 ml of hESC media into each insert, followed by dispensing 4 ml of hESC media under insert into each companion plate well.
13. Seed hESCs onto Matrigel-coated inserts.
    a. Slowly add 1 ml of resuspended hESCs dropwise into each insert, trying to distribute clusters evenly throughout membrane surface.
14. Dispense 4 ml of hESC media under insert into each companion plate well.

iii. Passage and Expansion of hESC Colonies on Membrane Inserts

1. Prepare a 1 mg/ml solution of collagenase IV in DMEM/F12 and filter sterilize.
2. Remove 6-well inserts (housed in a companion plate) containing hESC culture from incubator.
3. Aspirate media.
4. Add collagenase IV solution to each well.
    a. 0.5 ml per well for a 6-well insert.
5. Using a P200 pipette scrape colonies off of plastic with pipette tip.
    1. Gently triturate colonies by aspirating and dispensing the collagenase IV solution within each well.
6. Collect cell suspension with 5 ml pipette and transfer to a conical tube containing collagenase IV solution.
    1. Pipette gently to mix and further break up colonies, careful not to break colonies into single cells.

7. Pellet cells at 1000 rpm for 4 min.
8. After spinning, remove supernatant and resuspend hES cells in fresh hESC media.
    1. Gently mix cell suspension.
9. Seed hES cells on newly prepared (Matrigel treated) 6-well membrane inserts.
    a. Add 1 ml of hES cell suspension to each insert.
    b. Make sure cells are evenly dispersed throughout well.
    c. Add 4 ml of hESC media to each well of the 6-well companion plate.
10. *Repeat as desired when colony morphology dictates need to passage colonies (too big, fused colonies, etc.).*

D. Membrane-Based Differentiation of ES Cells

The ability to differentiate ESCs into various lineages holds great promise for their use for tissue engineering and the study of early developmental processes that would not normally be accessible otherwise. ESCs have the capability to form all cell types in the adult. The elucidation of the appropriate methodologies to differentiate these cells down specific lineages, however, is still very much in the early stages. It is apparent that in addition to soluble growth factors that can be added to the media, cell-to-cell induction processes are most certainly involved in the differentiation of the tissues both by direct and indirect influences.

1. Membrane-Based Cardiomyocyte Differentiation from ES Cells

Microporous membranes offer an important opportunity to understand the cellular environments required to differentiate specific tissues. They provide a means to facilitate the culture of 3-D structures by providing increased exposure to nutrients and diffusible factors. In the example, a membrane-based system is used to study the differentiation of ESCs. The example uses the spontaneous differentiation of mouse ESCs augmented by an initiation step of forming 3-D EBs (Wobus *et al.*, 1991) of defined size by the use of a hanging drop methodology (Fig. 4) (Dang *et al.*, 2002; Kurosawa *et al.*, 2003; Yamada *et al.*, 2002). In contrast to the propensity for mouse ES cells to differentiate into cardiomyocytes, human ES cells typically require further induction by the introduction of factors and/or by coculture with other cells (Passier *et al.*, 2005). For instance, Mummery *et al.* (2003) have shown that the coculture of human ES cells with visceral endoderm, typically present on the outer layers of EBs (Grabel *et al.*, 1998; Murray and Edgar 2001), can induce the formation of functional cardiomyocytes. The ability

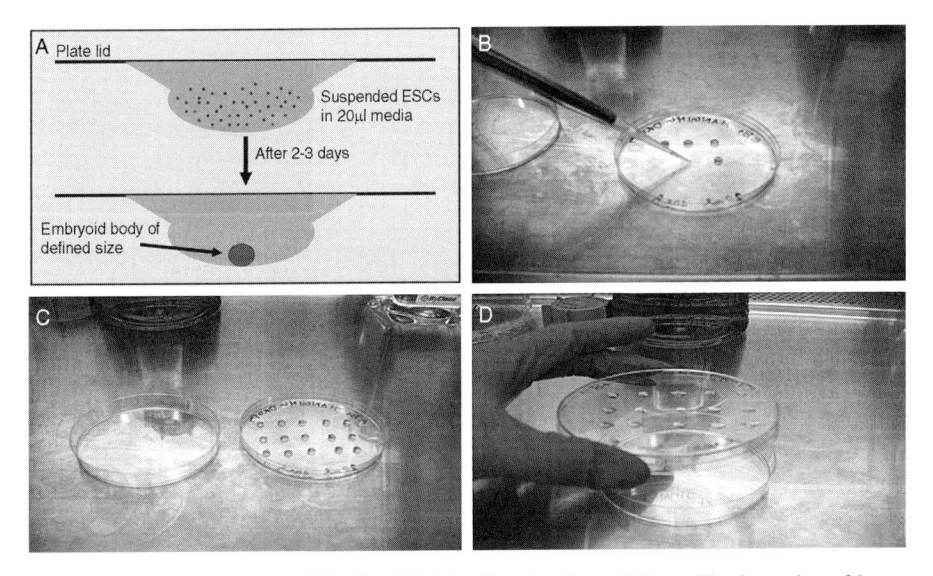

Fig. 4 Formation of Embryoid Bodies (EBs) by Hanging Drop Culture. The formation of homogeneously sized EBs was performed in hanging drop cultures. Schematic of hanging drop formation in inverted media droplets (A). Isolated ESC suspensions (see text) were placed onto inverted plate lids in 20 μl media drops (B and C). The lids were carefully flipped over right side up over a reservoir of PBS (D) and incubated for 2–3 days until EB formation was observed. (See Plate no. 4 in the Color Plate Section.)

for membrane-based coculture to dissect direct from indirect inducing influences is a powerful tool to enable the identification of diffusible factors that regulate cardiomyocyte differentiation.

The example below typifies the ability to perform differentiation studies on porous membrane substrates allowing for the ability to study indirect cell-to-cell influence on these processes. In addition to providing a suitable system for these studies, the ability to use single-well inserts, as used in this example, facilitates the ability to perform time studies as the inserts can be individually removed for downstream analysis such as immunocytochemistry or gene expression.

a. Materials

Cardiomyocyte Differentiation Media:

DMEM (5.6 mg/ml glucose)

10% FBS

2 mM glutamine

1 mM sodium pyruvate

100 units/ml penicillin/streptomycin

1% nonessential amino acids (from 100× stock)

b. Protocol

a. EB Formation by Hanging Drop Method

It is important for differentiation studies to remove the fibroblast feeder cells so that they do not incorporate themselves into the EBs. The separation of ES cells from the MEFs (as shown in step 2 below) is accomplished by the observation that MEFs typically begin to adhere to tissue culture treated plastic faster than do ES cells. Utilizing this observation, a series of transfers of ESC–MEF cell suspensions into new TC-treated flasks can result in the removal of the majority of MEFs. As an alternative to this time-intensive method, ES cells can be expanded by a membrane insert-based indirect coculture as described above, thus eliminating the need for separation since only ESCs are present on the top of the membrane by this method.

1. Expand mES cells on an MEF feeder layer in media containing 20% FBS and ESGRO supplement.
2. Enrichment of ES cells from MEF feeder layer cells by selective adhesion.
    a. Wash flasks 2× with prewarmed DPBS 10 ml per T75 flask (let them sit for 1–2 min per wash).
    b. Add 0.25% trypsin 3 ml per flask, which should take about 3 min, and monitor microscopically to prevent overdigestion.
    c. Triturate cells with 5 ml pipette until cells are in suspension.
    d. Add 12 ml ESC media to inactivate trypsin.
    e. Mix well and wash flask wall to remove all cells.
    f. Add cell suspension to new TC flask.
    g. Incubate at 37 °C for 30–45 min.
    h. Remove cell suspension and repeat steps f and g. Three incubations will typically remove most MEFs, leaving a majority of ES cells in the cell suspension.
3. Transfer ES cell suspension to a conical tube and pellet cells, 1000 rpm for 4 min.
4. Remove supernatant and resuspend cells in fresh media containing 10% FBS and without ESGRO.
5. Form EBs (Fig. 4).
    a. Transfer 20 µl drops containing 1500–2000 ES cells in 10% FBS (no ESGRO) to a non-tissue culture-treated plastic plate lid and carefully invert over a plate containing 15 ml of DPBS. DPBS will keep the drops from evaporating.
    b. Gravity will force the ES cells to pool to the bottom of the drop to form cell clusters (Fig. 4A).
    c. EBs should form within 1–2 days.

6. Coat 6-well plates/inserts with 0.1% gelatin in DPBS and incubate for at least 30 min.
   a. 1 ml for 6-well companion plates.
   b. 0.5 ml for 6-well hanging inserts.
7. Remove excess gelatin before attaching EBs.
8. Carefully flip hanging drop lid upside down. Harvest EBs with P200 pipette tip individually and transfer (~20–25 EBs) to treated 6-well plates/inserts in 10% FBS (no ESGRO®).
9. Cultivate EBs in 10% FBS (no ESGRO®) for 5–20 days with regular media replacement.

b. Gross Phenotypic Analysis of Cardiomyocyte Differentiation

Cardiomyocyte differentiation has the distinct phenotype that the EBs will show areas of rhythmic beating that can be seen under normal light microscopy. This phenotype, however, typically occurs most in the presence of fresh media and at physiological temperature. For these reasons, it is imperative that fresh, warm media is exchanged before visual observation and that plates are either viewed one at a time or viewed on a heated microscope stage to ensure temperature control.

1. Exchange fresh cardiomyocyte media 2 h before visualization and/or scoring of beating of attached EBs.
2. Cardiomyocyte beating is best observed after removal from incubator; visualize beating for scoring soon after removal from incubator, perhaps one plate at a time unless heated stage is available.

Note: Beating has been observed in some instances to be light sensitive as well. It is often observed that turning off the light source for a minute then turning it back on facilitates further beating. For these reasons, try to limit the exposure to the microscope light for observation purposes.

c. Immunochemical Analysis of Cardiac Marker Expression in Differentiated EBs (Fig. 5)

1. Wash EBs 3 times with 1 ml of DPBS.
2. Add 600 µl of ice cold 90% methanol for 5 min.
3. Wash EBs 3 times with 1 ml of DPBS.
4. Add 600 µl of 1° antibody solution [e.g., sarcomeric myosin heavy chain (β-MHC) in a DPBS solution of 0.1% saponin and 1% FBS].
5. Incubate at 37 °C, 95% RH, for 1 h.
6. Wash EBs 3 times with 1 ml of DPBS.
7. Add 600 µl of fluorescent labeled 2° antibody solution
8. Incubate at 37 °C, 95% RH, for 1 h.

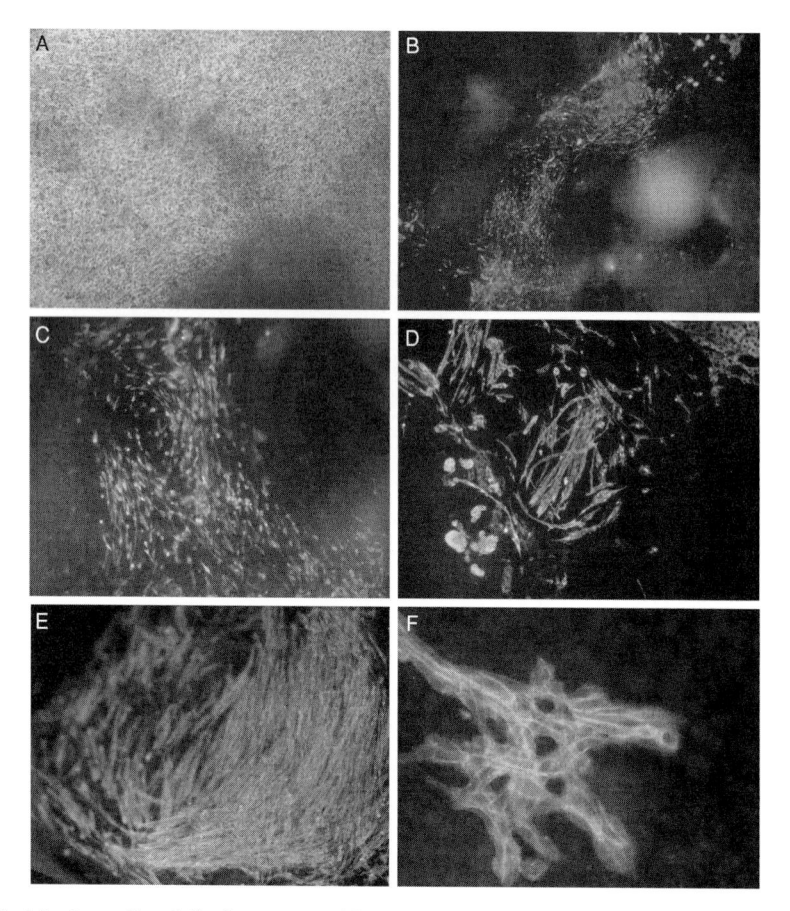

Fig. 5 Membrane-Based Cardiomyocyte Differentiation from Murine Embryonic Stem Cells. Differentiated embryoid bodies (EBs) were visualized after 12 days postattachment on 1-μm polyethylene terephthalate (PET) membranes by phase contrast (A, 50×) or immunflourescence (B–F). For immunocytochemical analysis, EBs were fixed with 90% methanol and permeablized with 0.1% saponin. Staining was performed for the identification of cardiac β-myosin heavy chain [(# MF-20, Developmental Studies Hybridoma Bank at the University of Iowa) (B, 50×, C, 100× and D, 400×)] or sarcomeric α-actinin [(Sigma # A7811) (E, 100× and F, 400×)]. Microscopy was performed through the PET membrane. (See Plate no. 5 in the Color Plate Section.)

9. Wash EBs 3 times with 1 ml of DPBS.
10. Add just enough mounting fluid to cover the membrane. See below for microscopic procedures.

E. Microscopic and Immunochemical Analysis Using Membrane Inserts

Perhaps the most basic and important ability in cell biology is to visualize cellular morphology and to tag specific markers (i.e., proteins) that indicate proliferation, function, differentiation, and cellular identity. Early work with microporous membrane supports for cell culture and assays were limited in their ability to perform high-definition imaging by their use of opaque materials such as cellulose acetate and polycarbonate. Developments in membranes made from optically clear materials, such as PET and PTFE (Biopore–Millipore CM), have made it possible to grow and differentiate cells on a substrate that readily facilitates microscopy and immunofluorescence studies. Additionally, these membranes can be easily removed from their plastic support in order to facilitate more advanced imaging techniques such as confocal and electron microscopy, techniques often quite difficult to perform in cumbersome tissue cultureware, such as plates and flasks or on fragile glass coverslips.

1. Membrane-Based Immunochemical Analysis

Immunochemical tagging of stem cell markers is an important set of techniques needed to define the pluripotency of the cells as well as identify tissue types after differentiation. Cellular markers can be typically divided by their location; these markers are typically extracellular (such as surface receptors or ligands) or intracellular (transcription factors, cytoskeletal proteins, etc.). Typically, the former can be stained on live cells directly, whereas the latter are typically fixed with appropriate fixative material and permeablized before incubation with primary antibody. Fixation is required to stabilize subcellular morphology and prevent degradation of antigens during subsequent staining procedures. Typically, cell preparations are submerged in a fixative solution (such as cold 90% methanol or 3.7% formaldehyde).

Procedure:

(Note: Reagents and buffers should be carefully pipetted down the side of the well in order to prevent disturbing cells adhered to the membrane)

1. Aspirate all media from inside filter plate wells and from companion plate wells or feeder trays. Fill filter plate wells with 1 ml of washing buffer (typically DPBS or HBSS). It is important to also fill the companion wells (28–32 ml for feeder trays, 1 ml for 24-well companion plates) with wash buffer to properly wash the underside of the membrane.
2. Incubate at room temperature for about 5 min and repeat wash 2 more times. Do not allow to dry.
3. If fixation is desired, add 200 µl of fixative solution to the inside of each well. For unfixed extracellular staining, proceed to step 5. It is not required to treat the underside of the membrane with fixative. Incubate according to protocol instructions. Generally, treatment is for ∼5 min. During this time, the solution should remain in the well and not leak through the membrane.
4. After treatment, aspirate fixative and fill filter and companion wells with washing buffer. Repeat steps 1 and 2 in order to fully remove the fixative solution from both sides of the filter membrane. Do not allow cells to dry.

5. Dilute primary antibody according to vendor recommendations. In order to obtain best results, it is recommended that optimal working dilutions be determined by the user. If permeabilization is required (such as for cytoplasmic or nuclear antigens), saponin can be added to the solution at a concentration of 0.1%.
6. Add 100 µl of antibody solution to each well, incubate at recommended temperature (typically room temperature or 4 °C) with mild shaking or rocking to ensure that solution wets out the entire filter surface. If antibody is fluorescently labeled (direct labeling), cover plate with foil to protect from light.
7. Aspirate antibody solution and wash both sides of membrane as indicated in steps 1 and 2 to remove all unbound antibody.
8. If performing indirect labeling with a secondary antibody, repeat steps 5 through 7 with secondary antibody. For fluorescent antibodies, continue to microscopy procedure. For enzyme-linked assays (HRP, etc.), follow vendor procedures for developing using 100 µl per well in each step.

2. Microscopy Procedure

Microscopic examination of samples can be performed in two modes: directly in the membrane insert or plate under low magnification, or on microscope slides for higher magnification using removed membrane.

a. Direct In-Plate Visualization Modes (Lower Magnifications)
1. Add 50 µl of mounting fluid to each sample well. If using fluorescence, it is recommended to use a mounting fluid that contains an anti-fade additive to prevent photobleaching

Viewing from below the plate (through PET membrane)

Optically clear membranes such as PET and Biopore–CM allow the visualization of cells through the membrane from below using typical inverted microscopes (Pitt *et al.*, 1987). Although inserts can be placed directly atop microscopes slides for viewing, multiwell membrane plates will typically require long working distance objectives of at least 2–3 mm to span from the bottom of the feeder tray to the top of the membrane in order to focus on the cells on the membrane if the feeder tray or companion plate is in place. Fixed cells that do not require to be visualized through media can be viewed directly without the feeder tray or companion plate but care should be taken not to contaminate the objective by liquid residue (media, mounting fluid) on membrane.

Viewing from above the plate (PET or PCF membrane)

Cells can be viewed in a conventional microscope directly from above. Cells can be visualized through the lid or with the lid removed. Working distances of the objective must be longer when reading from above compared to from below. Typically 5–10× objectives are used that have at least a (A) 13.6 mm or a (B) 18.0 mm working distance when viewing without or with the lid, respectively.

b. Visualizing Removed Membranes on Microscope Slides (for Higher Magnification or with Objectives with Short Working Distances)

The membrane can be removed from each well for microscopic evaluation. This allows for higher magnification examination and storage of the slides for future use.

Removal of membrane and mounting:

1. With a sharp scalpel, make a small incision in the edge of the membrane and carefully cut along the well side approximately one quarter of the diameters of the well. Using forceps, carefully hold the membrane while continuing to cut around well diameter to remove membrane. Use care to prevent membrane from curling.
2. Place membrane disk, cells facing up, onto a microscope slide. Add 50-μl mounting fluid to the membrane disk and allow wetting out in order to prevent bubbles under the disk. Slowly lower a cover slip at an angle to allow air bubbles to be removed.

3. Basic Sample Preparation for Electron Microscopy

In addition to benefiting from high magnification for light or fluorescent microscopy, the removal of the membrane offers a major advantage over cumbersome plastic tissue culture plates in the sample preparation for cell visualization by electron microscopy.

a. Materials
- Glutaraldehyde solution for electron microscopy, ~8% in H_2O
- 0.2 M sodium Cacodylate Buffer pH 7.2 (sodium cacodylate hydrate dissolved in H_2O)
- 2% glutaraldehyde in 0.1 M Cacodylate Buffer pH 7.2 with 0.1 M sucrose
- 1% osmium tetroxide in 0.1 M Cacodylate Buffer pH 7.2
- Prepared graded alcohol solutions: 25%, 50%, 75%, 95%, 100%

b. Sample Fixation Method
1. Replace growth media with 2% glutaraldehyde. Do this several times to ensure a complete transition from growth media to fixative. Fix for 2 h. Cells may be left in fixative overnight at 4 °C.
2. Rinse cells with 0.1 M Cacodylate Buffer, pH 7.2, with 0.1 M sucrose 3 times (let them sit for 5 min per wash)
3. Post fix with 1% osmium tetroxide in 0.1 M Cacodylate Buffer, pH 7.2, for 5–10 min
4. Rinse with 0.1 M Cacodylate Buffer, pH 7.2, 3 times (let them sit for 5 min per wash)

c. Dehydration and Membrane Removal
1. Dehydrate samples through graded ethanol series (25%, 50%, 75%, 95%, 3 × 100%), 5 min each.
2. In order to prevent leaking and to ensure complete dehydration, alcohol treatment steps should be performed with membrane inserts in companion plate wells with both sides of the membrane exposed to the alcohol.
3. Air dry inserts until membranes are completely dry.
4. With a sharp scalpel, make a small incision in the edge of the membrane and carefully cut along the well side approximately one quarter of the diameters of the well. Using forceps, carefully hold the membrane while continuing to cut around well diameter to remove membrane. Use care to prevent membrane from curling.
5. Place membrane disk, cells facing up, onto a microscope slide. Perform scanning electron microscope (SEM) according to microscope-specific procedure.

IV. Summary

ESCs offer the ability to simulate development in many ways not previously feasible *in vitro*. In order to model the *in vivo* environment, a variety of systems will prove to be essential in understanding the appropriate cues that direct tissue-specific differentiation. These cues will include factors and cellular interactions that are asymmetric and indirect.

The solid and nonpermissive nature of impermeable tissue culture plastic limits the *in vitro* growth and differentiation of cells by restricting the ability to be exposed to media components, nutrients, and growth factors to only one side. The featureless, planar nature of plastic supports also limits cells from their proper attachment to structured supports typically experienced *in vivo*.

For over five decades, microporous membranes have been pivotal in allowing cells to be grown and interact with factors and other cells on a structured support that simulates the *in vivo* 3-D environment. Experimentally, the ability for porous membrane-based systems to manipulate and measure cellular function from both sides of the culture has been essential to the understanding of factors that influence cellular function asymmetrically. In addition, the ability to separate cells via a porous separation has allowed for the dissection of direct and indirect effects of intercellular communication that has facilitated the understanding of diffusible factors during development.

The important contributions of microporous membranes that have accelerated the understanding of these effects in other systems throughout the years will prove critically important methods in the identification of the influences required for the maintenance and differentiation of ESCs.

Acknowledgments

The authors extend grateful thanks to Ajay Sharma, John Lynch, and Ken Ludwig (Millipore, Danvers, MA) for their critical manuscript review and recommendations. Also, the authors acknowledge Matthew Singer (Millipore, Temecula, CA) for initial feasibility studies and technical suggestions.

References

Akutsu, H., Cowan, C., and Melton, D. (2006). Human embryonic stem cells. *Methods Enzymol.* **418**, 78–92.

Albini, A., Iwamoto, Y., Kleinman, H., Martin, G., Aaronson, S., Kozlowski, J., and McEwan, R. (1987). A rapid *in vitro* assay for quantitating the invasive potential of tumor cells. *Cancer Res.* **47**(12), 3239–3245.

Amit, M., and Itskovitz-Eldor, J. (2006). Feeder-free culture of human embryonic stem cells. *Methods Enzymol.* **420**, 37–49.

Amit, M., Margulets, V., Segev, H., Shariki, K., Laevsky, I., Coleman, R., and Itskovitz-Eldor, J. (2003). Human feeder layers for human embryonic stem cells. *Biol. Reprod.* **68**(6), 2150–2156.

Baum, J., Mowat, A., and Kirk, J. (1971). A simplified method for the measurement of chemotaxis of polymorphonuclear leukocytes from human blood. *J. Lab. Clin. Med.* **77**(3), 501–509.

Boheler, K., Crider, D., Tarasova, Y., and Maltsev, V. (2004). Cardiomyocytes derived from embryonic stem cells. *Methods Mol. Med.* **108**, 417–436.

Boyden, S. (1962). The chemotactic effect of mixtures of antibody and antigen on polymorphonuclear leucocytes. *J. Exp. Med.* **115**, 453–466.

Byers, S., Hadley, M., Djakiew, D., and Dym, M. (1986). Growth and characterization of polarized monolayers of epididymal epithelial cells and Sertoli cells in dual environment culture chambers. *J. Androl.* **7**(1), 59–68.

Chen, H. (2005). Boyden chamber assay. *Methods Mol. Biol.* **294**, 15–22.

Chen, U., and Kosco, M. (1993). Differentiation of mouse embryonic stem cells *in vitro*: III. Morphological evaluation of tissues developed after implantation of differentiated mouse embryoid bodies. *Dev. Dyn.* **197**(3), 217–226.

Choo, A., Padmanabhan, J., Chin, A., and Oh, S. (2004). Expansion of pluripotent human embryonic stem cells on human feeders. *Biotechnol. Bioeng.* **88**(3), 321–331.

Cook, J., Crute, B., Patrone, L., Gabriels, J., Lane, M., and Van, B. R. (1989). Microporosity of the substratum regulates differentiation of MDCK cells *in vitro*. *In Vitro Cell. Dev. Biol.* **25**(10), 914–922.

Dang, S., Kyba, M., Perlingeiro, R., Daley, G., and Zandstra, P. (2002). Efficiency of embryoid body formation and hematopoietic development from embryonic stem cells in different culture systems. *Biotechnol. Bioeng.* **78**(4), 442–453.

Fujioka, T., Yasuchika, K., Nakamura, Y., Nakatsuji, N., and Suemori, H. (2004). A simple and efficient cryopreservation method for primate embryonic stem cells. *Int. J. Dev. Biol.* **48**(10), 1149–1154.

Fuller, S., and Simons, K. (1986). Transferrin receptor polarity and recycling accuracy in "tight" and "leaky" strains of Madin-Darby canine kidney cells. *J. Cell. Biol.* **103**(5), 1767–1779.

Gildea, J., Harding, M., Gulding, K., and Theodorescu, D. (2000). Transmembrane motility assay of transiently transfected cells by fluorescent cell counting and luciferase measurement. *Biotechniques* **29**(1), 81–86.

Goel, V., Accomazzo, M., DiLeo, A., Meirer, P., Pitt, A., and Pluskal, M. (1992). Deadend microfiltration: Applications, design, and cost. *In* "Membrane Handbook" (W. S. Winston Ho, and K. Sirkar, eds.), pp. 506–565. Springer, New York.

Grabel, L., Becker, S., Lock, L., Maye, P., and Zanders, T. (1998). Using EC and ES cell culture to study early development: Recent observations on Indian hedgehog and Bmps. *Int. J. Dev. Biol.* **42**(7), 917–925.

Grobstein, C. (1953). Morphogenetic interaction between embryonic mouse tissues separated by a membrane filter. *Nature* **172**(4384), 869–870.

Hovatta, O., Mikkola, M., Gertow, K., Stromberg, A., Inzunza, J., Hreinsson, J., Rozell, B., Blennow, E., Andang, M., and Ahrlund-Richter, L. (2003). A culture system using human foreskin fibroblasts as feeder cells allows production of human embryonic stem cells. *Hum. Reprod.* **18**(7), 1404–1409.

Hwang, N., Kim, M., Sampattavanich, S., Baek, J., Zhang, Z., and Elisseeff, J. (2005). The effects of three dimensional culture and growth factors on the chondrogenic differentiation of murine embryonic stem cells. *Stem Cells* **18**, 18.

Kleinman, H., McGarvey, M., Liotta, L., Robey, P., Tryggvason, K., and Martin, G. (1982). Isolation and characterization of type IV procollagen, laminin, and heparan sulfate proteoglycan from the EHS sarcoma. *Biochemistry* **21**(24), 6188–6193.

Kurosawa, H., Imamura, T., Koike, M., Sasaki, K., and Amano, Y. (2003). A simple method for forming embryoid body from mouse embryonic stem cells. *J. Biosci. Bioeng.* **96**(4), 409–411.

Lebkowski, J., Gold, J., Xu, C., Funk, W., Chiu, C., and Carpenter, M. (2001). Human embryonic stem cells: Culture, differentiation, and genetic modification for regenerative medicine applications. *Cancer J.* **7**(2), S83–S93.

Lee, J., Song, J., Lee, J., Park, J., Kim, S., Kang, S., Kwon, J., Kim, M., Roh, S., and Yoon, H. (2004). Available human feeder cells for the maintenance of human embryonic stem cells. *Reproduction* **128**(6), 727–735.

Liu, W., Li, Y., Cunha, S., Hayward, G., and Baskin, L. (2000). Diffusable growth factors induce bladder smooth muscle differentiation. *In Vitro Cell. Dev. Biol. Anim.* **36**(7), 476–484.

Miyamoto, K., Hayashi, K., Suzuki, T., Ichihara, S., Yamada, T., Kano, Y., Yamabe, T., and Ito, Y. (2004). Human placenta feeder layers support undifferentiated growth of primate embryonic stem cells. *Stem Cells* **22**(4), 433–440.

Mostov, K., and Deitcher, D. (1986). Polymeric immunoglobulin receptor expressed in MDCK cells transcytoses IgA. *Cell* **46**(4), 613–621.

Mummery, C., Ward-van, O. D., Doevendans, P., Spijker, R., van, d. B. S., Hassink, R., van, d. H. M., Opthof, T., Pera, M., De, l. R. A., Passier, R., and Tertoolen, L. (2003). Differentiation of human embryonic stem cells to cardiomyocytes: Role of coculture with visceral endoderm-like cells. *Circulation* **107**(21), 2733–2740.

Murray, P., and Edgar, D. (2001). The regulation of embryonic stem cell differentiation by leukaemia inhibitory factor (LIF). *Differentiation* **68**(4–5), 227–234.

Ng, E., Davis, R., Azzola, L., Stanley, E., and Elefanty, A. (2005). Forced aggregation of defined numbers of human embryonic stem cells into embryoid bodies fosters robust, reproducible hematopoietic differentiation. *Blood* **106**(5), 1601–1603.

Passier, R., Oostwaard, D., Snapper, J., Kloots, J., Hassink, R., Kuijk, E., Roelen, B., de, L. R. A., and Mummery, C. (2005). Increased cardiomyocyte differentiation from human embryonic stem cells in serum-free cultures. *Stem Cells* **23**(6), 772–780.

Pitt, A., Gabriels, J., Badmington, F., McDowell, J., Gonzales, L., and Waugh, M. (1987). Cell culture on a microscopically transparent microporous membrane. *Biotechniques* **5**(2), 162–171.

Robertson, E. (1987). Embryo-derived stem cell lines. *In* "Teratocarcinomas and Embryonic Stem Cells: A Practical Approach" (E. Robertson, ed.), pp. 71–112. IRL Press, Oxford, UK.

Rosler, E., Fisk, G., Ares, X., Irving, J., Miura, T., Rao, M., and Carpenter, M. (2004). Long-term culture of human embryonic stem cells in feeder-free conditions. *Dev. Dyn.* **229**(2), 259–274.

Rutten, A., Bequet-Passelecq, B., and Koeter, H. (1990). Two-compartment model for rabbit skin organ culture. *In Vitro Cell. Dev. Biol.* **26**(4), 353–360.

Smith, A., and Hooper, M. (1987). Buffalo rat liver cells produce a diffusible activity which inhibits the differentiation of murine embryonal carcinoma and embryonic stem cells. *Dev. Biol.* **121**(1), 1–9.

Thomson, J., Itskovitz-Eldor, J., Shapiro, S., Waknitz, M., Swiergiel, J., Marshall, V., and Jones, J. (1998). Embryonic stem cell lines derived from human blastocysts. *Science* **282**(5391), 1145–1147.

3. Membrane-Based Embryonic Stem Cell Culture

Verfaillie, C. (1992). Direct contact between human primitive hematopoietic progenitors and bone marrow stroma is not required for long-term *in vitro* hematopoiesis. *Blood* **79**(11), 2821–2826.

von, A. U., and Mackay, C. (2000). T-cell function and migration. Two sides of the same coin. *N. Engl. J. Med.* **343**(14), 1020–1034.

Wang, G., Zhang, H., Zhao, Y., Li, J., Cai, J., Wang, P., Meng, S., Feng, J., Miao, C., Ding, M., Li, D., and Deng, H. (2005). Noggin and bFGF cooperate to maintain the pluripotency of human embryonic stem cells in the absence of feeder layers. *Biochem. Biophys. Res. Commun.* **330**(3), 934–942.

Widdicombe, J., Sachs, L., and Finkbeiner, W. (2003). Effects of growth surface on differentiation of cultures of human tracheal epithelium. *In Vitro Cell. Dev. Biol. Anim.* **39**(1–2), 51–55.

Wobus, A., Wallukat, G., and Hescheler, J. (1991). Pluripotent mouse embryonic stem cells are able to differentiate into cardiomyocytes expressing chronotropic responses to adrenergic and cholinergic agents and Ca2+ channel blockers. *Differentiation* **48**(3), 173–182.

Xu, C. (2006). Characterization and evaluation of human embryonic stem cells. *Methods Enzymol.* **420**, 18–37.

Xu, C., Inokuma, M., Denham, J., Golds, K., Kundu, P., Gold, J., and Carpenter, M. (2001). Feeder-free growth of undifferentiated human embryonic stem cells. *Nat. Biotechnol.* **19**(10), 971–974.

Xu, C., Jiang, J., Sottile, V., McWhir, J., Lebkowski, J., and Carpenter, M. (2004). Immortalized fibroblast-like cells derived from human embryonic stem cells support undifferentiated cell growth. *Stem Cells* **22**(6), 972–980.

Yamada, T., Yoshikawa, M., Kanda, S., Kato, Y., Nakajima, Y., Ishizaka, S., and Tsunoda, Y. (2002). *In vitro* differentiation of embryonic stem cells into hepatocyte-like cells identified by cellular uptake of indocyanine green. *Stem Cells* **20**(2), 146–154.

CHAPTER 4

Culture of Rodent Spermatogonial Stem Cells, Male Germline Stem Cells of the Postnatal Animal

Hiroshi Kubota[*] and Ralph L. Brinster[†]

[*]Laboratory of Cell and Molecular Biology
Department of Animal Science
School of Veterinary Medicine
Kitasato University
Towada, Aomori 034-8628, Japan

[†]Department of Animal Biology
School of Veterinary Medicine
University of Pennsylvania
Philadelphia, Pennsylvania 19104

Abstract
I. Introduction
II. Rationale
    A. Basic Concept of SSC Culture
    B. Components of SSC Culture System
III. Methods
    A. Protocol for Mouse SSC Culture
    B. Protocol for Rat SSC Culture
IV. Materials
    A. Reagents for Cell Preparation
    B. Stock Solution of Reagents for SFM and Cell Culture
    C. Reagents for STO Feeder Layers
V. Discussion
    References

Abstract

Spermatogonial stem cells (SSCs), postnatal male germline stem cells, are the foundation of spermatogenesis, during which an enormous number of spermatozoa is produced daily by the testis throughout life of the male. SSCs are unique among stem cells in the adult body because they are the only cells that undergo self-renewal and transmit genes to subsequent generations. In addition, SSCs provide an excellent and powerful model to study stem cell biology because of the availability of a functional assay that unequivocally identifies the stem cell. Development of an *in vitro* culture system that allows an unlimited supply of SSCs is a crucial technique to manipulate genes of the SSC to generate valuable transgenic animals, to study the self-renewal mechanism, and to develop new therapeutic strategies for infertility. In this chapter, we describe a detailed protocol for the culture of mouse and rat SSCs. A key factor for successful development of the SSC culture system was identification of *in vitro* growth factor requirements for the stem cell using a defined serum-free medium. Because transplantation assays using immunodeficient mice demonstrated that extrinsic factors for self-renewal of SSCs appear to be conserved among many mammalian species, culture techniques for SSCs of other species, including farm animals and humans, are likely to be developed in the coming 5–10 years.

I. Introduction

Germ cells are specialized cells that pass the genetic information of an individual to the next generation. Production of functional germ cells is essential for continuation of the germline of the species. Spermatogenesis, the process of male germ cell production, takes place in the seminiferous tubules of the postnatal testis and is a highly productive system in the body. In the mammalian testis, more than 20 million sperms per gram of tissue are produced daily (Amann, 1986). The high productivity relies on spermatogonial stem cells (SSCs). Like other types of stem cells in adult tissues, SSCs self-renew and produce daughter cells that commit to differentiate throughout life of the male (Meistrich and van Beek, 1993). Furthermore, in mammals, SSCs are unique among stem cells in the adult body, because they are the only cells that undergo self-renewal and transmit genes to subsequent generations.

Stem cells are defined by their biological function; therefore, unequivocal identification of a stem cell requires a functional assay (Weissman *et al.*, 2001). A functional transplantation assay for SSCs was developed in mice a decade ago and made it possible to study the biological function of SSCs, including self-renewal and differentiation (Brinster and Avarbock, 1994; Brinster and Zimmermann, 1994). In the transplantation assay, donor cells are harvested from the testes of fertile mice and are microinjected into seminiferous tubules of recipient mice. A small subpopulation of transplanted testis cells, the SSCs, colonize the basement membrane of the seminiferous tubules and begin to proliferate. Individual donor SSCs eventually form spermatogenic colonies in the seminiferous tubules of the recipients. When testicular cells from transgenic mice that express the *Escherichia coli LacZ* gene, which encodes a β-galactosidase, or the green fluorescence protein (GFP) gene are transplanted, donor colonies can be identified and counted readily because of the transgene expression in the donor cells (Fig. 1). About 2 months after transplantation of mouse SSCs, differentiated germ cells in a colony have progressed to spermatozoa in the seminiferous tubules of the recipient animal. The reconstituted spermatogenic colonies continuously produce spermatozoa throughout the remaining life of the recipient males. Several lines of evidence indicate that each donor-derived spermatogenic colony arises from a single SSC (Dobrinski *et al.*, 1999b; Kanatsu-Shinohara *et al.*, 2006b; Zhang *et al.*, 2003), and the colonization efficiency of transplanted SSCs is estimated to be 5–12% (Nagano, 2003; Ogawa *et al.*, 2003). Following transplantation, the recipient males can

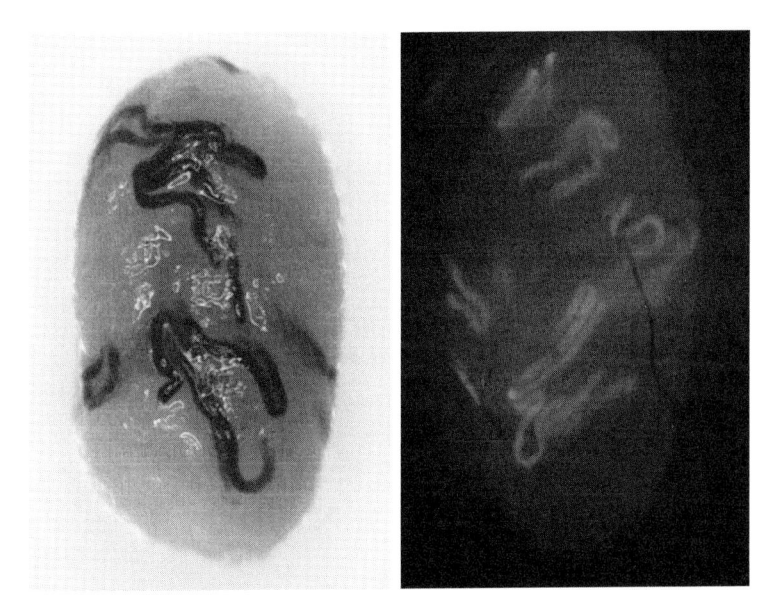

Fig. 1 Colony formation of donor-derived spermatogenesis in infertile recipient mouse testes. Testicular germ cells were isolated from transgenic mice that express reporter genes (*LacZ* or GFP) and injected into testes of mice treated with busulfan. Two months after transplantation, donor-derived spermatogenesis is reconstituted. Left: testis transplanted with *LacZ* expressing SSCs from transgenic mouse line B6.129S-Gt(ROSA)26Sor/J (ROSA, Jackson Laboratory). The testis was stained with 5-bromo-4-choloro-3-indolyl β-D- galactoside (X-gal). Right: testis transplanted with GFP-expressing SSCs from transgenic mouse line C57BL/6-TgNACTB-EGFP)1Osb/J (Jackson Laboratory). Each blue stretch or green stretch of cells in the testes represents a colony of spermatogenesis that arises from a single SSC. (See Plate no. 6 in the Color Plate Section.)

become fertile and produce progeny with donor cell haplotype, demonstrating normal function of the spermatozoa originating from transplanted germ cells (Brinster and Avarbock, 1994). Thus, it is clear that the spermatogenic colony-forming cells are SSCs, and the spermatogonial transplantation technique is a functional assay that allows quantitative evaluation of SSCs from a variety of sources.

Existence of a definitive functional assay to unequivocally identify SSCs provides an ideal experimental system to study stem cell biology. Using the functional assay, SSCs and the surrounding microenvironment, or the stem cell niche, in the seminiferous tubules have been studied (Brinster, 2002). Furthermore, by means of genetic modification of SSCs isolated from testes followed by transplantation, it has been shown that an SSC is a valuable vehicle to generate genetically modified animals (Hamra *et al.*, 2002; Nagano *et al.*, 2001a).

Because of the enormous potential of SSCs in basic research and applied science, including agriculture and medicine, development of an *in vitro* culture system for stem cells is extremely important. An early study demonstrated that SSCs could survive on STO (SIM mouse embryo-derived thioguanine and ouabain resistant) mouse embryonic fibroblast feeder layers for several months in culture (Nagano *et al.*, 1998). Recently, several methods to culture rodent SSCs for long periods have been reported (Hamra *et al.*, 2005; Kanatsu-Shinohara *et al.*, 2003; Kubota *et al.*, 2004b; Ryu *et al.*, 2005). To develop a long-term culture system for SSCs, one of the most crucial objectives is identification of extrinsic factors essential to promote self-renewal and expansion of SSCs *in vitro*. Previous studies using transgenic mice with gain-of-function and loss-of-function of glial cell line-derived neurotrophic factor (GDNF) indicated that this ligand is a key growth factor to control survival and proliferation of undifferentiated spermatogonia and perhaps SSCs *in vivo* (Meng *et al.*, 2000). Using a serum-free culture system, we clearly demonstrated that GDNF is indeed the primary growth factor for SSC self-renewal (Kubota *et al.*, 2004b). In the presence of GDNF, SSCs formed tightly packed clumps of cells and continuously proliferated. Clump-forming germ cells kept on expanding for more than 6 months in the defined serum-free medium supplemented with GDNF, and reconstituted long-term spermatogenesis following transplantation into recipient testes (Fig. 2). Furthermore, GDNF was found to be the crucial extrinsic factor for rat SSC self-renewal and proliferation (Ryu *et al.*, 2005). With slight modifications of the culture condition for mouse SSCs, rat SSCs can be expanded *in vitro*, and progeny derived from cultured rat SSCs could be generated. In this chapter, we describe a detailed method to culture mouse and rat SSCs. Since this is a technical and methodological chapter, neither the developmental biology of male germ cells nor the physiology of spermatogenesis will be reviewed. Excellent publications have described these processes in detail (de Rooij, 1998; Nagy *et al.*, 2003; Russell *et al.*, 1990; Zhao and Garbers, 2002).

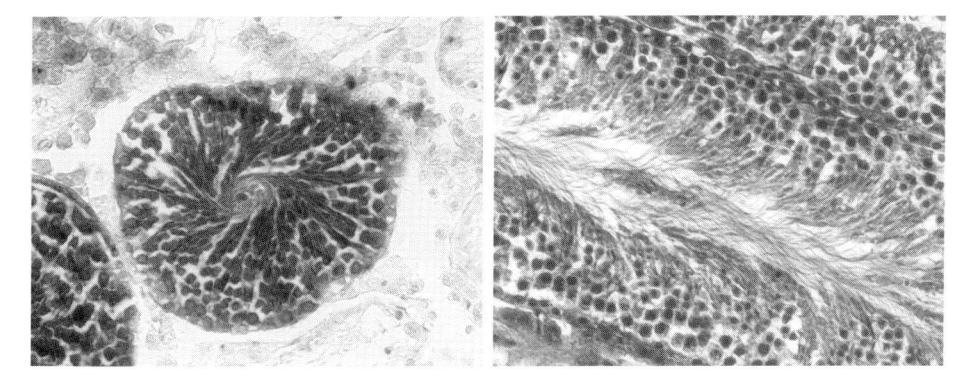

Fig. 2 Histological cross sections of seminiferous tubules of recipient W^{54}/W^r testes transplanted with cultured SSCs. Histological sections of testes transplanted with ROSA SSCs that were cultured for 6 months (left, stained with X-gal, counter stain; Nuclear Fast Red) and 7.5 months (right, stain; Hematoxylin-Eosin). (See Plate no. 7 in the Color Plate Section.)

II. Rationale

A. Basic Concept of SSC Culture

Development of new culture conditions for animal cells has been largely empirical. At present no unifying culture technique applicable to various tissue-specific stem cells has been developed. Despite early hopeful expectations by stem cell biologists, recent studies have indicated that characteristics of tissue-specific stem cells do not appear to be conserved (Kubota *et al.*, 2003). However, because the general process of spermatogenesis is believed to be conserved among many species, the self-renewal machinery of SSCs in different species might be similar. In fact, xenotransplantation experiments demonstrated that SSCs from all mammals examined, including rats, rabbits, dogs, pigs, cattle, horses, baboons, and humans, colonized and proliferated or were maintained in the seminiferous tubules of immunodeficient mice (Clouthier *et al.*, 1996; Dobrinski *et al.*, 1999a, 2000; Nagano *et al.*, 2001b, 2002; Oatley *et al.*, 2004). These results suggest that critical exogenous factors to promote self-renewal of SSCs are conserved among various species. Therefore, we employed a systematic approach to develop a SSC culture system using the mouse as a model because once a defined culture condition for mouse SSCs was developed, it would form the basis for other species.

Stem cells generally divide rarely in normal physiological conditions (Meistrich and van Beek, 1993; Morrison *et al.*, 1997; Potten and Morris, 1988). The microenvironment surrounding stem cells, which is called the stem cell niche, controls the timing of proliferation and differentiation of the stem cells (Spradling *et al.*, 2001). To maintain stem cells *in vitro*, reconstitution of the stem cell niche would be ideal, but likely not completely attainable. However, if one can maintain self-renewing stem cells *in vitro*, the culture condition probably provides essential signals that promote self-renewal found endogenously in the stem cell niche *in vivo*. Clearly, any knowledge about stem cell niche factors provides valuable information for the development of an *in vitro* culture system for maintaining stem cells.

To develop a defined culture condition for mouse SSCs, we chose a culture system that consists of a germ cell population enriched for stem cells, serum-free hormonally defined culture medium, and mitotically inactivated STO feeder cells (Kubota *et al.*, 2004a). Originally this culture system was developed for hepatic progenitors, hepatoblasts, in the rat (Kubota and Reid, 2000). A previous study clearly demonstrated that the culture system is useful to investigate stem/progenitor cells and the surrounding microenvironment, because the condition minimized unknown components in the culture (Kubota and Reid, 2000). Such an *in vitro* system will allow controlled and detailed investigation of factors involved in cell fate decisions. We first optimized the basic culture condition to allow mouse SSCs to survive for a short period by modifying the serum-free medium for hepatoblasts. The modified culture condition was able to maintain mouse SSCs without loss of the stem cell activity assessed by the transplantation assay for at least 1 week (Kubota *et al.*, 2004a). The availability of a functional assay made possible the evaluation of various conditions for survival of SSCs. Subsequently, we sought factors that increase stem cell number in the culture. After extensive screening utilizing the serum-free culture technique and the transplantation assay, we determined extrinsic factors essential for self-renewal proliferation of SSCs *in vitro* (Kubota *et al.*, 2004b). As described earlier, the culture system consists of three components, enrichment of SSCs, serum-free medium, and feeder cells. Because each element was important for the systematic development of the mouse SSC culture system, we describe the three components in the following sections.

B. Components of SSC Culture System

1. Spermatogonial Stem Cells

The number of SSCs in the testis is very low, presumably as few as 1 in 3000–4000 cells in the adult mouse testis (Tegelenbosch and de Rooij, 1993). Therefore, unfractionated testis cell suspensions are not ideal as a starting cell population for SSC culture. The great majority of unfractionated testicular cells in adult testes are differentiated or differentiating germ cells. These germ cells have limited proliferative activity; therefore, they die and disappear gradually in culture although the cell number is high at the beginning. In addition, since differentiating germ cells are nonadherent cells, they can be removed easily by changing the medium, while SSCs and spermatogonia generally adhere to the stromal feeder cells or extracellular matrices (Shinohara *et al.*, 1999). Although the differentiated germ cells showed limited proliferative activity in culture, there are two reasons to remove them from starting cell populations. First, the stem cells are rare in the testis cell suspension; therefore, it is almost impossible to identify the stem cells in culture. Because microscopic observation of cell divisions in culture is a valuable indication of stem cell proliferation, a large number of differentiated cells are an impediment to accurate assessment of stem cell behavior. Second, it is thought that differentiating germ cells may provide growth-inhibitory signals to undifferentiated spermatogonia, including stem cells (Bootsma and Davids, 1988; de Rooij et al., 1985; Meistrich and van Beek, 1993). Existence of such inhibitory signals or negative feedback systems is believed to occur in other stem cell systems as well (Loeffler and Potten, 1997). Although the identity of these signals is not known, it is desirable to eliminate any possibility of a detrimental effect on SSCs.

Somatic cells, such as Sertoli cells, Leydig cells, myoid cells, and fibroblasts, also exist in the testis, although they are not major cell populations in adult testes. It is believed that terminally differentiated cells do not proliferate; therefore, most somatic cells will disappear in culture during subculturing. However, because they could produce a variety of hormones, growth factors, or extracellular matrices, it is important to remove as many somatic cells as possible to minimize the complexity of the culture conditions. In addition, fibroblastic cells generally can proliferate in regular culture medium containing animal serum; even if they are isolated from adult tissues (see below). Based on our experience, SSCs from testes of immature animals are easier to culture, but removal of testicular somatic cells from these cell suspensions is even more critical. Rapid overgrowth of fibroblastic cells can be a major problem in primary cultures, and fibroblastic cells from young animals have a high proliferative ability. In the adult, the number of testicular fibroblastic cells is fewer. However, they can gradually become dominant if serum is added to the culture medium. Serum-supplemented culture conditions selectively expand fibroblasts, because serum contains a variety of trophic factors for fibroblasts (Sato et al., 1960). Use of a germ cell population enriched for SSCs is crucial to diminish the detrimental effects of fibroblastic cells and other somatic cells on SSC survival and proliferation.

2. Serum-Free Defined Medium

In general, addition of serum facilitates survival and proliferation of animal cells *in vitro*; however, several crucial drawbacks exist. First, serum contains complex materials, which have been as yet poorly defined or characterized. In addition, there is considerable batch variation depending on the physiological conditions, sex, and age of donors. Second, serum contains inhibitors of certain tissue-specific cells (Barnes and Sato, 1980b; Enat et al., 1984). Third, serum enriches growth factors for mesenchymal cells, such as platelet-derived growth factor or fibroblast growth factors; therefore, mesenchymal cells, particularly fibroblasts, selectively overgrow in serum-supplemented medium (Sato et al., 1960). These fibroblasts produce factors to inhibit proliferation of other cell types.

Serum-free hormonally defined medium was developed by Gordon Sato's group in the 1970s (Barnes and Sato, 1980b; Bottenstein et al., 1979; Hayashi and Sato, 1976). This series of studies revealed that one of the major functions of serum is to provide hormones or growth factors which stimulate replication of cells. In serum-free media supplemented with specific hormones or growth factors, many mammalian cells were able to be maintained without loss of the cell lineage-specific and developmental stage-specific characteristics (Barnes and Sato, 1980a,b; Bottenstein et al., 1979). Although there was no report of long-term cultures to maintain functional germ cells in such serum-free hormonally defined medium, early studies culturing testicular cells in serum-free medium clearly demonstrated that using defined medium is a powerful approach to study the physiology of testicular somatic cells, such as Sertoli cells or Leydig cells (Mather, 1980; Mather et al., 1981). Serum-free culture is a prerequisite to identify hormones or growth factors essential for self-renewal of SSCs.

3. Feeder Cells

Generally, when a small number of cells are placed in a culture dish, they do not grow well. Conditioned media or feeder cells commonly have been used to culture cells at low cell densities (Ham, 1963). While conditioned media support cell growth by soluble factors, feeder cells are able to stimulate cells cocultured not only by soluble factors but also by insoluble signals through direct cell–cell contact or via extracellular matrices. Originally, nonmultiplying irradiated feeder cells were used to supply conditioning factors for colony formation from single HeLa cells (Puck and Marcus, 1955). The most commonly used coculture system with a feeder layer technique was first developed for human epidermal keratinocyte culture (Rheinwald and Green, 1975). The original culture system consisted of irradiated 3T3 cells as mesenchymal feeders with serum-supplemented medium, and it supported colony formation and serial cultivation of human keratinocytes (Rheinwald and Green, 1975). A coculture system for clonal growth of rat hepatic progenitors using STO feeder cells and a defined serum-free medium (Kubota and Reid, 2000) is an evolved form of the human keratinocyte culture.

Although SSCs can be enriched by several methods (Kubota et al., 2003, 2004a; Shinohara et al., 1999, 2000), it is still laborious to obtain a large number of stem cells. In addition, the slow proliferation rate is a common characteristic of tissue stem cells. Therefore, a coculture system using feeder cells is reasonable for cultivation of stem cells. Early studies with hematopoietic stem cell culture have shown that coculture using stromal monolayers derived from hematopoietic tissues was able to support self-renewal of hematopoietic stem/progenitor cells for several months (Dexter et al., 1977). It is worthwhile to point out that primary mouse embryonic fibroblast (MEF) feeders or STO feeder cells have generally been used for culture of germline-derived pluripotent stem cells, such as embryonic carcinoma (EC) stem cells, embryonic stem (ES) cells, or embryonic germ (EG) cells (Evans and Kaufman, 1981; Martin, 1981; Martin and Evans, 1975; Matsui et al., 1992; Resnick et al., 1992).

One drawback to using feeder cells is that they produce factors that are not defined. Although our culture system uses feeder cells, they were prepared from a well-established mouse cell line, STO cells, instead of primary MEFs. Recent studies have demonstrated that fibroblasts from different anatomic locations are significantly different, and substantial heterogeneity of embryonic fibroblasts exists (Chang *et al.*, 2002). Using feeders of an established cell lines minimizes the variability of the unknown contribution of the feeder cells to the culture conditions.

III. Methods

A. Protocol for Mouse SSC Culture

1. Preparation of Testis Cell Suspension and Enrichment of SSCs

Identification of surface antigens on SSCs is necessary for immunoselection. Using fluorescence-activated cell sorting (FACS) in conjunction with the transplantation assay, the surface phenotype of mouse SSC was determined to be Thy-1$^+$ αV-integrin$^{-/dim}$ α6-integrin$^+$ c-kit$^-$ and major histocompatibility complex class I (MHC-I)$^-$ (Kubota *et al.*, 2003). Thy-1 is a surface marker expressed on SSCs in neonatal, pup, and adult testes of the mouse. Following identification of the surface antigen for SSC by FACS, a method to enrich mouse SSCs using microbeads conjugated with a Thy-1 antibody was developed (Kubota *et al.*, 2004a). The method, magnetic-activated cell sorting (MACS), using Thy-1 antibody cannot completely purify the Thy-1$^+$ αV-integrin$^{-/dim}$ α6-integrin$^+$ c-kit$^-$ MHC-I$^-$ cell population, but the technique requires less time and less tissues to obtain an SSC-enriched cell population. The SSC content of the cell population (MACS Thy-1 cells) depends on the age of animal. Usually, testis cells from younger mice result in cell populations with a greater percentage of SSCs than do cells from adult testes, because fewer Thy-1$^+$ somatic cells are present in younger testes. Gonocytes (pre-spermatogonia) can be enriched by MACS from neonatal testes; however, they are quiescent in the testis and appeared to take more time to start proliferating at the beginning of the SSC culture than do MACS Thy-1 cells from pup testes (unpublished observation). Based on our experience, MACS Thy-1 cells from pup testis at 5–8 days postpartum (dpp; day of birth is 0 dpp) are the best population for SSC culture in the mouse. The following procedure is for cell preparation of MACS Thy-1$^+$ cells from 4 to 8 pup testes (2–4 pups) at the age of 5–8 dpp. The procedure consists of three steps: digestion of pup testes; Percoll fractionation; and MACS separation. The procedure to enrich Thy-1$^+$ cells by MACS was performed according to the manufacture's protocol. Details for preparing each reagent in the following procedures are described in the *Materials* section.

Testis cell preparation

1. Remove testes from pups with fine forceps using sterile procedures and collect the testes in a 35-mm petri dish in 3 ml of Hank's balanced salt solution (HBSS).

2. Transfer the testes to a second dish of HBSS and remove tunica albuginea under a dissecting microscope.
3. Using a p200 pipette, transfer the testis tissue without tunica to a 15 ml conical centrifuge tube containing 0.5 ml of 7 mg/ml DNase I solution and 4.5 ml of 0.25% Trypsin-EDTA.
4. Pipette up and down with p1000 pipette to disperse seminiferous tubules.
5. Incubate the tissues at 37 °C for 5 min.
6. After pipetting with p1000 pipette several times, incubate the tube at 37 °C for an additional 3 min. At this point, the cell suspension will be viscous.
7. Add 0.7 ml of fetal bovine serum (FBS) (one tenth volume) to stop enzymatic digestion, also add 0.5 ml of 7 mg/ml DNase to digest genomic DNA from dead cells.
8. Pipette well to make a single cell suspension. If cells remain clumped, add another 0.5 ml of 7 mg/ml DNase and pipette again.
9. Filter the cell suspension through a 40-μm pore nylon cell-strainer (BD Biosciences 352340) followed by washing the cell-strainer with HBSS.
10. Centrifuge the cell suspension at $600 \times g$ for 7 min at 4 °C.
11. Remove supernatant and resuspend cells in 10 ml of PBS-S (Dulbecco's PBS supplemented with 1% FBS, 10 mM HEPES, 1 mg/ml glucose, 1 mM pyruvate, 50 units/ml penicillin, and 50 μg/ml of streptomycin).
12. Count cells. Typical testicular cell number obtained by this method is 1×10^6 to 1.3×10^6 cells/testis at 5 dpp and 1.7×10^6 to 2×10^6 cells/testis at 8 dpp.

Percoll fractionation

1. Slowly overlay 5 ml of cell suspension on 2 ml of 30% Percoll solution in a 15 ml conical centrifuge tube. Do not put more than 2×10^7 cells in the 15 ml tube.
2. Centrifuge at $600 \times g$ for 7 min at 4 °C without using the centrifuge brake.
3. Carefully remove the cells and debris at the interface between the HBSS and the 30% Percoll solution. Then, remove all aqueous phases containing HBSS and 30% Percoll solution. Leave the pellet at the bottom of the tube.
4. Resuspend the pellet of cells in 2 ml of PBS-S and transfer the cell suspension into a 5 ml polypropylene tube (BD Biosciences 352063). Count cell number. The range of cell recovery is 40–70%.
5. Centrifuge cell suspension at $600 \times g$ for 7 min at 4 °C.
6. Resuspend the pellet in 90 μl of PBS-S and pipette well to make a single cell suspension.

MACS separation

1. Add 10 μl of magnetic microbeads conjugated with anti-Thy-1 antibody (Miltenyi Biotec 130–049–101, Auburn, CA) into the 90 μl of the cell suspension and mix well. Ten microliter of Thy-1 microbeads is for one separation column. The typical range of the cell number for 10 μl of Thy-1 microbeads is 3×10^6 to 6×10^6 cells. Use 20 μl of Thy-1 microbeads when cell number is more than 6×10^6 cells.

2. Incubate the cell suspension containing Thy-1 microbeads for 20 min at 4 °C. Mix gently by tapping every 10 min. Longer incubation (for 30–40 min) occasionally increases the recovery rate of Thy-1$^+$ cells.

3. Add 2 ml of PBS-S to the tube to dilute Thy-1 microbeads and centrifuge at $600 \times g$ for 7 min at 4 °C. Remove the supernatant completely and resuspend in 1 ml of PBS-S.

4. Place a separation column (MS Column; Miltenyi Biotec 130-042-201) in the magnetic field of the mini MACS Separation Unit (Miltenyi Biotec 130-142-102) and rinse with 0.5 ml of PBS-S.

5. Apply the cell suspension to the column. After the cell suspension has passed through the column and the column reservoir is empty, wash the column with 0.5 ml of PBS-S three times.

6. Remove the column from the MACS Separation Unit and elute the magnetically retained cells slowly into a 5 ml polypropylene tube (BD Biosciences 352063) with 1 ml of serum-free culture medium (see below) using the plunger supplied with the column.

7. Centrifuge the tube containing the cells at $600 \times g$ for 7 min at 4 °C and resuspend the cell pellet with 1 ml of mouse serum-free medium (SFM) for rinsing. Repeat this step once.

8. After the final rinsing step, resuspend cells in 0.5 ml of SFM and count the cell number. The recovery of Thy-1$^+$ cells from one MS Column is 1.6×10^5 to 2.2×10^5 cells.

2. Serum-Free Medium

The composition of SFM for mouse SSCs is shown in Table I. To prepare mouse SFM, bovine serum albumin (BSA) powder and antibiotics are added to MEMα medium and stored at 4 °C overnight, because it takes several hours to dissolve BSA completely. The following day, stock solutions of transferrin, free fatty acid (FFA) mixture (Chessebeuf and Padieu, 1984, see Table II), selenium (Na$_2$SeO$_3$), freshly prepared 2-mercaptoethanol (2-ME), insulin, HEPES, and putrescine are added to the MEMα containing BSA and antibiotics. Preparation of stock solutions for each supplemental component is described in the *Materials* section. SFM is sterilized by filtration using a 0.2 μm membrane filter and stored at 4 °C.

Table I
Components in Mouse Serum-Free Medium (SFM) and Modified Mouse SFM

	Manufacturer	Catalog number	Mouse SFM[a]	Modified mouse SFM[b]
MEMα	Gibco/Invitrogen	12561	Basal media	Basal media
Penicillin	Gibco/Invitrogen	15140	50 units/ml	50 units/ml
Streptomycin	Gibco/Invitrogen	15140	50 μg/ml	50 μg/ml
BSA	MP Biomedicals/ICN	See Note[c]	0.2%	0.6%
Transferrin	Sigma	T 1283	10 μg/ml	100 μg/ml
FFA mixture	Sigma	See Table II	7.6 μeq/L	15.2 μeq/L
Na$_2$SeO$_3$	Aldrich/Sigma	481815	3×10^{-8} M	6×10^{-8} M
L-glutamine	Gibco/Invitrogen	25030	2 mM	2 mM
2-ME	Sigma	M 7522	50 μM	100 μM
Insulin	Sigma	I 5500	5 μg/ml	25 μg/ml
HEPES	Sigma	H 0887	10 mM	10 mM
Putrescine	Sigma	P 5780	60 μM	120 μM

[a] The concentration of each component in serum-free medium for mouse SSC culture is indicated.

[b] Modified mouse SFM is made by increasing several components of mouse SFM. Serum-free medium for rat SSC culture is modified mouse SFM plus 10% DDW (10 ml of water is added to 100 ml of modified mouse SFM). Thus, the final concentrations of each supplemental component in rat SFM is about 90% of that in modified mouse SFM.

[c] Note: The source and lot of BSA are important. BSA of MP Biochemicals (formerly ICN), catalog number 810661 (lot number 2943C) and 194774 (lot number R14550), supported long-term *in vitro* proliferation of mouse SSCs. BSA from Sigma, catalog number A3803 (lot numbers 064K0720, 025K1497, and 124K0729), could be used for mouse SSC culture. For rat SSC culture, BSA of MP Biochemicals, catalog number 810661 (lot number 2943C and 4561H), was effective.

Table II
Preparation of Free Fatty Acid Mixture

Free fatty acid (FFA)	Manufacturer	Catalog number	Stock solution[a]	100 meq/l FFA mixture[b]
Linolenic acid	Sigma	L 2376	1 M	5.6 μl (5.6 mM)
Oleic acid	Sigma	O 1008	1 M	13.4 μl (13.4 mM)
Palmitoleic acid	Sigma	P 9417	1 M	2.8 μl (2.8 mM)
Linoleic acid	Sigma	L 1012	1 M	35.6 μl (35.6 mM)
Palmitic acid	Sigma	P 0500	1 M	31.0 μl (31.0 mM)
Stearic acid	Sigma	S 4751	151 mM	76.9 μl (11.6 mM)
Absolute ethanol				834.7 μl
Final volume				1000 μl (100 meq/l)

[a] Ethanol is used to make stock solutions of each FFA. One molar stock solutions of linolenic acid, oleic acid, palmitoleic acid, and linoleic acid are liquid at room temperature. One molar palmitic acid and 151 mM stearic acid are solid. These two FFA stock solution need to be heated to 45–50 °C to dissolve.

[b] Seventy-six microliter of 100 meq/l FFA mixture is added to 1000 ml of serum-free medium. This makes 7.6 μeq/l FFA mixture at the final concentration in the medium. Numbers in parentheses indicate final concentration in serum-free medium.

3. Preparation of Feeder Cells

STO cells (SNL76/7 cells) (McMahon and Bradley, 1990) were obtained from Dr. A. Bradley (The Wellcome Trust Sanger Institute, London). STO cells are routinely grown in 10 cm dishes in Dulbecco's modified Eagle's medium (DMEM) supplemented with 7% FBS (serum-supplemented medium for STO cells, SSM/STO). Although the culture methods are based on previous reports using newborn calf serum for the growth medium (Robertson, 1987), STO cells in the medium with FBS appeared to grow faster. Cultures of STO cells should be passaged promptly when they reach confluency because it is possible to accumulate noncontact-inhibited cells in the population. In addition, continuous passage of STO cells may also select clones that cannot support SSC culture. Therefore, it is advisable to expand STO cells at early passages and cryopreserve immediately in several low-passage aliquots.

1. Add mitomycin C at a final concentration of 10 μg/ml in culture medium for confluent 10 cm dishes of STO cells. Incubate the dishes for 3–4 h at 37 °C. Avoid more than 5 h exposure to mitomycin C.
2. Remove the medium containing mitomycin C from the STO cells and wash plates three times with 10 ml of HBSS.
3. Digest the cells with trypsin-EDTA solution for 3 min at 37 °C and mix with SSM/STO to stop the trypsin activity. Collect digested cells and centrifuge at $600 \times g$ for 7 min at 4 °C. Resuspend the cells in SSM/STO at 12×10^6 cells/ml. Add to the cell suspension an equal volume of 2× freeze medium [20% dimethyl sulfoxide (DMSO) in SSM/STO] slowly at 4 °C. The final cell concentration will be 6×10^6 cells/ml in 10% DMSO in SSM/STO.
4. Aliquot 1 ml of the STO cell suspension into each cryotube and store at −70 °C using a freezing container (Nalgene 5100–0001).
5. For preparation of monolayers of STO cell feeders, thaw one vial of mitomycin C-treated STO cells at 37 °C and put the cell suspension into a 50 ml conical centrifugation tube.
6. Add slowly 9 ml of SSM/STO on ice to dilute the freeze medium and mix gently. Centrifuge at $600 \times g$ for 7 min at 4 °C.
7. Resuspend the cell pellet with SSM/STO to make a cell suspension at a cell concentration of 2×10^5 cells/ml. The cell viability is usually more than 90%.
8. Seed the cells on to gelatinized plates at a concentration of 5×10^4 cells/cm^2. Typically put 1 ml of the cell suspension (2×10^5 cells/ml) into one well of a 12-well plate. Gelatinized plates are prepared by precoating plates with 0.1% gelatin for 1 h at 37 °C. STO monolayers can be created with fresh mitomycin-C treated STO cells as well.
9. Use the STO feeder cells within 4 days. If older STO feeder cells (5–7 days after seeding) are used for SSC culture, change the medium on the fourth day.

4. Culture of Mouse SSCs

While the culture of SSCs is dependent on the presence of appropriate growth factors, other elements can influence cell proliferation. After placing freshly isolated MACS Thy-1 cells in the primary culture, initiation of germ cell clump formation is an important indication of likely successful continuous culture of SSC (Fig. 3A) (Kubota et al., 2004b). Each culture shows some difference in the number of germ cell clumps and the growth speed. Probably the age of the testes, purity of Thy-1$^+$ cells, and damage caused by cell preparation are factors influencing initial and continuous clump formation. One of the most influential factors observed is the mouse strain used. While SSCs from DBA/2 mice proliferate easily in the presence of GDNF alone, stem cells from other mouse strains such as C57BL/6 or 129/SvCP require soluble GDNF family receptor alpha 1 (GFRα1) and basic fibroblast growth factor (bFGF or FGF2) for continuous proliferation *in vitro*

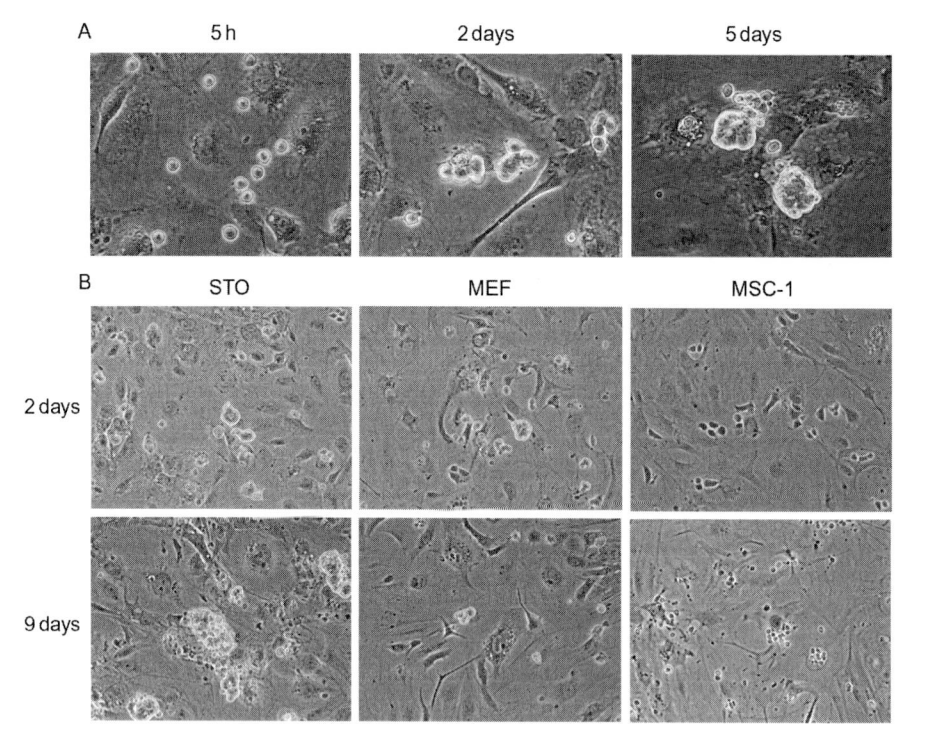

Fig. 3 Microscopic observation of cultured SSCs. (A) development of germ cell clumps from 129/SvCP Thy-1$^+$ pup testis cells isolated by MACS. MACS Thy-1 cells were placed on STO feeder cells in serum-free defined medium supplemented with GDNF. Single Thy-1$^+$ cells at 5 h in culture (left). Initiation of cell clump formation at 2 days (middle). Growth of germ cell clumps at 5 days (right). (B) Comparison of feeder cells for their ability to support germ-cell clump formation and growth. C57BL/6-derived SSCs were placed on STO, MEF, or MSC-1 feeders in the presence of GDNF, GFRα1, and bFGF, and cultured for 9 days. (See Plate no. 8 in the Color Plate Section.)

(Kubota *et al.*, 2004b). The growth advantage of DBA/2-derived SSCs was reported earlier (Kanatsu-Shinohara *et al.*, 2003). Cellular responses to GDNF are mediated by a multicomponent receptor complex consisting of RET receptor tyrosine kinase and a glycosil phosphatidylinositol-anchored ligand-binding subunit, GFRα1 (Sariola and Saarma, 2003). Because RET stimulation by soluble GFRα1 potentiates downstream signaling (Paratcha *et al.*, 2001), addition of soluble GFRα1 may play an important role for *in vitro* proliferation of SSCs in most mouse strains. bFGF has been shown to be a potent growth factor for *in vitro* proliferation of primordial germ cells (Matsui *et al.*, 1992; Resnick *et al.*, 1992). Therefore, bFGF may also provide a critical stimulus to support SSC replication in culture. Moreover, recent studies suggest that bFGF may play an important role for SSC proliferation in human testes (Goriely *et al.*, 2003).

From our experience, SSCs from DBA/2 appear to proliferate faster in the SFM supplemented with GDNF, GFRα1, and bFGF than in GDNF alone. In the presence of three factors, SSCs expand in culture easily from DBA/2 mice, with more difficulty from C57BL/6 mice and the most difficulty from 129/SvCP mice.

Primary culture of Thy-1$^+$ germ cells

1. Remove culture medium from STO monolayer cultures and rinse the plates with HBSS twice to wash out residual medium containing serum.
2. Place 5×10^4 to 10×10^4 MACS Thy-1 cells in wells of 12-well plates containing STO monolayers. Two to four days after seeding of STO feeder cells in gelatin-coated wells is optimal for SSC primary culture.
3. Add recombinant human GDNF, rat GFRα1, and human bFGF at a final concentration of 20 ng/ml, 150 ng/ml, and 1 ng/ml, respectively.
4. Maintain cells at 37 °C in a humidified 5% CO_2 atmosphere. Change the medium every other day. By 48 h in culture, Thy-1$^+$ germ cells form small clumps with tight intercellular contacts (Fig. 3A). They continuously proliferate and form large clumps (Fig. 3A).

Recently, several studies reported SSC culture using feeder cells. Kanatsu-Shinohara *et al.* used MEF for long-term culture of SSCs derived from DBA/2 gonocytes (Kanatsu-Shinohara *et al.*, 2003). Hamra *et al.* reported that feeder cells from a mouse Sertoli cell line (MSC-1) to be superior to STO feeder cells, although germ cells were cultured on the feeder cells for only a short period (Hamra *et al.*, 2004, 2005). Both culture conditions contain FBS. MSC-1 was established from transgenic mice carrying a fusion gene composed of human Müllerian inhibitory substance transcriptional regulatory sequences linked to the SV40 T-antigen gene (Peschon *et al.*, 1992). Using our culture system, MEF feeders supported initiation of germ cell clumps; however, the growth was not as fast as STO cell feeders (Fig. 3B). MSC-1 feeder cells maintained germ cell survival and proliferation poorly (Fig. 3B). By 9 days after initial plating, most germ cells on MSC-1 feeder layers died.

Subculture and establishment of long-term culture

The first subculture is performed 6–9 days after initial culture of MACS Thy-1 cells on STO feeders. The second and subsequent subcultures also are performed at a similar interval (6–9 days). Timing of subculture must be determined subjectively on the basis of the number of proliferating germ cells and testicular fibroblasts.

1. Remove culture medium and add 0.5 ml of trypsin-EDTA solution per well of a 12-well plate to digest the germ cell clumps.
2. Incubate cultures for 3–5 min at 37 °C and add 0.1 ml of FBS to stop digestion. Pipette digested cells gently with p1000 pipette several times. In some instances, genomic DNA from dead cells may cause cell aggregation and clumping, especially when many germ cells are cultured for more than 1 week. In that case, add 0.1 ml of 7 mg/ml DNase solution with the FBS.
3. Collect digested cells from a well of the 12-well plate in a 15 ml conical centrifugation tube and dilute the cell suspension with 2 ml of mouse SFM. Centrifuge at $600 \times g$ for 7 min at 4 °C.
4. Resuspend the cell pellet with 2 ml of mouse SFM and centrifuge at $600 \times g$ for 7 min at 4 °C. Cells are resuspended in 3 ml of SFM and plated onto fresh STO cell feeders. The split ratio of the first subculture is 1:2.
5. Sometimes testicular fibroblastic cells outgrow clump-forming germ cells. Because clump-forming germ cells are attached on STO feeders weakly, they can be removed from the feeders by pipetting gently with p1000 pipette (Ryu *et al.*, 2005). Following gentle pipetting on the surface of feeder cells, collect the culture medium containing the detached germ cell clumps and centrifuge at $600 \times g$ for 7 min at 4 °C.
6. Resuspend the pellet in mouse SFM and pipette gently several times to break up cell clumps. Place cells onto fresh STO feeder cells in the same size well as the original culture (1:1).

Usually, 1 month after initial plating, clump-forming germ cells constantly proliferate on the feeder in mouse SFM supplemented with GDNF, GFRα1, and bFGF (Fig. 4A, phase contrast). Once cultures reach that stage, subculture can be conducted every 4–7 days at the split ratio of 1:2–1:4.

5. Characteristics of Cultured Mouse SSCs

The clump-forming germ cells express germ cell markers, such as germ cell nuclear antigen 1 (GCNA1) or mouse vasa homologue (MVH) (Enders and May, 1994; Fujiwara *et al.*, 1994; Kubota *et al.*, 2004b; Oatley *et al.*, 2006). More importantly, they show very similar characteristics with freshly isolated SSCs. Previous studies using transplantation assays and FACS demonstrated that the phenotype of SSCs in pup testes is αV-integrin$^{-/dim}$ α6-integrin^{+} Thy-1$^{lo/+}$

4. Spermatogonial Stem Cell Culture

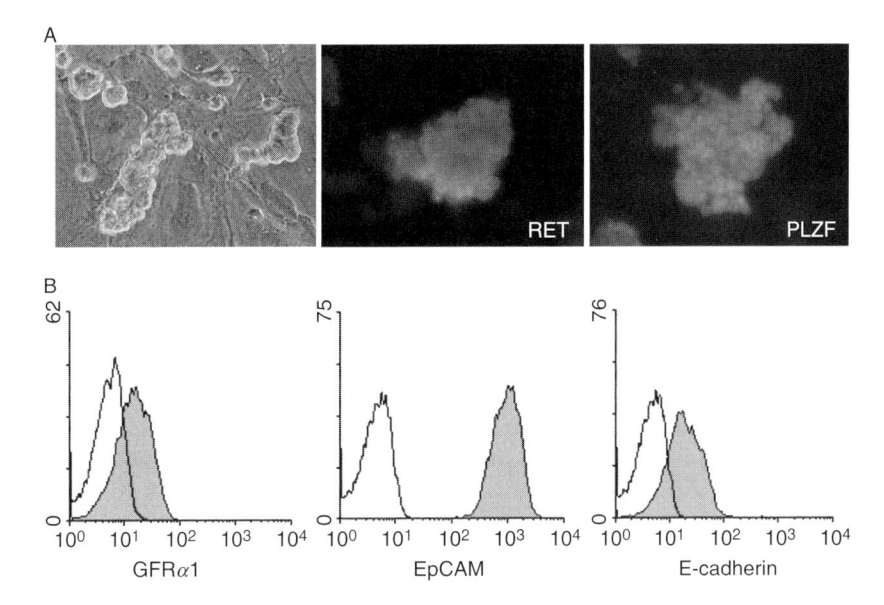

Fig. 4 Phenotypic characteristics of cultured SSCs. (A) SSCs from C57BL/6 pup testes continuously proliferate on STO feeder cells in serum-free defined medium supplemented with GDNF, GFRα1, and bFGF. A phase contrast image of SSCs cultured for 1 month (left) and immunocytochemistry of RET (middle) and PLZF (right). Clump-forming germ cells were stained with antibodies against RET or PLZF. (B) Cell surface molecules on clump-forming germ cells were analyzed by flow cytometry. Cultured SSCs express GFRα1, EpCAM, and E-cadherin. Filled histograms represent stained cells with antibodies indicated. Open histograms indicate unstained cells. (See Plate no. 9 in the Color Plate Section.)

(Kubota *et al.*, 2004a). Flow cytometric analysis showed that continuously cultured germ cells are also αV-integrin$^{-/dim}$ α6-integrin$^+$ Thy-1$^{lo/+}$, and the surface phenotype of cultured germ cells was constant during a 3-month culture period (Kubota *et al.*, 2004b). Transplantation assays indicated that the stem cell activity of clump-forming germ cells during a 3-month culture period and fresh αV-integrin$^{-/dim}$ α6-integrin$^+$ Thy-1$^{lo/+}$ cells isolated by MACS was similar (Kubota *et al.*, 2004b). Usually, the stem cell activity is represented by donor-derived colony number generated in the recipient testes per 10^5 transplanted cells. In our studies, the values of stem cell activity of freshly isolated cells and cultured cells were both ~500 colonies per 10^5 αV-integrin$^{-/dim}$ α6-integrin$^+$ Thy-1$^{lo/+}$ cells transplanted (Kubota *et al.*, 2004b). Since the colonization efficiency is ~5%, 1 in 10 clump-forming germ cells was an SSC. These results indicate that the culture conditions efficiently support proliferation of αV-integrin$^{-/dim}$ α6-integrin$^+$ Thy1$^{lo/+}$ cells. All clump-forming germ cells expressed RET tyrosine kinase, which is the signal transducer of GDNF (Fig. 4A) (Kubota *et al.*, 2004b). Although soluble GFRα1 was added to the culture medium, cultured germ cells expressed GFRα1 on the cell surface (Fig. 4B) (Kubota *et al.*, 2004b). In addition to RET and GFRα1 expression, clump-forming germ cells are positive for several spermatogonia markers including promyelocytic leukaemia zinc finger (PLZF) protein, epithelial cell adhesion molecule (EpCAM, CD326), and E-cadherin (Fig. 4A and B) (Buaas *et al.*, 2004; Costoya *et al.*, 2004; Tokuda *et al.*, 2007; van der Wee *et al.*, 2001).

The cultured SSCs shared several characteristics of undifferentiated ES cells, such as POU5F1 (previously known as Oct-3/4) expression and alkaline phosphatase activity (Kubota *et al.*, 2004b). However, significant differences exist. For example, although FBS supported ES cells in culture, the constituents of FBS are detrimental to SSC proliferation (Kubota *et al.*, 2004b). When SSCs were cultured in medium containing only 0.1% FBS, proliferation of SSCs was dramatically decreased compared with that of SSCs in SFM (Kubota *et al.*, 2004b). Cultured SSCs do not generate tumors when transplanted to immunocompromised mice (Kubota *et al.*, 2004b), whereas ES cells produce teratocarcinomas when injected into mice (Evans and Kaufman, 1981; Martin, 1981). In addition, SSCs are negative for Nanog, which is highly expressed in undifferentiated ES cells (Chambers *et al.*, 2003; Mitsui *et al.*, 2003; Oatley *et al.*, 2006).

B. Protocol for Rat SSC Culture

1. Enrichment of Rat SSCs

SSCs from rat pup testis can be enriched by immunoselection using an EpCAM antibody (Ryu *et al.*, 2004). In the rat pup testis (8 dpp), ~8% of testis cells are EpCAM$^+$. The transplantation assay indicated that almost all SSCs express EpCAM (Ryu *et al.*, 2004). EpCAM is a homophilic cell–cell adhesion molecule and has been used for identification and purification of primordial germ cells or spermatogonia in rodents (Anderson *et al.*, 1999; Moore *et al.*, 2002; van der Wee *et al.*, 2001). Although rat SSCs express Thy-1 antigen, a subpopulation of the testicular somatic cells also express this surface molecule; therefore, MACS using Thy-1 antibody did not enrich rat SSCs efficiently. To obtain an SSC-enriched cell population for culture, we developed a protocol for isolation of EpCAM$^+$ cells using MACS with anti-EpCAM antibody (MACS EpCAM cells). The protocol is for 4–8 testes (2–4 pups) from 8–12 dpp rat pups.

Testis cell preparation and MACS separation

1. Prepare testes for digestion as described in the mouse section.
2. Put testes into 10 ml of 1 mg/ml collagenase solution and incubate 5–8 min. Tap the bottom of the tube gently every 2–3 min.
3. Centrifuge at 600 × g for 1 min at 4 °C to collect loose seminiferous tubules.
4. Remove supernatant and add 10 ml of HBSS to loosen tubules by inverting the tube. Repeat steps 3 and 4 for rinsing.
5. Remove supernatant and add 1.5 ml of 7 mg/ml DNase and 6 ml of trypsin-EDTA.

6. Digest the tissues as described in the mouse protocol.
7. Add 0.9 ml of FBS and 1 ml of 7 mg/ml DNase and pipette well using a p1000 pipette.
8. Filter the cells through a 40 μm pore nylon cell strainer and spin down cells at 600 × g for 7 min at 4 °C.
9. Resuspend cells in 14 ml of MEMα containing 1% FBS and count the cell number. Typical cell recovery is 10×10^6 to 13×10^6 cells/testis at 9–10 dpp.
10. Employ Percoll fractionation as described in the mouse protocol.
11. For MACS separation, resuspend 50×10^6 cells of Percoll-fractionated cells in 5 ml of PBS-S.
12. Add 0.3–0.5 μg of anti-rat EpCAM antibody (Clone: GZ1) per 10^6 cells for labeling with primary antibody and incubate for 20 min on ice.
13. After rinsing the cells twice with 10 ml of PBS-S, resuspend cells in 0.4 ml of PBS-S. Add 0.1 ml of goat anti-mouse IgG microbeads (Miltenyi Biotec, 130–048–402) to label magnetically EpCAM$^+$cells. Incubate cells for 20 min on ice.
14. Add 10 ml of PBS-S and centrifuge at 600 × g for 7 min at 4 °C.
15. Remove the supernatant and resuspend in 2 ml of PBS-S.
16. Separate EpCAM $^+$ cells using two MS column according to the manufacture's protocol. For detail, see the mouse protocol described earlier.

Approximately 4–6% of cells applied to MS columns are recovered. For example, 2×10^6 to 3×10^6 EpCAM$^+$ cells are obtained from 50×10^6 cells (see step 11).

2. Culture of Rat SSCs

The basic culture condition for rat SSC culture consists of SFM, STO feeder cells, and a growth factor combination that are the same as for mouse SSC culture. Although components of SFM for rat SSCs are identical to those of mouse SFM, the concentrations of several components in the SFM for rat SSCs were increased. The SFM that contains increased concentrations of several supplements was designated modified mouse SFM. The concentration of each supplement in the modified mouse SFM is indicated in Table I. In addition, reduction of medium osmolarity by the addition of water resulted in an increase in clump-forming germ cells (Ryu et al., 2005). The modified mouse SFM diluted with 10% (vol/vol) distilled water was designated rat SFM.

To prepare rat SFM, BSA powder, transferrin, and antibiotics are added to MEMα, and the medium is stored at 4 °C overnight. The next day, stock solutions of FFA, Na$_2$SeO$_3$, freshly prepared 2-ME, insulin, HEPES, and putrescine are added to the MEMα containing BSA and transferrin followed by sterilization by filtration through a membrane filter of 0.2 μm pore size.

Because the concentration of several nutrients is increased in the rat SFM, the medium supports somatic cell growth better than mouse SFM. As a result, we observed during cultures a gradual increase in the number of testicular somatic cells contaminating the original MACS EpCAM cell population. Because testicular somatic cells interfered with SSC maintenance and replication (Kubota *et al.*, 2004a), their number was decreased when necessary by removing germ cell clumps at the time of subculture using gentle pipetting of medium across the surface of the feeder layer and recovery of the detached clumps with the culture medium. The collected clumps are digested with a 1:25 dilution of trypsin/EDTA (final 0.01% trypsin/40 nM EDTA) for 1 min or longer if necessary. Enzymatic digestion is stopped by FBS. Following gentle pipetting, clump-forming germ cells become small clumps, but not single cells. After rinsing the germ cells twice, place them on fresh STO feeder layers. During the first 2 months after initial plating of MACS EpCAM cells, the split ratio for subculture should be 1:1 or 1:1.5 at an interval of every 7–10 days. When testicular fibroblasts outgrow germ cells in culture, germ cell clumps can be collected by gentle pipetting, even though the sizes of clumps are small. In that case, enzymatic digestion is not necessary, because trypsin treatment damages clump-forming cells and results in a decrease in cell recovery. Once clump-forming germ cells constantly proliferate in culture, subculture can be conducted at the split ratio of 1:2. Low atmospheric oxygen (5%) was used for long-term culture of rat SSCs (Ryu *et al.*, 2005). Low atmospheric oxygen has shown a beneficial effect on long-term proliferation of several types of mammalian cells *in vitro* (Ezashi *et al.*, 2005; Parrinello *et al.*, 2003).

3. Characteristics of Cultured Rat SSCs

The surface phenotype of freshly isolated rat SSCs in pup testis is EpCAM$^+$ Thy-1^{lo} β3-integrin$^-$ (Ryu *et al.*, 2004). The surface antigen profile of rat SSCs cultured for more than 10 months was basically similar to that of fresh SSCs (Ryu *et al.*, 2005). However, the surface expression of these antigens increased slightly in cultured cells. POU5F1 expression and alkaline phosphatase activity are present in cultured rat SSCs, indicating that mouse and rat SSCs share these characteristics found in ES cells or PGCs. Rat SSCs also express RET and GFRα1, which are the receptors for GDNF (Ryu *et al.*, 2005).

IV. Materials

A. Reagents for Cell Preparation

DNase: DNase I (Sigma DN25) solution is prepared at 7 mg/ml in HBSS and sterilized by filtration.

Collagenase: Collagenase type IV (Sigma C5138) solution is prepared at 1 mg/ml in HBSS and sterilized by filtration.

Trypsin-EDTA: 0.25% trypsin and 1 mM EDTA (Gibco/Invitrogen 25200).

Antibiotics: 10,000 units/ml penicillin and 10,000 µg/ml of streptomycin (Gibco/Invitrogen 15140).

FBS: FBS (Hyclone) is treated at 56 °C for 30 min. Aliquots are stored at −20 °C.

Pyruvate: Stock solution is prepared at 100 mM sodium pyruvate (Sigma P2256) in distilled deionized water (DDW), sterilized by filtration using 0.2 μm membrane filter, and stored at 4 °C.

PBS-S: Dulbecco's PBS supplemented with 1% FBS, 10 mM HEPES (Sigma H0887), 1 mg/ml glucose (Sigma G6152), 1 mM pyruvate, 50 units/ml penicillin, and 50 μg/ml streptomycin sterilized by filtration using 0.2 μm membrane filter, and stored at 4 °C.

Percoll solution: 30% (vol/vol) Percoll (Sigma P4937) is prepared in Dulbecco's PBS containing 1% FBS, 50 units/ml penicillin, and 50 μg/ml streptomycin. Percoll solution is sterilized by filtration using a 0.2 μm membrane filter and stored at 4 °C.

B. Stock Solution of Reagents for SFM and Cell Culture

Transferrin: Stock solution is prepared at 10 mg/ml in Dulbecco's PBS and stored at −20 °C.

FFA mixture: Detailed information is described in Table II. Make a small volume aliquot (e.g., 50–100 μl/0.5 ml tube) of FFA mixture. For storage of stock solutions of each FFA or FFA mixture aliquot, it is advised to flush vials or tubes with nitrogen before closing to prevent oxidation of FFAs. The tubes are sealed with parafilm and stored at −20 °C.

Na_2SeO_3: Stock solution is prepared at 3×10^{-4} M in DDW and stored at −20 °C.

2-ME: 100 mM solution is prepared freshly each time medium is made. 100 mM solution is prepared by adding 7 μl of 14.4 M 2-ME (Sigma M7522) in 1 ml of MEMα medium.

Insulin: Stock solution is prepared at 10 mg/ml in 10 mM HCl and stored at −20 °C.

Putrescine: Stock solution is prepared at 100 mM in Dulbecco's PBS and stored at −20 °C.

GDNF: Human recombinant GDNF (R&D Systems, 212-GD-010) stock solution is prepared at 20 μg/ml in Dulbecco's PBS containing 0.1% BSA and stored at −70 °C.

GFRα1: Rat recombinant GFRα1/Fc Chimera (R&D Systems, 560-GR-100) stock solution is prepared at 100 μg/ml in Dulbecco's PBS containing 0.1% BSA and stored at −70 °C.

bFGF: Human recombinant bFGF (BD Biosciences, 354060) stock solution is prepared at 10 μg/ml in Dulbecco's PBS containing 0.1% BSA and stored at −70 °C.

C. Reagents for STO Feeder Layers

SSM/STO (serum-supplemented medium for STO cells): The medium for STO cell culture is Dulbecco's modified Eagle's medium (DMEM, Gibco/Invitrogen 11965) supplemented with 7% FBS, 100 μM 2-ME, 10 mM HEPES, 2 mM glutamine, 50 units/ml penicillin, and 50 μg/ml streptomycin. SSM/STO is sterilized by filtration using a 0.2 μm membrane filter and stored at 4 °C.

Mytomycin C: Stock mitomycin C (Sigma M4287) solution is prepared at 200 μg/ml in Dulbecco's PBS and stored at −70 °C.

Freezing medium (2×): 20% dimethyl sulfoxide (DMSO, Sigma D2650) in SSM/STO.

Gelatin: 0.1% gelatin (porcine skin type A, Sigma G2500) in DDW is autoclaved for sterilization and stored at room temperature.

V. Discussion

An exciting practical aspect of the development of culture systems for SSCs is that the technique establishes a foundation for sophisticated genetic manipulation, including targeted modification, of the species from which the stem cells were isolated (Kanatsu-Shinohara et al., 2006a). The conservation of GDNF signaling in mouse and rat SSCs as the essential pathway to stimulate in vitro self-renewal and the previously demonstrated ability of SSC of many species to maintain and proliferate in mouse seminiferous tubules suggest that a similar culture system can be developed to obtain continuous proliferation of SSCs of many mammalian species, including humans. In addition, because a large number of SSCs can be generated in culture, they represent a powerful resource for gene analysis to elucidate the mechanisms governing self-renewal and differentiation of the stem cells (Oatley et al., 2006). Moreover, continuous in vitro proliferation of SSCs of any species lays the foundation for the development of systems to support germ cell differentiation in vitro. Modulating culture conditions that support differentiation processes of male germ cells resulting in production of functional gametes in vitro will create a valuable model for studying the molecular and cellular biology of male germ cell differentiation. Such an in vitro experimental system may allow development of new therapeutic strategies for infertility (Brinster, 2007; Kubota and Brinster, 2006). The progressive development of culture systems for SSCs of other mammalian species including humans is imminent and will provide the basis for a wide range of studies on the biology of the stem cell, in vitro differentiation of germ cells, and modification of germlines.

Acknowledgments

Research support has been provided by the National Institutes of Health and Robert J. Kleberg, Jr. and Helen C. Kleberg Foundation.

References

Amann, R. P. (1986). Detection of alterations in testicular and epididymal function in laboratory animals. *Environ. Health Perspect.* **70,** 149–158.

Anderson, R., Schaible, K., Heasman, J., and Wylie, C. (1999). Expression of the homophilic adhesion molecule, Ep-CAM, in the mammalian germ line. *J. Reprod. Fertil.* **116,** 379–384.

Barnes, D., and Sato, G. (1980a). Methods for growth of cultured cells in serum-free medium. *Anal. Biochem.* **102,** 255–270.

Barnes, D., and Sato, G. (1980b). Serum-free cell culture: A unifying approach. *Cell* **22,** 649–655.

Bootsma, A. L., and Davids, J. A. (1988). The cell cycle of spermatogonial colony forming stem cells in the CBA mouse after neutron irradiation. *Cell Tissue Kinet.* **21,** 105–113.

Bottenstein, J., Hayashi, I., Hutchings, S., Masui, H., Mather, J., McClure, D. B., Ohasa, S., Rizzino, A., Sato, G., Serrero, G., Wolfe, R., and Wu, R. (1979). The growth of cells in serum-free hormone-supplemented media. *Methods Enzymol.* **58,** 94–109.

Brinster, R. L. (2002). Germline stem cell transplantation and transgenesis. *Science* **296,** 2174–2176.

Brinster, R. L. (2007). Male germline stem cells: From mice to men. *Science* **316,** 404–405.

Brinster, R. L., and Avarbock, M. R. (1994). Germline transmission of donor haplotype following spermatogonial transplantation. *Proc. Natl. Acad. Sci. USA* **91,** 11303–11307.

Brinster, R. L., and Zimmermann, J. W. (1994). Spermatogenesis following male germ-cell transplantation. *Proc. Natl. Acad. Sci. USA* **91,** 11298–11302.

Buaas, F. W., Kirsh, A. L., Sharma, M., McLean, D. J., Morris, J. L., Griswold, M. D., de Rooij, D. G., and Braun, R. E. (2004). Plzf is required in adult male germ cells for stem cell self-renewal. *Nat Genet.* **36,** 647–652.

Chambers, I., Colby, D., Robertson, M., Nichols, J., Lee, S., Tweedie, S., and Smith, A. (2003). Functional expression cloning of Nanog, a pluripotency sustaining factor in embryonic stem cells. *Cell* **113,** 643–655.

Chang, H. Y., Chi, J. T., Dudoit, S., Bondre, C., van de Rijn, M., Botstein, D., and Brown, P. O. (2002). Diversity, topographic differentiation, and positional memory in human fibroblasts. *PNAS* **99,** 12877–12882.

Chessebeuf, M., and Padieu, P. (1984). Rat liver epithelial cell cultures in a serum-free medium: Primary cultures and derived cell lines expressing differentiated functions. *In Vitro* **20,** 780–795.

Clouthier, D. E., Avarbock, M. R., Maika, S. D., Hammer, R. E., and Brinster, R. L. (1996). Rat spermatogenesis in mouse testis. *Nature* **381,** 418–421.

Costoya, J. A., Hobbs, R. M., Barna, M., Cattoretti, G., Manova, K., Sukhwani, M., Orwig, K. E., Wolgemuth, D. J., and Pandolfi, P. P. (2004). Essential role of Plzf in maintenance of spermatogonial stem cells. *Nat. Genet.* **36,** 653–659.

de Rooij, D. G. (1998). Stem cells in the testis. *Int. J. Exp. Pathol.* **79,** 67–80.

de Rooij, D. G., Lok, D., and Weenk, D. (1985). Feedback regulation of the proliferation of the undifferentiated spermatogonia in the Chinese hamster by the differentiating spermatogonia. *Cell Tissue Kinet.* **18,** 71–81.

Dexter, T. M., Allen, T. D., and Lajtha, L. G. (1977). Conditions controlling the proliferation of hemopoietic stem cells *in vitro*. *J. Cell. Physiol.* **91,** 335–344.

Dobrinski, I., Avarbock, M. R., and Brinster, R. L. (1999a). Transplantation of germ cells from rabbits and dogs into mouse testes. *Biol. Reprod.* **61,** 1331–1339.

Dobrinski, I., Avarbock, M. R., and Brinster, R. L. (2000). Germ cell transplantation from large domestic animals into mouse testes. *Mol. Reprod. Dev.* **57,** 270–279.

Dobrinski, I., Ogawa, T., Avarbock, M. R., and Brinster, R. L. (1999b). Computer assisted image analysis to assess colonization of recipient seminiferous tubules by spermatogonial stem cells from transgenic donor mice. *Mol. Reprod. Dev.* **53,** 142–148.

Enat, R., Jefferson, D. M., Ruiz-Opazo, N., Gatmaitan, Z., Leinwand, L. A., and Reid, L. M. (1984). Hepatocyte proliferation *in vitro*: Its dependence on the use of serum-free hormonally defined medium and substrata of extracellular matrix. *Proc. Natl. Acad. Sci. USA* **81,** 1411–1415.

Enders, G. C., and May, I. I. (1994). Developmentally regulated expression of a mouse germ cell nuclear antigen examined from embryonic day 11 to adult in male and female mice. *Dev. Biol.* **163**, 331–340.

Evans, M. J., and Kaufman, M. H. (1981). Establishment in culture of pluripotential cells from mouse embryos. *Nature* **292**, 154–156.

Ezashi, T., Das, P., and Roberts, R. M. (2005). Low O2 tensions and the prevention of differentiation of hES cells. *Proc. Natl. Acad. Sci. USA* **102**, 4783–4788.

Fujiwara, Y., Komiya, T., Kawabata, H., Sato, M., Fujimoto, H., Furusawa, M., and Noce, T. (1994). Isolation of a DEAD-family protein gene that encodes a murine homolog of Drosophila vasa and its specific expression in germ cell lineage. *Proc. Natl. Acad. Sci. USA* **91**, 12258–12262.

Goriely, A., McVean, G. A., Rojmyr, M., Ingemarsson, B., and Wilkie, A. O. (2003). Evidence for selective advantage of pathogenic FGFR2 mutations in the male germ line. *Science* **301**, 643–646.

Ham, R. G. (1963). An improved nutrient solution for diploid Chinese hamster and human cell lines. *Exp. Cell Res.* **29**, 515–526.

Hamra, F. K., Chapman, K. M., Nguyen, D. M., Williams-Stephens, A. A., Hammer, R. E., and Garbers, D. L. (2005). Self renewal, expansion, and transfection of rat spermatogonial stem cells in culture. *Proc. Natl. Acad Sci USA* **102**, 17430–17435.

Hamra, F. K., Gatlin, J., Chapman, K. M., Grellhesl, D. M., Garcia, J. V., Hammer, R. E., and Garbers, D. L. (2002). Production of transgenic rats by lentiviral transduction of male germ-line stem cells. *Proc. Natl. Acad Sci USA* **99**, 14931–14936.

Hamra, F. K., Schultz, N., Chapman, K. M., Grellhesl, D. M., Cronkhite, J. T., Hammer, R. E., and Garbers, D. L. (2004). Defining the spermatogonial stem cell. *Dev. Biol.* **269**, 393–410.

Hayashi, I., and Sato, G. H. (1976). Replacement of serum by hormones permits growth of cells in a defined medium. *Nature* **259**, 132–134.

Kanatsu-Shinohara, M., Ikawa, M., Takehashi, M., Ogonuki, N., Miki, H., Inoue, K., Kazuki, Y., Lee, J., Toyokuni, S., Oshimura, M., Ogura, A., and Shinohara, T. (2006a). Production of knockout mice by random or targeted mutagenesis in spermatogonial stem cells. *Proc. Natl. Acad. Sci. USA* **103**, 8018–8023.

Kanatsu-Shinohara, M., Inoue, K., Miki, H., Ogonuki, N., Takehashi, M., Morimoto, T., Ogura, A., and Shinohara, T. (2006). Clonal origin of germ cell colonies after spermatogonial transplantation in mice. *Biol. Reprod.* **75**, 68–74.

Kanatsu-Shinohara, M., Ogonuki, N., Inoue, K., Miki, H., Ogura, A., Toyokuni, S., and Shinohara, T. (2003). Long-term proliferation in culture and germline transmission of mouse male germline stem cells. *Biol. Reprod.* **69**, 612–616.

Kubota, H., Avarbock, M. R., and Brinster, R. L. (2003). Spermatogonial stem cells share some, but not all, phenotypic and functional characteristics with other stem cells. *Proc. Natl. Acad. Sci. USA* **100**, 6487–6492.

Kubota, H., Avarbock, M. R., and Brinster, R. L. (2004a). Culture conditions and single growth factors affect fate determination of mouse spermatogonial stem cells. *Biol. Reprod.* **71**, 722–731.

Kubota, H., Avarbock, M. R., and Brinster, R. L. (2004b). Growth factors essential for self-renewal and expansion of mouse spermatogonial stem cells. *Proc. Natl. Acad. Sci. USA* **101**, 16489–16494.

Kubota, H., and Brinster, R. L. (2006). Technology insight: *In vitro* culture of spermatogonial stem cells and their potential therapeutic uses. *Nat. Clin. Pract. Endocrinol. Metab.* **2**, 99–108.

Kubota, H., and Reid, L. M. (2000). Clonogenic hepatoblasts, common precursors for hepatocytic and biliary lineages, are lacking classical major histocompatibility complex class I antigen. *Proc. Natl. Acad. Sci. USA* **97**, 12132–12137.

Loeffler, M., and Potten, C. S. (1997). Stem cells and cellular pedigrees-a conceptual introduction. *In* "Stem Cells" (C. S. Potten, ed.), pp. 1–27. Academic Press, San Diego.

Martin, G. R. (1981). Isolation of a pluripotent cell line from early mouse embryos cultured in medium conditioned by teratocarcinoma stem cells. *Proc. Natl. Acad. Sci. USA* **78**, 7634–7638.

Martin, G. R., and Evans, M. J. (1975). Differentiation of clonal lines of teratocarcinoma cells: Formation of embryoid bodies *in vitro*. *Proc. Natl. Acad. Sci. USA* **72**, 1441–1445.

Mather, J. P. (1980). Establishment and characterization of two distinct mouse testicular epithelial cell lines. *Biol. Reprod.* **23**, 243–252.

Mather, J. P., Saez, J. M., and Haour, F. (1981). Primary cultures of Leydig cells from rat, mouse and pig: Advantages of porcine cells for the study of gonadotropin regulation of Leydig cell function. *Steroids* **38**, 35–44.

Matsui, Y., Zsebo, K., and Hogan, B. L. (1992). Derivation of pluripotential embryonic stem cells from murine primordial germ cells in culture. *Cell* **70**, 841–847.

McMahon, A. P., and Bradley, A. (1990). The Wnt-1 (int-1) proto-oncogene is required for development of a large region of the mouse brain. *Cell* **62**, 1073–1085.

Meistrich, M. L., and van Beek, M. E. A. B. (1993). Spermatogonial stem cells. *In* "Cell and Molecular Biology of the Testis" (C. Desjardins, and L. L. Ewing, eds.), pp. 266–295. Oxford University Press, New York.

Meng, X., Lindahl, M., Hyvonen, M. E., Parvinen, M., de Rooij, D. G., Hess, M. W., Raatikainen-Ahokas, A., Sainio, K., Rauvala, H., Lakso, M., Pichel, J. G., Westphal, H., *et al.* (2000). Regulation of cell fate decision of undifferentiated spermatogonia by GDNF. *Science* **287**, 1489–1493.

Mitsui, K., Tokuzawa, Y., Itoh, H., Segawa, K., Murakami, M., Takahashi, K., Maruyama, M., Maeda, M., and Yamanaka, S. (2003). The homeoprotein Nanog is required for maintenance of pluripotency in mouse epiblast and ES cells. *Cell* **113**, 631–642.

Moore, T. J., Boer-Brouwer, M., and Dissel-Emiliani, F. M. (2002). Purified gonocytes from the neonatal rat form foci of proliferating germ cells *in vitro*. *Endocrinology* **143**, 3171–3174.

Morrison, S. J., Shah, N. M., and Anderson, D. J. (1997). Regulatory mechanisms in stem cell biology. *Cell* **88**, 287–298.

Nagano, M., Avarbock, M. R., Leonida, E. B., Brinster, C. J., and Brinster, R. L. (1998). Culture of mouse spermatogonial stem cells. *Tissue Cell* **30**, 389–397.

Nagano, M., Brinster, C. J., Orwig, K. E., Ryu, B. Y., Avarbock, M. R., and Brinster, R. L. (2001a). Transgenic mice produced by retroviral transduction of male germ-line stem cells. *Proc. Natl. Acad. Sci. USA* **98**, 13090–13095.

Nagano, M., McCarrey, J. R., and Brinster, R. L. (2001b). Primate spermatogonial stem cells colonize mouse testes. *Biol. Reprod.* **64**, 1409–1416.

Nagano, M., Patrizio, P., and Brinster, R. L. (2002). Long-term survival of human spermatogonial stem cells in mouse testes. *Fertil. Steril.* **78**, 1225–1233.

Nagano, M. C. (2003). Homing efficiency and proliferation kinetics of male germ line stem cells following transplantation in mice. *Biol. Reprod.* **69**, 701–707.

Nagy, A., Gertsenstein, M., Vintersten, K., and Behringer, R. (2003). "Manipulating the Mouse Embryo: A Laboratory Manual" Cold Spring Harbor Laboratory Press, Cold Spring Harbor, New York, USA.

Oatley, J. M., Avarbock, M. R., Telaranta, A. I., Fearon, D. T., and Brinster, R. L. (2006). Identifying genes important for spermatogonial stem cell self-renewal and survival. *Proc. Natl. Acad. Sci. USA* **103**, 9524–9529.

Oatley, J. M., Reeves, J. J., and McLean, D. J. (2004). Biological activity of cryopreserved bovine spermatogonial stem cells during *in vitro* culture. *Biol. Reprod.* **71**, 942–947.

Ogawa, T., Ohmura, M., Yumura, Y., Sawada, H., and Kubota, Y. (2003). Expansion of murine spermatogonial stem cells through serial transplantation. *Biol. Reprod.* **68**, 316–322.

Paratcha, G., Ledda, F., Baars, L., Coulpier, M., Besset, V., Anders, J., Scott, R., and Ibanez, C. F. (2001). Released GFRalpha1 potentiates downstream signaling, neuronal survival, and differentiation via a novel mechanism of recruitment of c-Ret to lipid rafts. *Neuron* **29**, 171–184.

Parrinello, S., Samper, E., Krtolica, A., Goldstein, J., Melov, S., and Campisi, J. (2003). Oxygen sensitivity severely limits the replicative lifespan of murine fibroblasts. *Nat. Cell Biol.* **5**, 741–747.

Peschon, J. J., Behringer, R. R., Cate, R. L., Harwood, K. A., Idzerda, R. L., Brinster, R. L., and Palmiter, R. D. (1992). Directed expression of an oncogene to Sertoli cells in transgenic mice using mullerian inhibiting substance regulatory sequences. *Mol. Endocrinol.* **6**, 1403–1411.

Potten, C. S., and Morris, R. J. (1988). Epithelial stem cells *in vivo*. *J. Cell. Sci. Suppl.* **10**, 45–62.

Puck, T., and Marcus, P. (1955). A rapid method for viable cell titration and clone production with hela cells in tissue culture: The use of X-irradiated cells to supply conditioning factors. *Proc. Natl. Acad. Sci. USA* **41,** 432–437.

Resnick, J. L., Bixler, L. S., Cheng, L., and Donovan, P. J. (1992). Long-term proliferation of mouse primordial germ cells in culture. *Nature* **359,** 550–551.

Rheinwald, J. G., and Green, H. (1975). Serial cultivation of strains of human epidermal keratinocytes: The formation of keratinizing colonies from single cells. *Cell* **6,** 331–343.

Robertson, E. J. (1987). Embryo-derived stem cell lines. *In* "Teratocarcinomas and Embryonic Stem Cells: A Practical Approach" (E. J. Robertson, ed.), pp. 71–112. IRS Press, Oxford, England.

Russell, L. D., Ettlin, R. A., Sinha-Hikim Amiya, P., and Clegg, E. D. (1990). "Histological and Histopathological Evaluation of the Testis" Cache River Press, Clearwater, FL.

Ryu, B. Y., Kubota, H., Avarbock, M. R., and Brinster, R. L. (2005). Conservation of spermatogonial stem cell self-renewal signaling between mouse and rat. *Proc. Natl. Acad. Sci. USA* **102,** 14302–14307.

Ryu, B. Y., Orwig, K. E., Kubota, H., Avarbock, M. R., and Brinster, R. L. (2004). Phenotypic and functional characteristics of spermatogonial stem cells in rats. *Dev. Biol.* **274,** 158–170.

Sariola, H., and Saarma, M. (2003). Novel functions and signalling pathways for GDNF. *J. Cell Sci.* **116,** 3855–3862.

Sato, G., Zaroff, L., and Mills, S. E. (1960). Tissue culture populations and their relation to the tissue of origin. *Proc. Natl. Acad. Sci. USA* **46,** 963–972.

Shinohara, T., Avarbock, M. R., and Brinster, R. L. (1999). beta1- and alpha6-integrin are surface markers on mouse spermatogonial stem cells. *Proc. Natl. Acad. Sci. USA* **96,** 5504–5509.

Shinohara, T., Orwig, K. E., Avarbock, M. R., and Brinster, R. L. (2000). Spermatogonial stem cell enrichment by multiparameter selection of mouse testis cells. *Proc. Natl. Acad. Sci. USA* **97,** 8346–8351.

Spradling, A., Drummond-Barbosa, D., and Kai, T. (2001). Stem cells find their niche. *Nature* **414,** 98–104.

Tegelenbosch, R. A., and de Rooij, D. G. (1993). A quantitative study of spermatogonial multiplication and stem cell renewal in the C3H/101 F1 hybrid mouse. *Mutat. Res.* **290,** 193–200.

Tokuda, M., Kadokawa, Y., Kurahashi, H., and Marunouchi, T. (2007). CDH1 is a specific marker for undifferentiated spermatogonia in mouse testes. *Biol. Reprod.* **76,** 130–141.

van der Wee, K. S., Johnson, E. W., Dirami, G., Dym, T. M., and Hofmann, M. C. (2001). Immunomagnetic isolation and long-term culture of mouse type A spermatogonia. *J. Androl.* **22,** 696–704.

Weissman, I. L., Anderson, D. J., and Gage, F. (2001). Stem and progenitor cells: Origins, phenotypes, lineage commitments, and transdifferentiations. *Annu. Rev. Cell Dev. Biol.* **17,** 387–403.

Zhang, X., Ebata, K. T., and Nagano, M. C. (2003). Genetic analysis of the clonal origin of regenerating mouse spermatogenesis following transplantation. *Biol. Reprod.* **69,** 1872–1878.

Zhao, G. Q., and Garbers, D. L. (2002). Male germ cell specification and differentiation. *Dev. Cell.* **2,** 537–547.

CHAPTER 5

Characterization of Human Amniotic Fluid Stem Cells and Their Pluripotential Capability

Laura Perin, Sargis Sedrakyan, Stafano Da Sacco, and Roger De Filippo

Childrens Hospital Los Angeles
Saban Research Institute
Developmental Biology Program
Keck School of Medicine
University of Southern California

    Abstract
I. Introduction
II. Amniotic Fluid Cell Composition and Stem Cells
    A. Cell Population Within the Amniotic Fluid
    B. Isolation of Stem Cells from Amniotic Fluid and Their Characterization
III. *In Vitro* Potential of Amniotic Fluid Stem Cells
    A. Adipogenic Differentiation
    B. Myogenic Differentiation
    C. Osteogenic Differentiation
    D. Endothelial Differentiation
    E. Hepatocytes Differentiation
    F. Neurogenic Differentiation
IV. *Ex Vivo* and *In Vivo* Application of Amniotic Fluid Stem Cells
    A. *Ex Vivo* Application of Amniotic Fluid Stem Cells: Integration into Mouse Embryonic Kidneys
    B. *In Vivo* Application of Amniotic Fluid Stem Cells: Integration into Adult Mouse Kidney
V. Discussion
    References

Abstract

Over the past decade, there has been ever-increasing emphasis placed on stem cells and their potential role in regenerative medicine for reconstruction of bio-artificial tissues and organs. Scientists have looked at various sources for pluripotential cells ranging from embryonic stem cells to adult stem cells. Amniocentesis is a well-established technique for the collection of cells derived from the human embryo. In this chapter, we are going to describe how to isolate, maintain in culture, and characterize the pluripotential capabilities of stem cells derived from amniocentesis in an *in vitro* and *in vivo* system.

Cell samples are obtained from human pregnancies, and the progenitor cells are isolated from male fetuses with a normal karyotype in order to confirm the absence of maternal admixed cells. Progenitor cells express embryonic-specific cell markers, they show a high self-renewal capacity with 350 population doublings, and normal ploidy is confirmed by cell-cycle analyses. They maintain their undifferentiated state, pluripotential ability, clonogenicity, and telomere length over the population doublings.

The progenitor cells are inducible to different cell lineages (osteogenic, adipogenic, skeletal muscle, endothelial, neuronal, and hepatic cells) under specific growth conditions. The ability to induce cell-type-specific differentiation is confirmed by phenotypic changes, immunocytochemistry, gene expression, and functional analyses. In addition, we will describe an application of these cells in an *ex vivo* and *in vivo* system for potential in organ (renal) regeneration.

The progenitor cells described in this chapter have a high potential for expansion, and may be a good source for research and therapeutic applications where large numbers of cells are needed. Progenitor cells isolated during gestation may be beneficial for fetuses diagnosed with malformations and could be cryopreserved for future self-use.

I. Introduction

Over the past decade stem cell research has emerged as an area of major interest for its potential in regenerative medicine applications (Stocum, 2001). The possibility of repairing damaged organs or creating *in vitro* tissues for transplantation will most likely become a reality in the future.

Embryonic stem cells (ESCs), derived from blastocysts, propagate readily (Thomson *et al.*, 1998) and are capable of forming aggregates (embryoid bodies) that generate a variety of specialized cells including neural, cardiac, and pancreatic cells (Reubinoff, 2000). Embryonic stem cells seem to be the most reliable source for these efforts, but ethical concerns and their potential to develop teratomas may ultimately preclude their usefulness clinically (Takahashi *et al.*, 2006). In order to avoid these issues, scientists have looked to other sources for pluripotent cells.

Amniotic fluid has been used as a safe and reliable screening tool for genetic and congenital diseases in the fetus for many years. However, amniotic fluid also contains a vast repository of progenitor cells that may have a useful role in bioengineering applications. Streubel *et al.* (1996) reported using nonhematopoetic cells for the conversion of amniocytes into myocytes. Recently, a c-kit positive population of human and rodent amniotic fluid stem cells that give rise to cells of myogenic, neuronal, adipogenic, hepatic, osteogenic, and endothelial lineages has been reported and characterized (De Coppi *et al.*, 2007). The volume and the composition of the amniotic fluid changes during pregnancy, and by 8 weeks gestation, the fetal kidney begins fluid production that rapidly increases in volume during the second trimester (Mark *et al.*, 2005). Contact between amniotic fluid and compartments of the developing fetus seems to best explain the presence of different cell types. The presence of mature cell lines derived from all three germ layers has been identified (Gosden, 1983; Hoehn and Salk, 1992). Mesenchymal and hematopoetic progenitor cells have also been shown to exist before the 12th week of gestation in humans (Torricelli *et al.*, 1993). Cells expressing proteins and various genetic markers from specific tissue types including brain, heart, and pancreas have all been discovered but more investigation is necessary to completely categorize the various cell types that exist within the amniotic fluid (Bossolasco *et al.*, 2006; George *et al.*, 2004; McLaughlin *et al.*, 2006).

In this chapter, we will discuss the recent advances in the field of human amniotic fluid stem cells (hAFSC). We are going to describe the methods used to establish their pluripotential characteristics, their ability to differentiate *in vitro* in cell lines derived from all the three germ layers. In addition, we will report one specific *ex vivo* and *in vivo* application of hAFSC for renal regeneration.

II. Amniotic Fluid Cell Composition and Stem Cells

A. Cell Population Within the Amniotic Fluid

1. Culture of Amniotic Fluid samples

Amniotic fluid samples are usually donated from cytogenetics laboratories in which they routinely screen fetuses for genetic abnormalities. In these laboratories, the cells within the amniotic fluid are usually cultured in regular media (DMEM, 10% serum and 1% penicillin/streptomycin, Sigma) or in Amniomax (Gibco-Invitrogen), or a 1:1 mix of both for at least 3 weeks in cover glass. After this time, the cells are discarded by cytogenics laboratories and they are collected for characterization and isolation of stem cells in our laboratory. The samples analyzed in the laboratory usually are male samples, to avoid contamination from mother's cells, and samples obtained from a fetus with a normal karyotype. These samples are, therefore, processed with a gestational age ranging from 15 to 19 weeks. They are detached from the cover glass using a trypsin/EDTA (Gibco-Invitrogen) solution and then cultured in a petri dish with Chang Media [α-MEM (Gibco-Invitrogen), containing

15% ES-FBS (Gibco-Invitrogen), 1% glutamine (Gibco-Invitrogen), and 1% penicillin/streptomycin (Gibco-Invitrogen), supplemented with 18% Chang B (Irvine Scientific) and 2% Chang C (Irvine Scientific)] at 37 °C with 5% CO_2 atmosphere. During the entire culture period the cells are maintained at around 70–80% of confluency and without any feeder layer as described by De Coppi P. *et al.* (2007).

2. Different Types of Cells Within the Amniotic Fluid

The total population of cells within the amnioitc fluid culture, as previously described, represents a very heterogenous morphology as shown in Fig. 1.

Samples between the 15th and the 19th weeks of gestation present a different milieu of cells derived from all three germ layers and there is a variation in progenitors present.

The total population was investigated for endoderm-, mesoderm-, and ectoderm-marker expression as shown by reverse transcription-polymerase chain reaction (RT-PCR) (Fig. 2A) and Western-blotting techniques (Fig. 2B).

It was determined that expression of markers from endoderm and mesoderm is higher in samples of amniotic fluid from early gestational ages versus those samples taken from later gestational ages. In contrast, ectodermal markers were equally expressed in early and later samples of human amniotic fluid.

B. Isolation of Stem Cells from Amniotic Fluid and Their Characterization

A pluripotent stem cell population can be isolated from amniotic fluid, as shown by De Coppi P. *et al.* using a positive selection for cell membrane receptor c-kit.

The c-kit receptor is a protein-tyrosine kinase that is specific for stem cell factor. This complex has been suggested to be involved in embryogenesis as well as carcinogenesis. The c-kit receptor and its ligand are also involved in hematopoiesis and identify a specific hematopoietic progenitor cell. It has also been suggested by other authors that c-kit is also expressed in the heart (Beltrami *et al.*, 2003) and retina stem cells (Koso *et al.*, 2007), indicating that the c-kit receptor can identify a stem cell population within different organs.

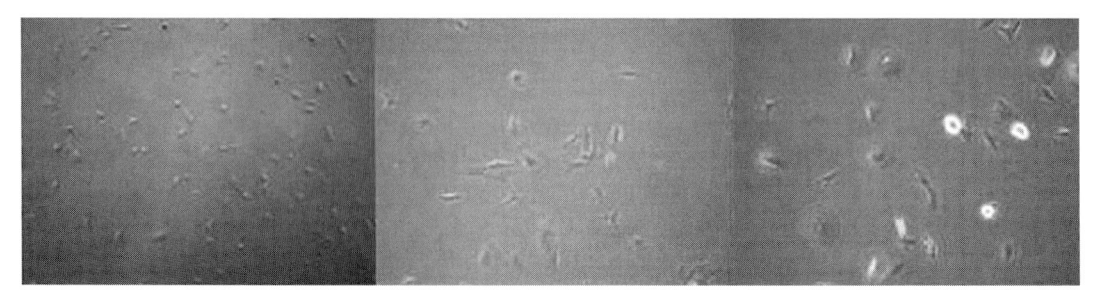

Fig. 1 Morphology of cells of the entire population of amniotic fluid (15 weeks of gestational age. The different morphology of the cells is evident. 5× (A, B), 10×(C).

5. Human Amniotic Fluid Stem Cells and Their Pluripotential Capability

A

	Weeks	Endoderm				Ectoderm			Mesoderm		Stem	
		AFP	Goosecoid	Sox-17	Cxcr-4	E-cadherin	Fgf-5	NCAM	Tal-1	Brachyury	Flk-1	Oct-4	C-kit
A	15		•		•	•	•	•			•	•	•
B	16	•			•		•	•		•	•	•	•
C	16	•	•			•	•	•					•
D	17	•					•	•					
E	17	•			•		•				•	•	•
F	17					•		•			•	•	
G	17												•
H	17	•					•	•					
I	18	•				•	•	•			•	•	
J	18			•		•	•	•				•	•
K	18	•				•	•	•			•	•	•
L	18					•					•		
M	18	•				•	•	•			•	•	•
N	19	•					•	•					
O	19					•	•	•	•		•		
P	20	•					•				•		

B

	Weeks	Endoderm				Ectoderm			Mesoderm		Stem	
		AFP	Goosecoid	Sox-17	Cxcr-4	E-cadherin	Fgf-5	NCAM	Tal-1	Brachyury	Flk-1	Oct-4	C-kit
A	15	•		•					•				
B	16	•		•	•	•	•				•	•	
C	16	•	•		•	•	•	•					
D	17												
E	17	•	•	•	•	•	•		•	•			
F	17	•		•			•	•					
G	17	•		•				•					
H	17												
I	18	•	•	•			•		•	•			
J	18	•	•	•					•	•			
K	18	•				•	•	•			•	•	
L	18												•
M	18			•			•	•				•	
N	19												
O	19			•			•		•				•
P	20			•									•

Fig. 2 (A) Results of the analysis of RNA expression through retrotranscription, amplification, and gel electrophoresis. Markers for all three germ layers and pluripotent characteristics were investigated. All the products of the amplification were confirmed through sequencing. Each of the 16 samples (from A to P) represent the entire amniotic fluid cell populations. (B) Western blotting analysis of samples (as described in Fig. 2A) from the entire population of amniotic fluid. Proteins from all three germ layers and pluripotential characteristics were analyzed.

The stem cell population is usually selected with the MACS (Magnetic Activated Cell Sorter) or FACS (Fluorescence Activated Cell Sorter) system only and 0.8–1% of the entire cell population expresses the surface marker. The c-kit[pos] cells are immediately cultured in petri dishes with no need of feeder layer in Chang Media, at 37 °C with 5% CO_2 atmosphere. For the first week, they maintain a round shape while afterwards they can become elongated and assume a fibroblast-like morphology. At this stage, if the cells are maintained at 70–80% of confluency, they can be cultured for many population doublings and split using a trypsin/EDTA solution every 48–72 h, maintaining a normal karyotype (Fig. 3A).

Using FACS sorting it was determined that amniotic fluid stem cells express some surface markers and transcription factors distinctive of embryonic stem cells such us OCT-4 and SSEA-4, confirming that they possess some important characteristics that embryonic stem cells also have, and suggesting that these cells retain pluripotential capability (Fig. 3D). In addition, they stained positively for a number of surface markers characteristic of mesenchymal and/or neural stem cells, including CD29, CD44 (hyaluronan receptor), CD73, CD90, and CD105.

Each sample showed positivity for Class I major histocompatibility (MHC) antigens (HLA-ABC), and some were weakly positive for MHC Class II (HLA-DR). The AFS cells were negative for markers of the hematopoietic lineage (CD45) and of hematopoietic stem cells (CD34, CD133).

The c-kit[pos] cells have a very high proliferative capacity exceeding Hayflick's limit. Their doubling time is around 36 h, with some variation between samples. Over the population doublings the cells maintain a normal karyotype, and they also present normal regulation of the control checkpoints of the cell cycle, in particular the G1 and G2, in preparation for chromosome replication and entrance into mitosis (Fig. 3B).

In addition, when cultured for more than 350 pd, the c-kit[pos] cells are able to maintain the telomeric length unchanged, indicating their ability to preserve this characteristic of undifferentiated stem cells (Fig. 3C).

Stem cells from amniotic fluid are easily infected with both lentivirus and retrovirus and show great capacity for expressing reporting markers such us

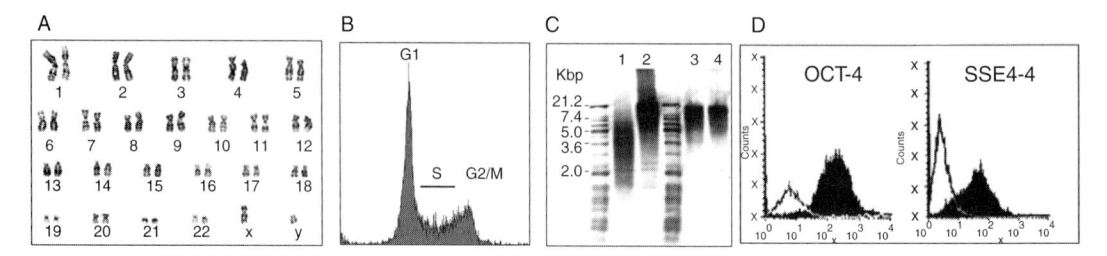

Fig. 3 (A) hAFSC show normal karyotype after 250 pds, (B) a control of the cell cycle, (C) Lane 1: short length telomere standards. Lane 2: high length telomere standards. Lane 3: telomere length of AFS cells early passage (20p.d.). Lane 4: telomere length of AFS cells late passage (250p.d.). maintenance of the telomeric length, and (D) hAFSC express embryonic markers like oct-4 and SSEA-4.

GFP (green fluorescent protein), Lac-Z, and luciferase, confirming their value for *in vitro* and *in vivo* application for studying tissue regeneration mechanisms.

C-kit[pos] cells showed a high clonal capability. Using techniques of clonal dilution together with retrovirus infection in order to identify the clones and the subclones, it was shown that the cells were able to give rise to a clonal population and maintain similar characteristics displaying the same pluripotential capacity of the original line.

III. *In Vitro* Potential of Amniotic Fluid Stem Cells

C-kit[pos] positive cells from amniotic fluid are able to differentiate into cell types derived from all three germ layers in an *in vitro* system if they are exposed to various molecules and growth factors that can stimulate change into differentiated phenotypes and their functional activity. The methods described are again mainly described in the article by De Coppi P. *et al.* (2007).

A. Adipogenic Differentiation

In order to stimulate the c-kit[pos] cells to differentiate into adipocyte-like cells, they are seeded at a density of 3000 cells/cm^2 and are cultured in DMEM low-glucose medium (Gibco-Invitrogen) with 10% FBS (Gibco-Invitrogen), 1% penicillin/streptomycin (Gibco-Invitrogen), and a different adipogenic supplement such as dexamethasone (1 µM, Sigma), 3-isobutyl-1-methylxanthine (1 mM, Sigma), insulin (10 µg/ml, Sigma), and indomethacin (60 µM, Sigma) as reported by Jaiswal *et al.* (1997).

After 1 week in culture, the cells change morphology from a typical elongated shape to a more rounded configuration, with some small granules within the cytoplam positive for the oil-o-red stain that identifies lipid production. In addition, the cells start expressing adipogenic genes, such as lipoprotein lipase and pparγ2, after 8 and 16 days of differentiation. These genes are expressed during adipogenic differentiation during development. The differentiation media is changed every 3 days. (Fig. 4).

B. Myogenic Differentiation

In order to stimulate the c-kit[pos] cells to differentiate into myocyte-like cells, the cells were seeded at a density of 3000 cells/cm^2 on plastic plates precoated with matrigel (incubation for 1 h at 37 °C at 1 mg/ml in DMEM) in DMEM low-glucose (Sigma) formulation containing 10% horse serum (Gibco-Invitrogen), 0.5% chick embryo extract (US Biological), and 1% penicillin/streptomycin (Gibco-Invitrogen) (Rosenblatt *et al.*, 1995). Twelve hours after seeding, 10 µM 5-aza-2′-deoxycytidine (Sigma) is added to the culture medium for 24 h. Incubation is then continued in complete medium lacking 5-azaC, with medium changes every 3 days. After 24 h in culture, it is possible to notice some multinucleated cells

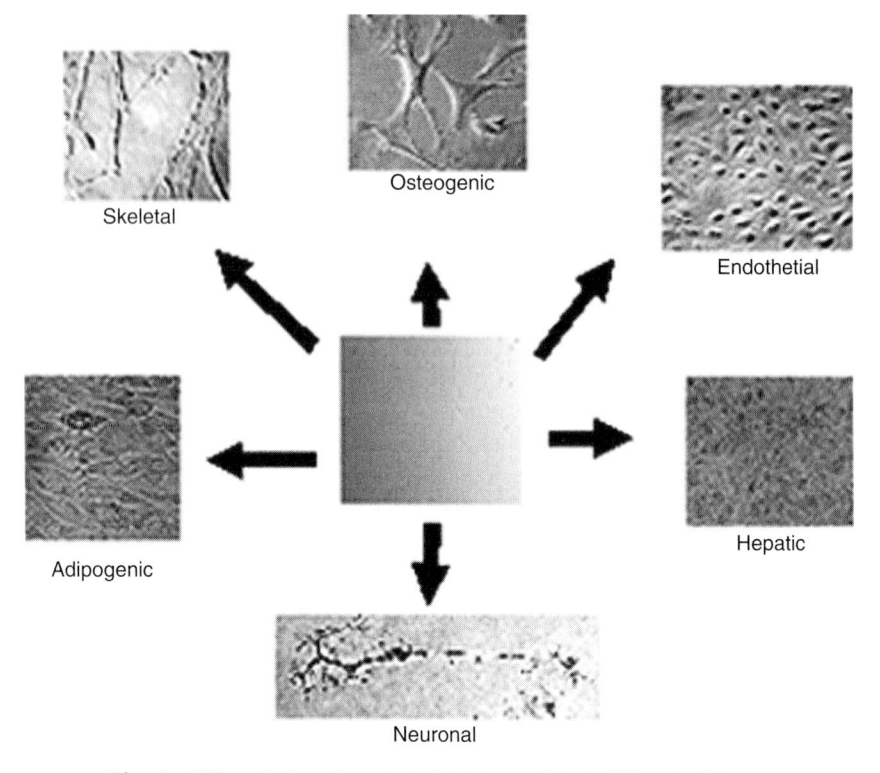

Fig. 4 Differentiation of amniotic fluid stem cells into different cell types.

that could be identified, after 10 days, as a skeletal myofiber. During the differentiation process the cells were able to express early muscle genes, such as MyoD and Myf-5, as well as sarcomeric tropomyosin and desmin. The differentiation media was changed every 3 days. (Fig. 4).

C. Osteogenic Differentiation

In order to stimulate the c-kit^{pos} cells to differentiate into osteocyte-like cells, they are seeded at a density of 3000 cells/cm^2 and were cultured in DMEM low-glucose medium (Gibco-Invitrogen) with 10% FBS (Gibco-Invitrogen), 1% penicillin/streptomycin (Gibco-Invitrogen), and osteogenic supplement such as dexamethasone (100 nM, Sigma), β-glycerophosphate (10 mM, Sigma), and ascorbic acid-2-phosphate (0.05 mM, Sigma) (Jaiswal et al., 1997) After 4 days in culture, the cells change morphology and then start to express, after 2 weeks in culture, genes usually expressed during osteocyte development ranging from early and late maturation of the cells such as Runx2 and Osteocalcin. In addition, the cells were able to produce a mineralized matrix, revealed with the Von Kossa staining (Jaiswal et al., 1997), and the production of calcium, demonstrated with the alkaline phosphatase colorimetric assay (Jaiswal et al., 1997). This expression is increased over time once the cells are stimulated with osteogenic supplements. The differentiation media was changed every 3 days (Fig. 4).

D. Endothelial Differentiation

In order to stimulate the c-kit^{pos} cells to differentiate into endothelial-like cells, the cells were seeded at a density of 3000 cells/cm^2 in endothelial basal media-2 (EBM-2, Clonetics BioWittaker) in gelatin-coated dishes. Every 2 days, bFGF (basic-fibroblastic growth factor, 2 ng/ml, Sigma) was added to the culture and after 1 week in culture the cells showed a change in morphology, reaching a tubular shape after 2 weeks. They express some early endothelial progenitor markers, such as CD31, as well as adult endothelial markers, like Factor VIII, von Willebrand Factor, and P1H12. After 1 month in culture, the cells were able to form capillary-like structures in the dish expressing CD31 and VCAM (vascular cell adhesion molecule). The differentiation media was changed every 3 days (Fig. 4).

E. Hepatocytes Differentiation

In order to stimulate the c-kit^{pos} cells to differentiate into hepatocyte-like cells, the cells were seeded at a density of 3000 cells/cm^2 in a matrigel-coated dish and cultured until they reached a 50–60% confluency. The cells were cultured in DMEM (Gibco-Invitrogen), 15% FBS (Gibco-Invitrogen), monothioglycerol (300 µM, Sigma), HGF (hepatocytes growth factor, 20 ng/ml, Sigma), oncostatin M (10 ng/ml, Sigma), dexamethasone (10^{-7} M, Sigma), FGF4 (fibroblastic growth factor, 100 ng/ml, Sigma), and 1× ITS (insulin-transferrin-selenium, Gico-Invitrogen) The cells were cultured in these conditions for at least 2 weeks, then trypsined and plated into a collagen sandwich gel. A layer of collagen was seeded on the plate, then the cells were plated above the layer, and finally another layer of collagen was added to the cells maintained up to 45 days. The differentiation media was changed every 3 days. During the differentiation process the cell changed morphology from elongated to the rounded shape typical of a hepotocyte-like cell. The differentiated cells were able to produce albumin and were positive for the production of urea after 3 weeks in culture. In addition, at early stages of differentiation, the cells express HNF4α (hepotocyte nuclear factor 4α), c-met, and MDR (multidrug resistance) (Fig. 4).

F. Neurogenic Differentiation

In order to stimulate the c-kit^{pos} cells to differentiate into neural-like cells, the cells were seeded at a density of 3000 cells/cm^2 in DMEM low-glucose (Gibco-Invitrogen), 1% penicillin/streptomycin (Gibco-Invitrogen), 2% DMSO (Sigma), butylated hydroxynisole (200 µM, Sigma), and NGF (neurogenic growth factor, 25 ng/ml, Sigma). After 2 days, the medium was replaced with Chang supplemented with NGF only.

In an alternative 2-step differentiation process for specific dopaminergic cells, the cells were seeded in a fibronectin (1 μg/ml, Sigma)-coated plate and cultured with DMEM/F12 (Gibco-Invitrogen) plus 1 × N2 (Gibco-Invitrogen) and bFGF (10 ng/ml, Sigma), and the bFGF was freshly replaced every 2 days (De Coppi *et al.*, 2007). The differentiated cells were able to express nestin and also glyceraldehyde-3-phosphate dehydrogenase and GIRK2. In addition, they were able to secrete the neurotransmitter glutamic acid (Fig. 4).

IV. *Ex Vivo* and *In Vivo* Application of Amniotic Fluid Stem Cells

A. *Ex Vivo* Application of Amniotic Fluid Stem Cells: Integration into Mouse Embryonic Kidneys

The differentiation ability of the AFS cells has further been tested in an *ex vivo* system (embryonic kidneys) in order to evaluate their potential application in kidney tissue regeneration. Perin *et al.* (2007) have shown for the first time the use of amniotic c-kit derived cells for kidney regeneration. The kidney primordia were obtained from timed-pregnant female mice, at embryonic days E12.5–E18 under a dissecting microscope. Once isolated, they were placed on a 0.4-μm pore size transwell membrane (Corning) and were cultivated at the medium-gas interphase in a humidified 37 °C incubator with 5% CO_2. The culture media was composed of Leibovitz's L-15 (Sigma) supplemented with 1% pen/strep (Gibco-Invitrogen) and 2% FBS (Gibco-Invitrogen). Perfusion channels were created with a 10-μm diameter glass needle under direct microscopy without conveying subsequent damage to the kidney but permitting homogenous perfusion and bathing in the culture media throughout the whole organ. Meanwhile, hAFSC labeled with GFP and Lac-Z were trypsinized, counted, and labeled with a cell-surface marker (CM-DiI, Molecular Probe).

Transduction with lentiovirus codifying for GFP and Lac-z was performed with standard protocols used in laboratories. Briefly two cycles of transduction were performed by removing the old medium and adding new virus supernatant and medium. Twenty-four hours after the initial transduction, cells are thoroughly washed 3 times with phosphate-buffered saline (PBS, Sigma) before *in vitro* or *in vivo* analysis. For the labeling with CM-DiI, the cells were trypsinized in 0.05 M trypsin/EDTA (Sigma) solution and centrifuged at 1500 rpm for 5 min and then were incubated with a working solution of CM DiI of 1 mg/ml for 5 min at 37 °C followed by an incubation of 15 min at 4 °C and 3 washes with PBS (Fig. 5A–C) and loaded into a 15-μm diameter transfer tip of a CellTram Oil injector (Eppendorf) with no serum. Each kidney received a single injection (1–2 μL) and was immediately placed on the top of a Transwell filter in the incubator for 2–10 days (Fig. 5D).

The survival and the differentiation capability of hAFSC after injection was analyzed by methods of histology, chromogenic *in situ* hybridization, PCR, and live imaging using standard protocols. H&E staining 5 days postinjection revealed

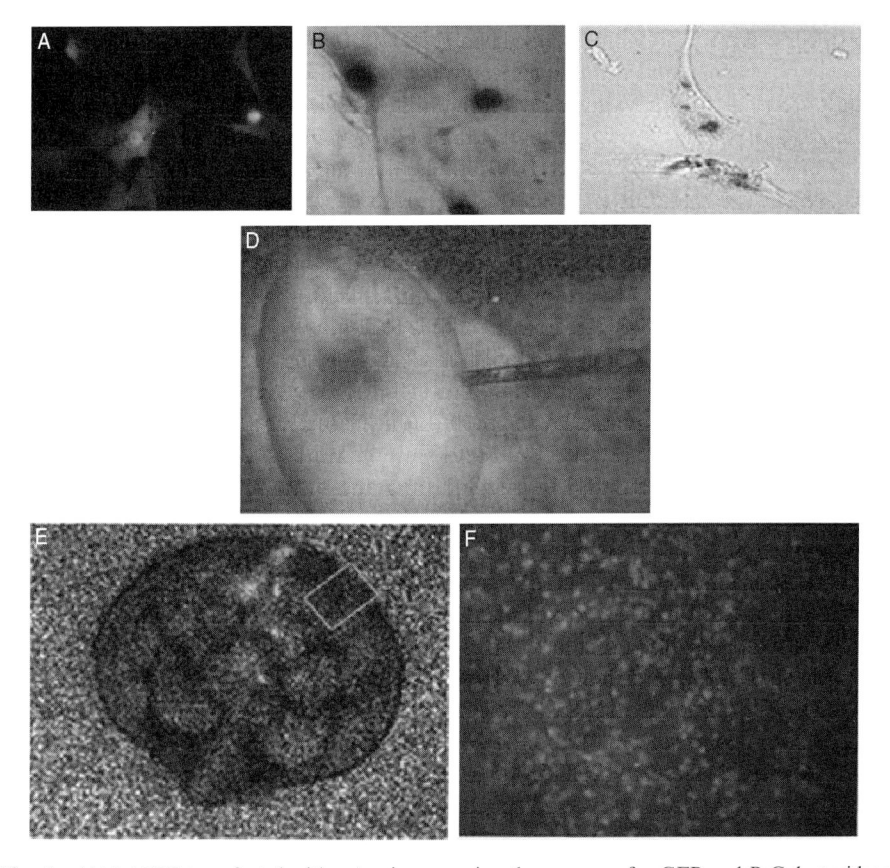

Fig. 5 (A) hAFSC transfected with retrovirus carrying the sequence for GFP and B-Galactosidase under fluorescent microscopy (40×). (B) Lac-z nuclear staining of hAFSC (40×). (C) hAFSC labeled pink under the light microscopy with cell surface marker CM-DiI (40×). (D) Microinjection of hAFSC labeled with CM-DiI under direct vision (4×) into the center of the embryonic kidney. (E) GFP labeled cells shown in the embryonic kidney under fluoroscopy (4×) at day 0 of injection. (F) Live imaging, at 4-h intervals, of the embryonic kidney after 4 days of culture demonstrates GFP labeled hAFSC multiplying and spreading throughout the entire organ from the center to the periphery. (See Plate no. 10 in the Color Plate Section.)

the cells injected into E13 kidneys integrated into C- and S- shape structures, stroma, and renal vesicles of the murine embryonic kidney. After 6 days of culture, CM-DiI labeled hAFSC were also detected in embryonic tubular and glomerular structures. These findings were further confirmed by chromogenic *in situ* hybridization. The integration of hAFSC into the metanephric structures was additionally confirmed by the migration of the injected cells from the site of injection, the center of the embryonic kidney, to the periphery, documented by both live imaging and by histological techniques (Fig. 5E and F). This observation strongly correlates to the centrifugal pattern of induction, morphogenesis, and differentiation of the metanephros, proceeding from the center to the periphery of the embryonic organ.

After 9 days of culture, expression of early embryonic kidney markers was detected using RT-PCR. The human-specific kidney genes expressed by the injected hAFSCs were Zona occludin-1 (ZO-1), claudin, and glial-derived neurotrophic factor (GDNF). These were negative in the controls (Fig. 6). The expression of GDNF is of particular importance because it represents direct evidence for successful hAFSC induction into kidney tissue. This is supported based on the fact that GDNF expression is usually seen in only very early stages during branching of the ureteric bud (UB) into the metanephric mesenchyme (MM) (Basson et al., 2006; Costantini and Shakya, 2006). After the branching ,GDNF expression is downregulated and other genes take over during the final development phase of the kidney. Thus, it was concluded that the strong human GDNF expression observed must arise from hAFSC that had been induced by the surrounding mouse embryonic environment, to undergo renal differentiation.

Overall, it was demonstrated that hAFSC are a potential source of stem cells capable of integrating into metanephric kidney structures during development and confirmed that these cells, when injected, follow all the steps of the nephrogenesis.

B. *In Vivo* Application of Amniotic Fluid Stem Cells: Integration into Adult Mouse Kidney

After having proven that hAFSC can survive and integrate into developing metanephric kidney structures when injected *in vitro*, the stem cells were tested in an *in vivo* animal model of renal injury. The animal model used for these experiments is acute

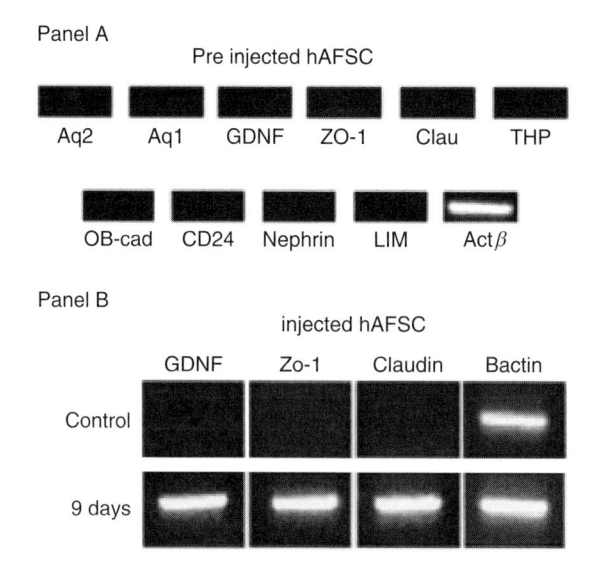

Fig. 6 Panel A: RT-PCR of hAFSC prior to injection demonstrates that these cells are negative for most kidney markers. Panel B: hAFSC, 9 days after injection into embryonic kidneys demonstrate early kidney gene markers. Zona occludens 1 (ZO-1, 760 bp), glial-derived neurotrophic factor (GDNF, 630 bp), and claudin (480 bp), when compared to no expression of kidney markers in control (hAFSC prior to injection).

tubular necrosis (ATN) in which the epithelial cells of tubules are damaged by a rhabdomyolosis process induced by intramuscular injection of glycerol (50% solution, 9 ml/Kg, Sigma) administered in equally divided doses into each upper hind limb, inducing renal dysfunction. The mice used for these experiments were female *nu/nu* [(Charles River) and also female wild type (C57/BL, Charles River)] as control. In the *in vivo* experiments, hAFSC labeled with luciferase and a cell surface marker (CM-Dil, Molecular Probe) were directly injected into damaged kidneys using an open surgery method. A small dorsal incision of about 10 mm was made to expose the kidney. 1×10^6 hAFSC labeled with a cell-tracking dye (CM Dil, as above described) were injected into both the damaged organs using a microinjector [(Eppendorf CellTram vario and Eppendorf transfer tips-RP (ICSI)]. Three (3) weeks after injection the integration of hAFSC into tubular structures was shown by histology, supported by expression of human-specific mature kidney markers with PCR (unpublished data).

At different times points (ranging from 1 day up to 6 months) the mice were sacrificed and the kidneys were collected and processed with standard histological and molecular methodologies in order to analyze the level of integration and differentiation of the stem cells.

These experiments demonstrate that the hAFSC were able to home into the damaged organ and survive in an *in vivo* environment. In addition, they integrate into damaged tubules, as demonstrated by immunofluorescence and by PCR. The integrated hAFSC were able to express kidney markers specific for tubular kidney-specific proteins (data not published).

V. Discussion

In this chapter, we described a stem cell population isolated from amniotic fluid. Amniocentesis is a relatively benign modality used for prenatal screening. These cells may offer an alternative method of access to fetal stem cells without harm to the fetus itself. hAFSC possess some characteristics in common with hESC. They express embryonic markers, they maintain their telomeric length, and they are able to give rise to clones. In addition, they can be maintained in culture for many population doublings. Once seeded *in vitro* and stimulated with different growth factors and molecules, hAFSC are able to give rise to different cell lines derived from all the three germ layers. They express mesenchymal stem cell markers and they appear to be more pluripotent than stem cells derived from bone marrow.

Some *ex vivo* and *in vivo* experiments in mice also confirm their ability to differentiate into different structures when stimulated in an environment that is different from an *in vitro* one. *Ex vivo* injection of the cells into mouse embryonic kidneys during gestational age E13–E18 has shown that hAFSC can not only survive but also integrate into the early structures of the metanephric kidney and follow the normal centrifugal pattern of kidney development. Expression of GDNF by hAFSC on the 9th day of culture is direct evidence for the involvement of these cells in the early stages of nephrogenesis. Direct *in vivo* injections of hAFSC into glycerol damaged kidneys revealed similar results, integrating themselves into the site of the injury, the renal tubules, after intramuscular glycerol injection. Thus, hAFSC seem to be a suitable source of stem cells, with a potential to be an alternative clinical treatment for a large number of kidney diseases in the future. A very important aspect of these *in vivo* applications is that hAFSC do not form teratomas when implanted into an *in vivo* environment. This fact is of great interest for their future applications to patients as therapeutics. However, there is still a lot of work to be done before these cells are shown safe and harmless for clinical applications. More accurate diagnostics need to be investigated not the least of which are their immunological propensity to exclude any potential for rejection once transplanted *in vivo*.

References

Basson, M. A., Watson-Johnson, J., Shakya, R., Akbulut, S., Hyink, D., Costantini, F. D., Wilson, P. D., Mason, I. J., and Licht, J. D. (2006). Branching morphogenesis of the ureteric epithelium during kidney development is coordinated by the opposing functions of GDNF and Sprouty1. *Dev. Biol.* **299**, 466–477.

Beltrami, A. P., Barlucchi, L., Torella, D., Baker, M., Limana, F., Chimenti, S., Kasahara, H., Rota, M., Musso, E., Urbanek, K., Leri, A., Kajstura, J., *et al.* (2003). Adult cardiac stem cells are multipotent and support myocardial regeneration. *Cell* **114**(6), 763–776.

Bossolasco, P., Montemurro, T., Cova, L., Zangrossi, S., Calzarossa, C., Buiatiotis, S., Soligo, D., Bosari, S., Silani, V., Deliliers, G. L., Rebulla, P., and Lazzari, L. (2006). Molecular and phenotypic characterization of human amniotic fluid cells and their differentiation potential. *Cell Res.* **16**(4), 329–336.

Costantini, F., and Shakya, R. (2006). GDNF/Ret signaling and the development of the kidney. *Bioessays* **28**, 117–127.

De Coppi, P., Bartsch, G., Siddiqui, M. M., Xu, T., Santos, C. C., Perin, L., Mostoslavsky, G., Serre, A. C., Snyder, E. Y., Yoo, J. J., Furth, M. E., Soker, S., *et al.* (2007). Isolation of amniotic stem cell lines with potential for therapy. *Nat. Biotechnol.* **25**(1), 100–106.

George, T., Rachel, W., Daniela, P., Gert, L., and Michael, F. (2004). The amniotic fluid cells proteome. *Electrophoresis* **25**, 1168–1173.

Gosden, C. M. (1983). Amniotic fluid cell types and culture. *Br. Med. Bull.* **39**(4), 348–354.

Hoehn, H., and Salk, D. (1992). Morphological and biochemical heterogeneity of amniotic fluid cells in culture. *Methods Cell Biol.* **26**, 11–34.

Jaiswal, N., Haynesworth, S. E., Caplan, A. I., and Bruder, S. P. (1997). Osteogenic differentiation of purified, culture-expanded human mesenchymal stem cells *in vitro*. *J. Cell. Biochem.* **64**(2), 295–312.

Koso, H., Satoh, S., and Watanabe, S. (2007). c-kit marks late retinal progenitor cells and regulates their differentiation in developing mouse retina. *Dev. Biol.* **301**(1), 141–154.

Mark, A., Underwood, M. D., William, M., Gilbert, M. D., Michael, P., and Sherman, M. D. (2005). Amniotic fluid: Not just fetal urine anymore. *J. Perinatol.* **25**, 341–348.

McLaughlin, D., Tsirimonaki, E., Vallianatos, G., Sakellaridis, N., Chatzistamatiou, T., Stavropoulos-Gioka, C., Tsezou, A., Messinis, I., and Mangoura, D. (2006). Stable expression of a neuronal dopaminergic progenitor phenotype in cell lines derived from human amniotic fluid cells. *J. Neurosci. Res.* **15**(7), 1190–1200.

Perin, L., Giuliani, S., Jin, D., Sedrakyan, S., Carraro, G., Habibian, R., Warburton, D., Atala, A., and De Filippo, R. E. (2007). Renal differentiation of amniotic fluid stem cells. *Cell Prolif.* **40**(6), 936–948.

Reubinoff, B. E., Pera, M. F., Fong, C. Y., Trounson, A., and Bongso, A. (2000). Embryonic stem cell lines from human blastocytes: Somatic differentiation *in vitro*. *Nat. Biotechnol.* **18,** 399–404.

Rosenblatt, J. D., Lunt, A. I., Parry, D. J., and Partridge, T. A. (1995). Culturing satellite cells from living single muscle fiber explants. *In Vitro Cell. Dev. Biol. Anim.* **31**(10), 773–779.

Stocum, D. L. (2001). Stem cell in regenerative biology and medicine. *Wound Rep.* **9,** 429–442.

Streubel, B., Martucci-Ivessa, G., Fleck, T., and Bittner, R. E. (1996). *In vitro* transformation of amniotic cells to muscle cells--background and outlook. *Wien. Med. Wochenschr.* **146**(9–10), 216–217.

Takahashi, K., Ichisaka, T., and Yamanaka, S. (2006). Identification of genes involved in tumor-like properties of embryonic stem cells. *Methods Mol. Biol.* **329,** 449–458.

Thomson, J. A., Itskovitz-Eldor, J., Shapiro, S. S., Waknitz, M. A., Swiergiel, J. J., Marshall, V. S., and Jones, J. M. (1998). Embryonic stem cells derived from human blastocysts. *Science* **282,** 1145–1147.

Torricelli, F., Brizzi, L., Bernabei, P. A., Gheri, G., Di Lollo, S., Nutini, L., Lisi, E., Di Tommaso, M., and Cariati, E. (1993). Identification of hematopoietic progenitor cells in human amniotic fluid before the 12th week of gestation. *Ital. J. Anat. Embryol.* **98**(2), 119–126.

CHAPTER 6

Method to Isolate Mesenchymal-Like Cells from Wharton's Jelly of Umbilical Cord

Kiran Seshareddy, Deryl Troyer, and Mark L. Weiss

Department of Anatomy and Physiology
College of Veterinary Medicine
Kansas State University
Manhattan, KS 66506

    Abstract
  I. Introduction
 II. Rationale
III. Methods
IV. Materials
 V. UCMSCs and Culture Characteristics
VI. Discussion
    References

Abstract

The umbilical cord is a noncontroversial source of mesenchymal-like stem cells. Mesenchymal-like cells are found in several tissue compartments of the umbilical cord, placenta, and decidua. Here, we confine ourselves to discussing mesenchymal-like cells derived from Wharton's Jelly, called umbilical cord matrix stem cells (UCMSCs). Work from several laboratories shows that these cells have therapeutic potential, possibly as a substitute cell for bone marrow-derived mesenchymal stem cells for cellular therapy. There have been no head-to-head comparisons between mesenchymal cells derived from different sources for therapy; therefore, their relative utility is not understood. In this chapter, the isolation protocols of the Wharton's Jelly-derived mesenchymal cells are provided as are protocols for their *in vitro* culturing and storage. The cell culture methods provided will enable basic scientific research on the UCMSCs. Our vision is that both umbilical cord blood and UCMSCs will be commercially collected and stored in the future for preclinical work, public and private banking services, etc. While umbilical cord blood banking standard operating procedures exist, the scenario mentioned above requires clinical-grade UCMSCs. The hurdles that have been identified for the generation of clinical-grade umbilical cord-derived mesenchymal cells are discussed.

I. Introduction

Mesenchymal stromal cells (MSCs), as defined by the International Society for Cellular Therapy, are plastic-adherent cells with a specific surface phenotype that have the capacity to self-renew and to differentiate into various lineages including bone, cartilage, and adipose (Dominici et al., 2006). Such cells can be derived from several different sources, such as trabecular bone, adipose tissue, synovium, skeletal muscle, dermis, pericytes, blood, and bone marrow (Tuan et al., 2003).

MSCs derived from bone marrow and adipose tissue have been studied extensively. MSCs derived from bone marrow can be differentiated into bone, cartilage, tendon, muscle, adipose tissue, and hematopoietic cell-supporting stroma (Baksh et al., 2004). Thus, they are candidates to treat patients suffering from bone disorders, heart failure, etc. Since MSCs can be isolated from adults in significant number, they have been examined closely for therapeutic utility. For example, MSCs support the ex vivo expansion of hematopoietic stem cells (Dexter et al., 1977; Friedenstein et al., 1974), act as immune modulators (Le, 2006), release cytokines and growth factors (Caplan and Dennis, 2006), and they home to sites of pathology (Studeny et al., 2002).

It is estimated that more than 50 clinical trials are ongoing using bone marrow-derived MSCs for a variety of indications, for example, acute myocardial infarction, stroke, and graft versus host disease. Nevertheless, there are limitations associated with MSCs derived from bone marrow for cell-based therapy. For example, collection of MSCs from bone marrow is an invasive and painful procedure. In normal aging, the marrow cavity fills with yellow fat. Thus, there may be difficulty in obtaining MSCs from older individuals. Along these lines, differences have been found between bone marrow-derived MSCs collected from the fetus versus adult-derived MSCs. For example, fetal MSCs have a longer life in vitro compared to adult-derived MSCs (Guillot et al., 2007): MSCs derived from adults have a useful lifespan in vitro of about five passages (Tuan et al., 2003).

In addition to bone marrow, MSCs may be derived from adipose tissue. While adipose-derived MSCs (ASCs) have been studied less than bone marrow-derived MSCs, ASCs may be induced to differentiate into osteocytes (Shen et al., 2006), cartilage (Jin et al., 2007a,b), and cardiomyocytes (Nakagami et al., 2006; Zhang et al., 2007), and they display both similar surface phenotype and immune properties to bone marrow-derived MSCs (McIntosh et al., 2006). While there is no shortage of the adipose material within the United States, the procurement of adipose tissue involves an invasive and painful surgical procedure. There is no comparison work done to evaluate ASCs from the fetus with adult-derived ASCs.

Our lab (Weiss *et al.*, 2006) and others (Conconi *et al.*, 2006; Lu *et al.*, 2006) have demonstrated that cells derived from the Wharton's Jelly in umbilical cords (so called umbilical cord matrix cells or UCMSCs) have properties of MSCs. While UCMSCs have surface phenotype (Weiss *et al.*, 2006), differentiation capability (Lu *et al.*, 2006), and immune properties similar to MSCs derived from bone marrow and adipose (Weiss *et al.*, unpublished data), UCMSCs are more similar to fetal MSCs in terms of their *in vitro* expansion potential. In contrast to bone marrow- and adipose-derived MSCs, UCMSCs are isolated from the umbilical cord following birth and may be collected following either normal vaginal delivery or cesarean section. As described below, UCMSCs are easily expandable *in vitro,* and may be cryogenically stored, thawed, and reanimated. While the collection process for human materials is elaborated on here, UCMSCs have also been isolated using modified protocols from dog, cat, rat, mouse, horse, bovine, and swine umbilical cord (protocols available upon request). Human UCMSCs grow as plastic-adherent cells, express a surface phenotype similar to other MSCs (Weiss *et al.*, 2006), and differentiate into multiple lineages (Wang *et al.*, 2004). Umbilical cord matrix cells have been safely transplanted and ameliorated symptoms in an animal model of Parkinson's disease (Medicetty *et al.*, 2003; Weiss *et al.*, 2006), neural damage associated with cardiac arrest/resuscitation (Wu *et al.*, 2007), retinal disease (Lund *et al.*, 2007), and cerebral global ischemia (Jomura *et al.*, 2007). Finally, UCMSCs that have been mitotically inactivated can be used as a feeder layer for embryonic stem cells (Saito, 2006; Toumadje and Auerbach, unpublished observations).

II. Rationale

Because of their physiological properties, and because of the ease of isolation, expansion, and banking capability, UCMSCs may be useful clinically and experimentally. These detailed protocols should enable further research on this interesting population.

III. Methods

- **Isolation of Cells**

    i. Use of umbilical cord tissue from human subjects requires Institutional Review Board (IRB) approval and a signed informed consent form. Umbilical tissue falls into an interesting niche. On the one hand, it is a discarded, (potentially) anonymous tissue, and thus may qualify for an IRB exemption. However, since DNA testing makes UCMSCs individually identifiable, an IRB may assign a protocol number and track the work. Once you secure IRB approval, you must find an obstetrics/gynecology physician and OB/GYN staff at a local hospital to assist; this is key to obtaining a steady supply of umbilical material. The informed consent outlines your project, and must be signed by the donor, and witnessed. The consent form is retained by your OB/

GYN collaborator to maintain donor confidentiality. We collect anonymous biographic information. For example, the sex of donor, weeks of gestation, normal or cesarean-section delivery, approximate cord length, preeclampsia, and twins are recorded. Cords are specifically excluded from individuals with questionable health status, for example, stillbirth, preeclampsia, infectious disease, STD, or hepatitis-positive mother.

ii. After the delivery of the baby, the umbilical cord is collected and stored in a sterile specimen cup containing 0.9% normal saline at 4 °C until processing.

iii. Typically, the cord is processed within 12–24 h of birth. See note 1.

iv. The cord is handled in an aseptic fashion and processed in a Type II Biological Safety Cabinet. See note 2.

v. The surface of the cord is rinsed with sterile phosphate buffered saline to remove as much blood as possible. See Fig. 1.

vi. The length of the cord is estimated. The cord is manipulated in a sterile 10 cm petri dish.

vii. The cord is cut into 3–5 cm long pieces using a sterile blade. See Fig. 3.

viii. Blood vessels are removed from each piece after incising the cord lengthwise. The remaining tissue is rinsed. See note 3. See Figs. 2, 4, 5 and 6.

ix. The cord tissue is placed into a sterile 15 ml centrifuge tube and incubated in 3 ml of enzyme solution for 1 h at 37 °C (see solution A below). See note 4.

x. After 1 h, the cord pieces are crushed using serrated thumb forceps to release as many cells as possible into the solution. See Fig. 7.

xi. The tissue is moved to a new sterile 10 cm^2 dish filled with phosphate buffered saline, swirled for 5 min, and moved to a new centrifuge tube containing enzyme solution (see solution B below). The tube is incubated for 30 min at 37 °C. See Fig. 8.

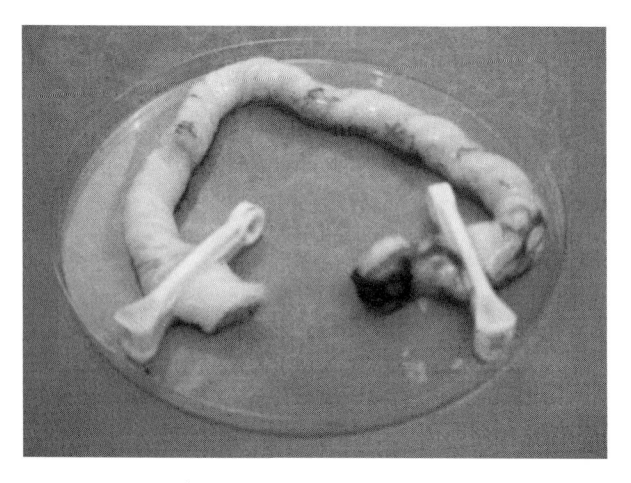

Fig. 1 Human umbilical cord.

6. Umbilical Cord Matrix Stem Cells (UCMSCs)

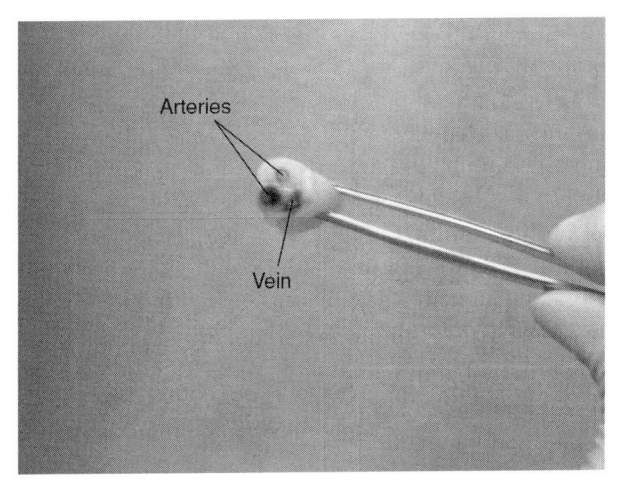

Fig. 2 Cross-section of human umbilical cord showing the two umbilical arteries and the umbilical vein (labeled by the arrow).

Fig. 3 Human umbilical cord cut into 3–5 cm long pieces.

xii. During this incubation, the centrifuge tube containing solution A is centrifuged at 1000 rpm for 5 min. The supernatant is discarded and 3 ml of medium (see the "medium" section) is added to the cell pellet. The cells are resuspended in medium by trituration with a 1000 µl pipette tip to minimize bubble formation and foaming, and the tube is placed in the incubator until the second enzymatic digestion is completed.

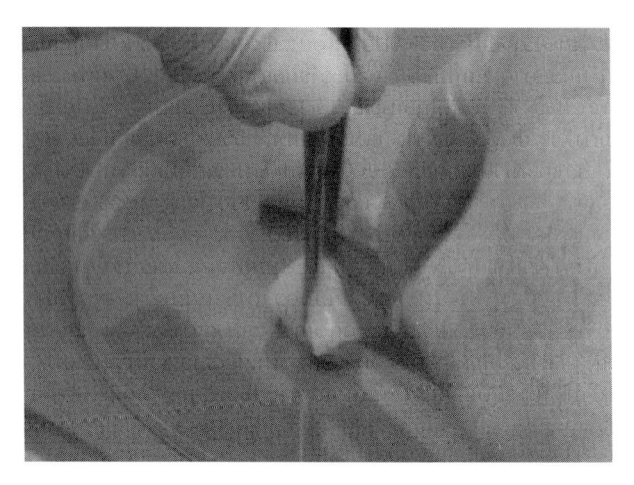

Fig. 4 Incise the amnion lengthwise, to expose the blood vessels.

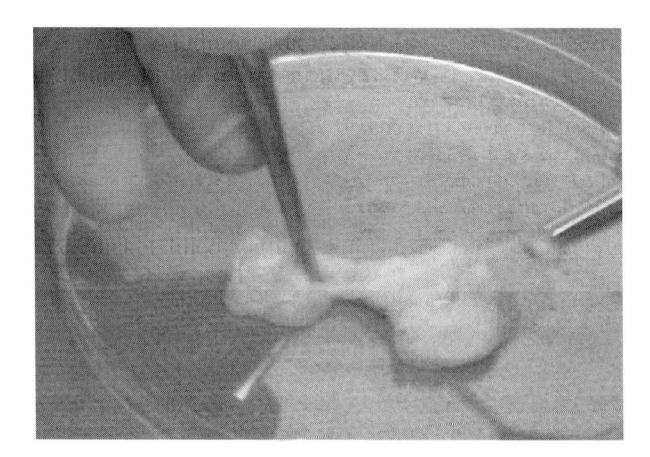

Fig. 5 Grasp each vessel with thumb forceps and tease it from the surround tissue.

xiii. After the second enzymatic digestion is complete, the cord pieces are squeezed in solution B to remove as many cells from Wharton's jelly as possible.

xiv. The tube is centrifuged at 1000 rpm for 5 min. The supernatant is discarded and 3 ml of medium is added to the cell pellet. The cells are resuspended in medium by titurition as in step XII.

xv. The cells from the two enzymatic digestion steps are combined. Optional step: Lyse red blood cells. See note 5.

xvi. The live cells are counted using a hemocytometer and plated in a 6-well tissue culture plate at a concentration of 30,000 cells per cm^2. See note 6.

xvii. Incubate plate at 37 °C, 5% CO_2 for 24–72 h.

6. Umbilical Cord Matrix Stem Cells (UCMSCs)

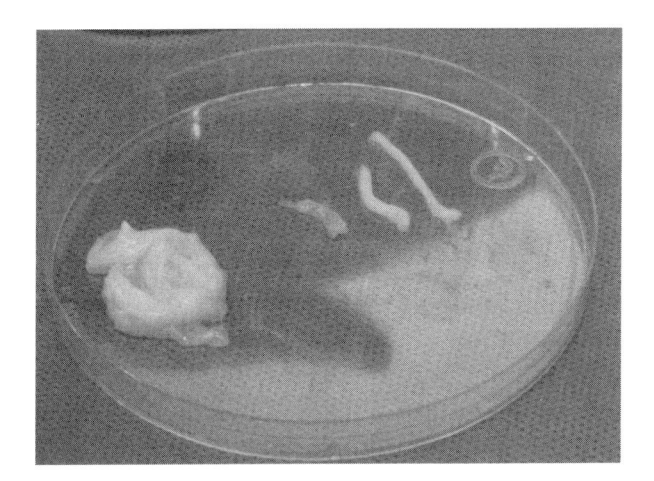

Fig. 6 Blood vessels removed from a piece of umbilical cord tissue.

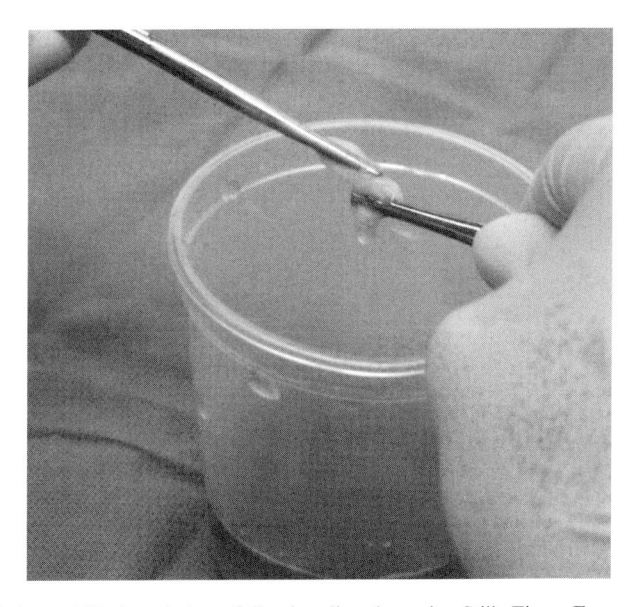

Fig. 7 Crush the umbilical cord tissue following digestion using Stille Tissue Forceps. This releases cells from the Wharton's Jelly into the solution.

xviii. After 24–72 h, the floating cells are transferred to a new plate to allow additional cells to adhere.

xix. The cells in the original plate are fed with fresh medium.

xx. The cells are fed by the removal/replacement of half the medium every 2–3 days till the cells reach approximately 80% confluence.

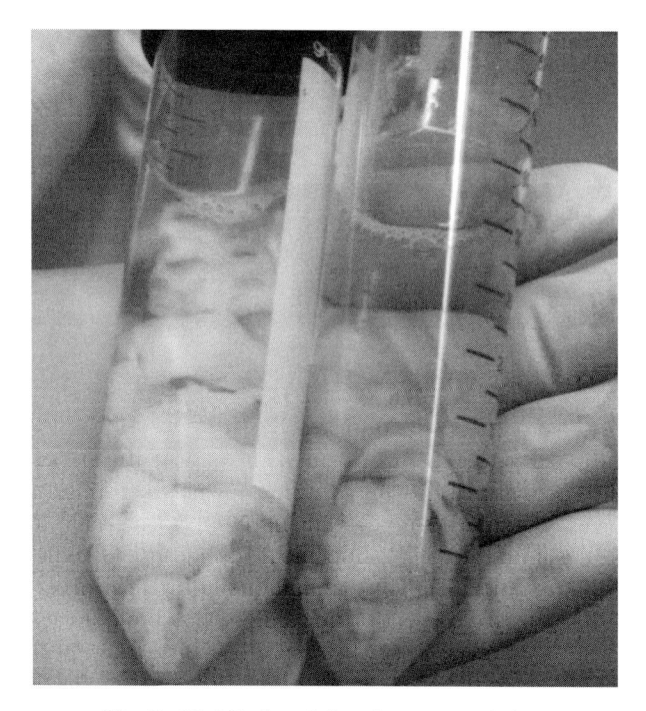

Fig. 8 Umbilical cord tissue in enzyme solution.

- **Passaging the cells:**
    i. The cells are passaged when they are 80–90% confluent. See note 7.
    ii. The medium is aspirated off and the cells are rinsed with sterile phosphate buffered saline (Ca^{2+} free).
    iii. A minimum amount of warmed, CO_2-equilibrated trypsin–EDTA (0.05%) is added to the plate and/or flask to cover the culture surface—0.5 ml to each well of a 6-well plate, 1 ml to a T-25 flask, and 2 ml to a T-75 flask.
    iv. The plate and/or flask is allowed to sit at room temperature for 1–2 min.
    v. Then the detachment of the cells is observed under a microscope and detachment facilitated by repeatedly tapping the plate and/or flask gently on a hard surface.
    vi. The cells are not allowed to be in contact with trypsin–EDTA for more than 5 min. See note 8.
    vii. The trypsinization reaction is neutralized by adding 2–3 times volumes of medium.
    viii. The solution containing the cells is transferred to a 15 ml sterile centrifuge tube and centrifuged at 1000 rpm for 5 min at room temperature.

ix. The supernatant is discarded and the cells are resuspended gently in fresh medium.
x. The cells are counted and transferred to a new plate or flask at a concentration of 10,000 cells per cm^2 in fresh medium. See note 9.
xi. The plates and/or flasks are incubated at 37 °C, saturating humidity and 5% CO_2.
xii. The plates and/or flasks are checked for confluence every day and the cells are fed every other day by removing half the medium and replacing it with fresh medium.
xiii. The amount of medium in a plate or flask is as follows:

>One well of a 6-well plate, 1.5 ml
>One T-25 flask, 4 ml
>One T-75 flask, 10 ml

- **Feeding the cells**

    i. The cells are fed every other day or every 3 days.
    ii. Half the medium in the plate or flask is aspirated off and is replaced with fresh medium.

- **Counting the cells**

    i. The cells are counted by trypan blue exclusion assay using a hemocytometer. This standard method will not be elaborated here (protocol provided upon request).

- **Cryopreservation**

    i. The cells collected for freezing should be in the growth phase.
    ii. The cells are lifted as described for passaging, except that 4 °C freezing medium is added to the cells rather than resuspending them in medium.
    iii. The cells in the freezing medium are transferred into a cryovial at 4 °C. See note 10.
    iv. The cryovial is transferred to a controlled rate cooler, like Mr. Frosty, maintained at 4 °C and placed in the coldest part of the −80 °C freezer.
    v. The cryovial is transferred to a liquid nitrogen tank in a day or two.
    vi. To thaw, the vials are removed from the LN_2 tank and plunged into a 37 °C water bath with gentle swirling.
    vii. Before the last ice crystal has melted, the vials are wiped down twice with 70% ethanol and moved to the Bio Safety Cabinet (BSC).
    viii. The contents of the vial are added slowly with gentle swirling to 5 ml of fresh medium and the vial rinsed twice with medium.

ix. The cells are pelleted and resuspended in 1 ml of medium and a live/dead count made.
x. Cells are plated at 20,000 cells/cm² for the first passage following thaw. See note 11.

- **Flowcytometry**

    i. The cells are lifted as described in passaging and are resuspended in phosphate buffered saline (Ca^{2+} free) at 1–2 million/ml.
    ii. Hundred microliters of this cell suspension is taken in each of 12×75 mm Falcon polystyrene FACS tubes.
    iii. The appropriate amount of conjugated antibody or isotype control is added to each FACS tube.
    iv. Tubes are incubated at room temperature in the dark for 15–20 min.
    v. After the incubation, the cells are washed with 2 ml of phosphate buffered saline (Ca^{2+} free).
    vi. The FACS tubes containing the cells are centrifuged at 1000 rpm for 3 min.
    vii. The supernatant is aspirated off.
    viii. The cells are resuspended in 200 µl of phosphate buffered saline (Ca^{2+} free) and ran through a flow cytometer.
    ix. Typically for each tube, 10,000 events are collected and the data are analyzed using Cell Quest software.

Antibody	Vendor	Catalog no.	Antibody per 100 µl cell suspension
PE isotype control(IgG1)	BD	555749	10 µl
FITC isotype control(IgG$_{2b}$)	BD	556655	10 µl
PE CD13(IgG1)	BD	555394	10 µl
PE CD44(IgG$_{2b}$)	BD	556655	10 µl
PE CD49e(IgG1)	BD	555617	10 µl
PE CD90(IgG1)	BD	555596	5 µl
PE CD105(IgG1)	Fitzgerald	RDI CD105SNPE	3 µl

IV. Materials

- **Enzyme solutions**

    a. Solution A: Collagenase and Hyaluronidase

        i. Collagenase Type I, 300 units/ml
        ii. Hyaluronidase from ovine testes, 1 mg/ml
        iii. Phosphate buffered saline with 3 mM $CaCl_2$

b. Solution B: Trypsin-EDTA (0.1%)

- **Growth Medium #1**

Recipe to make 100 ml:		Final concentration
Low glucose DMEM	56 ml	56%
MCDB 201, pH 7.4	37 ml	37%
Insulin-transferrin-selinium, 100×	1 ml	1×
ALBU-Max 1, 100× (0.15 g/ml)	1 ml	0.15 mg/ml
Dexamethasone, 10^{-4} M	1 µl	1 nM
Ascorbic acid-2 phosphate 10^{-2} M	1 ml	100 µM
Antibiotic/antimycotic	1 ml	1×
Fetal bovine serum (FBS)	2 ml	2%
Epidermal growth factor (100 µg/ml stock)	10 µl	10 ng/ml
Platelet-derived growth factor (50 µg/ml stock)	20 µl	10 ng/ml
Glutamine 200 mM (100×)	1 ml	2 mM

Medium is filter sterilized using a 0.22 µm filter and stored refrigerated. The medium is protected from light and prepared fresh weekly.

- **Growth Medium #2**

To make 100 ml:		
Low glucose DMEM	90 ml	
MSC-qualified FBS	10 ml	
	Vendors and Catalog Numbers	
Low glucose DMEM, pH 7.4	Invitrogen	11885
MCDB 201, pH 7.4	Sigma	M-6770
Insulin-transferrin selenium 100× (ITS-X)	Invitrogen	51500056
AlbuMax I, 100× = 0.15 g/ml	Invitrogen	111200021
Dexamethasone, 10^{-4} M in DMSO	Sigma	D-4902
Ascorbic acid 2-phosphate	Sigma	A-8960
Pen/strep 100× (Pen G 10,000 U/ml; strep 10000 µg/ml)	Invitrogen	15140
FBS, for growth medium #1	Atlanta Biologicals	S11150
FBS, for growth medium #2	Invitrogen	12662-029
EGF (stock 100 µg/ml)	R&D Systems	236-EG-200
PDGF-BB (stock 50 ug/ml)	R&D Systems	520-BB-050
Collagenase Type I	Invitrogen	17100-017
Hyaluronidase	Fisher	ICN15127202
Dulbecco's phophate buffered saline	Invitrogen	14190250
Trypsin-EDTA	Invitrogen	25200-106

- Tissue culture plastics:

a. 15 ml conical centrifuge tubes	ISC Bioexpress	C3317-1
b. 50 ml conical centrifuge tubes	ISC Bioexpress	C3317-6
c. 6-well plates	Fisher	12-565-73
d. Tissue culture flasks-25 cm^2	Fisher	10-126-28
e. Tissue culture flasks-75 cm^2	Fisher	10-126-37
f. 50 ml 0.22 μm filters	Fisher	SCGP 005-25

Notes

1. On average, more than 500,000 cells are isolated from umbilical cords (average length of 53 cm, see Table I). Cells have been isolated up to 48 h after birth. There are no significant differences in the numbers of cells isolated or cord length between sexes.

2. Cord material should be considered potentially infectious and be handled with appropriate precautions.

Table I
Table Showing the Yield of Matrix Cells from Umbilical Cords of Varying Lengths

Sample ID	Gender	Length of umbilical cord	No. of cells isolated (approximately)
HUC 32	Female	50 cm	600,000
HUC 35	Male	55 cm	200,000
HUC 40	Female	55 cm	600,000
HUC 41	Female	40 cm	450,000
HUC 42	Male	55 cm	500,000
HUC 43	Male	45 cm	500,000
HUC 44	Male	50 cm	550,000
HUC 45	Male	55 cm	600,000
HUC 46	Male	55 cm	600,000
HUC 47	Male	50 cm	550,000
HUC 48	Male	55 cm	600,000
HUC 49	Male	60 cm	650,000
HUC 50	Male	38 cm	450,000
HUC 51	Male	42 cm	400,000
HUC 52	Female	60 cm	600,000
HUC 53	Female	57 cm	500,000
HUC 54	Male	55 cm	600,000
HUC 55	Female	60 cm	600,000
HUC 56	Female	55 cm	500,000
HUC 57	Female	42 cm	450,000
HUC 58	Female	50 cm	500,000
HUC 59	Female	65 cm	600,000
HUC 60	Female	70 cm	700,000
HUC 61	Male	70 cm	750,000
HUC 62	Male	65 cm	600,000
HUC 63	Female	50 cm	550,000
HUC 64	Female	35 cm	400,000

Cords HUC 61 and HUC 62 were held in the refrigerator for more than 24 h and were processed within 48 h of collection. The remaining cords were processed within 24 h of collection.

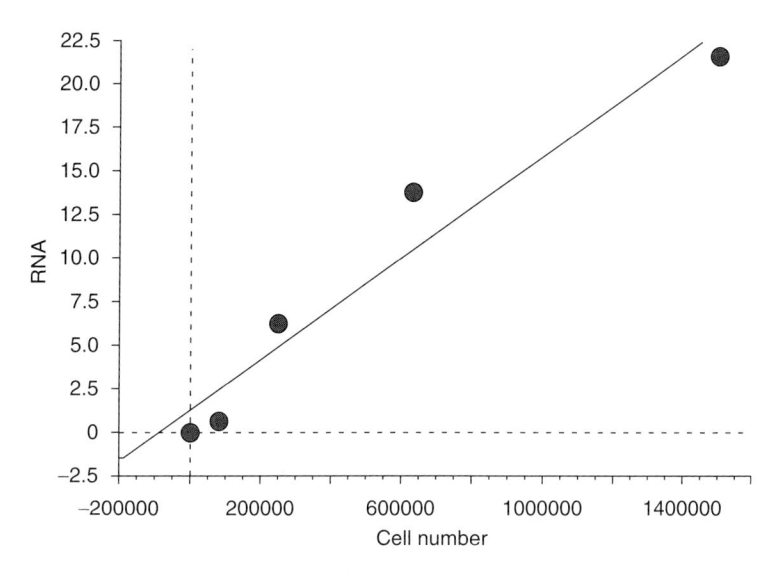

Fig. 9 Regression plot showing RNA yield from human umbilical cord matrix cells. RNA yield is in micrograms. $Y = 1.3 + 1.454E^{-5}X$; $R^2 = 0.941$.

3. For small animals, such as rat and mice, it is impractical to remove vessels. In this case, the vessels are not dissected from cord; rather the cord is torn to pieces using sharpened Dumont #5 forceps.

4. For small rodents, swine, bovine, and human, small (ca. 1 mm³) explants created by dicing the cord material with sterile blades may be used as starting material without enzymatic digestion. Start-up is slower than with enzymatically digested material.

5. Standard protocols for lysing red blood cells can be applied and may have a positive effect. This has not been quantified or investigated fully.

6. UCMSCs like high-cell density and start faster when grown at high density; 15,000 cells per square centimeter is recommended.

7. At passage, early passages are at splits of 1:2. Passages 4 and on may be at splits of 1:3 or 1:4. The smallest, round phase bright cells are important for maintaining the culture and are most sensitive to effects of enzyme and other mechanical damage.

8. The cells are sensitive to enzymatic treatment and to rough manipulation. Avoid exposure for more than 5 min to enzymes. If the cells do not lift rapidly, the UCMSCs may have differentiated or senesced. Preliminary work indicates that UCMSCs may prefer 2–5% oxygen tension for culture (K. E. Mitchell, University of Kansas, personal communication). This has not been verified in our lab but it agrees with our observation of the cells being sensitive to stressful treatment.

9. In early work, hyaluronic acid-coated plates were used. Subsequently, this was found not to be required.

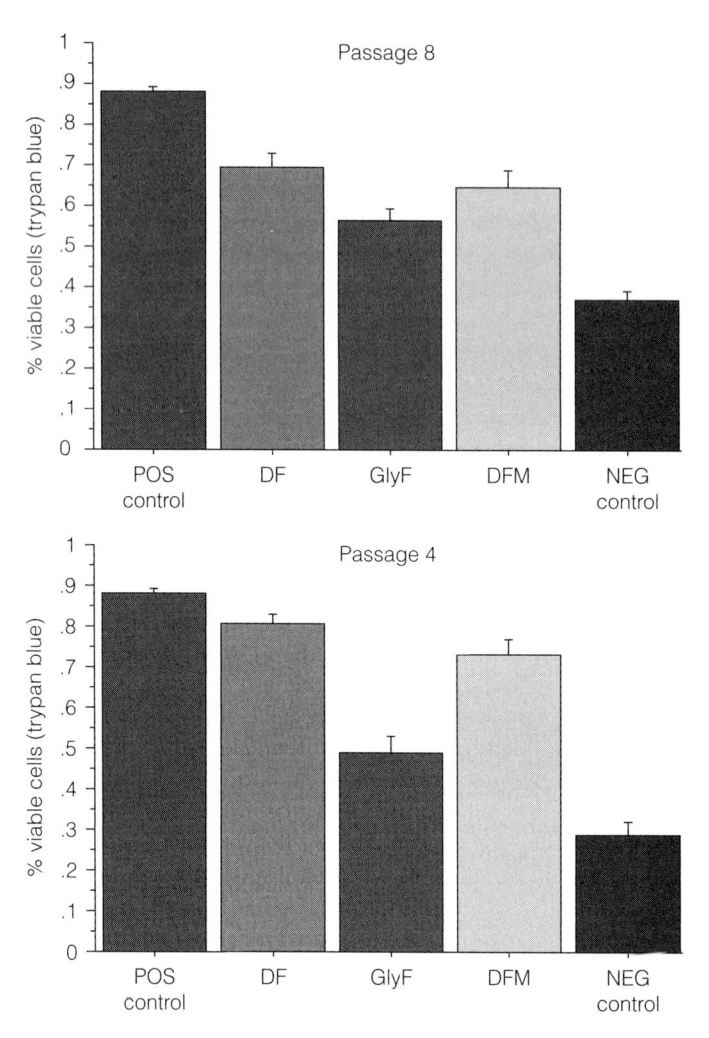

Fig. 10 Viability of umbilical cord matrix (UCM) cells at passage 4 (bottom) and passage 8 (top). The positive control was cell viability before freezing. The negative control is growth medium only. Abbreviations: DF: 90% FBS and 10% DMSO; GlyF: 90% FBS and 10% glycerol; DFM: 50% growth medium, 40% FBS, and 10% DMSO.

10. As mentioned in results, four different freezing medium have been tested. While good viability is obtained overall, significant variability has been noted at start-up after thawing. This seems to be due to factors involving the freezing (and not the thawing) procedure. The best way to ensure consistency and optimal viability is to prepare small batches of cells for freezing and avoid trying to process many plates at a time.

11. When plating cells after a thaw, if cells do not attach in 24 h and start growing, discard and try a second vial.

Surface markers (before freezing)

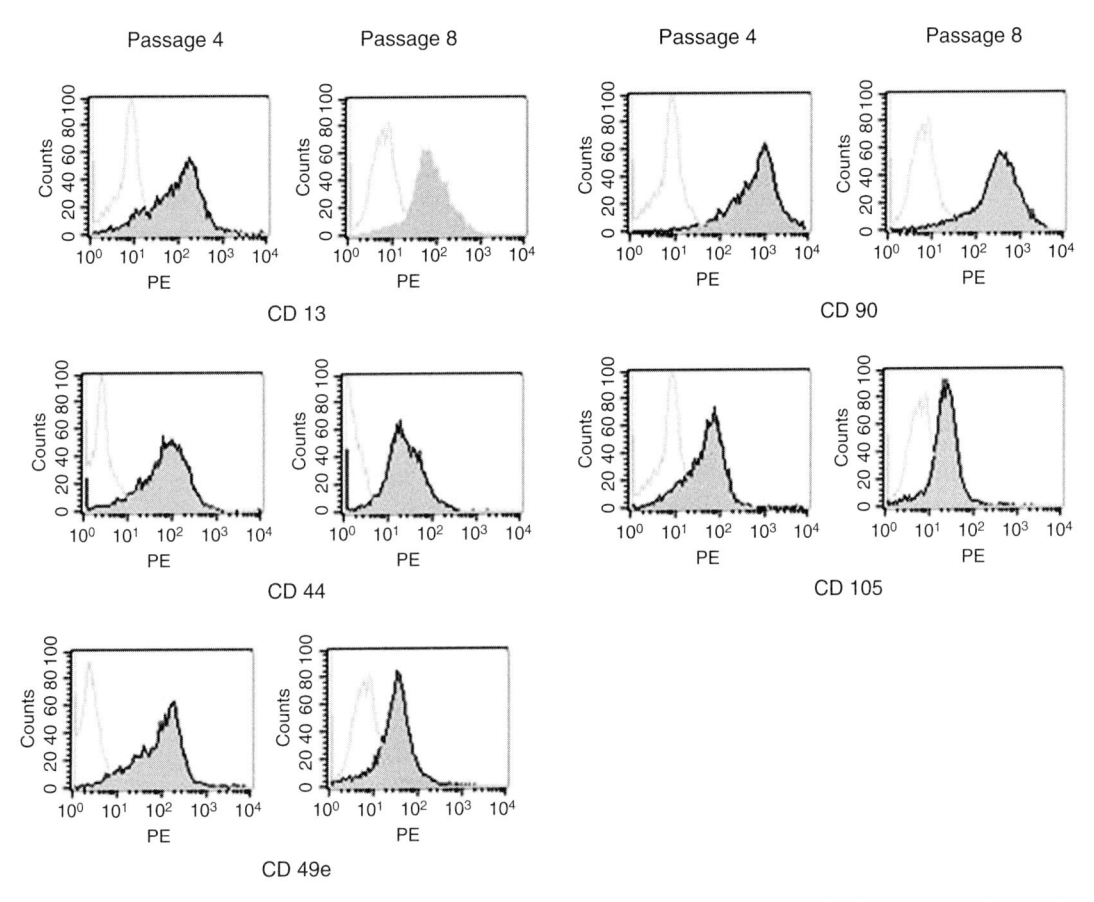

Fig. 11 Flow cytometry analysis–Histogram plots of expression of cell surface markers before freezing (frozen in 90% DMSO and 10% FBS) at passage 4 and passage 8.

12. The cells at early passage have typical fibroblast, that is, MSC-like morphology, whereas cells at late passages are elongated and start losing the characteristic morphology of MSCs. See Figs. 13 and 14.

V. UCMSCs and Culture Characteristics

1. Experiments were carried out to optimize the freezing protocol for human UCMSCs. The cells from nine umbilical cords were frozen at passage 4 and passage 8 in three different freezing media; growth medium served as a negative control. The results from viability at thaw are shown in Fig. 10. From this data, it

Surface markers (after freezing)

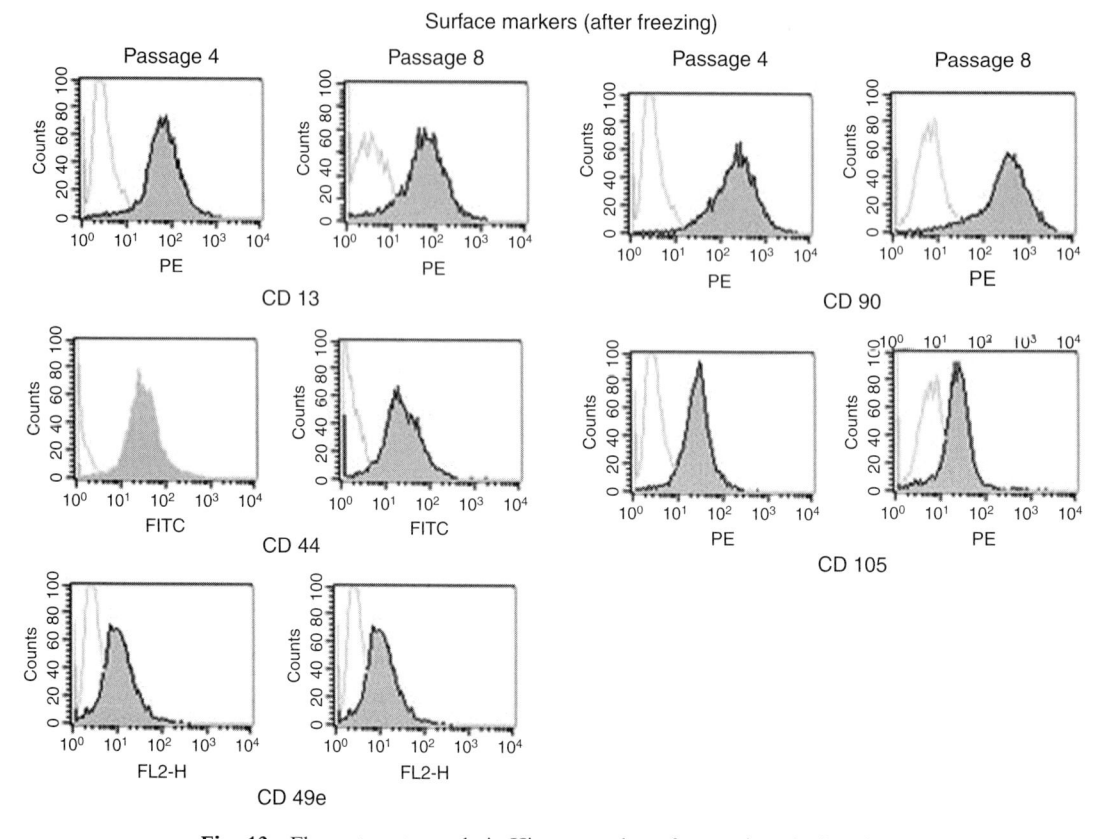

Fig. 12 Flow cytometry analysis–Histogram plots of expression of cell surface markers after freezing (frozen in 90% DMSO and 10% FBS) at passage 4 and passage 8.

appears that 90% FBS and 10% DMSO yields about 80% viability; this is more than for the other media tested.

2. Flow cytometry experiments were done to evaluate the effect of passage and freezing medium (90% FBS and 10% DMSO) on the expression of five cell surface markers. Histogram plots for before and after freezing, respectively, are shown in Figs. 11 and 12. From these results, it appears that there is no significant change in the expression of surface markers at early and late passages by a freeze–thaw cycle.

3. Total cellular RNA was obtained from cells of two UCMSCs isolates (isolates 46 and 47) at different passages using the Qiagen RNeasy kit. The yield is plotted as a regression line (data shown in Fig. 9).

VI. Discussion

The cells from Wharton's Jelly can be isolated using another method called the "Explant Method." For this method, the tissue is chopped into small pieces, about 1 mm^2, and plated with medium. The explants attach to the substrate and the cells

6. Umbilical Cord Matrix Stem Cells (UCMSCs)

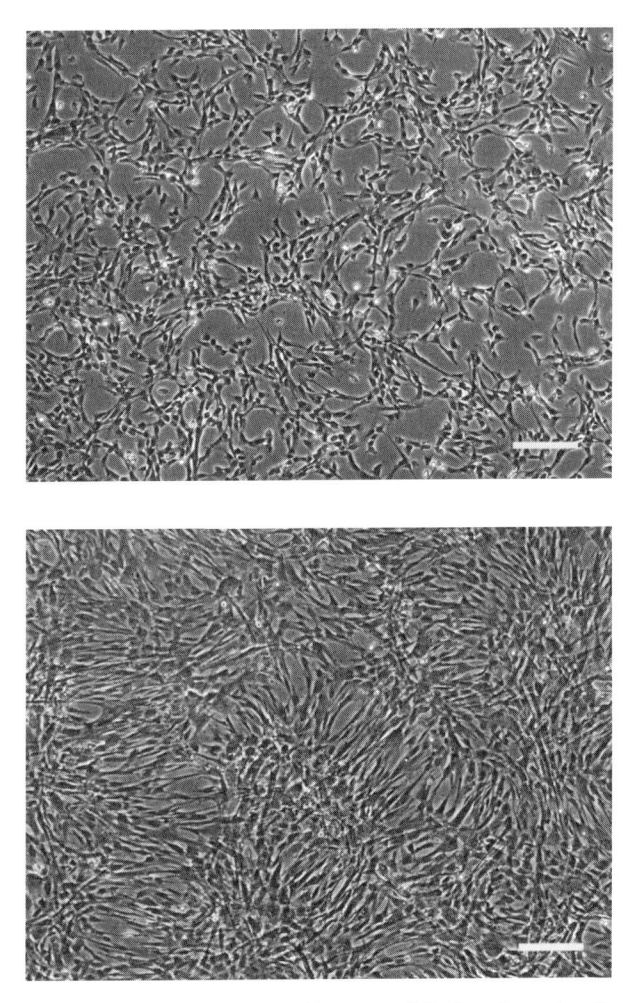

Figs. 13 and 14 Phase contrast micrograph of human umbilical cord matrix cells at passage 1 and passage 11. Calibration bar-100 μm. 100× magnification.

outgrow from the tissue. These cells are harvested and passaged. The shortcoming with this method is the inability to determine the number of cells that have been isolated from the cord at passage, because the cells continue to outgrow from the explants even after the cells have been harvested. With the enzymatic digestion method, the number of cells isolated from the umbilical cord is fixed and known (see Table I). The determination of population doubling of the cells is possible when the number of cells to start with is known.

Several hurdles remain before UCMSC culture is optimal to grow cells for use in human clinical trials. Current culture practices use FBS as a constituent of the medium for UCMSCs, which introduces some limitations. The cells might become infected because of transmission of pathogens (prions and virus) from the serum and/or adherence of animal protein molecules to the cell surface, causing immune reactions. There is a possibility for the cells to get contaminated during culture. If the patients' safety is compromised, the cells cannot be used to treat diseases in spite of their potential. Also, batch to batch variability may be seen even if the serum is procured from the same vendor. This variability may affect the growth of the cells and the time for which they remain in the undifferentiated state. Therefore, to use the cells in clinical trials, they have to be grown in an appropriate serum-free medium (Mannello and Tonti, 2007). The isolation of cells is done manually and is time consuming (3–4 h per cord). This is another hurdle in generating clinical grade cells. To be able to use the cells in clinics, the isolation procedure ideally has to involve minimal tissue handling and be less time consuming.

Acknowledgments

We thank our colleagues in the Midwest Institute for Comparative Stem Cell Biology at Kansas State University for their editorial assistance and James Hong for taking pictures and for assistance in the preparation of the chapter, specifically Dr. Duane Davis and Deryl Troyer. We thank Chad Mauer, Nathan Bammes, Katrina Fox, Satish Medicetty, Hong He, Barbara Lutjemeier, Mark Banker, Julie Hix, Marla Pyle, and Cameron Anderson for their contribution to this work. We are grateful for the help provided by Dr. Suzanne M. Bennett, Obstetrician and Gynecologist of the Women's Health group. The work was funded by the Midwest Institute of Comparative Stem Cell Biology and by NIH NS034160 to MLW. Some laboratory members were funded by the Terry C. Johnson Center for Basic Cancer Research, K-BRIN, Howard Hughes Medical Institute, the College of Veterinary Medicine Dean's office, and the Provost's Office of Kansas State University. MLW is working as a consultant for the Regenerative Medicine Institute (RMI, Las Vegas, NV). Kansas State University Research Foundation (KSURF) has submitted an application to the US Patent Office for intellectual protection for the umbilical cord matrix cells. Contact KSURF for information regarding licensing.

References

Baksh, D., Song, L., and Tuan, R. S. (2004). Adult mesenchymal stem cells: Characterization, differentiation, and application in cell and gene therapy. *J. Cell. Mol. Med.* **8**, 301–316.

Caplan, A. I., and Dennis, J. E. (2006). Mesenchymal stem cells as trophic mediators. *J. Cell. Biochem.* **98**, 1076–1084.

Conconi, M. T., Burra, P., Di, L. R., Calore, C., Turetta, M., Bellini, S., Bo, P., Nussdorfer, G. G., and Parnigotto, P. P. (2006). CD105(+) cells from Wharton's jelly show *in vitro* and *in vivo* myogenic differentiative potential. *Int. J. Mol. Med.* **18**, 1089–1096.

Dexter, T. M., Allen, T. D., and Lajtha, L. G. (1977). Conditions controlling the proliferation of haemopoietic stem cells *in vitro*. *J. Cell. Physiol.* **91**, 335–344.

Dominici, M., Le, B. K., Mueller, I., Slaper-Cortenbach, I., Marini, F., Krause, D., Deans, R., Keating, A., Prockop, D., and Horwitz, E. (2006). Minimal criteria for defining multipotent mesenchymal stromal cells. The International Society for Cellular Therapy position statement. *Cytotherapy* **8**, 315–317.

Friedenstein, A. J., Chailakhyan, R. K., Latsinik, N. V., Panasyuk, A. F., and Keiliss-Borok, I. V. (1974). Stromal cells responsible for transferring the microenvironment of the hemopoietic tissues. Cloning *in vitro* and retransplantation *in vivo*. *Transplantation* **17**, 331–340.

Guillot, P. V., Gotherstrom, C., Chan, J., Kurata, H., and Fisk, N. M. (2007). Human first-trimester fetal MSC express pluripotency markers and grow faster and have longer telomeres than adult MSC. *Stem Cells* **25**, 646–654.

Jin, X. B., Sun, Y. S., Zhang, K., Wang, J., Ju, X. D., and Lou, S. Q. (2007a). Neocartilage formation from predifferentiated human adipose-derived stem cells *in vivo*. *Acta Pharmacol. Sin.* **28**, 663–671.

Jin, X. B., Sun, Y. S., Zhang, K., Wang, J., Shi, T. P., Ju, X. D., and Lou, S. Q. (2007b). Ectopic neocartilage formation from predifferentiated human adipose derived stem cells induced by adenoviral-mediated transfer of hTGF beta2. *Biomaterials* **28**, 2994–3003.

Jomura, S., Uy, M., Mitchell, K., Dallasen, R., Bode, C. J., and Xu, Y. (2007). Potential treatment of cerebral global ischemia with Oct-4 + umbilical cord matrix cells. *Stem Cells* **25**, 98–106.

Le, B. K. (2006). Mesenchymal stromal cells: Tissue repair and immune modulation. *Cytotherapy* **8**, 559–561.

Lu, L. L., Liu, Y. J., Yang, S. G., Zhao, Q. J., Wang, X., Gong, W., Han, Z. B., Xu, Z. S., Lu, Y. X., Liu, D., Chen, Z. Z., and Han, Z. C. (2006). Isolation and characterization of human umbilical cord mesenchymal stem cells with hematopoiesis-supportive function and other potentials. *Haematologica* **91**, 1017–1026.

Lund, R. D., Wang, S., Lu, B., Girman, S., Holmes, T., Sauve, Y., Messina, D. J., Harris, I. R., Kihm, A. J., Harmon, A. M., Chin, F. Y., Gosiewska, A., and Mistry, S. K. (2007). Cells isolated from umbilical cord tissue rescue photoreceptors and visual functions in a rodent model of retinal disease. *Stem Cells* **25**, 602–611.

Mannello, F., and Tonti, G. A. (2007). No breakthroughs for human mesenchymal and embryonic stem cell culture: Conditioned medium, feeder layer or feeder-free; medium with foetal calf serum, human serum or enriched plasma; serum-free, serum replacement nonconditioned medium or ad-hoc formula? All that glitters is not gold!. *Stem Cells* **25**(7), 1603–1609.

McIntosh, K., Zvonic, S., Garrett, S., Mitchell, J. B., Floyd, Z. E., Hammill, L., Kloster, A., Di, H. Y., Ting, J. P., Storms, R. W., Goh, B., Kilroy, G., Wu, X., and Gimble, J. M. (2006). The immunogenicity of human adipose-derived cells: Temporal changes *in vitro*. *Stem Cells* **24**, 1246–1253.

Medicetty, S., Mitchell, K. E., Weiss, M. L., Mitchell, B. M., Martin, P., Davis, D., Morales, L., Helwig, B., Beerenstrauch, M., Abou-Easa, K., Hildreth, T., and Troyer, D (2003). Matrix cells from Wharton's jelly form neurons and glia. *Stem Cells* **21**, 50–60.

Nakagami, H., Morishita, R., Maeda, K., Kikuchi, Y., Ogihara, T., and Kaneda, Y. (2006). Adipose tissue-derived stromal cells as a novel option for regenerative cell therapy. *J. Atheroscler. Thromb.* **13**, 77–81.

Saito, S. Y. K. (2006). Bovine UCMC support equine embryonic stem cells. *In* "Embryonic Stem Cell Protocols." Vol. I, pp. 59–79. Human Press, NJ, USA.

Shen, F. H., Zeng, Q., Lv, Q., Choi, L., Balian, G., Li, X., and Laurencin, C. T. (2006). Osteogenic differentiation of adipose-derived stromal cells treated with GDF-5 cultured on a novel three-dimensional sintered microsphere matrix. *Spine J.* **6**, 615–623.

Studeny, M., Marini, F. C., Champlin, R. E., Zompetta, C., Fidler, I. J., and Andreeff, M. (2002). Bone marrow-derived mesenchymal stem cells as vehicles for interferon-beta delivery into tumors. *Cancer Res.* **62**, 3603–3608.

Tuan, R. S., Boland, G., and Tuli, R. (2003). Adult mesenchymal stem cells and cell-based tissue engineering. *Arthritis Res. Ther.* **5**, 32–45.

Wang, H. S., Hung, S. C., Peng, S. T., Huang, C. C., Wei, H. M., Guo, Y. J., Fu, Y. S., Lai, M. C., and Chen, C. C. (2004). Mesenchymal stem cells in the Wharton's jelly of the human umbilical cord. *Stem Cells* **22**, 1330–1337.

Weiss, M. L., Medicetty, S., Bledsoe, A. R., Rachakatla, R. S., Choi, M., Merchav, S., Luo, Y., Rao, M. S., Velagaleti, G., and Troyer, D. (2006). Human umbilical cord matrix stem cells: Preliminary characterization and effect of transplantation in a rodent model of Parkinson's disease. *Stem Cells* **24**, 781–792.

Wu, K. H., Zhou, B., Yu, C. T., Cui, B., Lu, S. H., Han, Z. C., and Liu, Y. L. (2007). Therapeutic potential of human umbilical cord derived stem cells in a rat myocardial infarction model. *Ann. Thorac. Surg.* **83**, 1491–1498.

Zhang, D. Z., Gai, L. Y., Liu, H. W., Jin, Q. H., Huang, J. H., and Zhu, X. Y. (2007). Transplantation of autologous adipose-derived stem cells ameliorates cardiac function in rabbits with myocardial infarction. *Chin Med. J. (Engl.)* **120**, 300–307.

CHAPTER 7

Isolation, Characterization, and Differentiation of Human Umbilical Cord Perivascular Cells (HUCPVCs)

Jane Ennis,[*] R. Sarugaser,[‡] A. Gomez,[‡] Dolores Baksh,[†] and John E. Davies[§]

[*]Tissue Regeneration Therapeutics,
Suite 512 Toronto, ON M5G IN8

[†]Organogenesis Inc.
Canton Massachusetts 02021

[‡]Institute of Biomaterials and Biomedical Engineering,
University of Toronto, Toronto,
ON M5S 3G9, Canada

[§]Faculty of Dentistry, Institute of Biomaterials and Biomedical Engineering,
University of Toronto, Ontario,
Canada M5S 3G9

I. Introduction
II. Methods
    A. Isolation and Culture of Human Umbilical Cord Perivascular Cells
    B. Cryopreservation of Dissected Umbilical Vessels
    C. Differentiation into Mesenchymal Phenotypes
    D. Culture of Single-Cell Seeded Clones
    E. Phenotypic Characterization of HUCPVCs
III. Concluding Remarks
    References

I. Introduction

There are currently over 36 clinical trials being performed, throughout the world, using both autologous and allogeneic mesenchymal stromal/stem cells (MSCs) for the treatment of a wide range of pathologies (http://clinicaltrials.gov/). These regenerative therapies can be broadly subdivided into the treatment of immune and inflammatory diseases; coadministration with other—particularly haematopoietic—stem cells; and engineering of the connective tissues. These cell therapeutic strategies are predominantly founded on the two most important characteristics of MSCs, their multilineage differentiation potential and their immunoprivileged and immunosuppressive phenotype, and herald an increasingly broad spectrum of potential clinical uses for mesenchymal stem cells.

Although MSCs have been identified in, and sourced from, various tissues, including bone marrow (Friedenstein et al., 1976; Kuznetsov et al., 1997), fat (Zuk et al., 2002), muscle (Lee et al., 2000), and placenta (Takahashi et al., 2004), the most common source employed clinically is bone marrow. However, while it is the most comprehensively researched source of MSCs, bone marrow can be obtained only by an elective and invasive surgical procedure undertaken on young individuals (commonly 18–35-year-old donors) and thus is of somewhat limited supply. In addition, only 1 in 100,000 nucleated cells in adult human bone marrow is an MSC, and it is known that this number decreases with advancing age (Kern et al., 2006; Stenderup et al., 2003). However, MSCs are now considered to be present in almost all organs of the body, are predominantly found in association with blood vessels, and are considered by many to be synonymous with pericytes. Indeed, bone marrow-derived MSCs are considered perivascular in origin (Shi and Gronthos, 2003), and the perivascular niche has been shown to be the source of mesenchymal progenitors in many organs (da Silva et al., 2006). Pericytes share the multidifferentiation potential of MSCs and have been shown to be capable of osteogenic, chondrogenic, and adipogenic differentiation both *in vitro* and in diffusion chambers *in vivo* (Doherty et al., 1998; Farrington-Rock et al., 2004). Such MSC-like cells are not the only cell type associated with capillaries since another vascular-associated cell type, called calcifying vascular cells (CVCs), is derived from the tissue of the vascular wall, and has also been shown to be capable of differentiating into bone, cartilage, smooth muscle, marrow stroma, and fat (Abedin et al., 2004; Tintut et al., 2003). Thus, there is a significant volume of evidence that tissue-resident MSCs are closely associated with the vasculature.

Given the increasing promise of the clinical utility of MSCs, together with the need to establish alternative, higher yield, sources of these cells to those currently obtained from marrow, we have explored the possibility of harvesting MSCs from the perivascular tissue of the human umbilical cord, a tissue which is discarded at birth. Since the umbilical cord is a rapidly growing organ comprising three blood vessels supported in a specialized connective tissue, we hypothesized that the cells of the latter, the so-called Wharton's Jelly, would be differentiated from an MSC source which, like other MSCs, should be present in the perivascular region of the organ (Sarugaser et al., 2005). Indeed, the cells of the perivascular region of the umbilical cord tissue had already been shown to have both a different cell phenotype and increased proliferative capacity, compared to the remaining tissue of the cord (Nanaev et al., 1997) and more recently others have recognized that the perivascular cells are distinctly different from those of the remainder of the Wharton's Jelly of the cord (Friedman et al., 2007), and possess the equivalent phenotype to bone marrow-derived MSCs (Baksh et al., 2007). It is not surprising, therefore, that there has recently been a surge of interest in harvesting mesenchymal cells from the human umbilical cord (Can and Karahuseyinoglu, 2007). Furthermore, since we have shown that the frequency of CFU-Fs in P0 populations of perivascular cells is 1:300 (Sarugaser et al., 2005), and since there are currently over 130 million births throughout the world annually, the perivascular tissue of the human umbilical cord represents a rich and unlimited alternative supply of MSCs for cell-based therapies. For these reasons we provide herein details of an improved extraction method for these perivascular cell populations, a method of cryogenically storing the umbilical vessels without the need for prior enzymatic removal of the perivascular cells, methods to demonstrate multilineage differentiation capacity and immunoregulatory phenotype, together with other methods of characterization of this unique MSC population.

II. Methods

A. Isolation and Culture of Human Umbilical Cord Perivascular Cells

The anatomy of the cord is simple. There are three large vessels running through the length of the cord (two arteries and one vein) with a surrounding gelatinous connective tissue called Wharton's jelly (WJ), which is enveloped in the amniotic epithelium, as shown in Fig. 1. While the structure is simple it is important to

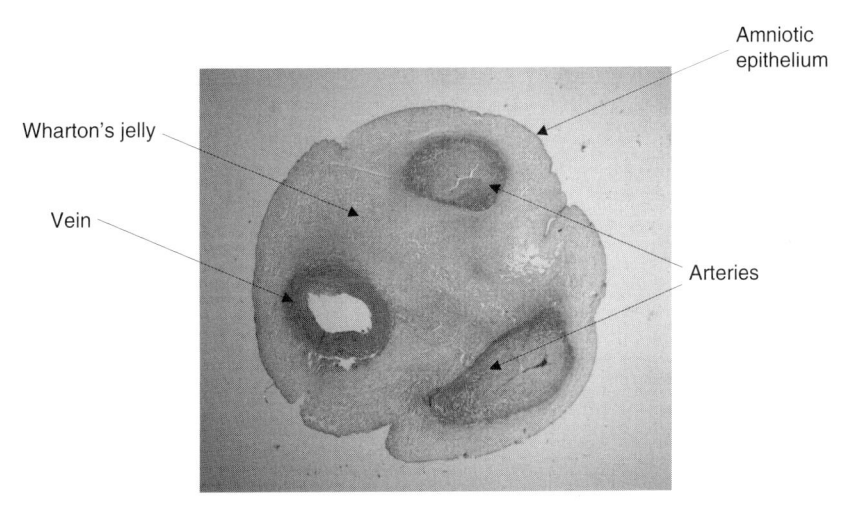

Fig. 1 Light microscope image of a human umbilical cord in cross-section, and stained with hematoxylin and eosin. Magnification 1.6×.

realize that this connective tissue is specialized for the specific physiological role of preventing kinking of the vessels during normal movements of the developing embryo/fetus. For this reason the majority of the cells of Wharton's Jelly are myofibroblasts, as demonstrated by Takechi *et al.* (1993) who also conjectured that they may differentiate from more primitive mesenchymal precursors in response to the pulsatile environment of the umbilical cord. Thus, "stromal cells" have been extracted from the WJ, or cord matrix, using a variety of techniques including mincing and/or enzymatically digesting the WJ tissue (Friedman *et al.*, 2007; Lund *et al.*, 2007; Wang *et al.*, 2004), and explant cultures (Mitchell *et al.*, 2003), but the following method was devised to specifically extract the mesenchymal precursors of these matrix cells from their perivascular niche.

Initially, having collected a 20–30 cm length of cord, in an 8 oz collection container filled with α-MEM culture medium and antibiotics, from the delivery suite at the hospital directly after birth, and transporting it at ambient temperature to the laboratory facility, the cord was placed on a sterile dissecting tray and cut into 4–5 cm segments. Part of the rationale for using such short segments (4–5 cm) of cord was that we were concerned that the vessels would break when being pulled from the bulk WJ. Each short fragment was then processed in the following manner: the epithelium was removed with forceps, the umbilical vessels were then teased apart including their associated perivascular tissue, and the vessels were then looped and tied using silk sutures to prevent hematopoietic or endothelial contamination. These vessel loops were then placed in 20 ml of Type I collagenase and digested overnight at 37 °C, following which the cells were removed from the supernatant, collected, washed, counted, and plated on tissue culture treated plastic at a density of 4000 cells/cm^2. Thus, rather than discarding the vessels and their proximal connective tissue, we employed the vessels as carriers of the perivascular tissue from which we could conveniently enzymatically digest the cells of interest. This extraction procedure results in a high yield of proliferating cells which are strongly adherent to tissue culture treated plastic (Fig. 2), and demonstrates that

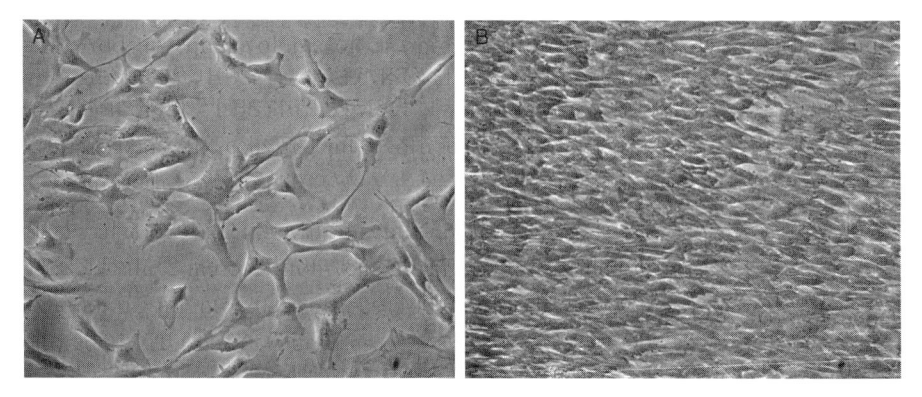

Fig. 2 Representative phase micrograph images of HUCPVCs (Human Umbilical Cord Perivascular Cells) in culture. (A) HUCPVCs as they would appear when plated for 3–5 days, (B) 12–14 days.

the procedure can be undertaken with very small samples of cord tissue (we have sometimes received only 3 cm from the delivery suite), yet still result in a rapidly expanding cell population.

After processing many cords in the above fashion we realized that the vessels are sufficiently strong to work with longer segments of cord, and we now process a complete length of cord without first creating shorter segments. To facilitate this we developed a custom-designed dissection table to hold the cord (Fig. 3). Thus, the entire cord is placed on the table and clamped at one end. Using a scalpel, a circumferential incision is made in the amniotic epithelium close to the clamped end of the cord. The epithelium is then separated from the bulk of the cord using blunt dissection around the entire incision, and grasped with paired forceps and pulled away from the clamped end. In this manner, the epithelium is generally removed in one long tubular piece, rather like a sock. If it does not all come off in one piece, the process can be repeated to remove the residual epithelium. The cord is then severed from the clamped end, and the vessels separated longitudinally. When separated, the vessels are looped and tied using suture silk, and placed in the collagenase digest as above. This procedure reduces the dissection time of a standard 30 cm piece of cord to ~15 min.

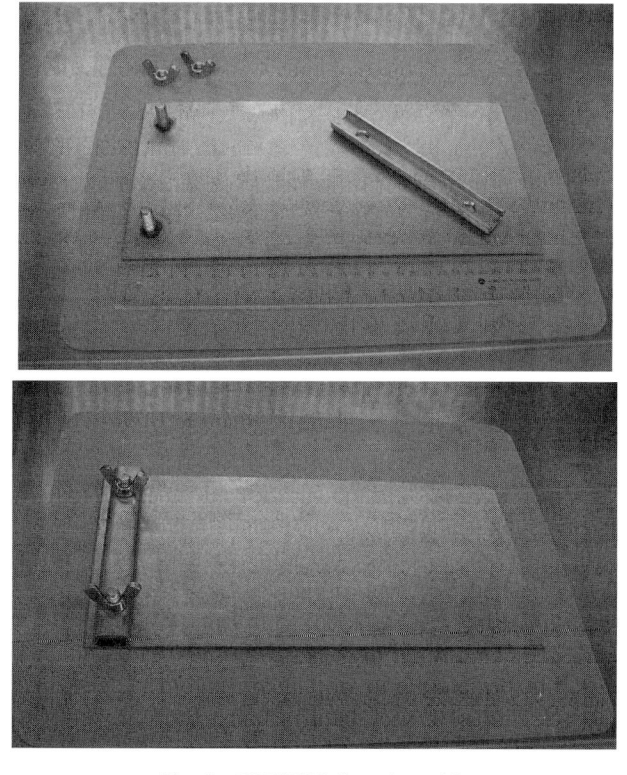

Fig. 3 HUCPVC dissection table.

To further reduce processing time, we have also reduced both the enzymatic digestion time and the time required to remove the vessels from the digestate. Previously, we reported an overnight digestion in collagenase contained in tubes held in a laboratory rotisserie to maintain some fluid flow. This resulted in a turbid digestate that still contained the vessels, and repeated washing was required to remove the vessels and reduce the viscosity of the solution enough to efficiently isolate the cells. This time can be considerably shortened by adding hyaluronidase to the collagenase and also by employing a laboratory rotator to create a more aggressive fluid flow. This has reduced the digest time to $\sim$4 h with a commensurate decrease in the technician time required for the entire harvesting procedure.

B. Cryopreservation of Dissected Umbilical Vessels

In the case where the HUCPVCs themselves are not needed immediately, it is not necessary to perform the entire dissection/digestion procedure. Once the vessels have been successfully separated from each other and are ready to be digested, the vessels are then immersed in 5–15 ml of cryoprotectant solution (such as CryoStor™), and incubated at 4 °C for 15–60 min. Following their incubation, the vessels are transferred to freezing vials or bags, and placed in a controlled-rate freezer to −80 °C. Once they have achieved −80 °C, they are placed directly in a vapor-phase liquid nitrogen storage facility. The vessels can be stored indefinitely at liquid nitrogen temperatures until the cells are needed. The tissue is thawed at 42 °C for 5–15 min, and immediately placed in the digest as per the usual protocol. This procedure allows for a much faster processing of cords, and further reduction in technician time.

C. Differentiation into Mesenchymal Phenotypes

A key characteristic of MSCs is their ability to differentiate into the mesenchymal phenotypes including bone, cartilage, and fat. Standard DMEM assays for the differentiation and detection of cells into these phenotypes are as follows:

1. **Osteogenesis**: Cells were grown in osteogenic growth medium [10 nM dexamethasone (DEX), 5 mM β-glycerophosphate, 50 µg/ml ascorbic acid (AA), and 10 nM 1,25-dihydroxy vitamin D_3]. On day 21, cultures were stained for alkaline phosphatase (ALP) activity and mineralization was assessed by von Kossa staining after 5 weeks. Quantification of ALP activity was determined by adding 4 mg/ml p-nitrophenyl phosphate in an alkaline buffer solution into each well, followed by incubation for 15 min at 37 °C, with activity measured using a microplate reader.

2. **Chondrogenesis**: Cells were grown in high-density pellet cultures (2.5×10^5 cells) for 21 days in serum-free medium (50 µg/ml AA, 40 µg/ml L-proline, 100 µg/ml sodium pyruvate, and 0.1 µM DEX), with and without 10 ng/ml recombinant human TGF-β3. On day 21, pellets were prepared for Alcian Blue staining to detect sulfated glycosaminoglycan (sGAG) and collagen content (Picrosirius Red staining).

3. **Adipogenesis**: Cells were cultured as monolayer in the presence of adipogenic supplements [1 μM DEX, 1 μg/ml insulin, and 0.5 mM IBMX (3-isobutyl-1-methylxanthine)]. On day 21, cultures were stained with Oil Red O stain and dye content quantified by isopropanol elution and spectrophotometry.

Using these techniques, HUCPVCs were shown to have the ability to differentiate down the osteogenic, chondrogenic, and adipogenic lineages; however, they can generate an osteogenic phenotype at a faster rate than BM-MSCs and demonstrate increased osteogenesis when compared to BM-MSCs. For example, mineralization is detected as early as day 10 in HUCPVCs versus 14 days in BM-MSCs cultured under osteogenic condition. By day 21, HUCPVCs express higher ALP activity relative to BM-MSCs (Baksh *et al.*, 2007). Furthermore, significantly greater numbers of bone nodules, as assayed by von Kossa staining, are detected in osteogenic cultures initiated with HUCPVCs than with BM-MSCs.

D. Culture of Single-Cell Seeded Clones

A unique feature of HUCPVCs is the extremely high frequency of progenitors at harvest that is maintained in continuing culture. This high frequency has enabled study of these cells at the single-cell level using clonal analysis. HUCPVCs are isolated as above, and passed through a 70 μm filter to ensure single-cell separation. Equal proportions of male and female cells are mixed together in suspension in order to aid in the assessment of clonality (see below). The suspension is then diluted to 1 cell/250 μl in culture media. Using a multichannel micropipette, 50 μl of the suspension (0.2 cells/well effective dilution) is deposited into individual wells of 96-well tissue culture plates, which are then incubated overnight. The following day, a further 50 μl of media is added to the wells. The media is then replaced every 5 days for 15 days. The plates are then observed by light microscopy for the presence of cells in individual wells. Only wells with cells are continued in culture with media replacement every 2–3 days until they reach 80–90% confluence. The cloned cells are then subcultured and transferred into individual wells of 6-well plates and then into T75 flasks for further expansion. The clonal purity of the population can be determined by FISH analysis for the absolute presence/absence of the Y-chromosome. These cloned cells can then be plated as required for analysis of mesenchymal differentiation potential as previously described.

By isolating and recloning clonal HUCPVC populations, we can determine if a single cell is capable of self-renewing and differentiating down one or more of the five mesenchymal lineages (fibroblast, bone, cartilage, adipose, and muscle). If a clonal population as well as its subclonal population can differentiate into all five mesenchymal lineages, then the original cell can be considered to be a true mesenchymal stem cell. Because of the significantly lower progenitor frequencies in other MSC sources, this culturing process has been prohibitive to date. Using this strategy, we have determined that ~9% of surviving clones are capable of differentiating down all five mesenchymal lineages, this ability is maintained in ~11% of surviving subclones, thus demonstrating the existence of a true mesenchymal stem cell (Sarugaser *et al.*, 2007).

E. Phenotypic Characterization of HUCPVCs

1. Flow Cytometry

HUCPVCs can be easily characterized by flow cytometric measurements of cellular markers. The cells are extracted as above, and the culture expanded to the required numbers. For each marker, 3×10^5 cells are used to ensure enough cells for gating and accurate measurements. The cells are washed using two rinses of phosphate buffered saline (PBS), and resuspended in 250 µl PBS and 5 µl of the conjugated (FITC or PE) antibody is added. The cells are incubated with the antibody for 45 min at room temperature, and then washed with PBS and run on the flow cytometer. The experimental samples were compared to an unstained control in order to correct for autofluorescence. If an unconjugated antibody was used, the experimental samples were also compared to cells incubated with only the 2° antibody to correct for nonspecific binding. The HUCPVC population is well-defined based on its size and granularity, and thus is easily gated for measurement. For intracellular markers, the cells must first be treated with methanol for 2 min to permeabilize the cells. This methodology has been used to illustrate that HUCPVCs are CD45−, CD34−, CD44+, CD49e+, CD73+, CD90+, and CD105+, indicating an MSC phenotype. In addition to the standard phenotypic markers, flow cytometric analysis revealed that ∼50% of HUCPVCs express CD146, which corresponds to ∼2.5-fold increase in CD146 expression relative to BM-MSCs. CD146 is characteristically expressed on circulating endothelial cells, which also express CD31, a classical endothelial marker (Duda *et al.*, 2006). The expression of CD146 is usually present in diseases such as inflammatory, immune, infectious, neoplastic, and cardiovascular disease (Blann *et al.*, 2005) and has thus been considered an ideal marker for a source of cells for therapeutic neovascularization and vascular repair (Wu *et al.*, 2005). Taken together, these results infer a potential novel role of HUCPVCs to circulate in biological fluid, similar to CD146+ endothelial cells and participate in the repair of diseased and/or damaged tissue. Importantly, the multilineage capacity of HUCPVCs is maintained in the CD146+ sorted HUCPVC population (Baksh *et al.*, 2007).

2. Wnt Pathway

Recently, Wnts have been implicated in regulating the differentiation potential of adult human BM-MSCs (Boland *et al.*, 2004a; De Boer *et al.*, 2004a,b; Etheridge *et al.*, 2004) and thus represent an important pathway regulating the maintenance of MSCs. To assess the Wnt signaling pathway's molecular profile, RNA is harvested and analyzed, following the manufacturer's recommended protocol on an Oligo GEArray® Human Wnt Signaling Pathway Microarray. In addition to expression, transduction of the Wnt signals can also be assessed. An optimized version of the TOP-FLASH luciferase reported plasmid (pLG3-OT) is used to assay canonical Wnt-responsive transcriptional activity. Cells are cotransfected by nucleofection with expression constructs of Wnt3a and Wnt5a and empty vector (control) and Renilla luciferase plasmid (1 µg) under the control of the CMV promoter and TOP-FLASH luciferase plasmid (2 µg). Luciferase activity is assayed using the Dual Luciferase Assay System according to the manufacturer's protocol (Promega Corp.), and reporter luciferase activity normalized to that of Renilla luciferase.

Using these methods, HUCPVCs not only express Wnt signaling genes but also possess the molecular machinery necessary to transduce signals through the Wnt signaling pathway. Taken together, these findings reveal that HUCPVCs may be regulated by the Wnt signaling pathway, similar to BM-MSCs (Boland *et al.*, 2004b) and thus, the tools, reagents, etc. used to study the Wnt signaling pathway already established for a variety of cell systems, including BM-MSCs, can be applied to HUCPVCs.

3. Telomerase Activity (TRAP Assay)

Of significant importance to stem cell research and clinical applications is their ability to self-renew. If cells are capable of infinite self-renewal, they may persist *in vivo*, and cause cancerous outcomes. Thus, it is important to assess cells for the presence of telomerase, as telomerase acts in the preservation of the telomeres, preventing their shortening as cells divide, thereby allowing the cell to theoretically proliferate indefinitely. Telomerase is assessed using the TRAPeze® XL Telomerase Detection Kit, and following the specific kit instructions. Using this methodology we determined that HUCPVCs had undetectable levels of telomerase activity; therefore, telomeres would shorten as HUCPVCs divide, thereby giving them a limited lifespan. These results were confirmed using a cancer gene chip, which included telomerase.

4. Cancer Gene Array

It has been demonstrated that HUCPVCs have a greater proliferative capacity when compared to BM-MSCs. However, after 20 days of adherent culture, HUCPVCs do not experience contact-inhibited cell growth, unlike BM-MSCs, but rather continue to grow by multilayering and forming aggregates overlying a layer of confluent cells. These cell aggregates contain live cells as determined by trypan blue exclusion, and when transferred to a new tissue culture flask, demonstrate the capacity to plate out and proliferate to colonize the culture substrate. A Cancer Pathway Finder gene array (Cat.#: EHS-033 human) can be used to detect changes in candidate genes associated with the cancer pathways. The Cancer Pathway Finder microarray contains 113 gene representative of 6 biological pathways involved in transformation and tumorigenesis, namely, cell cycle control and DNA damage repair, apoptosis and cells senescence, signal transduction molecules and transcription factors, adhesion, angiogenesis, and invasion and metastasis. Briefly, total RNA from cells grown under basal, undifferentiated conditions for 14 days is harvested using TRIzol reagent and purified on Qiagen's RNeasy mini columns. Three micrograms of purified total RNA is then converted to cDNA, and 4 μg of cRNA is hybridized to each microarray gene chip. cRNA labeling, amplification, overnight hybridization, and detection of gene expression was determined according to the manufacturer's protocol (SuperArray). Relative changes in gene expression are analyzed and quantified based on densiometry using the GEarray expression analysis suite software.

Using this array, we confirmed that HUCPVCs share a similar gene expression profile to normal BM-MSCs (Fig. 4). Of the genes tested, only 12.5% have different expression levels between the two cell types. This methodology can be utilized to assess gene expression in a range of available arrays, thereby providing a suite of information which can be compared to other cell types.

5. Immunocharacterization

As mentioned above, MSCs have an important immunoprivileged and immunosuppressive phenotype which allows them not only to be employed in mismatched recipients but also as a means to reduce an on-going immune reaction. Indeed, many studies have now shown that when MSCs are cultured with unmatched lymphocytes they do not cause alloreactivity, and that they suppress active immune reactions both *in vitro* and *in vivo* (Le Blanc *et al.*, 2003; Ringden *et al.*, 2006). These capacities can be tested *in vitro* using simple culture methodologies.

a. *In vitro* **alloreactivity**: Peripheral blood is obtained from consenting donors using heparinized syringes, and separated using a Ficoll-Paque™ density centrifugation (20 ml blood layered slowly over 20 ml Ficoll), and the leukocytes are removed from the buffy coat and counted. HUCPVCs are isolated and plated as per standard methods and plated in 96-well plates at 10^4 cells/well. Once attached, after approximately 2 h, 10^5 leukocytes are added to each well. The plates were incubated at 37 °C with 5% CO_2 air in RPMI-1640 media containing HEPES (25 mmol/L), L-glutamine (2 mmol/L), 10% fetal bovine serum, and antimicrobials. The cells were allowed to incubate for 6 days, after which they were stained with 5-bromo-2-deoxyuridine (BrdU, a base analog of thymidine), and assessed for BrdU incorporation using flow cytometry.

b. **Immunosuppression**: HUCPVCs are isolated, and plated as per standard methods, and plated in 96-well plates at 10^4 cells/well. Peripheral blood is obtained from two mismatched consenting donors using heparinized syringes, and separated using a Ficoll-Paque™ density centrifugation. The leukocytes are removed from the buffy coat and counted. Following this, both leukocyte populations are plated with the HUCPVCs at 10^5 cells/well. The plates are allowed to incubate for 6 days, counted on a ViCell-XR™ cell counter, and assessed for CD25 expression by flow cytometry. These cultures are also performed in which the leukocyte

RPS27A	(AKT1)	(ANGPT1)	(ANGPT2)	(APAF1)	(ATM)	(BAD)	(BAI1)
(BAX)	(BCL2)	(BCL2L1)	(BIRC5)	(BRCA1)	(BRCA2)	(CASP8)	(CASP9)
(CCND1)	(CCNE1)	(CD44)	(CDC25A)	(CDH1)	(CDK2)	(CDK4)	(CDKN1A)
(CDKN1B)	(CDKN2A)	(CFLAR)	(CHEK2)	(COL18A1)	(CTNNB1)	(E2F1)	(EGF)
(EGFR)	(ERBB2)	(ETS2)	(FGF2)	(FGFR2)	(FLT1)	FOS	(GZMA)
(HGF)	(HTATIP2)	(ICAM1)	(IFNA1)	(IFNB1)	IGF1	IL8	(ITGA1)
(ITGA2)	(ITGA3)	(ITGA4)	(ITGA5)	(ITGA6)	(ITGAV)	(ITGB1)	(ITGB3)
ITGB5	JUN	(KAI1)	(KISS1)	(MAP2K1)	(MAPK14)	(MCAM)	(MDM2)
(MET)	(MICA)	MMP1	MMP2	MMP9	(MTA1)	(MTA2)	(MTSS1)
MYC	(NCAM1)	NFKB1	NFKBIA	NME1	(NME4)	(PDGFA)	(PDGFB)
(PIK3CB)	(PIK3R1)	PLAU	PLAUR	(PNN)	(PRKDC)	(PTEN)	(RAF1)
(RASA1)	RB1	S100A4	(SERPINB2)	(SERPINB5)	(SERPINE1)	SNCG	(SRC)
(SYK)	(TEK)	(TERT)	(TGFB1)	(TGFBR1)	THBS1	THBS2	TIMP1
TIMP3	(TNF)	TNFRSF10B	TNFRSF1A	(TNFRSF25)	(FAS)	(TP53)	(TWIST1)
(UCC1)	(VEGF)	(PUC18)	(Blank)	Blank	AS1R2	AS1R1	AS1
GAPDH	B2M	HSPCB	HSPCB	ACTB	ACTB	BAS26	BAS2C

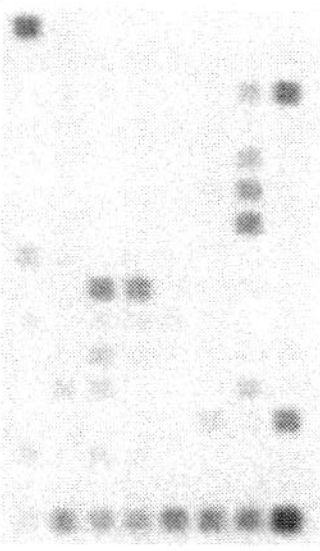

A

RPS27A	AKT1	(ANGPT1)	(ANGPT2)	(APAF1)	(ATM)	(BAD)	(BAI1)
(BAX)	(BCL2)	(BCL2L1)	(BIRC5)	(BRCA1)	(BRCA2)	(CASP8)	(CASP9)
(CCND1)	(CCNE1)	(CD44)	(CDC25A)	(CDH1)	(CDK2)	CDK4	CDKN1A
(CDKN1B)	(CDKN2A)	(CFLAR)	(CHEK2)	(COL18A1)	(CTNNB1)	(E2F1)	(EGF)
(EGFR)	(ERBB2)	(ETS2)	(FGF2)	(FGFR2)	(FLT1)	(FOS)	(GZMA)
(HGF)	(HTATIP2)	(ICAM1)	(IFNA1)	(IFNB1)	IGF1	IL8	ITGA1
(ITGA2)	ITGA3	(ITGA4)	(ITGA5)	(ITGA6)	(ITGAV)	ITGB1	(ITGB3)
ITGB5	(JUN)	(KAI1)	KISS1	(MAP2K1)	(MAPK14)	(MCAM)	(MDM2)
(MET)	(MICA)	MMP1	MMP2	MMP9	(MTA1)	(MTA2)	(MTSS1)
MYC	(NCAM1)	NFKB1	NFKBIA	NME1	(NME4)	(PDGFA)	(PDGFB)
(PIK3CB)	(PIK3R1)	PLAU	PLAUR	PNN	(PRKDC)	PTEN	(RAF1)
RASA1	RB1	S100A4	(SERPINB2)	(SERPINB5)	SERPINE1	SNCG	(SRC)
(SYK)	(TEK)	(TERT)	(TGFB1)	(TGFBR1)	THBS1	THBS2	TIMP1
TIMP3	(TNF)	TNFRSF10B	TNFRSF1A	(TNFRSF25)	(FAS)	(TP53)	(TWIST1)
(UCC1)	(VEGF)	(PUC18)	(Blank)	Blank	AS1R2	AS1R1	(AS1)
GAPDH	B2M	HSPCB	HSPCB	ACTB	ACTB	BAS2C	BAS2C

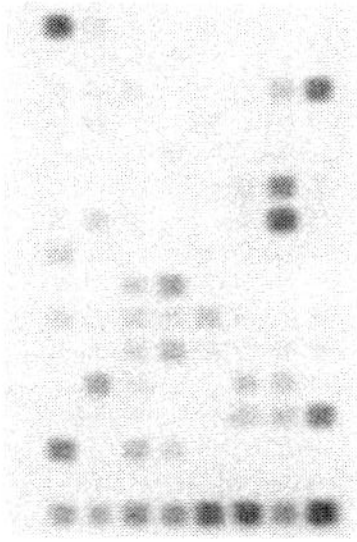

B

Fig. 4 (A) Results of BM-MSCs and (B) HUCPVCs when run on a Cancer Pathway Finder gene array. BM-MSCs and HUCPVCs share a similar genetic profile for those genes tested, with 87.5% of genes showing similar expression. All genes in parentheses are not expressed.

populations were separated from the HUCPVCs by a semipermeable TransWell® membrane insert. The HUCPVCs are cultured on the Transwell® surface, and the insert is transferred to a 24-well plate containing PBLs from two mismatched donors (as described above).

Using these methodologies, we have shown that the immunological profile of HUCPVCs is bioequivalent to that of BM-MSCs. HUCPVCs do not cause alloreactivity in mismatched lymphocytes *in vitro*, and suppress an active immune response (even when added at days 3 and 5 of a 6-day culture) (Fig. 5). In addition, HUCPVCs are capable of reducing the activation state of T lymphocytes as indicated by a reduction in CD25 expression (data not shown). These capacities are similar to those shown by BM-MSCs, and when compared directly are shown to be have the same level of effect and trends among doses (Ennis *et al.*, 2008).

6. Transduction

In order for HUCPVCs to be tracked for *in vivo* studies, especially on a long-term basis, they can be transduced with a fluorescent marker. To this end, we have modified existing Lentiviral Transfection protocols from "Cell Biology: A Laboratory Handbook" (Salmon and Trono, 2006) and the Ellis Lab (UHN, Toronto) for the modification of HUCPVCs with green fluorescent protein (GFP). The transduction process involves the transfection of 293 cells with the desired DNA plasmids (vector DNA, 10 μg gag/pol-expressing plasmid, 10 μg of rev-expressing plasmid, 10 μg of tat-expressing plasmid, 5 μg of VSV-G-expressing plasmid with

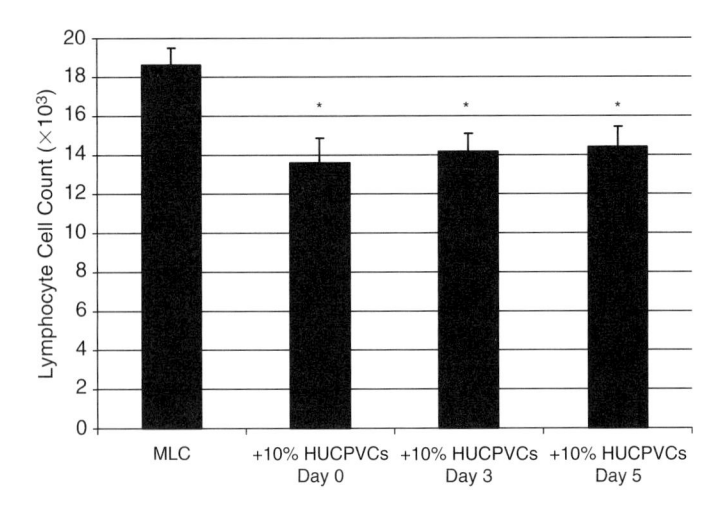

Fig. 5 Upon the inclusion of 10% HUCPVCs in a two-way mixed lymphocyte culture (MLC), a significant reduction in lymphocyte cell number is seen; regardless of when the cells were added (day 0, 3, or 5) in a 6-day culture. * $p < 0.05$, $n = 3$.

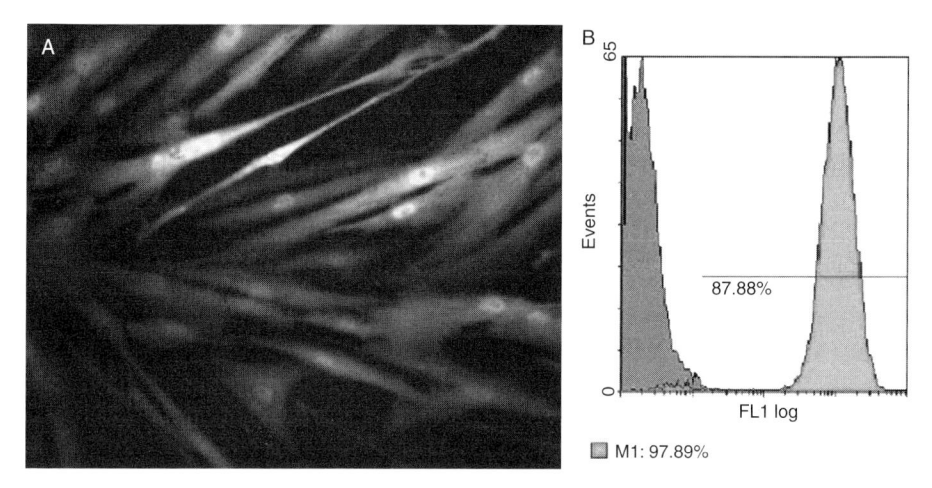

Fig. 6 (A) Fluorescent micrograph of HUCPVCs (Human Umbilical Cord Perivascular Cells) lentivirally transfected with Green Fluorescent Protein (GFP) (field width = 0.3 mm), (B) Flow cytometry measurement of GFP expression, measuring 97.89% of HUCPVCs positively expressing the protein, gated on control. (See Plate no. 11 in the Color Plate Section.)

2.5 M $CaCl_2$). These are allowed to incubate overnight, after which the medium is changed. Cells are left with this medium for three more days, after which the supernatant of the cells is collected and filtered through a 0.45 mm filter. The viral supernatant is then concentrated by ultracentrifugation (50,000 × g for 90 min) or using an Amicon Ultra-15 Centrifugal Filter device (100,000 MWCO; Millipore). When this process is complete, the viral supernatant can be combined with the HUCPVCs at a concentration determined by titering the concentrated virus, and allowed to incubate at 37 °C overnight with polybrene (4 mg/ml). The following day, more medium is added, and the cells incubate for 6 more hours before changing the media. This methodology can result in a transduction efficiency of 97% (Fig. 6), with maintenance of good proliferative rates. While we initially focused on GFP as our transfected protein for *in vivo* localization, with a functioning lentiviral transduction protocol, it is possible to transduce the cells with any protein.

III. Concluding Remarks

This chapter has provided a broad overview of the techniques necessary to extract and culture human umbilical cord perivascular cells (HUCPVCs), and also methods to determine their functionality. Our results to test for mesenchymal functionality through differentiation, immunologic phenotype, gene expression, and ability to be transfected have indicated that HUCPVCs are bioequivalent to BM-MSCs and further corroborate our previous findings (Baksh *et al.*, 2007;

Ennis *et al.*, 2008). We have improved the cell harvesting procedures since our original description (Sarugaser *et al.*, 2005), and have made the process more reliant on technology than technique and, concomitantly, reduced the laboratory time required. With these streamlined tissue processing methods, and the ready availability of human umbilical cords, the perivascular tissue of the human umbilical cord clearly represents a rich and potentially unlimited alternative, ethically noncontroversial, supply of MSCs for cell-based therapies.

Acknowledgments

The authors thank Elaine Cheng for the cord dissections, and the staff of Women's College Hospital Toronto for their support. Also, we thank the Ontario Research and Development Challenge Fund (grant to JED).

References

Abedin, M., Tintut, Y., and Demer, L. L. (2004). Mesenchymal stem cells and the artery wall. *Circ. Res.* **1,** 671–676.

Baksh, D., Yao, R., and Tuan, R. (2007). Comparison of proliferative and multilineage differentiation potential of human mesenchymal stem cells derived from umbilical cord and bone marrow. *Stem Cells* **45**(67), 99–106.

Blann, A. D., Woywodt, A., Bertolini, F., Bull, T. M., Buyon, J. P., Clancy, R. M., Haubitz, M., Hebbel, R. P., Lip, G. Y., Mancuso, P., Sampol, J., Solovey, A., *et al.* (2005). Circulating endothelial cells. Biomarker of vascular disease. *Thromb. Haemost.* **93**(2), 228–235.

Boland, G. M., Perkins, G., Hall, D. J., and Tuan, R. S. (2004a). Wnt 3a promotes proliferation and suppresses osteogenic differentiation of adult human mesenchymal stem cells. *J. Cell. Biochem.* **93**(6), 1210–1230.

Boland, G. M., Perkins, G., Hall, D. J., and Tuan, R. S. (2004b). Wnt 3a promotes proliferation and suppresses osteogenic differentiation of adult human mesenchymal stem cells. *J. Cell. Biochem.* **93**(6), 1210–1230.

Can, A., and Karahuseyinoglu, S. (2007). Concise review: Human umbilical cord stroma with regard to the source of fetus-derived stem cells. *Stem Cells* **25**(11), 2886–2895.

da Silva, M. L., Chagastelles, P. C., and Nardi, N. B. (2006). Mesenchymal stem cells reside in virtually all post-natal organs and tissues. *J. Cell Sci.* **119**(Pt. 11), 2204–2213.

De Boer, J., Siddappa, R., Gaspar, C., van Apeldoorn, A., Fodde, R., and Van Blitterswijk, C. (2004a). Wnt signaling inhibits osteogenic differentiation of human mesenchymal stem cells. *Bone* **34**(5), 818–826.

De Boer, J., Wang, H. J., and Van Blitterswijk, C. (2004b). Effects of Wnt signaling on proliferation and differentiation of human mesenchymal stem cells. *Tissue Eng.* **10**(3–4), 393–401.

Doherty, M. J., Ashton, B. A., Walsh, S., Beresford, J. N., Grant, M. E., and Canfield, A. E. (1998). Vascular pericytes express osteogenic potential *in vitro* and *in vivo*. *J. Bone Miner. Res.* **13**(5), 828–838.

Duda, D. G., Cohen, K. S., di Tomaso, E., Au, P., Klein, R. J., Scadden, D. T., Willett, C. G., and Jain, R. K. (2006). Differential CD146 expression on circulating versus tissue endothelial cells in rectal cancer patients: Implications for circulating endothelial and progenitor cells as biomarkers for antiangiogenic therapy. *J. Clin. Oncol.* **24**(9), 1449–1453.

Ennis, J., Gotherstrom, C., Le Blanc, K., and Davies, J. (2008). *In vitro* Immunologic Properties of Human Umbilical Cord Perivascular Cells. *Cytotherapy*. In Press.

Etheridge, S. L., Spencer, G. J., Heath, D. J., and Genever, P. G. (2004). Expression profiling and functional analysis of wnt signaling mechanisms in mesenchymal stem cells. *Stem Cells* **22**(5), 849–860.

Farrington-Rock, C., Crofts, N. J., Doherty, M. J., Ashton, B. A., Griffin-Jones, C., and Canfield, A. E. (2004). Chondrogenic and adipogenic potential of microvascular pericytes. *Circulation* **110**(15), 2226–2232.

Friedenstein, A. J., Gorskaja, J. F., and Kulagina, N. N. (1976). Fibroblast precursors in normal and irradiated mouse hematopoietic organs. *Exp. Hematol.* **4**(5), 267–274.

Friedman, R., Betancur, M., Boissel, L., Tuncer, H., Cetrulo, C., and Klingemann, H. (2007). Umbilical cord mesenchymal stem cells: Adjuvants for human cell transplantation. *Biol. Blood Marrow Transplant.* **13**(12), 1477–1486.

Kern, S., Eichler, H., Stoeve, J., Kluter, H., and Bieback, K. (2006). Comparative analysis of mesenchymal stem cells from bone marrow, umbilical cord blood, or adipose tissue. *Stem Cells* **24**(5), 1294–1301.

Kuznetsov, S., Krebsbach, P., Satomura, K., Kerr, J., Riminucci, M., Benayahu, D., and Robey, P. (1997). Single-colony derived strains of human marrow stromal fibroblasts form bone after transplantation *in vivo*. *J. Bone Miner. Res.* **12**(9), 1335–1347.

Le Blanc, K., Tammik, L., Sundberg, B., Haynesworth, S., and Ringden, O. (2003). Mesenchymal stem cells inhibit and stimulate mixed lymphocyte cultures and mitogenic responses independently of the major histocompatibility complex. *Scand. J. Immunol.* **57**, 11–20.

Lee, J. Y., Qu-Petersen, Z., Cao, B., Kimura, S., Jankowski, R., Cummins, J., Usas, A., Gates, C., Robbins, P., Wernig, A., and Huard, J. (2000). Clonal isolation of muscle-derived cells capable of enhancing muscle regeneration and bone healing. *J. Cell Biol.* **150**(5), 1085–1100.

Lund, R. D., Wang, S., Lu, B., Girman, S., Holmes, T., Sauvé, Y., Messina, D. J., Harris, I. R., Kihm, A. J., Harmon, A. M., Chin, F. Y., Gosiewska, A., et al. (2007). Cells isolated from umbilical cord tissue rescue photoreceptors and visual functions in a rodent model of retinal disease. *Stem Cells* **25**(3), 602–611.

Mitchell, K. E., Weiss, M. L., Mitchell, B. M., Martin, P., Davis, D., Morales, L., Helwig, B., Beerenstrauch, M., Abou-Easa, K., Hildreth, T., Troyer, D., and Medicetty, S. (2003). Matrix cells from Wharton's jelly form neurons and glia. *Stem Cells* **21**(1), 50–60.

Nanaev, A. K., Kohnen, G., Milovanov, A. P., Domogatsky, S. P., and Kaufmann, P. (1997). Stromal differentiation and architecture of the human umbilical cord. *Placenta* **18**(1), 53–64.

Ringden, O., Uzunel, M., Rasmusson, I., Remberger, M., Sundberg, B., Lönnies, H., Marschall, H. U., Dlugosz, A., Szakos, A., Hassan, Z., Omazic, B., Aschan, J., et al. (2006). Mesenchymal stem cells for treatment of therapy-resistant graft-versus-host disease. *Transplantation* **81**(10), 1390–1397.

Salmon, P., and Trono, D. (2006). Practical Course on HIV-derived Vectors—Handbook for the design, production, and titration. *In* "Cell Biology: A Laboratory Handbook." p. 437. Elsevier, San Diego.

Sarugaser, R., Lickorish, D., Baksh, D., Hosseini, M. M., and Davies, J. E. (2005). Human umbilical cord perivascular (HUCPV) cells: A source of mesenchymal progenitors. *Stem Cells* **23**(2), 220–229.

Sarugaser, R., Hanoun, L., Keating, A., Stanford, W. L., and Davies, J. (2007). Mesenchymal stem cells self-renew and differentiate according to a deterministic hierarchy. International Society for Stem Cell Research, 5th Annual Meeting.

Shi, S., and Gronthos, S. (2003). Perivascular niche of postnatal mesenchymal stem cells in human bone marrow and dental pulp. *J. Bone Miner. Res.* **18**, 696–704.

Stenderup, K., Justesen, J., Clausen, C., and Kassem, M. (2003). Aging is associated with decreased maximal life span and accelerated senescence of bone marrow stromal cells. *Bone* **33**(6), 919–926.

Takahashi, K., Igura, K., Zhang, X., Mitsuru, A., and Takahashi, T. A. (2004). Effects of osteogenic induction on mesenchymal cells from fetal and maternal parts of human placenta. *Cell Transplant.* **13**(4), 337–341.

Takechi, K., Kuwabara, Y., and Mizuno, M. (1993). Ultrastructural and immunohistochemical studies of Wharton's jelly umbilical cord cells. *Placenta* **14**, 235–245.

Tintut, Y., Alfonso, Z., Saini, T., Radcliff, K., Watson, K., Bostrom, K., and Demer, L. L. (2003). Multilineage potential of cells from the artery wall. *Circulation* **108**(20), 2505–2510.

Wang, H. S., Hung, S. C., Peng, S. T., Huang, C. C., Wei, H. M., Guo, Y. J., Fu, Y. S., Lai, M. C., and Chen, C. C. (2004). Mesenchymal stem cells in the Wharton's jelly of the human umbilical cord. *Stem Cells* **22**(7), 1330–1337.

Wu, H., Riha, G. M., Yang, H., Li, M., Yao, Q., and Chen, C. (2005). Differentiation and proliferation of endothelial progenitor cells from canine peripheral blood mononuclear cells. *J. Surg. Res.* **126**(2), 193–198.

Zuk, P. A., Zhu, M., Ashjian, P., De Ugarte, D. A., Huang, J. I., Mizuno, H., Alfonso, Z. C., Fraser, J. K., Benhaim, P., and Hedrick, M. H. (2002). Human adipose tissue is a source of multipotent stem cells. *Mol. Biol. Cell* **13**(12), 4279–4295.

CHAPTER 8

Hepatic Stem Cells and Hepatoblasts: Identification, Isolation, and *Ex Vivo* Maintenance

Eliane Wauthier,[*,1] **Eva Schmelzer,**[*,1] **William Turner,**[†,1]
Lili Zhang,[*,1] **Ed LeCluyse,**[∥,1] **Joseph Ruiz,**[¶,1] **Rachael Turner,**[†,1]
M. E. Furth,[§,1] **Hiroshi Kubota,**[*,1] **Oswaldo Lozoya,**[†,2]
Claire Barbier,[*,2] **Randall McClelland,**[*,3] **Hsin-lei Yao,**[†,3]
Nicholas Moss,[*,3] **Andrew Bruce,**[¶,3] **John Ludlow,**[¶,3] **and**
L. M. Reid[*,†,‡,§,1]

[*]Departments of Cell and Molecular Physiology
UNC School of Medicine
Chapel Hill, North Carolina 27599

[†]Join Department of Biomedical Engineering at NCSU and
UNC School of Medicine
Chapel Hill, North Carolina 27599

[‡]Program in Molecular Biology and Biotechnology
UNC School of Medicine
Chapel Hill, North Carolina 27599

[§]Institute for Regenerative Medicine
Wake Forest Baptist Medical Center
Winston Salem, North Carolina 27157

[¶]Vesta Therapeutics
801–8 Capitola Drive, Suite 801
Durham, North Carolina 27713

[∥]CellzDirect, Inc.
Pittsboro, North Carolina 27312

---

[1] Authors contributed technologies and helped prepare the review.
[2] Authors helped with the preparation of the review.
[3] Authors established technologies described in the review but did not help with the preparation of the review.

I. Introduction
II. Epithelial–Mesenchymal Relationship
III. Stem Cells and Lineage Biology
    A. Maturational Lineage Stages from Zygote to Mature Liver Cells
    B. Strategies Influenced by the Knowledge of Lineage Biology
IV. Liver Processing
    A. General Comments
    B. Liver Perfusions in Mice and Rats—Readily Adaptable to Any Mammalian Liver (Other than Human)
    C. Fetal Human Livers (Livers of Gestational Ages 14–20 Weeks)
    D. Liver Perfusions: Neonatal, Pediatric, and Adult Human Livers
V. Fractionation of Liver Cell Subpopulations
    A. Analysis of Ploidy
    B. Size
    C. Cell Density
    D. Antigenic Properties
VI. Feeder Cells
    A. Preparation of Mesenchymal Feeders from Fetal Livers
    B. Preparation of Feeders from STO Cells: A Murine Embryonic Cell Line
    C. Feeders of hUVECs Feeders
    D. Replacement of Feeders with Defined Paracrine Signals
VII. Extracellular Matrix
    A. Paradigms in the Construction of Extracellular Matrix
    B. Rules for Cultures on Matrix Substrata
    C. Protocols for Preparing Extracellular Matrix Components
VIII. Culture Media
    A. The Basal Medium
    B. Serum
    C. Lipids
    D. Soluble Signals: Autocrine, Paracrine, and Endocrine Factors (HDM)
    E. Influence of Serum-Free HDM on Epithelial Cells in Culture
IX. Monolayer Cultures of Cells
    A. Monolayer Cultures in a Growth State with Defined Conditions
    B. Monolayer Cultures in a Differentiated State with Defined Conditions
X. Three-Dimensional Systems
    A. Spheroids
    B. Hyaluronan Hydrogels as Three-Dimensional Scaffolding for cells
        Appendix 1: Nomenclature, Glossary, and Abbreviations
        Appendix 2: Details on the Preparation of Buffers/Reagents
        Appendix 3: Commercial Sources of Reagents
        Appendix 4: Sources of Primary and Secondary Antibodies
        for Immunoselection and for Characterization of the Cells
    References

I. Introduction

This review is an update of a methods review published previously (Macdonald *et al.*, 2002) combined with some information taken from a lengthy review on hepatic stem cells (HpSCs) (Schmelzer *et al.*, 2006b) and then providing current protocols on HpSCs, especially human hepatic stem cells (hHpSCs). The conditions needed for a subpopulation of cells from one species are, with rare exceptions, similar to those needed for the subpopulation derived from all other species. Therefore, though the protocols we are listing have been developed with studies on rodents or human cells, they should work as is, or with minor modifications, on cells from any species.

Some of the detailed protocols available in the earlier review (Macdonald *et al.*, 2002) are not repeated here. This includes how to develop a serum-free, hormonally defined medium or HDM; how to prepare tissue extracts enriched in extracellular matrix (e.g. Matrigel, Biomatrix, ECM); and how to design biodegradable, polylactide scaffoldings or microcarriers in ways appropriate for progenitors and use of bioreactors. Even more detailed discussions on bioreactors are given in another review (Macdonald *et al.*, 1999). Also, protocols for factors or reagents that are commercially available are not given. Some of the especially critical ones are listed in the Appendices with notation of their commercial sources.

II. Epithelial–Mesenchymal Relationship

The liver is similar to all metazoan tissues in that the epithelial–mesenchymal relationship constitutes the organizational basis for the tissue (Gittes *et al.*, 1996; Lemaigre and Zaret, 2004; Reid and Jefferson, 1984; Young *et al.*, 1995). Epithelia (hepatic, lung, pancreas, etc.) are wed to a specific type of mesenchymal cells (endothelia, stroma, stellate cells, smooth muscle cells). Dynamic interactions between the two are mediated by a set of soluble signals, autocrine and paracrine, and a set of insoluble signals found on the lateral borders of homotypic cells, the lateral extracellular matrix, and on the basal borders between heterotypic cells, the basal extracellular matrix (Brill *et al.*, 1994; Furthmayr, 1993; Kallunki and Tryggvason, 1992; Martinez-Hernandez and Amenta, 1993c, 1995; Nimni, 1993; Reid *et al.*, 1992; Stamatoglou and Hughes, 1994; Terada *et al.*, 1998). Furthermore, endocrine (i.e. systemic) regulation and modulation is achieved, in part, by regulating some aspect of the epithelial–mesenchymal relationship (Bernfield *et al.*, 1992). Normal epithelial cells will not survive for long and will not function properly unless epithelial cells and appropriate mesenchymal cells are cocultured (Bhatia *et al.*, 1996; LeCluyse *et al.*, 1996; Liu and Chang, 2000; Reid and Jefferson, 1984). To escape from the need for cocultures, one can place the cells of interest in contact with appropriate extracellular matrix components and in medium containing the soluble signals (either defined components or "conditioned medium") produced by interactions between epithelia and mesenchymal cells (Jefferson et al., 1984a; Reid et al., 1992; Tillotson et al., 2001; Xu et al., 2000).

In recent years, there has been recognition that the epithelial–mesenchymal relationship is lineage dependent (Cheng et al., 2007; Reid et al., 1992; Schmelzer et al., 2006b; Sigal et al., 1993). Epithelial stem cells are partnered with mesenchymal stem cells, and their differentiation is co-ordinate (Jung et al., 1999; Kubota et al., 2007; Lemaigre and Zaret, 2004). In the liver, the lineages begin with the HpSCs paired with their mesenchymal partners, angioblasts (Schmelzer et al., 2007) that interact with multiple forms of paracrine signals (Sicklick et al., 2005). These two give rise to descendents in a stepwise, lineage-dependent fashion, and their descendents remain in a partnership throughout differentiation (Table I). Tissue engineering involves the mimicking of the liver's epithelial–mesenchymal relationship with recognition of the lineage-dependent phenomena (Schmelzer et al., 2006b; McClelland et al., 2004; Xu et al., 2000).

III. Stem Cells and Lineage Biology

Quiescent tissues (liver, brain), as well as rapidly renewing tissues (skin, intestine), are organized as precursor cell populations (i.e. stem cells and committed progenitors) that yield daughter cells undergoing a maturational process ending in apoptotic cells (Cheng et al., 2007; Fuchs and Byrne, 1994; Fuchs and Raghavan, 2002; Reid, 1990; Sigal et al., 1992, 1995; Temple and Alvarez-Buylla, 1999). The different maturational stages of cells within the maturational lineages are distinct phenotypically and in their requirements for ex vivo maintenance (Macdonald et al., 2002). Therefore, it is ideal to purify specific subpopulations of cells at distinct maturational stages in order to have cells that behave uniformly under specified culture conditions. Purification of the cells involves enzymatic dissociation of the tissue followed by fractionation methods that can include immunoselection technologies (Kubota and Reid, 2000; Kubota et al., 2007; LaGasse, 2007; Sigal et al., 1994, 1999). Most of the culture conditions required for the different maturational stages of liver cells have been defined and involve the use of entirely purified soluble signals and extracellular matrix components (Macdonald et al., 2002; McClelland et al., 2008; Schmelzer et al., 2006b; Schmelzer et al., 2007; Turner et al., 2006; Turner et al., 2008). Certain conditions can be used to maintain the cells with reproducible growth properties and others used to put the cells into a state of growth arrest with expression of particular tissue-specific genes (Table II).

Each cellular subpopulation at a specific maturational stage (e.g. stem cells, committed progenitors, diploid adult cells, polyploid adult cells) has a unique phenotype (antigenically, biochemically, morphologically) with its own distinct

Table I
Lineages of Epithelial–Mesenchymal Partners in Liver (Known or Assumed)

Lineage stage (#): Epithelia	Mesenchymal cells
Stem cell niche: *Ductal plates (fetal/neonatal livers)/canals of Hering (pediatric/adult livers)*	
(1) Hepatic stem cells [pluripotent]	Angioblasts and hepatic stellate cell precursors
(2) Hepatoblasts (HBs) [bipotent]	Activated angioblasts and hepatic stellate cells
Immediately outside the stem cell niche	
(3A) Committed hepatocytic progenitors [unipotent]	Endothelial cell precursors
(3B) Committed biliary progenitors [unipotent]	Stromal cell precursors
Liver Acinus-Zone 1 Parenchyma:	
(4A) "Small" hepatocytes [diploid]	Continuous endothelia
(4B) Intra-hepatic bile duct epithelia [diploid]	Stroma, smooth muscle cells
Liver Acinus-Zone 2 Hepatocytes	
(5A) Midacinar hepatocytes	Continuous and fenestrated endothelia (ploidy profile level depends on the species)
Liver Acinus-Zone 3 Hepatocytes	
(6A) Pericentral hepatocytes	Fenestrated endothelia (ploidy profile depends on the species)
(7A) Apoptotic Hepatocytes	Fenestrated endothelia (polyploidy but with DNA fragmentation)
Extrahepatic Biliary Epithelial Cells (Cholangiocytes)	
(5B) Bile duct epithelium	Stroma, smooth muscle cells (ploidy profile not known; hypothesized to be the cholangiocytes of the extrahepatic bile duct from liver to gall bladder)
(6B) Bile duct epithelium	Stroma, smooth muscle cells (ploidy profile not known; hypothesized to be the Cholangiocytes of the extrahepatic bile duct from gall bladder to gut)
(7B) Common duct epithelium	Stroma, smooth muscle cells (ploidy profile not known; hypothesized to be the epithelia of the Common duct—that shared with the pancreatic duct)

It is hypothesized that there are maturational lineage stages for the intra- and extra-hepatic bile duct epithelia, but these lineage stages have yet to be defined. The lineage stages for the hepatocytes and their associated mesenchymal cells are known and have been defined by studies on zonation of hepatocyte cell size, ploidy, antigen and gene expression, and morphology and by more recent studies in differentiation of stem cells to mature cells (Schmelzer *et al.*, 2006, 2007; Turner *et al.*, 2007).

set of conditions for *ex vivo* growth versus differentiation. For example, HpSCs must be co-cultured with their natural partners, angioblasts and hepatic stellate cells precursors (Schmelzer *et al.*, 2007). If the paracrine signals mediating the epithelial–mesenchymal relationship are known, they can be used to prepare completely defined culture conditions. If they are not known, then co-culturing of the epithelia and mesenchymal subpopulations is required for full functions.

Table II
Maturational Lineages Varying in Kinetics

Rapidly regenerating tissues (rapid kinetics)	Quiescent tissues (slow kinetics)
Turnover in days to weeks	*Turnover* in months to years
% polyploid cells low (e.g. 5–10%)	*% polyploid cells* intermediate (e.g. 30%) to high (e.g. 95%)
Representative tissues	*Representative tissues*
• Hemopoietic cells	• Lung, liver, pancreas, other internal organs
• Epidermis	• Blood vessels
• Intestinal epithelia	• Skeletal muscle
• Hair	• Nerve cells—including the brain
	• Heart muscle

Hypothesis: Kinetics of lineage inversely correlated with extent of polyploidy. References: Anatskaya *et al.* (1994), Brodskii *et al.* (1983), Brodsky and Uryvaeva (1977), Epstein and Gatens (1967), Matturri *et al.* (1991).

A. Maturational Lineage Stages from Zygote to Mature Liver Cells

[Pluripotent = cells that can give rise to daughter cells of more than one fate. Stem Cells = cells that are pluripotent and can self-replicate (self-renew) via symmetric division or can undergo differentiation by asymmetric division.]

Totipotent stem cells = zygote or cells of the morula. Pluripotent stem cells that can produce both the extraembryonic tissues (placenta, amnion) as well as the tissues derived from all 3 germ layers (ectoderm, mesoderm, endoderm).

Pluripotent embryonic stem (ES) cells = ES cells are pluripotent cells found in early embryos and that give rise to the tissues formed by the three germ layers (ectoderm, mesoderm, and endoderm) but not the extraembryonic tissue (e.g. placenta, amnion) (Martin, 1981; O'Shea, 1999; Smith *et al.*, 1988; Xu *et al.*, 2002).

Pluripotent cells in liver comprise two populations: hepatic stem cells (HpSCs) and possibly also hepatoblasts (HBs). The data thus far confirm that the HpSCs are stem cells; the data are incomplete in defining whether the HBs are stem cells.

HpSCs are pluripotent cells in fetal and postnatal livers that give rise to hepatoblasts (HBs) and to committed biliary progenitors (Schmelzer *et al.*, 2007; Sicklick *et al.*, 2005; Turner *et al.*, 2006). Their known antigenic profile is summarized in brief in Table II and is given in more detail elsewhere (Schmelzer *et al.*, 2006a). Unique defining markers include EpCAM, NCAM, Claudin 3, and CK19. They can be expanded *ex vivo* only when co-cultured with angioblasts or early stage endothelia or with appropriate matrix and soluble signals mimicking the HpSC–angioblast relationship (McClelland *et al.*, 2008; Turner *et al.*, 2006). These conditions have been defined and include a serum-free medium, "Kubota's Medium" (KM), designed for hepatic progenitors (Kubota and Reid, 2000), that can be used in combination with matrix components that include type III collagen (McClelland *et al.*, 2008a,b) and hyaluronans (Turner *et al.*, 2006; Turner et al, 2008). The doubling time under optimal conditions has been found to be a doubling every ~26 h.

Hepatoblasts (bipotent) are derived from HpSCs. They have a phenotype that overlaps extensively with that of the HpSCs (Schmelzer *et al.*, 2006a) with the qualifiers that they express ICAM-1 but not NCAM, fetal forms of P450s (e.g. human P450–3A7), higher levels of AFP, elevated levels of albumin, and have lost expression of claudin 3. The doubling time for HBs is estimated to be ∼40–50 h under the conditions tested to date.

Committed progenitors are unipotent precursors for a single cell type. They include biliary committed progenitors (CK19+, ALB−) and hepatocytic committed progenitors (CK19−, ALB+). The culture conditions for these cells overlap extensively with that of the HpSCs and HBs. The division rate for these is not known.

Adult diploid cells: All fetal and neonatal tissues contain only diploid cells, and these cells are able to undergo complete cell divisions both *in vivo* and *in vitro*. The number of divisions possible for adult diploid cells is estimated to be 5–8 divisions (Kubota and Reid, 2000; Mitaka *et al.*, 1993, 1995, 1998). The clonogenic expansion conditions that are successful with HpSCs and HBs are permissive for colony formation of diploid hepatocytes though the cells divide much more slowly, doubling time ∼60 h (Kubota and Reid, 2000), and show limited ability to be passaged.

Adult polyploid cells: Polyploid cells appear in liver tissue within a few weeks of postnatal life in mice or rats or by teenage years in humans (Gerlyng *et al.*, 1993; Liu *et al.*, 2003a; Mossin *et al.*, 1994; Saeter *et al.*, 1988; Seglen *et al.*, 1986). The majority of the human liver's polyploid cells are tetraploid, whereas higher levels of ploidy have been observed in murine hepatocytes (up to 32N) and rat hepatocytes (up to 8N) (Liu *et al.*, 2003; Higgins *et al.*, 1985; Severin *et al.*, 1984). The percentage of polyploid cells increases with age (Anatskaya *et al.*, 1994; Brodsky *et al.*, 1980; Carriere, 1967; Liu *et al.*, 2003a; Sigal *et al.*, 1999a) and after partial hepatectomy (Sigal *et al.*, 1999a) and is achieved by cells undergoing nuclear division without cytokinesis [62]. By ∼6 weeks of age, adult rats have livers in which the extent of polyploidy is at least 90% comprising 80% tetraploid cells and 10% octaploid cells.

Published data on ploidy in human liver varies greatly, and the interpretations to date are suspect due in some studies to the methods used for analyzing ploidy, in others to the sampling methods used, and in all of them to the sparse number of human liver samples analyzed. Prior to the 1970s, some investigators assumed that cells with multiple nuclei are polyploid and those with a single nucleus are diploid. The level of ploidy was defined using that as a guide in the analyses of histological sections of tissue. Since it is now well known that polyploid cells can be mononucleated, the data from these older studies must be re-evaluated. In more recent studies using valid methods for analyzing ploidy, the authors have used a small piece of the liver or even a needle aspirate (Anatskaya *et al.*, 1994; Anti *et al.*, 1994; Kudryavtsev *et al.*, 1993; Saeter *et al.*, 1988; Watanabe and Tanaka, 1982). Considering the heterogeneous nature of liver cells (Burger *et al.*, 1989; Gaasbeek Janzen *et al.*, 1987; Gumucio, 1989; Traber *et al.*, 1988), such small samples are unlikely to be representative of the tissue as a whole. All studies on human liver are made difficult by the extremely limited supply of reasonable

Table III
Age Effects on Ploidy Profiles of Parenchymal Cells

Rodents	Humans[a]
Fetal and neonatal: entirely diploid	*Fetuses, neonates, and children up to teenage years:* entirely diploid
Young adults: 4–5 weeks of age: 10% diploid; 80% tetraploid; 10% octaploid	*Young adults:* 20–40 years of age mostly diploid with perhaps up to 30% tetraploid
6 months and older: <5% diploid; >95% polyploid; polyploid cells are a mix of mononucleated and binucleated cells	*Older adults:* steady increase of polyploid cells with age; polyploid cells are mostly (entirely?) binucleated, tetraploid cells

[a]Estimates of ploidy profiles in human parenchyma vary due to the effects of ischemia: polyploid cells are lost selectively with ischemia, especially warm ischemia (Reid *et al.*, 2000).

quality samples and usually of only certain ages. Altogether these difficulties may explain the wide variations in findings on the extent of polyploidy in adult human livers. Some have claimed as few as 10%, others about 30%, and the highest numbers reported are up to 60% of the cells as polyploid. All agree that the percentage of polyploid liver cells increases with age, especially after 60 years of age (Anatskaya *et al.*, 1994; Gerlyng *et al.*, 1993; Kudryavtsev *et al.*, 1993; Saeter *et al.*, 1988; Stein and Kudryavtsev, 1992; Watanabe and Tanaka, 1982).

In summary, a rigorous analysis of polyploidy in human liver and the data correlated with age of donor has yet to be done (Table III). The properties of the diploid versus polyploid cells are quite distinct. Whereas diploid cells have been found capable of colony formation (Kubota and Reid, 2000; Mitaka *et al.*, 1995, 1998; Ogawa *et al.*, 2004; Tateno and Yoshizato, 1999), σ tetraploid cells are able to go through only one or two complete cell divisions as indicated by studies on routine primary cultures of adult liver cells (Gebhardt, 1988; Overturf *et al.*, 1999). Tetraploid cells are thought incapable of clonogenic growth, an assumption that has yet to be tested rigorously. Instead, they are ideal for an analysis of highly differentiated functions, some of which are expressed only in the polyploid cells such as a few of the cytochrome P450s (Hamilton *et al.*, 2001) (Tables IV and V).

B. Strategies Influenced by the Knowledge of Lineage Biology

Although all forms of progenitors are found in embryonic tissues, most adult tissues contain only determined stem cells and/or committed progenitors (Potten *et al.*, 2006). The known properties of the different maturational lineage stages of cells help to define their potential in academic, clinical, or industrial programs. Of the types of progenitors studied (e.g. mesenchymal, neuronal, muscle, epidermal, hepatic), all have been found to be readily cryopreserved (Chen, 1992; Ek *et al.*, 1993; Westgren *et al.*, 1994) and expanded *ex vivo* (Deans and Moseley, 2000;

Table IV
Summary of Lineage Stages within Liver for Representative Species

Maturational lineage stage (location)	Mouse	Rat	Human
1. *Stem cells (HpSCs)* (ductal plate/canals of Hering) ~7–9 μm, diploid	Class I MHC negative	Class I MHC negative	EpCAM+, NCAM+, claudin 3+, class I MHC antigens negative, AFP−, Hedgehog proteins and their receptor, patched
2. *Hepatoblasts* (HBs) (probable transit amplifying cells) parenchyma of fetal/neonatal livers; tethered to ends of canals of Hering in pediatric and adult livers. ~10–12 μm, diploid	Class I MHC negative, ICAM+, AFP++	Class I MHC negative, ICAM+, AFP++	EpCAM+, ICAM+, class I MHC antigens negative, AFP++, Hedgehog proteins and patched; P450 3A7
3. *Unipotent progenitors* (rare except in fetal/neonatal livers) ~12–15 μm/diploid	*Committed biliary progenitors:* CK19+, ALB−; *committed hepatocytic progenitors:* CK19−, ALB+		
4. *Diploid hepatocytes* ("small hepatocytes") ~18 μm	~5% of adult hepatocytes all in zone 1	~10% of adult parenchyma, all in zone 1	Zone 1 and 2 hepatocytes. Gluconeogenesis; Connexins 26, 32
5. *Diploid cholangiocytes:* Zone 1 of liver acinus	Aquaporins, DPPIV, various pumps associated with bile production (e.g., MDR3)		
6. *Tetraploid hepatocytes*	Zone 2: Parenchyma	Zone 2: The majority of the parenchyma (~80%)	Zone 3: 10–50% of adult parenchyma depending on donor age; express late genes such as P450 3A1
6B. *Polyploid biliary epithelia* (?)	Unknown if they exist, or if they do, then their location is unknown. We hypothesize that they do exist and are probably in the extrahepatic bile duct		
7. Hepatocytes of higher ploidy (8N to 32N)	Zone 2/3 (~10–20% of adult parenchyma)	Zone 3 (~10% of adult parenchyma)	Not present
8. Apoptotic cells	Next to central vein. Markers for apoptosis evident.		

Table V
Markers for Hepatic Stem/Progenitors

Marker	Species	Comments/references
A. Hepatic stem cell/progenitor markers that are cloned and sequenced		
Albumin	All species	Found weakly expressed in HpSCs; steady increase in amount expressed in hepatoblasts (HBs) and with the differentiation of cells of the hepatocytic lineage (Baumann et al., 1999b; Schmelzer et al., 2007b; Zaret, 2002).
α-Fetoprotein (AFP)	All species	AFP is *not* expressed by HpSCs but rather by HBs (Schmelzer et al., 2006a, 2007; Zaret, 1999). A variant form of α-fetoprotein is expressed by hemopoietic progenitors (Kubota et al., 2002) and is identical to that in hepatic cells except for exon-1 encoded sequences.
Cytokeratin 7/19	All species	Cytokeratins 7/19 are found in the HpSCs, the HBs, and biliary epithelia but not the mature hepatocytic parenchyma (Crosby et al., 2001; Schmelzer et al., 2007; Shiojiri, 1994; Tanimizu et al., 2003; Theise et al., 1999; Van Eyken et al., 1988).
Claudin 3	Humans	Present in HpSCs but not in HBs or later hepatocytic stages (Schmelzer et al., 2006a).
CD133/1 (prominin)	Humans	A transmembrane protein found on hepatic, endothelial, and hemopoietic stem cells (Corbeil et al., 1999; Dhillon et al., 2001; Schmelzer et al., 2007; Weigmann et al., 1997).
E-cadherin	All species	Transmembrane protein that is coupled to β-catenin. It is expressed in the cells in the earliest stages of embryogenesis.
Epithelial cell adhesion molecule (EpCAM)	All species	Present on HpSCs, HBs, and some committed progenitors but not on mature hepatocytes (LaGasse, 2007; Schmelzer and Reid, 2008; Schmelzer et al., 2007; Sicklick et al., 2005).
CD44H (hyaluronan receptor)	Rats and humans	Present on rat HBs and on hHpSCs and hHBs (Kubota and Reid, 2000; Turner et al., 2006).
MDR1 (multidrug resistance gene)	Rats	Present on HBs (Joseph et al., 2003; Ros et al., 2003). [not yet tested in humans but hypothesized to be present in hHpSCs and hHBs].
Intercellular cell adhesion molecule (ICAM1)	Rats and humans	Present on HBs, committed progenitors and mature parenchymal cells; not expressed by HpSCs (Kubota and Reid, 2000; Schmelzer et al., 2007).
Neuronal cell adhesion molecule (NCAM)	Humans	Present on HpSCs but not HBs (Schmelzer et al., 2007). NCAM was first identified in neuronal cells; its functions in neurons include interactions with the fibroblast growth factor receptor (FGFR) through the p59Fyn signaling pathway. Its functions in HpSCs are unknown at present.

Delta-like/preadipocyte factor-1 (DLK-Pref-1)	Mice	Transmembrane protein that contains epidermal growth factor-like repeats related to ones in the notch/delta/serrate family of proteins (Jensen et al., 2004; Tanimizu et al., 2003, 2004).
Wnt/β-catenin pathway	All species	A pathway that appears to be generic for stem cell populations (Austin et al., 1997; Plescia et al., 2001).

B. Markers found on tightly associated "companion cells" (angioblasts or hepatic stellate cell precursors)

CD117 (c-kit)	All species	Receptor for stem cell factor. Expressed by progenitors of mesodermal lineages. Originally thought to be on hepatic progenitors, but the data increasingly suggest that it is on angioblasts tightly associated with HpSCs (Baumann et al., 1999a; Dhillon et al., 2001; Fujie et al., 1994; Matsusaka et al., 1999; Omori et al., 1997).
CD 146	Human	Antigen expressed by activated endothelia and HpSTCs. Unknown functions (Schmelzer et al., 2007).
KDR	All species	VEGF receptor (Flt1) present on angioblasts and endothelia (Kearney et al., 2004; Lecouter et al., 2004; Roberts et al., 2004).
CD34	Rodents, mice, human	Expression of CD34 has been reported to be on hepatic progenitors but the data are questionable with more recent realization that the marker is on tightly associated mesenchymal cells. Thus, rigorous sorts for CD34+ cells do not yield clonogenic populations capable of liver reconstitution or of forming liver tissue *in vitro* (Crosby et al., 2001; Dan et al., 2006; Kubota et al., 2007; Omori et al., 1997; Petersen et al., 2003; Schmelzer et al., 2007; Suskind and Muench, 2004).

C. Transcription factors

Prox1	All species	Homeobox gene defining pancreatic and liver fates (Burke and Oliver, 2002; Sosa-Pineda et al., 2000).
Hex	All species	Homeobox gene found in early liver (Keng et al., 2000; Martinez Barbera et al., 2000).
HLX	All species	Gene expressed in mesenchyme and required endoderm to migrate into the cardiac mesenchyme (Hentsch et al., 1996).
HNF-1, HNF-3, HNF-4, HNF-6	All species	(Monaghan et al., 1993; Rollini and Fournier, 1999; Runge et al., 1998).
C/EBP	All species	(Timchenko et al., 1999; Tomizawa et al., 1998; Van den Hoff et al., 1994; Wang et al., 1995).
DBP	All species	(Van den Hoff et al., 1994).
c-jun proto-oncogene	All species	Defining transcriptional element for liver development (Eferl et al., 1999; Hilberg et al., 1993; Passegue et al., 2002).

(continues)

Table V (continued)

Marker	Species	Comments/references
D. Markers defining epithelial cells		
E-cadherin	All species	Cell adhesion molecule on parenchymal cells but not on mesenchymal cell types (Aberle et al., 1996; Nitou et al., 2002; Terada et al., 1998).
Cytokeratins 8/18	All species	Cytokeratins evident in all forms of epithelia (Germain et al., 1988).
E. Oval cell antigens		
Oval cell antigens	All species	Identified in the livers of hosts following injury, drug or radiation treatment. Present on both hepatic and hemopoietic cells. (Dunsford and Sell, 1989; Hixson et al., 1990; Grisham, 1997; Thorgeirsson et al., 2004).
A6	Murine	(Cantz et al., 2003; Engelhardt et al., 1993).
OC2 and OC3	Rat	(Eghbali et al., 1991; Hixson et al., 1990; Petzer et al., 1996; Sigal et al., 1994, 1999).
Cloned and sequenced markers not found in/on hepatic progenitors		
CD 45	All species	Common leucocyte antigens (Kubota and Reid, 2000; Minguet et al., 2003; Oren et al., 1999).
Glycophorin A	All species	Protein on red blood cells (Kubota and Reid, 2000; Sigal et al., 1994, 1995).
CD14	All species	Expressed on monocytes and macrophages (Yin et al., 2001).
CD38	All species	Antigen on various hemopoietic cells (e.g., B lymphocytes) (Crosbie et al., 1999; Petzer et al., 1996).
OX43	Rats	Antigen found on all vascular endothelial cells except those in brain. Also recognizes peritoneal macrophages (Fiegel et al., 2003a,b).
OX44 (also called CD53)	Rats	Cell surface glycoprotein found on all leucocytes but absent on red blood cells and platelets (Paterson et al., 1987).

Gage, 1994; Kubota and Reid, 2000; McClelland *et al.*, 2008). However, the embryonic stem cells are especially notable for their ability to survive freezing and to expand without differentiating if maintained under precise culture conditions (Chen, 1992; Cowan *et al.*, 2004; Levenberg *et al.*, 2004; Maltsev *et al.*, 1993; Resnick *et al.*, 1992). By contrast, attempts to cryopreserve adult liver cells (predominantly the polyploid cells) have met with limited success. Best results have been achieved only by embedding the cells in alginate or a form of extracellular matrix (Guyomard *et al.*, 1996; Koebe *et al.*, 1990, 1996; Watts and Grant, 1996). Significant *ex vivo* expansion and the ability to subculture adult liver cells have been observed only with the so-called "small hepatocytes" (Mitaka *et al.*, 1992, 1998; Ogawa *et al.*, 2004), a diploid subpopulation of liver cells. Typically, the mature cells undergo one or two rounds of division and then survive for a matter of days in culture, or for a few weeks when supplied with the appropriate extracellular matrix and medium conditions (Hamilton *et al.*, 2001; Macdonald *et al.*, 2002).

The ability of embryonic stem cells to give rise to all, or almost all, possible adult fates makes them appealing as a possible "one serves all" source for cell therapies (Beardsley, 1999; Brinster, 1974; Levenberg *et al.*, 2004; Schuldiner *et al.*, 2000). However, their use in cell transplantation for patients is complicated by their tumorigenic potential (Martin, 1981; Mendiola *et al.*, 1999; Nozaki *et al.*, 1999; Pedersen, 1999). The tumorigenicity of ES cells when injected at ectopic sites is being investigated extensively (Mendiola *et al.*, 1999; Pedersen, 1999), in a hope that it can be controlled to enable ES cells to reach their full potential both industrially and clinically. Until this is solved, the ES cells will become an excellent choice as a research tool for pharmaceutical/biotechnology companies and for bioartificial organs but cannot be used for cell therapies.

While determined stem cells are more restricted in their adult cell fates, they have not been found to be tumorigenic, making them the first choice for clinical programs in cell transplantation or for bioartificial organs (Beardsley, 1999; Cheng *et al.*, 2007; Susick *et al.*, 2002). Bone marrow and cord blood transplants, which represent the first forms of progenitor cell therapies, have been performed for years (Gluckman *et al.*, 1989; Rubinstein *et al.*, 1993). More recently, other forms of progenitor cell therapies are being tested in clinical trials; these include mesenchymal progenitor cells (Bruder *et al.*, 1994; Caplan, 1994; Dean and Moseley, 2000), neuronal progenitor cells (Cattaneo and McKay, 1991; Gage, 1994), and fetal pancreatic islet cell transplants (Reinholt *et al.*, 1998; Shumakov *et al.*, 1983). The early data from these trials are very encouraging for the future of progenitor cell therapies as a class.

The problems with determined stem cells include: (a) availability of tissues, a particular problem for organs that until now have derived only from brain-dead, but beating-heart donors; (b) the need for the development of purification schemes for isolation of the cells; (c) identification of optimal cryopreservation conditions; and (d) definition of the *ex vivo* expansion and differentiation conditions.

IV. Liver Processing

A. General Comments

Liver cells can be isolated readily using a standard perfusion process (Berry and Friend, 1969; Freshney, 2000). The method has many variations but in all its forms, it involves (a) anesthetizing an animal or obtaining an isolated liver or portion of a liver (human tissue), (b) catheterizing one or more major blood vessels leading into or from the liver (e.g. portal vein and/or the vena cava), (c) perfusing for 10–15 min with a chelation buffer containing EGTA to reduce the calcium concentrations in the liver, (d) perfusing with a buffer containing collagenase and a protease(s) (e.g. elastase) in a calcium-containing buffer for 10–15 min (mice, rats) to ~30 min (human) and (e) mechanically dissociating the liver by pressing the digested liver through cheese cloth or raking it with combs or pressing it through sieves (metal, plastic, cloth) of narrowing mesh size.

Enzymes Used in Liver Perfusions

Most liver perfusions are done with collagenase preparations that are partially purified. Different companies indicate the degree of purification of a collagenase preparation with company-specific nomenclature. For example, Sigma indicates the crudest collagenase preparation as type I, and the pure collagenase as type VIII, whereas Boehringer Mannheim refers to the crudest preparation as type A and the pure collagenase as type D. One must read the company's literature to learn the details of the nomenclature and its implications for the extract or purified factor(s) being sold. Generally, the liver perfusions are done with a preparation that is intermediate in purity (e.g. type IV in Sigma's series or type B or C in the Boehringer Mannheim series), since both collagenase and one or more proteases are required for optimal liver digestion. It has been learned that the most effective liver perfusions are achieved with a mixture of purified collagenase and purified elastase at precise ratios (Gill *et al.*, 1995; Olack *et al.*, 1999). A commercial preparation of purified collagenase and elastase, called *"Liberase"* (Boehringer Mannheim), designed for liver digestion is, available. However, its use has been limited due to its high cost. It is the enzyme mix of choice especially for preparations that are to be used clinically.

B. Liver Perfusions in Mice and Rats—Readily Adaptable to Any Mammalian Liver (Other than Human)

Prepare stock solutions a day or two ahead of time. Use highly purified, sterile water such as prepared using a Milli-Q purification system (Millipore) and then sterilize by autoclaving or by filtration through a Nalgene filtration unit. The water so prepared will be referred to as MQ water. The stock solutions are prepared as described in Appendix 2 and an those established by prior investigators (Berry and Friend, 1969; Seglen, 1976)

1. Supplies and Set Up

   1. Autoclave:
      - surgical tools: large forceps, small and large scissors, two hemostats
      - two glass funnels, Erlenmeyer flask (250 ml), two sheets of nytex, thread or suture, swabs (cotton tipped)
   2. Prepare buffers (see Appendix 2): 10× Leffert's, 1 M CaCl$_2$, 0.2 M EGTA, and buffers A, B, and C (see formulations below), culture medium. Isotonic Percoll [Leffert is given as representative of the buffers that can be used]
   3. Culture plates, Petri dishes, 50 cc conical tubes, animal rack, pan to accommodate rack/animal, ice buckets
   4. Fast rat overnight (optional)
   5. O$_2$ (tank), ethanol, collagenase (Sigma type IV), ketamine, bovine serum albumin (BSA), phosphate buffered saline (PBS), trypan blue

2. Perfusion Set Up

   - 42 °C water bath
   - Perfusion pump with stop cock three-way valve on two inlet lines (with 5 ml pipettes for bottles)
   - Single outlet line (to connect to catheter) that goes through the water bath
   - Flow rate at 30 ml/min
   - O$_2$ tank feeding two outlet tubes with screw clamps to control gas flow
   - 37 °C water bath
   - Low speed centrifuge
   - Good light source for surgery

3. Prepare (in Biological or Laminar Flow Hood)

   - Turn on water bath (42 °C). Rinse pump tubing with ethanol for 5 min, dH$_2$O 5 min.
   - *Add collagenase and BSA to buffers B and C.* Aliquot 2 × 50 cc conical tubes of *buffer C* on ice.
   - Put *buffers A and B* in 42 °C bath (weigh down with donuts) and insert lines.
   - Fill inlet with *buffer B* up to stopcock, then fill entire line with *buffer A*, eliminating any bubbles.
   - Insert O$_2$ lines into bottles, adjust to a low flow rate (take care that bubbles form above buffer inlets).

4. Surgery

- Weigh animal and anesthetize with ketamine (200 g rat, dose: 0.2 ml).
- Set up rack (to support animal) and pan (to accommodate rack/animal and collect fluids), instruments, two sutures (or pieces of thread), Petri dish, and five pieces of tape.
- Restrain animal limbs with tape. Open abdominal skin and cut up centerline. Cut laterally and spread.
- Cut through abdominal muscles up centerline to expose lower end of sternum, but do not open chest cavity. Take care not to nick the liver! Cut laterally and spread sides.
- Use sterile swab to move/sweep intestines out via the right side, exposing vena cava and portal vein.
- Attach hemostat to sternum and flip it upward.
- Use a swab to maneuver liver up against the diaphragm.
- Put a suture around inferior vena cava between renal and hepatic veins, and another around portal vein. Loop each for tying.
- Use a hemostat on intestine proximal to portal vein to increase tension, then catheterize portal vein.
- Tie off suture, turn on pump, and attach outlet to catheter.
- Watch liver swell and blanch, and cut inferior vena cava below suture.
- Tape down line to catheter for stability.

5. Perfusion

- Pump *buffer A* for 2 min to rinse liver well (removal of calcium).
- Push liver back down from diaphragm, then cut through diaphragm.
- Cut open the heart, then tie off suture on inferior vena cava. Open chest cavity for drainage.
- Cut small ligaments that attach liver to diaphragm. Push liver back up against diaphragm to increase flow pressure.
- Pump *buffer A* for about 5 min (150 ml).
- Switch to *buffer B*. Flow until liver is well digested (300–500 ml). Liver should swell and then spread out.
- Finally tissue structure breaks down visibly below the capsule. During this time, free liver from attaching membranes.

6. Cell Isolation

- Remove liver to Petri dish and add *buffer C* (room temperature, RT), using ~10–15 ml.
- Comb gently to break capsule, then shake liver with a forcep at the junction of the lobes to dislodge cells. Flip liver over and repeat. Ideally, only white connective tissue will remain. Do not comb resistant areas (often at margins

of lobes) since this will decrease overall viability. If the liver is not completely reduced to connective tissue, proceed with what shakes out.
- Filter cell suspension into glass funnel with a double layer of nytex membrane sitting into an Erlenmeyer flask. Rinse Petri dish with *Buffer C* and add to filtrate. Agitate nytex membrane to increase flow rate, but do not assist from above.
- Blow O_2 into flask gently then cover with parafilm and put into 37 °C bath for 5–10 min (at a slow flow rate, it will take 10 min; at a faster flow rat— 5 min). Swirl flask gently to resuspend cells, then incubate in ice for 10 min.
- Repeat nytex filtration into 50 cc conical tubes on ice in hood.
- Divide filtrate between 2 × 50 cc tubes then spin down at low speed centrifugation (50 × g) for 1 min.
- Aspirate supernatant (which is cloudy). Add 20 ml per tube of cold *buffer C* and gently resuspend cells.
- Spin down cells again for 2 min at 60 × g and resuspend as before. Repeat for third spin, again at 2 min at 60 × g. Resuspend in a *combined* volume of 20 ml.

7. Culture

- Count cells at 1:5 dilution in trypan blue on a hemacytometer.
- Calculate viability and concentration.
- Calculate for plating; plate cells in any of a number of basal media (DME, F12, DME/F12, RPMI 1640, Williams E) available commercially. For plating, the medium is supplemented with 5–10% serum, referred to as serum supplemented medium (SSM). Serum is required briefly (~30 min) for inactivation of enzymes. Thereafter, it can be minimized or even avoided altogether for plating if one uses one of the serum-free, hormonally defined media (HDM) (see tables below) with supplementation (for plating only) of 10–15% knockout serum replacement (Invitrogen).
- Incubate for 1–2 h at 37 °C (the lower the viability, the less the time).
- Remove media, wash with 1× PBS, then add either SSM or a serum-free HDM depending upon experimental needs.
- Incubate in CO_2 tissue culture incubator maintaining cultures with daily media changes.

C. Fetal Human Livers (Livers of Gestational Ages 14–20 Weeks)

1. General Comments

The protocols are those established by Reid and associates (Schmelzer *et al.*, 2006a, 2007; Turner *et al.*, 2006). Fetal livers comprise both hepatopoietic and hemopoietic progenitors. Indeed, in the gestational ages at which erythropoiesis peaks, the vast majority of cells are hemopoietic cells. Therefore, the focus of the protocol is to separate the hemopoietic cells (including progenitors) away from the parenchymal progenitor cells. Most of the hemopoietic cells are free-floating and not bound to each other by cell adhesion molecules or by extracellular matrix molecules. This feature became the focus of the protocol overcoming the fact that the hemopoietic progenitors and hepatic progenitors are similar in size and granularity. The tissue is only partially digested with enzymes allowing the parenchymal cells to remain as clumps that can be centrifuged by very low speed spins, leaving the supernatants to contain the free-floating hemopoietic cells. After debulking the tissue from the hemopoietic cells, the clumps of parenchyma can be subjected to further enzymatic digestions to yield either small aggregates or single cell suspensions optimal for immunoselection for the stem cell and hepatoblasts subpopulations.

2. Buffers

Cell Wash, Kubota's Medium, free fatty acid mixture, digestion buffer. See Appendix 2 for details on buffer preparation.

3. Liver Processing

1. Prepare ice bucket filled with ice and disposal container filled with 10% Clorox.
2. Unpack the liver and spin in the 50 ml Falcon tube for 5 min at 1100 rpm, 4 °C (Sorval RT7 and RTH swinging bucket).
3. Fill out documentation papers for the liver (record age, size, color, tissue integrity, sender's assigned number, and our assigned processing number).
4. Determine the liver volume using the milliliter graduations listed on the Falcon tube. Aspirate supernatant and place the liver into Petri dish.
5. Physically dissociate the liver with forceps and scalpels by lightly scraping the liver, pulling out any part of the biliary tree or connective tissue.
6. Add 25 ml of digestion buffer (per every 3 ml of liver volume) into the Petri dish. Using a pipette, transfer 3 ml liver fractions (which is now a ~25 ml solution) into 50 ml Falcon tubes ("tubes 1").
7. Recap the tube and parafilm the top to prevent leakage.
8. Place in bag and put in 37 °C water bath for 30–60 min, depending on grade of digestion (visual check), and shaking the tubes every 5 min.
9. Remove tube from water bath, place in holder, and remove ("tubes 1") supernatant down to ~7.5 ml. Place supernatant into another Falcon tube ("tubes 2") and keep on ice until the second digestion process is done.

10. Reconstitute the pellet from "tubes 1" with 30 ml of digestion buffer and shake as before in 32 °C water bath for 5–30 min.
11. Spin "tubes 1" and "tubes 2" at 1100 rpm at 4 °C (250 RCF, Sorvall RT7, and RTH swinging bucket).
12. Discard supernatant.
13. Pool pellets (each from tubes 1 and tubes 2, so you have the same amount of tubes as before) and resuspend in Cell Wash Media at 40 ml per tube.
14. Perform a slow spin at 300 rpm for 5 min at 4 °C (20 RCF, Beckman centrifuge GS-6R, and GH-3,7 swinging bucket). Carefully remove supernatant leaving 7.5 ml of media above the pellet.
15. Add Cell Wash to a total volume of 30 ml.
16. Redo steps 14–15 no more than two times (until red blood cells are minimal).
17. Remove supernatant, add 20 ml of Cell Wash solution per tube and agitate gently to resuspend the cells.
18. Filter through 75 µm mesh (nylon, PGC Scientific, Maryland) using pipette and gently pushing cells through mesh. (Mesh is placed on glass funnel that is put in the beaker).
19. Check cell number and viability (trypan blue).
20. Plate at ∼300,000 cells/100 mm tissue culture plate, in Kubota's medium supplemented with 5% fetal bovine serum (FBS) for 30–60 min.
21. Replace the medium with fresh serum-free Kubota's Medium.

D. Liver Perfusions: Neonatal, Pediatric, and Adult Human Livers

1. Sourcing of Human Livers from Postnatal Donors

Pediatric and adult human livers are from brain-dead-but-beating-heart donors, since the donor organ is procured for organ transplantation, and the liver's exquisite sensitivity to ischemia necessitates that the procurement process occur at the moment of death. The organ is removed from the donor and placed into transport buffer (typically University of Wisconsin solution, "UW" solution; also called Viaspan available commercially from UpJohn). In the United States, only 1–2% of the deaths are those who have undergone brain death prior to heart arrest. Thus, the number of donor organs/year is very small, on average ∼5000–6000 per year. Over 95% of these are used successfully for organ transplantation. The remaining 5% of the donor organs, or up to ∼250 livers per year, are livers rejected for organ transplantation for a variety of reasons including infections that result in the liver going to investigators studying that type of infection or ischemia, high percentage of fat, or other conditions resulting in the liver going to diverse academic or industrial investigators. The rejected livers are shunted to federal agencies that handle the distribution process to researchers. These livers, ranging in weight from 1500 to 2500 g, can be shipped as an intact organ to groups of investigators who can afford them or, more commonly due to the costs, shipped as sections of liver, partitioned by federal agency staff members to maximize the number of researchers receiving samples. The sample is shipped to the investigators within ~10–20 h from the time of removal from the donor or the "clamp time." The samples arrive flushed with the transport buffer, bagged and on ice. If one receives a portion of a partitioned liver, one receives a piece that is usually ~100–200 g and that must be perfused through cut blood vessels exposed on the surface of the sample. The conditions prior to death and the cold ischemia associated with the transport conditions of the liver or portion of a liver can result in the deterioration of the sample. Thus, the quality of the starting material is extremely variable.

For donor organs, the overall organ integrity and functions begin to deteriorate at significant rates after 18 h post-clamp; such organs will not be used for transplantation after this time. This cut-off timing for transplantation is under extensive investigation by groups trying to prolong the time, and, therefore, increase the numbers of organs that might be transplanted. In our experience, the quality of the cells prepared from donor organs that have been procured >18–20 h reflect this general phenomenon of deterioration, and lower yields and viability of the polyploid cell populations are observed compared with fresher organs or tissue. We have also observed that, in general, organs received more than 24 h after clamp time often do not yield cells of adequate quality; nor are the cells able to efficiently attach to culture substrata. However, the time threshold after which a particular organ cannot produce cells of adequate quality is affected by several factors including age of the donor, proficiency of organ preservation, the quality of the tissue perfusion, and disease state of the organ (e.g. extent of cirrhosis and steatosis). For the most part, organs should be a uniform tan or light brown color when received; organs that appear "bleached" or dark brown should not be used and generally yield only non-viable or CYP450-depleted cells. The medium containing phenol red in contact with liver cells isolated from damaged organs often has a characteristic pink color, especially when mixed with Percoll®, which is believed to reflect the depletion of certain macromolecules from damaged cells.

Mature parenchymal cells in pediatric and adult livers are very sensitive to ischemia, even cold ischemia, and begin dying soon after cardiac arrest. With every hour after death, more mature liver cells die such that by the latest time points tested, the only cells left are the stem cells and other early progenitors, the subpopulations most tolerant of ischemia (Reid et al., 2000). Although the stem cells can survive many hours, the dying mature cells release lytic enzymes that can damage the stem cells. Empirically, one can find HpSCs and other early progenitors from livers of asystolic donors for hours after death (>10 hours) They are recognizable by their expression of stem cell markers such as epithelial cell adhesion molecule, EpCAM (Schmelzer and Reid, 2008; Schmelzer et al., 2007). The EpCAM+ cells obtained from such livers are viable and will attach and grow in culture if the correct culture conditions are used for them (see below). However, the studies on them to date have been very limited. So, it is unknown if they have the potential to differentiate to fully mature parenchymal cells. Needed are studies defining the extent of ischemia (cold or warm) to which they can be subjected and still leave the stem cells with full differentiation potential.

Neonatal livers are from infants who die within the first year of life. It is not possible to define brain death in a neonate, since the posterior skull of a neonate does not close until 8 weeks and the anterior for up to 18 months after birth. Consequently, neonates can suffer significant brain damage resulting in swelling of the brain and yet recover. For them, death is defined always as cardiac arrest resulting in the fact that neonatal livers are always from asystolic donors. Since the neonatal liver is comprised predominantly of stem cells and progenitors, the entire organ as an organ survives for hours (>10 hours). Therefore, the stem cells/progenitors can survive even longer than those in adult livers given that the extent of mature cells dying is minimal (so, low levels of enzymes released). Consequently, neonatal livers are an ideal source of highly viable parenchymal cells for some hours after death (Reid *et al.*, 2000; Schmelzer *et al.*, 2007). Procurement of neonatal livers by organ procurement organizations (OPOs) began in 2001 through our efforts and continue to be handled by investigators affiliated with a biotechnology company, Vesta Therapeutics (Durham, NC). At present, it is the only company procuring and processing neonatal livers, though surely this will change in the coming years. There are rough estimates that at least one neonate dies on a medical center's neonatal intensive care unit (NICU) every week, and there are many such NICU units within the United States. Even conservative estimates suggest several thousand neonatal deaths/year in the United States, and, at present, only a handful of these neonates have been donors for tissue/organs procured by OPOs. Thus, there is considerable potential for tissue and organs from neonates who have died to become a major new source of high quality tissue for use for both research and clinical programs (Table VI).

2. Processing of Sections of Human Liver Tissue and Liver Cell Isolation

As stated above, the quality of the liver cell preparation is a reflection of the quality of the starting tissue. As such, the best sources of tissue for the isolation of liver cells are freshly resected biopsy samples, freshly preserved donor organs (<12 h from clamp time), and neonatal livers. Of course, these sources of tissue are rarely available to the average academic or industrial investigator. Therefore, strict guidelines must be in place to assure that the quality of the cells is adequate and that proper controls are introduced into every experiment. There are a number of methods described in the literature for the isolation of human liver cells from partial biopsy segments and whole lobes (LeCluyse, 2001; LeCluyse *et al.*, 2000; Li *et al.*, 1992b; Strom *et al.*, 1996). These approaches are essentially modified versions of the original two-step perfusion methods developed by Seglen and others (Berry and Friend, 1969; Seglen, 1976) for the isolation of liver cells from

Table VI
Sourcing of Human Livers

- *Fetal livers (14–20 weeks gestation)*
  - High percentage of stem cells, hepatoblasts (HBs), and committed progenitors
  - Ease in isolation
  - Ability to obtain and use them depends on legal issues and on political and cultural attitudes
- *Liver resections*
  - Neonatal, pediatric, and adult livers
  - Difficult to obtain; highly variable quality of tissue; small amounts
- *Organ donors ("brain-dead but beating heart donors"): cold ischemia*
  - ~1–2% of deaths; ~5000–6000 per year in United States
  - Pediatric and adult livers
  - Most used for transplantation; must compete for the small numbers of rejected livers ~100–200 per year
  - Highly variable quality of tissue
- *Cadaveric livers (asystolic donors): warm and cold ischemia*
  - All neonatal deaths and 98–99% of pediatric and adult deaths
  - Neonatal, pediatric, and adult livers
  - Cannot be used for transplantation, so all are available for research and cell therapy programs
  - Pediatric and adult livers—mature liver cells die within 1–2 hrs of death; stem cells (EpCAM+ cells) survive for >10 hrs but with increasing damage to the stem cells due to enzymes released by dying cells
  - Neonatal livers are ideal since so rich in stem cells and progenitors. Can isolate viable cells from neonatal livers for up to >10 hrs after death.

rat liver. Given below are protocols for processing of sections or chunks of human liver tissue as described by LeCluyse and associates (LeCluyse, 2001; Hamilton et al., 2001).

a. Buffers/Reagents/Supplies

See Appendix 2 for Kubota's Medium, chelation buffer, collagenase preparation.

Surgical glue: Loctite instant adhesive, Medical Grade (Loctite Corporation, Rocky Hill, Connecticut).

b. Processing

Remove liver sample from transport bag and place in a sterile tray.

1. Add some sterile Kubota's Medium (KM) to liver to keep it wet during the phase of connecting catheters.

2. The liver surface has been cut during the division of the original liver into samples, revealing many exposed blood vessels. Test for candidate blood vessels that will yield perfusion of a reasonable percentage of the liver by catheterizing, pumping buffer through them, and observing if the liver tissue swells and where the fluid emerges.

3. Once a blood vessel has been chosen, insert a catheter that fits snugly into the vessel and glue it into place with surgical glue. Secure the catheter in place by adding more glue at the cannula/tissue interface.

4. Seal all other openings on the cut surface using glue. For the larger openings, it may be necessary to seal them using a cotton-tipped applicator. Either the wooden part only can be used or the cotton tip. The applicator can be reduced to fit in the opening size and the wooden stick should be reduced in length so that no more than a centimeter protrudes. Secure applicators in place by making a glue collar around the edge.

5. Dab dry the cured surface of the liver, and cover the entire cut surface with a thin layer of glue, apply using a sterile cotton tipped applicator.

6. Once the glue has dried, place the liver on a weighing boat and connect perfusion tubing to the catheter and slowly start the perfusion. If no major leaks are observed, carefully place the liver tissue in the perfusion tank containing the chelation buffer. Submerge the liver sample with sufficient buffer so that it floats.

7. Slowly increase the flow rate for the chelation buffer. Perfuse with chelation buffer for 10–15 min. The flow rate and backpressure will vary with the size of the tissue, and how well it is sealed (On average, the initial flow rate and backpressure should be between 25–50 ml/min and 20–40 mm Hg, respectively).

8. Remove chelation buffer and replace with the collagenase buffer. Perfuse with collagenase buffer until liver is digested (~15–30 min). The digestion time will vary depending upon the activity of the collagenase used and the size of the liver. Outflow pressure should not exceed 30–40 mm Hg throughout the perfusion procedure. The back pressure will normally decrease as digestion progresses, and the tissue begins to break down. Do not over digest the liver as this will lead to overt cell damage and symptoms of oxidative stress.

The quality and integrity of the resulting cell material is dependent on a number of factors including the time of perfusion (preferably $\leq$30 min total perfusion time), flow rate and back pressure values, the quality of the starting material, and the quality of the crude collagenase (non-specific protease contamination varies from batch to batch).

9. When the perfusion is complete, remove the liver from the tank and place in a sterile bowl/dish of adequate size and then transfer to a sterile hood.

10. Add sufficient plating medium to cover the liver tissue.

11. Remove the dried capsule of glue and catheter from the surface of the tissue. Using tissue forceps and scissors, gently break open the outer Glisson's capsule. With the aid of the tissue forceps, mechanically dissociate the liver by raking the liver with sterile flea combs (purchased from any veterinary supply) or equivalent instrument and gently release the liver cells into the medium, leaving behind the connective tissue and any undigested material.

12. The digested material in the culture medium is then filtered through a series of Teflon mesh filters:

$$1000 \, \mu m \text{ mesh} \rightarrow 500 \, \mu m \text{ mesh} \rightarrow 100 \, \mu m \text{ mesh}$$

(Use large funnels, and filter in sterile 500–600 ml beakers).

Scrape off any remaining clumps and aggregates from the filters and transfer to a Petri dish. Remove the filter and dispose it. The clumps and aggregates removed from the filter can be rinsed with Leffert's buffer and frozen for biochemical analyses on the freshly isolated cells or they can be rinsed and plated for preparing feeder layers (see below) or discarded.

13. Transfer the cells that passed through the filters into 50–200 ml centrifuge tubes (the size of the centrifuge tubes will vary according to the amount of material), and wash by low speed centrifugation ($70 \times g$ for 4 min).

14. The resulting *pellet is enriched for the polyploid liver cells* that are typically above 22–25 μm in size. Remove and save the supernatant. Gently resuspend the pellet in a small amount of medium. At this point, pellets may be pooled if warranted, and smaller tubes (50 ml) may be used if the initial centrifugation was performed with larger tubes.

15. The *supernatant* can be kept for the isolation of non-parenchymal cells (e.g. hepatic stellate cells) (Kubota *et al.*, 2007), diploid parenchymal cells, and/or HpSCs—in the case of hHpSCs, cells expressing EpCAM (Schmelzer *et al.*, 2007). Spin the supernatant at $200 \times g$ for 4–5 min to pellet some of the remaining cells. Resuspend the pellet in the Kubota's Medium with 5% fetal bovine serum (FBS), count and store appropriately. Spin the supernatant again but now at $300 \times g$. Again resuspend the pellet in the Kubota's Medium with 5% fetal bovine serum (FBS), count the cells and store appropriately. Spin the supernatant at $500 \times g$ for 4–5 min to collect the final remaining cells. Resuspend the pellet in the Kubota's Medium with 5% fetal bovine serum (FBS). With increasing gravitational force, smaller and smaller cells will pellet and will be increasingly enriched for the diploid subpopulation of cells from the liver. The final spin will include cell debris but will also contain progenitors.

16. Resuspend the pellets (step 15; those enriched for the polyploid cells) in culture medium plus 90% isotonic Percoll, the ratio of volumes should be 3 parts culture medium to 1 part Percoll (e.g. 30 ml of cells in medium + 10 ml of 90% isotonic Percoll) (see reagents section for details on Percoll). *Note:* if the liver has a high fat content, then the Percoll must be altered accordingly by reducing the volume of Percoll used.

17. Centrifuge the Percoll suspension at $100 \times g$ for 5 min and then remove the supernatant [*Note:* this supernatant can be treated similarly to the initial supernatant (see step 16) to collect the diploid subpopulations and other smaller cells]. The top layer of the supernatant contains the dead cells; care should be taken not to disrupt this layer or contaminate the pellet with these cells. Gently resuspend the pellet in Kubota's Medium with 5% fetal bovine serum (FBS). Centrifuge for a final time at $70 \times g$ for 3 min.

18. Resuspend the final cell pellet in 4–5 ml of culture medium per ml of packed cell pellet.

19. Place the cell suspension on ice and count the cells in the final suspensions after centrifugation and/or fractionation. Use trypan blue exclusion assay or its equivalent to assess the percentage of viable cells.

3. Processing of Entire Human Livers (See Appendix 2 for Details of Buffer Preparation)

Donor organs rejected from transplant programs can be obtained through Organ Procurement Organizations (OPOs) that are part of United Organ Sharing Organizations (www.Unos.com). The organs are expensive (>$8000–$10,000/organ—prices vary depending on the OPO) given that one must reimburse the OPOs for the time needed for staff members to interface with families of the donors and for the costs of surgical removal of the organ from the donor. The organs rejected from transplant programs are rejected due to reasons that affect significantly the quality of the cells to be isolated and that include ischemia, some abnormality in the vasculature, excessive fat, and evidence of pathogens. Moreover, the need to use as many organs as possible for transplant patients results in prolonged maintenance of organs in transport buffers after removal from the donor such that all of them have some degree of at least cold ischemia. The procedure for isolating cells from whole adult and neonatal livers is similar to that outlined for processing sections or chunks of human livers with the following modifications:

For whole adult livers, the gall bladder is drained and sealed with an umbilical clamp. Umbilical clamps are used to seal the vena cava on the inferior side of the liver (i.e. same side as the gall bladder) as well. The appropriate-sized catheters (depends on the size of the liver) are inserted and sutured into both the portal vein and the hepatic artery. Between 500 and 1000 ml of UW solution, also called ViaSpan^{TM}, (DuPont Pharmaceuticals, Delaware), is perfused by hand, using a 60-ml syringe, through the catheters to flush any blood remaining within the organ out of the vena cava on the superior side of the liver. The organ is flushed until the UW buffer is clear. Then the vena cava is sealed with an umbilical clamp. The organ is then placed into the perfusion tank. Three liters of chelation buffer are required for the 10–15 min chelation perfusion step with flow rates of ~100 and 80 ml/min for the portal vein and hepatic infusion routes, respectively. The flow rate and backpressure will vary with the size of the tissue, and can be adjusted accordingly to prevent backpressure. The digestion perfusion step will require 3 L of digestion buffer. After the digestion (~30 min), the liver is perfused with 1 L of plating media lacking phenol red and containing 5% fetal bovine serum (FBS) for 5 min. Then steps 9–19 of the liver section/liver chunk processing protocol are followed as described above.

For neonatal livers, the gall bladder is sealed with a hemoclip while the vena cava on the inferior side of the liver is sealed with an umbilical clamp. A catheter is inserted into the vena cava on the superior side of the liver and fixed in place using surgical glue. All perfusion steps are performed in a tissue culture hood at RT with all buffers prewarmed to 37 °C. Approximately, 250–500 ml of UW solution is perfused by hand, using a 60-ml syringe, through the catheter to flush any blood remaining within the organ. The organ is flushed until the UW becomes clear. The portal vein and hepatic artery are then clamped using hemoclips. The organ is placed into sterile bowl then perfused by hand through the catheter, using a 60-ml syringe, with 1 L of prewarmed chelation buffer for 15 min. 1 L of digestion buffer is then perfused by hand through the catheter for 10–30 min; digestion is stopped when the cells begin to dissociate. After digestion, the liver is perfused with 1 L of plating media lacking phenol red containing 5% fetal bovine serum (FBS) for 5 min. Steps 9–19 of the liver section/liver chunk processing protocol are followed as described above.

V. Fractionation of Liver Cell Subpopulations

Liver cells have properties that are maturational lineage dependent. One can fractionate the cell suspensions into subpopulations with more uniform properties with regard to any of these properties. Some of the phenotypic properties for which fractionation has proven especially useful include those listed below.

A. Analysis of Ploidy

Ploidy is a critical variable for many of the functions of cells under consideration. Polyploid cells have less growth potential than do the diploid cells but express critical genes such as the mature forms of CYP450; conversely, the diploid cells have the greatest growth potential but express an overlapping but somewhat distinct set of genes from those expressed by the polyploid cells. Therefore, different experiments may demand cells of distinct ploidy. Ploidy can be analyzed on viable cells and fixed cells, whole cells, or isolated nuclei. The dye used to stain viable cells is Hoechst 33342 (Molecular Probes), and its use has been especially well characterized in lymphocytes. It is membrane permeable and DNA specific. A dye that works well with fixed cells is propidium iodide (Sigma). Analysis of liver cells is more complicated than for lymphocytes, since so many of the cells are polyploid, and many are multinucleated. For example, flow cytometric analyses and sorts cannot distinguish between mononucleated tetraploid cells and binucleated cells in which each nucleus is diploid. Also, analyses of liver cells are dramatically influenced by the quality of cell preparation.

The actual analysis of the cells, whether with Hoechst 33342 dye or propidium iodide, can be by flow cytometer, fluorescence microscopy, confocal microscope, or automated fluorescence image cytometer. The Hoechst dye has a maximal excitation wavelength at 350 nm and emission wavelength at 461 nm. The propidium iodide has a maximal excitation wavelength at 535 nm and emission wavelength at 617 nm.

The most accurate measurements of ploidy are using flow cytometry and automated fluorescence image cytometer. Obviously, morphology in combination with ploidy can be done only by fluorescence microscopy and confocal microscopy and is the only way to distinguish binucleated cells from mononucleated cells.

a. Method for fixed cells

- Use a single cell suspension as obtained by liver perfusion.
- Fix cells with methanol:acetic acid (ratio of 3:1) for 5 min.
- Wash with PBS 3 × 5 min each.
- Digest cells with RNase A (50 µg/ml, Sigma) in PBS for 30 min. at 37 °C.
- Add propidium iodide (Sigma) to a final concentration of 50 µg/ml.
- DNA content of cells is then analyzed by flow cytometry (e.g. FACScan).

b. Method for isolation of nuclei from freshly isolated cells

- Freshly isolated cells are washed twice with 5.0 ml of cold saline GM (g/l; glucose 1.1; NaCl 8.0; KCl 0.4; $Na_2HPO_4 \cdot 12H_2O$ 0.39; KH_2PO_4 0.15) containing 0.5 mM EDTA for chelating free calcium and magnesium ions.
- Cell pellets are then resuspended in 4.0 ml RSB swelling solution, containing 10 mM NaCl, 1.5 mM $MgCl_2$, 10 mM Tris (pH 7.4), and 0.5 mM EDTA.
- The cell pellets are then vortexed vigorously for 30 s and allowed to stand on ice for 5 min.
- After addition of 0.5 ml of 10% Nonidet P-40 (NP-40) detergent, the samples are vortexed again vigorously for 30 s and put on ice for 30–60 min.
- The samples are then centrifuged at 1000 rpm and the pellets are then washed with PBS twice.
- Check with a microscope; the pellets should be purified nuclei with little cytoplasmic contamination (Crissman and Hirons, 1994).

c. Ploidy analysis on nuclei

- Wash nuclei with PBS 3 × 5 min.
- Digest cells with RNase A (50 µg/ml, Sigma) in PBS for 30 min at 37 °C.
- Add propidium iodide (Sigma) to a final concentration of 50 µg/ml.
- DNA content of the nuclei is then analyzed by flow cytometry (e.g. FACScan analysis).

d. Ploidy analysis on viable cells

- Prepare freshly isolated cells.
- Stain cells with a solution of 5 µg/ml of Hoechst 33342 (Molecular Probes) in RPMI 1640 (GIBCO) supplemented with FBS (2–10% depending on cells and experimental plans; Hyclone), EGTA (0.5 mM), DNase (0.006%) (Sigma).

- Stain at 37 °C for 30 min.
- Put on ice and keep at 4 °C until analysis.
- Analyze by flow cytometry or by fluorescence microscopy.

B. Size

Cells are sieved through filters to enrich for cells of a desired size. For example, filter through a 20 μm filter to obtain the "small hepatocytes", cells averaging 17–18 μm in diameter and that are known for their ability to form colonies in culture (Ito *et al.*, 1996; Mitaka *et al.*, 1992).

Fractionation of Adult Liver Cells That Can Be Expanded at Low Seeding Densities

1. Prepare freshly isolated liver cell suspension.
2. Filter the cells through a 30-μm sieve to remove large cells and aggregated cells. Cellular viability should be >90% as measured by trypan blue exclusion.
3. If one uses the cells that pass through the 30 μm sieve to seed the plates, the efficiency of colony formation is ∼200 colonies per 1000 cells seeded or about 20%.
4. If higher efficiencies of colony formation are desired, filter the cells again through sieves (or fractionate with density gradients) and use cells less than 18 μm in diameter. This yields the so-called "small hepatocyte" fraction.
5. Plate the cells onto STO feeders and into the serum-free HDM for expansion.
6. Under the conditions specified, colonies of adult cells form at a lower growth rate than for the HpSCs but are easily visible within a week of culture. By about 20 days of culture, the adult colony-forming cell can generate 130 daughter cells (by contrast, the stem cells in the same time frame will yield a colony with 3000–4000 cells).

C. Cell Density

- Layer over Ficoll-Paque (Pharmacia Biotech) or Percoll (Sigma) and follow company instructions for their use. Collect the cells of desired density, rinse, and suspend in the plating medium.
- Centrifugal elutriation: cells are fractionated very gently by centrifuging them in a special rotor in which the buffer in which the cells are suspended is pumped upwards while the centrifuge is spinning. This yields two opposite forces: the centrifugal force in opposition to the direction of the fluid force and when carefully calibrated can yield cell fractions of very precise cell densities. See elsewhere for these procedures as used for liver cells, for example (Overturf *et al.*, 1999).

D. Antigenic Properties

One can immunoselect cells using panning, flow cytometry, columns, magnetic beads, and so on to isolate the cells having a desired antigenic profile (e.g. cells expressing connexin 32). Three examples are given: (1) immunoselection to eliminate red blood cells by panning; (2) immunoselection by flow cytometry to isolate purified rodent hepatoblasts; (3) immunoselection by magnetic bead selection for hHpSCs.

1. Immunoselection to Eliminate Specific Cell Types

Elimination of red blood cells is given as an example.

- Coat 150 mm *Petri dishes* (not tissue culture dishes) with species-specific monoclonal antibody to rabbit anti-rat red blood cell IgG —Let sit overnight at 4 °C—(100 µl per plate + 10 ml media to coat each plate).
- Keep buffer on plate to keep from drying out—BUT remove before adding cells to be panned.
- Add cells to the panning plates at 4 °C. After 5 min, gently transfer non-attached cells to 50 ml sterile tube. Repeat the process 2–4 times depending on the extent of contamination by the cells to be eliminated. For example, rat livers at embryonic days 16–18 are replete with red blood cells. One must pan them 4–5 times to eliminate the large numbers of red blood cells present. Fewer rounds (once or twice) of panning are required for eliminating red blood cells from adult tissues prepared by perfusion.
- Take a count of remaining cells in solution to establish how many cells you have and to arrive at the correct cell density for plating; check dishes to assess non-specific losses.

2. Use of Immunoselection to Purify Rodent HBs

Methods for identification and purification of hepatic progenitors have been developed in rodent systems using multiparametric flow cytometry in combination with multiple fluoroprobe-labeled monoclonal antibodies to purify cells of a defined antigenic profile (Kubota and Reid, 2000; Sigal *et al.*, 1994). Profile for immunoselection of rodent HBs: *Negative for CD45, CD34, and class I MHC antigens; positive for ICAM1 and CD44H; forward scatter: ~10 µm; low side scatter or granularity.*

Even when unique antigens are not known, one can enrich significantly for the cells of interest by doing a "negative sort" using fluoroprobe-labeled antibodies to markers on cells not of interest and then separating the population into those expressing those markers and those not. Secondarily, one can use side scatter, a measure of cytoplasmic complexity caused by mitochondria, ribosomes, and so on. The less mature cells are "agranular" or lower in granularity, whereas the more mature cells are more granular enabling one to enrich for cell populations of given granularity.

1. Rat Embryonic livers of a given gestational stage (e.g. E13 fetal livers) are isolated and digested with 800 U/ml collagenase (Sigma) and 20 U/ml thermolysin (Sigma) followed by further digestion with Trypsin-EDTA solution (Sigma).
2. Treat the cell suspension with 200 U/ml DNase I (Sigma).
3. The cell suspension should have a cell viability >90% by trypan blue exclusion.
4. Purification of the hepatoblast is by multiparametric flow cytometry as given below.
5. Cells are stained with monoclonal antibodies (mAbs) and sorted using a flow cytometer such as Cytomation's MoFlo or Becton Dickenson FACStarplus.
   - Block background staining with 20% goat serum and 1% teleostean gelatin (Sigma).
   - Stain cells with a monoclonal antibody of the class Ia antigen such as anti-RT1A^{a,b,l} B5 (Pharmingen) for Fisher 344 strain of rats, anti-RT1A OX18 (Pharmingen), anti-rat ICAM-1 1A29 (Pharmingen). (*Note: the antibodies must be chosen with respect to the species and strain of animal being used as the source of the liver tissue; this is especially true for identifying the class Ia and class Ib MHC antigens; for rats of the Fisher 344 strain, the antibody is against RT1A^{a,b,l} B5, and an antibody identifying both class Ia and class Ib antigen in this species and strain is OX 18*).
   - Sort for cells that are RT1A$^-$, OX18^{dull}, ICAM-1$^+$, side scatter^{high}. *Note:* the use of high side scatter is relative to that of the major contaminant in fetal livers: hemopoietic cells. In general, the parenchymal cell lineage has a higher level of side scatter or granularity than that of the hemopoietic subpopulations. By contrast, the HpSCs are relatively agranular or have low side scatter in comparison to that in mature parenchymal cells but have higher side scatter than the hemopoietic subpopulations.
   - The sorted cells should be plated on Mitomycin C-treated, mouse, embryonic stromal feeders (STO cells) and in Kubota's Medium (see Appendix 2).

3. Immunoselection of hHpSCs

Immunoselection can be done by any extant immunoselection technology (panning, beads, flow cytometry, etc.) (Schmelzer et al., 2006a, 2007). However, it is quite important to remember that the human progenitor cells require specific buffer conditions for survival, are *very* sensitive to shear forces, and present significant biohazards for the investigators doing the sorts due to the formation of aerosols during flow cytometric procedures. Therefore, one can do flow cytometry only in facilities with appropriate protective equipment and in ones willing to accommodate the need for specific buffer conditions. In our early studies, we had hoped to use the robust and precise immunoselection possible by multiparametric flow cytometry technologies, as had been used with rodent cells only to realize that few cells survived the process.

With respect to buffers, the immunoselection should take place in buffers optimal for the cells and with the generic requirements of nutrients (e.g. a basal medium), low levels of calcium (0.3–0.4 mM), lipids (mixture of free fatty acids bound to purified albumin—see Appendix 2), and insulin (∼1–5 µg/ml). The cells die rapidly in PBS, the most common buffer used as a sheath fluid in flow cytometers. Therefore, if flow cytometry is used, one must change the shealth fluid to a buffer permitting survival of the cells. We have found Kubota's Medium (Appendix 2) (Kubota and Reid, 2000) to be ideal.

The shear forces inherent in flow cytometry are harmful to the hHpSCs and hHBs. We have not done rigorous studies to define precisely the shear forces the stem cells will tolerate, since we converted to magnetic bead technologies to avoid both this problem and the limited access in our area to cytometers that might be used with viable human cells (requires special equipment for control of biohazards from aerosols). Only Dr. Eric Lagasse and his former associates at StemCells Inc. (Menlo Park, CA) have used flow cytometry successfully to isolate hHpSCs and hHBs. Their methods are published in a patent just issued (Lagasse, 2007). They also note the sensitivity of the human hepatic progenitors to shear forces and were successful only when they made use of a large cytometer nozzle (200 µm) and flow speeds under 2000 cells/s.

An alternative is to make use of magnetic bead technologies that consist of magnetic beads onto which are bound antibodies to specific antigens; the beads with antibodies are added to the cells, and the cells are fractionated into subpopulations that have the antigen versus those that do not. The purity of the cells with respect to the antigen being used for selection is nowhere near that achieved by flow cytometry. Yet the ease of use, the speed of the fractionation (minutes), the gentleness of the process, and the ability to use buffers optimal for the cells makes this immunoselection technology very economical and advantageous for most experimental needs. An especially useful version of magnetic bead technologies is that made available commercially from Miltenyi Biotec (Auburn, CA; www.miltenyibiotec.com) that uses microbeads (∼50 nm) superparamagnetic particles that are coupled to specific antibodies. Fractionation of the cells is done by subjecting the cells to a magnetic field after pouring them into columns of varying sizes from Mini-MACs (∼10^7 cells), MidiMACs (10^8 cells), AutoMACs (2×10^8 cells), and up to the SuperMACS (10^9 cells).

Below we describe fractionation of suspensions of human fetal liver cells using the Mini- and MidiMACS systems from Miltenyi™ (Miltenyi Biotec, Auburn, CA). The protocol was used for a publication on phenotypes of hepatic progenitors (Schmelzer et al., 2006a). General protocols and a detailed explanation of the principle of magnetic separation are available at the company's web page (www.miltenyi.com). In general, cells can be selected for the expression of a surface molecule (positive enrichment) or cell populations can be depleted of cells expressing a certain antigen (depletion). Further subselection of both, the positive or negative population, is possible by doing a second round using a different antibody.

Single cell human liver suspensions were obtained as described above. To avoid re-aggregation of single cells to clusters we added accutase (Innovative Cell Technologies, San Diego, CA), a mild enzyme, during some steps of the separation procedure. Volumes are given for the use of Miltenyi "MS" mini columns (cat. #130–042–201) but can be scaled up easily for the use of "LS" midi or larger columns. The given protocol is for the selection of human hepatic progenitors for those expressing EpCAM (CD326). As an example of how one can do multiple, successive separations, we describe the fractionation of suspensions of human fetal livers into four populations using EpCAM and ICAM (CD54). Thus, we obtain four populations: EpCAM+ICAM+, EpCAM−ICAM+, EpCAM+ICAM−, and EpCAM−ICAM−. If selection is done for only one antigen, cells can be selected using microbeads to which are directly coupled EpCAM (CD326) antibody (Miltenyi, cat. #130–061–101) or a primary EpCAM antibody which will be targeted by microbeads directly via IgG- or fluorochrome-mediated binding, thus, there is no need to implement a multisort kit for a single sort.

1. Prepare a cell suspension as described above for human fetal livers.
2. Count cell numbers. Do not use more than 10×10^6 cells per 500 μl volume and per mini-column.
3. Prepare separation buffer:
   a. For very short fractionation processes (ones taking a few minutes—therefore, a single antigen sort), one can use PBS containing 0.5% BSA and 2 mM EDTA (Sigma, St. Louis, MO). As soon as the fractionation process is completed, transfer immediately into Kubota's Medium.
   b. For fractionation processes taking more than 5 min, one can use modified Kubota's Medium with no calcium and with 2 mM EDTA. As soon as the fractionation is completed, the cells must be placed into Kubota's Medium.
4. Incubate cells in 10% accutase (Innovative Cell Technologies, San Diego, CA) in separation buffer at RT for about 15 min until cells are in single cell suspension.
5. Pellet cells by centrifugation (1100 rpm, 5 min, 4 °C).
6. Re-suspend cells in 500 μl separation buffer.
7. Stain cells with primary antibodies; incubate cells in the dark at 4 °C:
   a. Add 20 μl anti-ICAM PE antibody (Becton Dickinson PharMingen, cat. #55511) per 1×10^6 cells in 500 μl buffer, mix well, incubate for 30 min.
   b. After 10 min add 50 μl anti-EpCAM FITC antibody per 10×10^6 cells in 500 μl buffer, mix well, incubate for 20 min.
8. Wash cells by adding 10 ml separation buffer containing 10% accutase.
9. Centrifuge (1100 rpm, 5 min, 4 °C), discard supernatant completely.

10. Resuspend cells in 90 μl buffer and add 10 μl anti-FITC multisort microbeads (use anti-FITC multisort kit from Miltenyi, cat. #130-058-701).
11. Mix well and incubate for 20 min at 4 °C.
12. Wash cells by adding 10 ml separation buffer.
13. Centrifuge (1100 rpm, 5 min, 4 °C), discard supernatant completely.
14. Apply 500 μl separation buffer (without cells) on a MS column which is placed in a magnetic separator (Miltenyi, MiniMACS cat. #130-090-312, or OctoMACS cat. #130-042-109) and let run through completely, discard effluent, and use new collection tube.
15. Resuspend cells in 500 μl separation buffer containing 10% accutase, pipette cells onto column.
16. Collect effluent as negative EpCAM fraction; wash column three times with 500 μl separation buffer (also collect to negative fraction).
17. Remove column from magnetic separator.
18. Apply 1 ml separation buffer on the column, flush out EpCAM+ fraction using the plunger, discard column.

Further subselection of the negative and positive EpCAM cell fractions can be achieved. For the negative EpCAM fraction go on as follows:

19. Wash cells by adding 10 ml separation buffer containing 10% accutase.
20. Centrifuge (1100 rpm, 5 min, 4 °C), discard supernatant completely.
21. Resuspend cell pellet in 80 μl separation buffer per 10×10^6 total cells. For fewer cells, use same volume, for more increase volume accordingly.
22. Add 20 μl MACS Anti-PE MicroBeads (Miltenyi, cat. #130-048-801) per 10×10^6 total cells, for fewer cells use same volume.
23. Mix well and incubate for 20 min at 4 °C.
24. Wash cells carefully by adding 10 ml separation buffer.
25. Centrifuge (1100 rpm, 5 min, 4 °C), discard supernatant completely.
26. Resuspend cell pellet in 500 μl separation buffer.
27. Apply 500 μl separation buffer without cells on a new mini column placed in magnetic separator and let buffer run trough, discard effluent, and use new collection tube.
28. Apply cells onto the column.
29. Wash column three times with 500 μl separation buffer.
30. Collect cells and flow through from washes as negative ICAM fraction (which is also EpCAM negative).
31. Remove column from magnetic separator.
32. Apply 1 ml separation buffer on the column, flush out cells using the plunger to remove the positive ICAM fraction (which is also EpCAM negative).

To subselect the positive EpCAM fraction go on as follows:

33. Incubate cells in separation buffer with 20 µl per 1 ml MACS multisort release reagent for 15 min at 4 °C.
34. Wash cells carefully by adding 10 ml separation buffer containing accutase.
35. Centrifuge (1100 rpm, 5 min, 4 °C), discard supernatant completely.
36. Resuspend cell pellet in 50 µl separation buffer, add 30 µl MACS multisort stop reagent, mix well.
37. Add 20 µl anti-PE MicroBeads per 10×10^6 total cells, for fewer cells use same volume.
38. Mix well and incubate for 20 min at 4 °C.
39. Wash cells carefully by adding 10 ml separation buffer.
40. Centrifuge (1100 rpm, 5 min, 4 °C), discard supernatant completely.
41. Resuspend cell pellet in 500 µl separation buffer per 10×10^6 total cells.
42. Apply 500 µl separation buffer without cells on a mini column placed in magnetic separator and let run through, discard effluent, and use new collection tube.
43. Apply cells onto the column.
44. Wash column three times with 500 µl buffer.
45. Collect cells and flow through from washes as negative ICAM fraction (that is also positive for EpCAM).
46. Remove column from separator.
47. Apply 1 ml buffer on the column, flush out cells using the plunger to remove positive ICAM fraction (that is also positive for EpCAM).

VI. Feeder Cells

Cocultures with mesenchymal feeder cells facilitate survival, *ex vivo* expansion, and/or differentiation of hepatic parenchymal cells (Bhatia *et al.*, 1996; Gebhardt *et al.*, 1996; Mizuguchi *et al.*, 2001; Reid and Jefferson, 1984). The type of feeder influences the behavior (growth, gene expression) of the parenchyma. For example, the feeders of cells with properties associated with endothelia (e.g. human umbilical vein embryonic endothelial cell, hUVECs) influence the parenchyma distinctly from those that are stroma (e.g. 3T3, STO cells) (Kubota and Reid, 2000). Shown below is the protocol for preparation of stromal or endothelial cell feeders. Primary cultures used as feeders are, by far, the most biologically active, but must be freshly prepared and are useful, typically, for only 7–10 days.

A. Preparation of Mesenchymal Feeders from Fetal Livers

Fetal liver-derived stromal feeders work for colony formation of rodent hepatic progenitors but have been tested only in high density cocultures where 50,000–100,000 hepatic progenitor cells seeded onto matrix-coated transwell are placed over feeders in 24-well dishes or in 35 mm dishes (Reid *et al.*, 1994; Sigal *et al.*, 1994). They have not been tested in clonogenic assays. They are more demanding in their preparations than STO cells but possibly may offer some influence to the cells that are distinct from that by the STO cells or other feeder cell lines.

a. For rodent cultures, use rodent fetal livers from gestational ages 14 to 16 (E14–E16). Earlier embryonic ages also yield active signals but are impractical, since the liver is so tiny as to be difficult to remove surgically, and the number of cells in the liver is small. Thus, E14–E16 rodent fetal livers are sufficiently large to yield a number of feeder cultures and still produce the critical paracrine signals. A conditioned medium prepared with mesenchymal feeders in Kubota's Medium described below works well as long as the stem cells are seeded onto a porous surface (e.g. a transwell) coated with an embryonic matrix substratum (type IV collagen and laminin). Stromal feeders prepared from livers past embryonic day 17 but before E19 work also but tend to yield lower numbers of colonies and yield a conditioned medium that is variably successful. Feeders from neonatal or adult rodent livers give little to no biological activity.

b. Isolate 10–20 embryonic livers of the requisite gestational age (E14–E16) and digest the tissue with 800 U/ml collagenase (Sigma) and 20 U/ml thermolysin (Sigma) followed by further digestion with Trypsin-EDTA solution (Sigma).

c. Plate the unfractionated cell suspension onto tissue culture dishes (35 or 60 mm) at a seeding density of 10^6 cells/60 mm dish and into a basal medium (e.g. RPMI 1640; GIBCO/BRL) to which is added 5% FBS (GIBCO/BRL), 5 µg/ml insulin (Sigma), 10 µg/ml iron-saturated transferrin (Sigma), 4.4×10^{-3} M nicotinamide (Sigma), 0.2% BSA (Sigma), 5×10^{-5} M 2-mercaptoethanol (Sigma), 7.6 µEq/l free fatty acid, 2×10^{-3} M glutamine (GIBCO/BRL), 3×10^{-8} M Na_2SeO_3 and antibiotics.

d. Culture the cells, with daily medium changes, for ~1 week to permit stromal cells to expand.

e. Disperse the attached cells with trypsin-EDTA or with collagenase, rinse and replate at 200,000 cells/well of a 24-well plate or at 5×10^5 cells/35 mm dish or at 10^6 cells/60 mm dish.

f. Add the plating medium described above and culture cells overnight to permit the cells to attach and spread.

g. Progenitor cell populations can be seeded directly onto the feeders if the feeders are treated with Mitomycin C (as described above) or can be seeded on transwells coated with type IV collagen and laminin and placed over the feeders.

h. Once the progenitors are added, whether directly to the feeders or via transwells, culture the feeders and the progenitor cells in Kubota's Medium (Kubota and Reid, 2000) (refer to Appendix 2).

i. Change the medium daily or every other day with freshly prepared medium.

B. Preparation of Feeders from STO Cells: A Murine Embryonic Cell Line

STO cells are murine embryonic stromal cells and have been established as a cloned cell line from mouse SIM (Sandos Inbred Mice) embryonic fibroblasts (Martin and Evans, 1975). One can obtain stocks from the American Tissue Culture Collection, ATCC. Overall the procedure is similar to the original protocol described in the references at the ATCC and in our publication (Kubota and Reid, 2000). It should be noted that STO feeders are supportive of extensive expansion of rodent HBs but cause slowing of growth and differentiation with hHpSCs and hHBs (Schmelzer et al., 2007). For rodent HBs, we found several modifications useful. A subclone, STO5, was isolated from the parent STO cells, and found to have superior ability to maintain hepatic progenitors *in vitro* (Kubota and Reid, 2000). STO5 cells are routinely cultured in DMEM/F12 supplemented with 7.5% FBS and 1% DMSO on 10 cm tissue culture dishes. When confluent, the cells are passaged by trypsin-EDTA treatment and diluted 1:3 into fresh plates. Ideally, a confluent 10 cm dish of STO5 cells should yield $\sim 2 \times 10^6$ –4×10^6 cells.

1. Preparation of Cryopreserved STO5 Feeder Stocks

   1. Remove medium from confluent 100 mm plates of STO5 cells.
   2. Add 4 ml of 5 µg/ml Mitomycin C-containing medium per 100 mm plate.
   3. Incubate the plates for between 3 and 4 h.
   4. Aspirate the Mitomycin C medium from the STO5 cells. Wash each plate twice with 5 ml of HBSS.
   5. Trypsinize the cells and collect the cells in DMEM/F12 containing 7.5% FBS.
   6. Spin down the cells at 1200 rpm for 5 min. Resuspend the cells in DMEM/F12 plus 7.5% FBS to prepare a cell suspension of 6×10^6 cells/ml.
   7. Add an equal volume of cryopreservation buffer (cryopreservation buffer stock: 80% FBS + 10% DMSO) dropwise on ice to make a final concentration of 40% FBS, 10% DMSO.
   8. Dispense the cells at 1 ml/cryotube (3×10^6 cells/tube).
   9. Store the cells at –80 °C.

The vial contains 2 mg Mitomycin C and 48 mg NaCl.
Dissolve the 2 mg of Mitomycin C in 10 ml of distilled water.
Aliquot 0.5 ml into 20 cryotubes (100 µg/tube).
Stored at –20 or –80 °C until use.

　*The basic protocol of SIGMA recommends 10 µg/ml of Mitomycin C (Sigma; Catalog M-4287).

2. Preparation of STO5 Feeder Layers

1. Coat tissue culture plates with a 0.1% (w/v) solution of gelatin (SIGMA G-1890) in water. Sterilize the gelatin solution by autoclaving. Preheat (at 65 °C) the gelatin solution by microwaving before coating the plates.
2. Flood the plates with a few milliliters of the gelatin solution and leave for at least 10 min at RT. Aspirate.
3. Take one tube of frozen Mitomycin C-treated STO5 and thaw at 37 °C. Transfer the 1 ml cell suspension into a conical tube on ice. Add 9 ml of DMEM/F12 containing 7.5% FBS dropwise.
4. Dispense the suspension into the gelatin-coated plates at a dilution of ~4×10^4 cells/cm^2.
5. Change the medium to one containing FBS. Ideally, the feeder layers should be used within 3–5 days after their preparation. However, they last for quite a long time as a monolayer, so they can be used for up to 10 days. Aspirate the medium and rinse with HBSS twice prior to use.

C. Feeders of hUVECs Feeders

The hHpSCs expand on feeders of hUVECs and remain undifferentiated or minimally differentiated as long as serum is avoided.

a. Stocks of hUVECs can be obtained from several companies: PromoCell, Astarte Biologics, and AdvanCell.
b. Stocks of hUVECs can be plated onto tissue culture dishes in endothelial growth medium (EGM-2) plus 1–2% FBS (Cambrex; Walkersville, MD) and used as a feeder when confluent.
c. The feeders must be rinsed with serum-free Kubota's Medium repeatedly to eliminate serum residues when the hHpSCs are plated onto them.

D. Replacement of Feeders with Defined Paracrine Signals

Different embryonic feeders elicit different effects with specific stem cell populations. Some feeders are permissive for self-replication of the stem cells; some induce lineage restriction but with maintenance of growth; some induce extensive differentiation. Clearly, the reasons for the biological variations in effects are the paracrine signals (matrix and/or soluble ones) being produced by the feeders. Efforts to define those signals are ongoing. Thus far, the known soluble signals from feeders include leukemia inhibitory factor (LIF), a form of fibroblast growth factor (FGF), hedgehog proteins (Sonic, Indian), regulators of telomerase and of notch proteins, hepatocyte growth factor (HGF), and stromal-derived factor (SDF) (Fleming, 1998; Hernandez *et al.*, 1992; Hoffmann and Paul, 1990; Kubota *et al.*, 2007; Lewis, 1998; Malik *et al.*, 2002; Schmelzer *et al.*, 2007; Sicklick *et al.*, 2005; Walker *et al.*, 1999).

A major challenge of current investigators that should be met within the near future is the complete defining of matrix and soluble signals to enable feeder-free cultures of stem cells. Such an achievement will enable one to regulate stem cells precisely.

VII. Extracellular Matrix

Extracellular matrix is essential, especially for normal cells to survive and function (Brill *et al.*, 1993, 1994; Furthmayr, 1993). The chemistry and physical features of extracellular matrix are known to regulate gene expression and influence cell morphology in all tissues including liver (Brill *et al.*, 1994; Fujita *et al.*, 1986; Maher and Bissell, 1993; Reid *et al.*, 1992; Runge *et al.*, 1997; Suzuki *et al.*, 2003; Zvibel *et al.*, 1991). The extracellular matrix is a complex mixture of molecules between and around cells made insoluble by crosslinking. Excellent reviews are available on its chemistry and functions (Hamilton *et al.*, 2001; Maher and Bissell, 1993; Martinez-Hernandez and Amenta, 1993c).

It is important to note that there are many types of matrices with distinct chemical composition. Each cell type secretes and is associated with a specific type of matrix. Furthermore, the matrix chemistry changes when the cell is growing or quiescent or when it is in some pathological condition (Enat *et al.*, 1984; Reid, 1993). Thus, in order for the cell type of interest to mimic its *in vivo* counterpart in a specific physiological or pathological state, the associated matrix chemistry must be identified. One is helped by the understanding that there is a paradigm to how all types of extracellular matrix are made, and by the numerous studies identifying the matrix chemistry in various tissues.

A. Paradigms in the Construction of Extracellular Matrix

All cells produce an extracellular matrix, and the extracellular matrix inbetween any given set of cells contains components derived from all the cell types in contact with the matrix. The matrix components present on the lateral borders between homotypic cells includes cell adhesion molecules or "CAMs" (Fujimoto *et al.*, 2001; Stamatoglou and Hughes, 1994; Terada *et al.*, 1998) and proteoglycans (Bernfield *et al.*, 1992; Ruoslahti and Yamaguchi, 1991; Stow *et al.*, 1985; Vongchan *et al.*, 2005) and will be referred to as the "lateral matrix". That between the epithelium and a mesenchymal cell partner will be referred to as the basal matrix. It consists of basal adhesion molecules (laminins, fibronectins) that bind the cells to a collagen scaffolding; in addition there are other matrix molecules such as proteoglycans that are attached in many ways to the plasma membrane surface, to the basal adhesion molecules, and/or to the collagens.

Although the matrix chemistry is unique in each tissue, there are some components that are present in all matrices, and the chemistry is dependent upon the location within the liver acinus and with the maturational stage of the cells (Gressner and Vasel, 1985; Reid et al., 1992; Vongchan et al., 2005; McClelland et al., 2008). For example, the periportal zone of the liver acinus is comprised of type III and IV collagen, laminin, hyaluronans, and forms of heparan sulfate proteoglycans and chondroitin sulfate proteoglycans. Within the Space of Disse, the region between the endothelial cells and parenchymal cells, the matrix chemistry transitions stepwise from portal triad to central vein such that the matrix chemistry associated with mature cells around the central veins is comprised of stable forms of fibrillar collagens (e.g. type I collagen), and highly sulfated proteoglycans such as heparin proteoglycans (Martinez-Hernandez and Amenta, 1993a,b; Reid et al., 1992; Vongchan et al., 2005).

B. Rules for Cultures on Matrix Substrata

1. Cells attach quickly on an appropriate matrix substratum. Therefore, when plating cells on any matrix substrata, add the cells in sufficient volume of medium to allow equal distribution over the plates.

2. Since matrix is a mixture of proteins and carbohydrates, use DNA or RNA stains rather than protein or carbohydrate stains for assessing plating efficiency or clonal growth efficiency of cells on matrix.

3. Cells on complex matrices (Matrigel, Biomatrix) do not detach readily. To do growth curves of cell on complex matrices, you may have to scrape the plates, isolate the DNA, and determine DNA content.

4. Antibodies (and many other reagents) will non-specifically stick to matrices. For antibody staining of cells on matrices, expose the cells to a control antiserum or to blocking reagents first and then to the specific antiserum.

5. Cells can achieve a very high density if cultured on matrix substrata and especially if proteoglycans or glycosaminoglycans are used. For example, typical saturation densities for normal cells on a 100-mm tissue culture dish are 7×10^6–10×10^6 whereas in the presence of appropriate proteoglycans there can be greater than 10^8 cells/100 mm dish.

6. Extraction protocols of RNA or DNA from cell cultures on collagens, proteoglycans, or complex matrices *should avoid* first detaching the cells enzymatically unless (1) the enzymes are guaranteed to be free of nucleases and (2) detachment does not result in alteration of the RNA levels. Rather, use guanidine or guanidinum solution on the cultures and extract everything on the plate, both cells and matrix. Then use a purification protocol for the RNA or DNA that will eliminate the matrix components (e.g. phenol chloroform extractions and/or centrifuging through cesium chloride gradient). The purification protocol for glycosaminoglycans (GAGs) is so close to that for nucleic acids that the GAGs and nuclei acids can be co-purified; if this happens, then the GAGs can readily be eliminated by use of highly purified glycosidases (e.g. heparinase, heparitinase).

C. Protocols for Preparing Extracellular Matrix Components

Many of the extracellular matrix components can now be purchased commercially including many of the collagens (type I, type IV), basal adhesion proteins (laminins, fibronectins), proteoglycans, and tissue extracts enriched in matrix (e.g. Matrigel). Therefore, the protocols for preparing these will not be given here. Many publications and reviews of protocols include these protocols (Macdonald *et al.*, 2002). In Appendix 3 are listed sources for those matrix components that are available commercially and that are used routinely for cultured cells. Tissue extracts enriched in extracellular matrix ("ECM", Biomatrix, Matrigel, amniotic matrix) are prepared with protocols that take advantage of the relative insolubility of extracellular matrix components (Macdonald *et al.*, 2002). See the prior method review for their preparation.

VIII. Culture Media

For detailed descriptions and discussions of classical cell culture methods, see recently published books and reviews such as that referenced (Freshney, 2000). All methods of preparing cells for culture start with the disruption of the tissue and its dispersal into chunks or single-cell suspensions. In classical cell culture, the dispersed tissue or cells are plated onto an inflexible plastic substratum that has been exposed to a cationic ionizing gas making the polystyrene in the dishes polarized to reveal a negatively charged layer. Cells attach to the dishes via that negatively charged surface and subsequently form a more complicated adhesion surface with secreted forms of extracellular matrix complemented by matrix components from serum. The cells are suspended in or covered with a liquid medium consisting of a basal medium of salts and nutrients supplemented with a biological fluid such as serum. We have prepared multipled forms of serum-free media for liver cells and have found that some requirements are generic for all the maturational lineage stages and others are lineage-stage specific. Below is provided general information about the media requirements and then are given tables of serum-free conditions for *ex vivo* maintenance of liver cells.

A. The Basal Medium

The basal media that are commonly used (e.g. RPMI, DME, BME, Waymouth's) were developed originally for cultures of fibroblasts (Freshney, 2000). Although most of the constituents in these basal media are requirements for both epithelial cells and fibroblasts, some aspects of the basal media have been redefined for epithelia (Freshney, 2000). Two examples are trace elements and calcium levels. Highly differentiated epithelia require various trace elements or other factors that act as cofactors for the enzymes associated with their tissue-specific functions (McKeehan *et al.*, 1990; Sato, 1984; Taub and Sato, 1980). With respect to the calcium, its level in many of the commercially available media is above 1 mM, concentrations permissive for growth of fibroblasts, but that in recent years have been shown to be inhibitory to most epithelial cell types (Eckl *et al.*, 1987). This problem is exacerbated by culturing the cells in serum-supplemented media, since serum also contributes significantly to the calcium level. Most normal epithelial cells can grow in calcium concentration of ~0.4 mM (Eckl *et al.*, 1987; Santella, 1998).

B. Serum

For more than 60 years, the primary form of biological fluid used for supplementation of the basal media has been serum obtained from animals taken to commercial slaughterhouses. Some investigators utilize serum autologous to the cell types to be cultured. However, it is more common that the serum derives from animals that are slaughtered routinely for commercial usage, such as cows, horses, sheep, or pigs (Freshney, 2000). Fibroblasts (stroma and other mesenchymal cell types) do well in serum-supplemented media (SSM). By contrast, epithelial cell types such as liver parenchymal cells dedifferentiate rapidly, within hours to a few days, in SSM and then die, usually within 5–7 days when on culture plastic or within 7–14 days when on various matrix substrata (Jefferson *et al.*, 1984a). Over the last about 20 years, the need for serum supplementation has been reduced or eliminated through efforts by many investigators. The serum supplementation has been replaced with mixtures of defined and purified hormones and growth factors as proposed by the pioneering strategies of Gordon Sato (McKeehan *et al.*, 1990; Sato, 1984; Taub and Sato, 1980).

C. Lipids

All cells require lipids, but mature hepatocytes are capable of converting linoleic acid into most lipid derivatives. This is not true for progenitors. They must be provided with a mixture of free fatty acids, bound to an appropriate carrier molecule (e.g. albumin) and supplied also with high density lipoprotein (10 μg/ml) for long-term management of uptake and release of lipids by the cells. The preparation of a mixture of free fatty acids is as described originally by Chessebeuf and Padieu (1984) and is given in Appendix 2. The mixture can be combined with albumin (bovine, human, etc.) to prevent the free fatty acids from being toxic. The fatty acids are a strict requirement for progenitor subpopulations. For commercial sources of purified fatty acids, see Appendix 3.

D. Soluble Signals: Autocrine, Paracrine, and Endocrine Factors (HDM)

An approach to defining the soluble signals from cell to cell interactions has been to replace the serum supplements in medium with known and purified hormones and growth factors to yield a serum-free, hormonally defined medium or "HDM" as summarized in recent reviews (Macdonald *et al.*, 2002; Xu *et al.*, 2000).

Such media have been developed for many cell types enabling investigators to have greater control over cells being maintained *ex vivo*. Use of HDM results in selection of the epithelial cell type of interest from primary cultures containing multiple cell types. Almost all of the published HDM are optimized for cell growth. To observe optimal expression of differentiated functions, the HDM must be retailored (Reid, 1990; Reid and Luntz, 1997; Macdonald *et al.*, 2002). Each tissue-specific function requires a discrete set of hormones and growth factors, often at concentrations that differ from those required for rapid cell growth. For example, insulin levels required for growth are typically ~1–5 μg/ml, whereas those needed for optimal expression of connexins are 50–100 ng/ml (Fujita *et al.*, 1986; Rosenberg *et al.*, 1992; Spray *et al.*, 1987). Thus, some of the hormones conducive to growth can markedly inhibit tissue-specific functions. A rule of thumb is to develop an HDM for growth of cells and then use it as a starting point for identifying the conditions needed for differentiation of those cells. Detailed protocols for the development of an HDM for a cell type at a given lineage stage have been described elsewhere and will not be presented here (Macdonald *et al.*, 2002).

Purified hormones and growth factors are prepared individually and aliquoted as 1000× stocks. Storage conditions depends on the factor. Most are frozen at −20 °C. They are added to the basal medium and then filtered through a sterilization filtration unit that is low protein binding. In the tables below are the serum-free conditions (media used in combination with feeders or with matrix substrata) found useful for hepatic progenitors and their associated mesenchymal companion cells.

The basal media given in the tables below is RPMI 1640. It can be replaced with any of a number of basal media (DME/F12, William's, etc.) with the qualifier that the medium must be chosen to be rich in amino acids, with a calcium concentration below 0.5 mM for growth versus greater than 0.5 mM for differentiation, and with copper avoided for stem cell cultures that are being kept in self-replicative mode (Tables VII–IX).

E. Influence of Serum-Free HDM on Epithelial Cells in Culture

Serum-free, HDM have been found to select for parenchymal cells even when the cells are on tissue culture plastic (Kubota and Reid, 2000; Taub and Sato, 1980; Xu *et al.*, 2000; Schmelzer *et al.*, 2007). This results, within a few days, in cultures that are predominantly the cell type for which the HDM was developed. However, if the cultures are plated onto tissue culture plastic and in HDM, the life span of the primary cultures has been found to be ~1 week, at which time, the cells peel off the plates in sheets. Achievement of longer culture life spans under serum-free conditions is dependent upon using substrata of matrix components or extracts enriched in extracellular matrix in combination with the serum-free, hormonally defined medium (Xu *et al.*, 2000) (Table X).

Tissue-specific gene expression is dramatically improved in cultures under serum-free conditions and especially with serum-free medium supplemented with only the specific hormones needed to drive expression of a given tissue-specific gene

Table VII
Serum-Free Conditions for Expansion of HpSCs

Basal media	RPMI 1640 + nicotinamide (4.4 mM) + L-glutamine (2 mM)
Lipids	High density lipoprotein, HDL (10 μg/ml) + free fatty acids bound to purified human albumin (0.2% w/v) and at 7.6 μEq (Appendix 2)
Hormone requirements	Insulin (5 μg/ml), transferrin/Fe (5 μg/ml)
Trace elements	Selenium: 3×10^{-10} M; zinc sulfate: 5×10^{-11} M
Calcium	~0.3 mM
Feeders (these can be replaced, in part, by the matrix substrata given below)	Angioblasts/endothelia for self-replication
	Activated hepatic stellate cells for expansion with lineage restriction to hepatoblasts (HBs)
	STO cells for expansion of rodent HBs and for differentiation of human HpSCs and HBs
Matrix substrata	Type III collagen or hyaluronans for self-replication (McClelland et al., 2008; Turner et al., 2006); type IV collagen and/or laminin for lineage restriction to HBs
Known soluble	Angioblasts: LIF (Lin et al., 2000)
Signals from feeders	Activated hepatic stellate cells: HGF (Kubota et al., 2007)

Table VIII
Serum-Free Conditions for Differentiation of HpSCs (or of hepatoblasts, HBs)

Basal media	RPMI 1640 + nicotinamide (4.4 mM) + L-glutamine (2 mM)
Lipids	High density lipoprotein, HDL (10 μg/ml) + free fatty acids bound to purified human albumin (0.2% w/v) and at 7.6 μEq (see Appendix 2)
Shared hormone requirements	Insulin (5 μg/ml), transferrin/Fe (5 μg/ml)
Trace elements	Selenium: 3×10^{-10} M; zinc sulfate: 5×10^{-11} M; copper sulfate: 10^{-10} M
Calcium	~0.6 mM
Hormones/growth factors	EGF (50 ng/ml), T3 (10^{-9} M), hydrocortisone (10^{-8} M)
Heparins or heparin proteoglycans	10 μg/ml (heparins) or 1–5 ng/ml (heparin PGs)
Feeders	For hHpSCs, one can use STO feeders; unknown for rodent or murine HpSCs or HBs
Matrix substrata	Plating on or embedding in type I collagen gels (McClelland et al., 2008)

(Enat et al., 1984; Muschel et al., 1986). However, mRNA synthesis of tissue-specific genes is not restored by serum-free medium or by any known combination of hormones and growth factors; rather, the improved tissue-specific gene expression in serum-free media or in HDM is due to post-transcriptional regulatory mechanisms, often an increase in stability of specific mRNA species (Jefferson et al., 1984b). Restoration of mRNA synthesis occurs only with serum-free media or with a serum-free, HDM containing the specific hormones or factors found to influence a given gene and presented *in combination* with tissue-specific forms of heparin proteoglycans or their glycosaminoglycan chains, heparins (Spray et al., 1987; Zvibel et al., 1991, 1995). Heparins are often bleached in commercial

Table IX
Serum-Free Conditions for Mature Parenchymal Cells (Brill *et al.*, 1994; Macdonald *et al.*, 2002; Xu *et al.*, 2000)

Components	Concentrations
Basal media	RPMI 1640 + nicotinomide (4.4 mM) + L-glutamine (2 mM)
Trace elements	Selenium: 3×10^{-10} M; zinc sulfate: 5×10^{-11} M; copper sulfate: 10^{-10} M
Lipids	High density lipoprotein (10 μg/ml)
Free fatty acids	Linoleic acid, 2.7×10^{-6} M [mature parenchyma can survive on linoleic acid alone; however, the cells do better if given the entire mixture of free fatty acids as described in Appendix 2]
Calcium	0.6 mM
Hormones/growth factors	Insulin (5 μg/ml), epidermal growth factor (50 ng/ml), tri-iodothyronine or T3 (10^{-9} M), hydrocortisone (10^{-8} M)
Matrix substrata	Ideal is to let the cells be three-dimensional (spheroids). Partial effects are observed with cells embedded in type I collagen especially if combined with heparin proteoglycan

Table X
Serum-Free Conditions Used for Hepatic Stellate Cells (Kubota *et al.*, 2007)

Basal medium	RPMI 1640 (GIBCO) + l-glutamine (1 mM, GIBCO)
Lipids	High density lipoprotein (10 μg/ml) + mixture of free fatty acids bound to albumin
Free fatty acids	See preparation of these in Appendix 2
Trace elements	Selenium: 3×10^{-10} M; zinc sulfate: 5×10^{-11} M; copper sulfate: 10^{-10} M
Hormones and growth factors	Insulin (5 μg/ml), transferrin/Fe (5 μg/ml); EGF (10 ng/ml; Pepro Tech, Inc.); LIF (10 ng/ml; StemCell Technologies, Inc.)
Antimicrobials	Antibiotics from GIBCO
Feeders	STO Feeders (Mitomycin C-treated)

processes to eliminate the brown coloration that is due to iron deposits; the bleaching process is destructive to heparin's biological activity on gene expression; so use unbleached fractions. At present one cannot obtain commercially the most potent matrix components regulating tissue-specific gene expression: the tissue-specific forms of proteoglycans. Efforts are ongoing in a number of laboratories to isolate and characterize them. It is hoped that they will become available commercially in the near future. (Tables XI-XIII).

IX. Monolayer Cultures of Cells

A. Monolayer Cultures in a Growth State with Defined Conditions

Under completely defined conditions, cells will behave quite reproducibly. Those capable of extensive growth even at clonal seeding densities are the stem cells, committed progenitors, and diploid adult cells and under the growth

Table XI
Ex Vivo Growth Potential for the Known Lineage Stages

Hepatic stem cells (HpSCs)	Division rates of ~1/26 h under optimal conditions; can be subcultured repeatedly. One cell can generate >40,000 daughter cells in ~15 days (McClelland *et al.*, 2008b; Schmelzer *et al.*, 2007)
Hepatoblasts (HBs) and committed progenitors	Division rates of ~1/40–50 h; rodent HBs have been found to undergo ~12 divisions in 3 weeks and with one HB generating 4000–5000 daughter cells in 3 weeks (Kubota and Reid, 2000). The division rates possible for human HBs are not known though estimated to be similar to that of the rodent HBs. Addition of various mitogens (e.g. HGF) can dramatically affect their division potential.
Diploid adult cells ["small hepatocytes"]	One cell yields ~130 daughter cell in 3 weeks (~5–7 divisions total); limited ability to be subcultured (Kubota and Reid, 2000)
Polyploid adult cells	Attach, survive; DNA synthesis but limited or no cytokinesis (Liu *et al.*, 2003)

Table XII
Requirements for Tissue-Specific Gene Expression (Both Transcriptional and Posttranscriptional)

Requirements for all genes	Basal media: nutrient rich. One option = RPMI 1640 (GIBCO) + L-glutamine (1 mM, GIBCO)	
	Calcium	~0.6 mM
	Trace elements	Selenium: 3×10^{-10} M; zinc sulfate: 5×10^{-11} M; copper sulfate: 10^{-10} M
	Lipids	High density lipoprotein (10 μg/ml) + mixture of free fatty acids bound to albumin
	Free fatty acids	See details for preparation of these in Appendix 2
	Hormones	Insulin (5 μg/ml), tri-iodothyronine or T3 (10^{-9} M), hydrocortisone (10^{-8} M)
	Proteoglycan/GAGs	Heparin proteoglycan (ideally liver specific) (1–5 ng/ml) or can see partial effects with bovine lung heparin (5–10 μg/ml); intestinal heparin is weaker (15–20 μg/ml)
Additional requirements for representative genes	Albumin	Epidermal growth factor (50 ng/ml)
	Connexins	Epidermal growth factor (50 ng/ml); glucagon (10 μg/ml)
	IGF II	Prolactin (2 mU/ml), growth hormone (10 μU/ml)
References	(Spray *et al.*, 1987; Zvibel *et al.*, 1995, 1991)	

conditions will express "early" genes. By contrast, the mature, polyploid cells will show limited growth, even under optimal growth conditions, require higher seeding densities even for survival, and yet will provide the highest levels of certain of the adult-specific functions such as the CYP450s.

1. Hepatic stem cells (HpSCs)
    a. Isolate the HpSCs by the protocols given above.
    b. Use a basal medium that has a low calcium concentration, 0.3–0.4 mM and little or no copper.

Table XIII
Feedback Loop: Relevance to Studies Tissue Engineering of Liver whether *Ex Vivo* or *In Vivo*

Findings	Hypothesis	Predictions
Stem cells or progenitors do not grow *ex vivo* when cocultured with mature parenchymal cells or with conditioned medium from the mature cells	Mature parenchymal cells (e.g. polyploid cells) produce soluble signals constituting feedback loop that regulates stem cell compartment	• The signals do not exist in peritoneum; site is permissive for expansion and maturation of human liver cells • Other hosts (e.g. sheep, pig) that have higher proportion of diploid cells will be better models for studies of human hepatic progenitors. • Strategies for clinical programs must take feedback loops into account • Transplant purified human HpSCs or progenitors (therefore avoiding feedback loop from mature human cells) • Hosts with high polyploidy will require liver injury to mature cells (zones 2/3)

    c. Seed the cells at the desired densities onto or into purified matrix substrata (type III or type IV collagen with or without laminin; hyaluronan hydrogels) or onto feeder cells (for rodents: STO cells; for human: feeders of angioblasts/endothelia such as hUVECs) and into Kubota's Medium.

    d. If you want the cells to become HBs, plate onto type IV collagen and laminin and in Kubota's Medium supplemented with HGF (10 ng/ml).

    e. If you want to drive the colonies predominantly or entirely to the hepatocytic fate, add also 20 ng/ml EGF (Collaborative Biomedical Products).

    f. Thereafter, change the medium every day to every 3 days.

    g. Plating rodent HBs (purified by flow cytometry) at clonogenic seeding densities onto STO feeders and in KM results in colonies of cells visible by 2–3 days and yielding 3000–4000 cells/colony within 20 days.

    h. hHpSCs demonstrate clonogenic expansion rates that differ depending on whether they are on culture plastic (Schmelzer *et al.*, 2007) or on type III collagen (McClelland *et al.*, 2008a; McClelland *et al.*, 2008b). On culture plastic and in KM, the cells divide every 4–5 days, whereas on substratum of type III collagen and in KM, they divide every $\sim$26 h. Seeding the hHpSCs on type III collagen and in KM results in $\sim$2000 cells/35 mM dish by day 5 and $\sim$40,000 cells/35 mm by day 15, respectively.

Passaging of HpSC colonies or HB colonies

    a. *Passaging with PBS or by enzymatic means*: a "cloning" ring is placed around each colony, the medium is aspirated and replaced with buffers and/or reagents (enzymes).

- PBS (no calcium): the tightly packed morphology of the colony start showing separation between the cells (cells rounding) after a few minutes in PBS; after 30 min, gentle aspiration of the PBS will lift aggregates of colony cells that can be transferred to plastic or various matrices. The success rate for this is quite low, since most of the cells die in PBS.
- Most enzymatic methods tested do not work well for passaging of the stem/progenitor cells. This includes trypsin, acutase, hyaluronidase, heparinase, and heparitinase. Only one protocol tested worked Collagenase (500 U/ml) (or libinase) Chondroitinase ABC in KM. Lift the whole colonies after 30 min at 37 °C. The colonies can be transferred to plastic or other substrata.

  b. *Mechanical passaging methods proved better than must enzymatic ones or than calcium depletion ones (PBS) for passaging*:

Method 1:

  a. Lift colonies by gently nudging them off the plate (pushing along the perimenter of the colony, since it is not attached or minimally attached at its center) using a Gilson Pipetman and 200 μl tips.
  b. Place in digest buffer (collagenase, DNAse, prepared in Kubota's Medium at the same concentrations as fetal liver dissociation buffer) and let sit while picking (the cells survive for some time, so one can pick a number of colonies before adding mechanical disruption methods). After picking is done, close the tube and shake to break the colonies into smaller aggregates.

Method 2:

  a. Lift colonies with stem cell knives (Reubinoff *et al.*, 2000), whole or sections.
  b. Transfer the entire colony to new culture dish.
  c. *Replating*:
  The best results are obtained by transferring to dishes coated with type III or IV collagen and in Kubota's Medium with as little serum as possible (1–2%) or, ideally, serum-free but using knockout serum (Invitrogen) (10–15%). Very slow attachment if on culture plastic. More efficient reattachment if on matrix, whether for growth (type III or IV collagen) or for differentiation (Matrigel or type I collagen)

2. Normal mature parenchymal cells
   a. Isolate the cells by standard protocols for the cell type you are using.
   b. Use a basal medium that has a low calcium concentration, ~0.4 mM.
   c. Plate the cells at densities of 3×10^5–4×10^5/60 mm or at least 10^6/100 mm. Use dishes coated with type IV collagen, a medium that contains both the hormones and growth factors used in the HDM, and some serum

(1–2%) to permit them to attach and to inactivate any enzymes that might have been used in isolating the cells. You can avoid the serum altogether if you use laminin with the collagen and if you use no enzymes in isolating the cells (or have effective ways of inactivating those enzymes without providing a condition that is toxic to cells).
   d. Incubate the cells for 4–6 h at 37 °C or until they are firmly attached.
   e. Rinse the plates, removing debris and floating cells, and feed with the serum-free HDM. Change the medium every day, or at the very least, every other day (prepare the HDM fresh).
   f. The cells will grow and will survive up to a month under these conditions.
   g. If the cells are stem cells, you can expect to be able to subculture and clone the cells under these conditions. If the cells are adult cells, you can expect several rounds of division and then growth arrest. You will not be able to clone or subculture them or you will have limited ability to do so. Polyploid cells will not grow under these conditions, but may survive for some weeks.

B. Monolayer Cultures in a Differentiated State with Defined Conditions

1. Hepatic stem cells
   a. Isolate the stem cells by standard protocols for the cell type you are using.
   b. For epithelial stem cells, it is likely that you must use a basal medium with a calcium concentration above 0.5 mM and copper, critical for aspects of differentiation, must be added.
   c. Prepare feeder cells by using STO (ATCC) feeders or stromal feeders prepared from embryonic tissue from which the stem cells are derived. For example, use embryonic liver stroma for HpSCs.
   d. Overlay the feeders with a fibrillar collagen or with matrigel or plate the stem cells onto collagen-coated or Matrigel-coated transwells.
   e. Plate cells at desired density (high density of subconfluent to confluent) onto the collagen embedded feeders or onto the collagen-coated transwells and over feeders.
   f. Use a basal medium supplemented with insulin (5 µg/ml) and transferrin (5 µg/ml) and with 1–2% serum for plating.
   g. After 4–6 h or after the cells are firmly attached, switch to a serum-free medium supplemented with insulin, transferrin, and other hormones or factors known to drive differentiation and to enhance specific genes (e.g. glucagons for the connexins; EGF, insulin, and glucocorticoids for albumin). See Table IX.
   h. Add unbleached, bovine lung-derived heparin (Sigma; 20–50 µg/ml) or liver-derived heparin proteoglycans (~10 µg/ml).
   i. The timing for full differentiation will vary with the cell type and must be empirically defined. But should occur within a few days to a week.

2. Normal mature cells (e.g. diploid or polyploid mature liver cells)

   a. Isolate the cells as a single-cell suspension by standard methods.

   b. Plate the liver cells at high density, 6×10^6–8×10^6/100 mm dish, into HDM/SSM and onto a flexible and porous scaffolding (e.g. filter, swatch of nylon, transwell, biodegradable microcarrier) coated with fibrillar collagen, ideally type I collagen mixed with fibronectin, or with matrigel.

   c. The cells should attach within 4–6 h. Then rinse the cells with PBS and give them the serum-free HDM retailored to optimize tissue-specific gene expression. Tailor the medium (as described earlier) to suit whichever genes are to be expressed optimally.

   d. The medium fed to the cells after 4–6 h (*not* the plating medium) should be supplemented with lung-derived, *unbleached* heparin (Sigma; at 20–50 μg/ml) or ideally the liver-derived (not available commercially) heparin or its corresponding heparin proteoglycan (at ∼10 μg/ml).

   e. Note that the cells will not grow, but will survive, should be three-dimensional and highly differentiated. The cultures will survive for perhaps 4–6 weeks and will retain most of their differentiated functions. If you are able to add the heparin proteoglycan, you will have near normal transcription rates for most of the tissue-specific genes. If you use the GAGs, unbleached heparins, you can observe normal or near normal transcription rates of some but not all tissue-specific genes. The proteoglycans and GAGs can be tissue specific. So, ideally use the proteoglycan or GAG from the tissue being cultured (e.g. liver-derived heparin proteoglycan for liver cells). With the limited availability of unbleached heparins (and even more so for proteoglycans), you will probably have to screen the available ones and use the one that is most active on your cells. See Table XII for summary of the requirements for tissue-specific gene expression.

X. Three-Dimensional Systems

Two forms of three-dimensional culture systems are described: (A) spheroid cultures; and (B) cultures in hyaluronan hydrogels. Please see the prior review for cultures on microcarriers or in bioreactors (Macdonald *et al.*, 1999, 2002).

A. Spheroids

The observations by Landry and others (Landry and Freyer, 1984; Landry *et al.*, 1985) that freshly isolated adult rat liver cells cultured under the appropriate conditions could aggregate and form spheroids, opened the way for the development of spheroid cultures as a model for the long-term *ex vivo* maintenance of adult liver cells. In spheroid cultures, liver cells are prevented from attaching to the substratum and remain in suspension forming three-dimensional multicellular aggregates that demonstrate an extent of differentiation close to that for liver tissue *in vivo* (Koide *et al.*, 1989, 1990; Li *et al.*, 1992a,b; Matsushita *et al.*, 1991).

Spheroid cultures provides a three-dimensional configuration for liver cells allowing the maintenance of an *in vivo*-like cuboidal cell shape and distribution of the cytoskeleton as well as enhanced cell–cell contacts. It has been shown that deposition of extracellular matrix within spheroids takes place, re-creating a more *in vivo*-like microenvironment (Landry *et al.*, 1985). In addition to re-establishment of an *in vivo*- like morphology and ultrastructure, hepatocytes in spheroid culture also express many differentiated functions over long-term culture including secretion of albumin (Landry *et al.*, 1985; Wu *et al.*, 1996a) and transferrin, tyrosine amino transferase induction by glucocorticoids (Landry *et al.*, 1985), ammonium metabolism, urea synthesis, and gluconeogenesis (Wu *et al.*, 1996a). The main application of the system has been the development of bioartificial livers (McClelland and Coger, 2000; Macdonald *et al.*, 1999; Wu *et al.*, 1996b). The majority of the research on liver spheroids has focused on rat and pig liver cells; studies of human liver spheroid cultures have appeared more recently.

1. Factors Affecting Spheroid Formation and Morphology

The spheroid culture model is a technically demanding model, since there are many variables that can affect the outcome, and since media changes are cumbersome given that the spheroids are floating. If all aspects of the culture conditions and methods utilized are not optimal then the quality of the spheroids formed and thereby the results obtained are compromised. Outlined below are some of the most critical factors for spheroid culture.

a. Choice of Method for Spheroid Culture

Several different methods have been described in the literature for the generation and culture of spheroids. Typically, hepatocytes are either cultured on bacteriological dishes and placed on a rotary shaker at a constant speed (Li *et al.*, 1992a; Yamada *et al.*, 1998), or they are cultured on a non-adherent substratum in static culture (Koide *et al.*, 1990; Landry *et al.*, 1985). When large-scale spheroid formation is required, spinner flasks or large Erlenmeyer flasks may be substituted for the smaller tissue culture vessels. The main factors affected by the method used are the size and shape of the spheroids obtained, and the time required for spheroid formation.

A novel method that combined elements from the shaking and the static methods for spheroid formation has been described (Dilworth *et al.*, 2000; Hamilton *et al.*, 1996). The method involves the culture of liver cells on 6-well tissue culture plates coated with a non-adherent polymer, p-HEMA [poly(2-hydroxyethyl methacrylate)], and shaking of the cultures on a rotary shaker at a constant speed of 90 rpm. Using this novel combination, the efficiency of spheroid formation is improved, the time required to obtain fully formed spheroids is reduced, and the spheroids formed are more homogenous in shape and size. A common concern with the use of spheroid culture is the potential for the formation of necrotic centers. If the diameter of spheroids exceeds 250–300 μm, the lack of oxygen and nutrient diffusion will cause cell death at the center of the spheroids. However, the use of this novel method prevents the generation of spheroids that exceed a 250-μm diameter, thereby preventing the formation of necrotic centers, a major improvement to the model.

The combination of p-HEMA coated 6-well plates on a rotary shaker is the method of choice, and found to yield optimal results in combination with the use of SSM/HDM and a seeding density of 5×10^5 viable cells/ml (2 ml/well). Experience with p-HEMA coating of plates for spheroid formation has shown that the coating must be smooth and even on the plates and also fully dry prior to use. The p-HEMA must also be filtered prior to use to remove any undissolved p-HEMA or particulate impurities in the solution, as these will cause an uneven surface once the ethanol evaporates from the plates. Uneven coating of the plates due to impurities or poor evaporation of the ethanol will cause poor spheroid formation. The shaking speed will also affect the size of the spheroids formed, increasing speeds resulting in smaller spheroids (Moscono, 1961). However, sheer stress forces become an issue at speeds exceeding 110–120 rpm. Previous studies determined 90 rpm to be optimal for spheroid formation using p-HEMA-coated plates (Dilworth *et al.*, 2000; Hamilton *et al.*, 1996).

Liver cells in spheroid culture do not attach and flatten as is observed in conventional monolayer culture; instead liver cells remain in suspension and retain a rounded cell shape. Using the method described herewith, rat liver cells form initial spheroids, irregular in shape and size, after only 2 h in culture. Some unaggregated single cells are also present at this stage, but these are removed from the culture through subsequent medium changes. Spheroids gradually increase in size over the first week of culture, from small aggregates of cells with a diameter of ∼30–50 μm to fully formed compact spheroids with a diameter of ∼200 μm. The increase in diameter results from the fusion and restructuring of the initial spheroids to form mature spheroids. Spheroids become more homogenous in shape and size with time in culture. When fully formed, spheroids appear as tightly packed dense structures with a distinct smooth outer lining; this outer lining consists of a single layer of flattened liver cells mixed with extracellular matrix.

An interesting observation was made when culturing human liver cells as spheroids. The formation of spheroids takes longer with human hepatocytes than with rat hepatocytes, requiring ∼2–3 extra days to complete spheroid formation. We hypothesize that this is due to the higher level of diploid cells in adult livers; livers from young rodents are greater than 90% polyploid, whereas those from young humans are estimated to be mostly diploid (>70%). Therefore, if the spheroid formation is a property predominantly of the polyploid cells, then it may take longer for human liver cell suspensions to form the spheroid, and factors that drive the differentiation to the polyploid state, such as heparins, may facilitate the kinetics of the process; preliminary evidence with heparins suggest, indeed, that this hypothesis may be true. The hypothesis is supported also by a previous report that addition of proteoglycans speeded the formation of rat spheroids (Koide *et al.*, 1990). An alternative is the use an artificial polymer, Eudragit, that has been shown to enhance spheroid formation (Yamada *et al.*, 1998).

b. Culture Medium

The composition of the culture medium used has been shown to have a major influence on the spheroid formation as it is on morphology and function of cultured liver cells (LeCluyse *et al.*, 1996). In previous work carried out to study the effects of different media on spheroid formation and morphology, it was determined that the best ones contained stable forms of insulin, transferrin/Fe, and selenium. One that was found especially useful is "Hepatocyte Medium" (HM) in which these factors are added to a modified Leibovitz-15 medium that has been specially buffered (Hamilton *et al.*, 1996). It is now commercially available from Sigma. HM needs to be supplemented with 2% FBS for at least the first 3 days of culture in order to obtain optimal spheroid formation and hepatocyte function; serum has been shown to be essential for the first few days of culture for optimal spheroid formation on an orbital shaker (Yagi *et al.*, 1993). However, after formation of the spheroids, the serum must be eliminated for optimal expression of tissue-specific genes, especially the P450s that are adversely affected in cultures with serum supplementation (see Tables IX and XII).

Preliminary studies with rat liver cells indicate that the use of HDM accelerated the rate of spheroid formation. The potential application of this culture medium for improving the efficiency and speed of spheroid formation in human hepatocytes is currently under investigation.

c. Cell Viability and Quality

A major factor affecting spheroid formation is the quality of the liver cells seeded, with preparations having a 90% or greater viability being optimal for spheroid formation. Dead cells release DNA which is "sticky". Therefore, if there is a high percentage of dead cells and cell debris, it results in clumping of the small initial aggregates to yield very large spheroids, often above 300 μM. Spheroids from a hepatocyte preparation with low viability will often be irregular in shape and size, as the dead cells and clumps will get incorporated into the spheroids.

After liver cell isolation, the viability of the cells is determined by trypan blue exclusion. Although this is a useful guide to the overall viability of the cells, it does not provide any information on the functional aspects of the cells. So even if the viability is 90%, as indicated by trypan blue exclusion, the cells may still not provide good spheroids. The initial stages of spheroid formation are thought to require the expression of function-specific cell surface receptors and adhesions sites (Chow and Poo, 1982; Garrod and Nicol, 1981), *de novo* RNA and protein synthesis, as well as active cellular processes such as the re-establishment of junctional complexes (Landry and Freyer, 1984). Healthy liver cells in good condition and able to undertake complex cellular processes are required.

2. Protocol for Establishing Spheroids

Supplies:

- Falcon 6-well tissue culture plates (Becton Dickinson Labware, Franklin Lakes, NJ)
- p-HEMA (Sigma)
- HM (Sigma)
- FBS (Hyclone)
- Inova 2000 Orbital Shaker (New Brunswick Scientific, Edison, NJ)
- Sterile transfer pipets (Fisher Scientific)

(a) 6-well tissue culture plates are coated with of 2.5% pHEMA solution in 95% ethanol (2 ml per well). The ethanol is allowed to evaporate leaving a clear, polymer coating. On the surface of each well. The plates can be left overnight in a class II safety hood with no lids on for a more rapid evaporation, or placed in a dry incubator at 37 °C (lids on) for 3–4 days. Although slower, the latter method has been found to yield improved results.

(b) Freshly isolated hepatocytes are resuspended in HM supplemented with 2% FBS, 0.1 µM dexamethasone, and antibiotics (1× penicillin/streptomycin). These are seeded onto p-HEMA-coated plates at a density of 5×10^5 viable cells/ml of medium (2 ml of medium per well).

(c) Place the plates on an orbital shaker at 90 rpm and in an incubator at 37 °C and 5% CO_2/95% O_2.

(d) The medium is replaced after 24 h of culture and every 2 days thereafter. This is done by tilting the 6-well plates on a sharp angle and allowing the spheroids to settle to the bottom edge of the plate. The medium is then slowly and carefully aspirated off using a sterile plastic transfer pipette. Particular care must be taken on the initial days of culture when the spheroids are small, as loss of spheroids can occur during the medium changes. After the first medium change, spheroids often required separation from each other by gentle pipeting with a sterile plastic transfer pipette.

B. Hyaluronan Hydrogels as Three-Dimensional Scaffolding for cells

Hyaluronans are a unique linear polysaccharide, found at various molecular weights throughout most tissues of the body (Fraser et al., 1997) and unique among the families of glycosaminoglycans (GAGs) in that they have no core protein. As for all GAGs, they are negatively charged and have many active groups for which chemical modifications can be made. In nature, no acetylated or sulfated versions have been found (Mahoney et al., 2001; Shah and Barnett, 1992; Shu et al., 2002). The stabilized subunits of glucuronic acid and glucosamine in the hyaluronan chain lend themselves to mechanical and physiological properties helpful in maintaining matrix chemistry and physiological homeostasis. Hyaluronans are essential for lubrication, homeostasis, filtering, and embryonic development (Fraser and Laurent, 1989; Fraser et al., 1997; Knudson and Knudson, 1991; Kobayashi et al., 1994). Chemical reactions with the sugar groups of the hyaluronans through hydrogen bonding help facilitate their physiological effects. Their role in development and differentiation applies to the maturation lineages of the developing organism (Laurent and Fraser, 1986), and most likely to the developing organ systems in times of tissue regeneration as seen in the liver. There is a high correlation between the amount of hyaluronans present during mitotic waves and cell proliferation. In the case of cell migration, the cylindrical formations of water tunnels allows cells which possess appropriate receptors to go through self-mediated guiding down the tunnels, to other destinations (Laurent and Fraser, 1986; Turley, 1989). Large hyaluronan molecules at low density can be favorable for tunnels, whereas at high densities create an intertwined mesh as seen in synovial fluids. The molecule has been studied for biological capacities in areas of water homeostasis, transport, and cellular interactions. Within the body, the polysaccharide is quickly catabolized locally and has been shown to be eliminated by liver endothelial cells (Laurent and Fraser, 1991). This compatibility with degradation without large-scale immune responses makes the molecule an appropriate one for grafts for transplantation of cells and for delivering drugs. It is the physiological properties of the hyaluronan that can lend to creating a matrix sufficiently strong and variable to advance tissue engineering.

1. Choices of Hydrogels for Culturing HpSCs or HBs

The hepatic progenitors, HpSCs and HBs, have receptors for hyaluronans (Turner et al., 2006). Yet even cells without such receptors can be maintained in hyaluronan hydrogels by making mixtures of the hyaluronans with other extracellular matrix components. Indeed, commercially available hyaluronan hydrogels (Glycosan Biosciences, Salt Lake City, Utah) include combinations of hyaluronans with gelatin (type I collagens) or with other matrix molecules. A useful version is a hydrogel that is 95% hyaluronans and 5% type I collagen (gelatin).

2. Preparation of the Hyaluronan Hydrogels

The hyaluronans are stabilized by crosslinking (Fraser and Laurent, 1986; Fraser et al., 1997; Shu et al., 2002). The method of crosslinking affects many aspects of the hydrogels with respect to their uses for cells in three-dimensional culture formats. For example, aldehyde crosslinking provides hydrogels that are very stable and can be handled readily with forceps. Yet these hydrogels are so robust that they resist enzymatic digestion with hyaluronidase and resist dissolution with even high molar guanidine buffers (Turner et al., 2006). Therefore, cells cultured in such hydrogels may survive and grow but will grow slowly since the cells' own enzymatic machinery is incapable of modifying the gels. A more useful form of crosslinking is disulfide bridges utilizing technologies developed by Prestwich and associates (Shu et al., 2002). These hydrogels can be modified to vary in mechanical properties (rigidity, stiffness), in the matrix components bound (collagens, basal adhesion molecules), and/or in growth factors or hormones (e.g. VEGF) complexed within the hydrogel (Tunner et al., 2000). Below is the description for 2–4% hydrogels utilizing hyaluronans available commercially from Glycosan Biosciences, Utah. In creating 8 ml of hyaluronan-based solution (Glycosan, SLC), the final composition of a given gel may be 2 parts medium, and 3 parts HA solutions.

1. Purchase hyaluronans and crosslinker reagent from a commercial source (e.g. Glycosan Biosciences, Utah). The hyaluronans are sold as lyophilized, crystal-like powders. The crosslinker is polyethylene glycol diacrylate (PEGDA).
2. Weigh out desired amount of the hyaluronans and place into a 50-ml tube.
3. Add appropriate amount of medium (e.g. Kubota's Medium) to achieve the desired percentage (weight/volume). Typical percentages used are 2–4%.
4. Place into a 37 °C water bath to dissolve.
5. Check the pH of the solution. Adjust pH to 7.4.
6. Filter through a 0.45-mm sterile filter syringe.
7. Mix the PEGDA crosslinker at 3% in the same medium and sterile filter. There is no need for a water bath, since it dissolves at room temperature.

Portions for various gels:

Crosslinked gel (1 part crosslinker:1 medium:3 parts hyaluronan solution)
Example: For a total volume of 7.30 ml of crosslinked gel add:
1.46 ml crosslinker:1.46 medium:4.38 ml HA solution

Prior to mixing of the solutions, the final number of cells for embedding into a hydrogel should be contained within the volume of the medium.

3. Sterilization of the Hyaluronan Hydrogels

Sterilization of the HA hydrogels can be performed prior to crosslinking or afterwards by exposure to a Cesium source (e.g. JL Shepard Mark I Model 68 Cesium Irradiator) with a deliverable dosage of 40 Gray (40 J/kg), over a 10-min period (Turner et al., 2006). If cells are to be embedded within the hydrogels, then the hyaluronans should be sterilized prior to the crosslinking step.

4. Cultures in Hyaluronan Hydrogels

HA hydrogels are placed into culture wells. For example, one can use either 6-well culture dishes (Falcon–Beckton–Dickinson, Franklin Lakes, NJ), or for the smaller sized hydrogel matrices, chambered coverglass culturing slides (Lab-Tek–Nunc, Naperville, IL). Smaller hydrogels require no manipulation (priming) prior to inoculation with freshly isolated cells other than a pre-soak with medium. The larger hydrogels require slight manipulation to insure the removal of air bubbles from the sponges. In most cases, addition of 3 mL of medium onto the hydrogel will trap air bubbles; removal of air bubbles can be accomplished by slight compression–relaxation of the hydrogel, forcing air from the lateral sides. After priming, suspensions of cells can be seeded onto large HA hydrogels at 2×10^6–3×10^6 cells/hydrogel in medium and at 2×10^5–3×10^5 cells per small hydrogel. After 16 h initial incubation at 37 °C in a CO_2 incubator (Forma Scientific, Baton Rouge, LA), the medium is changed to reduce or eliminate dead cells or debris. The working volume for a 6-well plate is 3 ml and for the two-chambered wells is 2 ml. Cells can be cultured for weeks to months on the hydrogels with changes of the media every 2–3 days. Importantly, the cells can be grown on the surface of the hydrogel or embedded in the hydrogels (as spheroids). One can collect surface colonies by adding PBS that removes the calcium required for cadherin binding to the gels and then gently lifting the colonies with a small Pipetman. Care must be taken with this approach, since the hydrogel fragments can get swept up with the desired colonies. However, the easiest way to collect the cells is simply by dissolving the hydrogel with a combination of purified enzymes and dithiothreitol as described below.

Cell recovery solution for dissolving the scaffolds crosslinked by disulfide bridges:

Serum-free basal medium (e.g. for the HpSCs and HBs, use Kubota's Medium)	5.0 ml
Dithiothreitol (DTT)	0.231 g
Hyaluronidase (Sigma)	5.0 mg
Collagenase (Sigma)	3.0 mg
DNase (Sigma)	1.5 mg

If using a media that contains serum for culturing purposes, make certain to wash the hydrogels with a serum-free medium before starting the digestion process. After the process is completed, you may remove broken scaffolding from the solution by gentle centrifugation through a density gradient (e.g. Percoll).

1. Wash hydrogels with the Cell Wash (see Appendix 2) for 2–3 times for 1 h each to clear any serum (fewer washes required if cells are maintained in a serum-free medium).
2. Add a minimum of 2 ml of cell recovery solution per 0.5 ml of disulfide-linked hydrogel.

3. Incubate at 37 °C until gels are digested (digestion time is ~1 h).
4. Collect cells.

5. Fixation and Sectioning of Cells in the Hydrogels

Hydrogels with cells can be removed from the medium and then fixed using 3 ml of 4% paraformaldehyde to each well. After 1 h incubation, the fixative is removed and replaced with serum-free basal medium and prepared for cryopreservation. Prior to freezing, the basal medium is removed, and the matrix lifted using a flat edge spatula. It is placed into Optimal Cutting Temperature Compound (OCT) (Tissue-Tek, Torrance, CA), in 2 ml conical sample cups (Fischer Science, Pittsburgh, PA) and snap frozen using liquid nitrogen emersion. The preserved specimens can be kept at −80 °C until sectioned. Slides are made with 5 μm sections through the hydrogels. If sectioning is very difficult, one can go up to 10 μm sections, though the resolution with microscopy will be less.

Protocol for BrdU staining of cells within hydrogels:
(Roche Staining Kit—Cat No. 11 296 736 001)

1. Transfer hydrogel to new dish.
2. Aspirate the culture media.
3. Add the BrdU labeling medium (aliquoted as to instructions).
4. Wash three times with washing buffer or 1× PBS with 0.01% Triton 100-X.
5. Ethanol-fix the cells for 20 min at −20 °C.
6. Remove from freezer and wash 3× with washing buffer.
7. Incubate for 1 h in Anti-BrdU at 1:20 dilution in buffer at RT.
8. Wash three times with washing buffer.
9. Incubate for 30 min at 37 °C with secondary antibody 1:20 dilution with 0.1% Triton 100-X in PBS.
10. Wash three times with washing buffer.
11. Add DAPI for 2 min.
12. Wash three times with washing buffer.
13. Resuspend in Kubota's Medium.
14. Image.

Trizol resolution of RNA protocol:
Cell collection and preparation:

1. Remove the hydrogels from the culture plate and place into 2 ml Eppendorf tubes.
2. Spin at 12,000 RCF in a microcentrifuge.
3. Remove and discard supernatant (alternatively, save the supernatant for assays for secreted factors).
4. Remove and discard medium from culture well.

5. Add 1 ml Trizol each to the well and to the Eppendorf tube.
   – Continually pipette to insure break up of cells pelleted in the centrifugation stage.
   – When working with monolayers, you can use 1 ml for every 10 cm^2 plate area.
6. Collect and transfer contents of the well into a second Eppendorf.

Phase separation of RNA and protein:

1. Incubate sample for 5 min at RT.
2. Add 0.2 ml chloroform for every 1 ml Trizol used in the previous step.
3. Shake the tube vigorously for 15 s.
4. Incubate for 3 min at RT.
5. Centrifuge the tubes ($<12,000 \times g$) for 15 at 4 °C.
   – RNA stays in the upper, clear aqueous phase.
6. Remove the aqueous phase and place into a new Eppendorf

RNA precipitation:

1. Add 0.5 ml isopropyl alcohol per 1 ml Trizol used in the cell collection.
2. Incubate samples for 10 min at RT.
3. Centrifuge the tubes ($<12,000 \times g$) for 10 min at 4 °C.
4. Remove the supernatant and do not disturb pellet.

RNA wash:

1. Add 1 ml of 75% ethanol for every 1 ml Trizol used in the cell collection.
2. Vortex the sample for 5 s.
3. Centrifuge the sample ($<7500 \times g$) for 5 min between 2 and 8 °C.
4. Remove the supernatant from the pellet.

Redissolving the RNA pellet:

1. Air-dry the pellet briefly (do not let the pellet dry out entirely).
2. Dissolve the RNA in RNase free MQ water.
3. Incubate the sample for 10 min between 55 and 60 °C.
   [1] Determine concentration of RNA.

8. Hepatic Stem Cells and Hepatoblasts: Identification, Isolation, and *Ex Vivo* Maintenance

Appendix 1: Nomenclature, Glossary, and Abbreviations

For the sake of brevity, we have used abbreviations for words or phrases used very frequently. The terms defining cell populations are keyed to Figs. 1–14

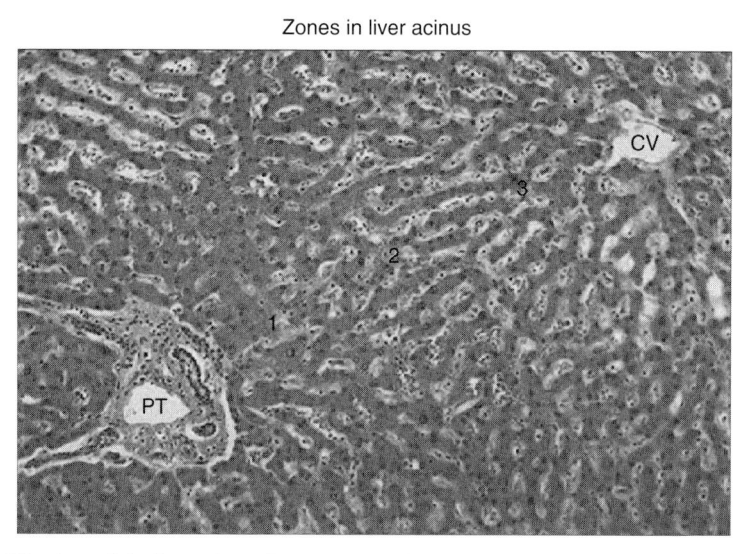

Fig. 1 Histology of the liver acinus. Shows the zonation from portal triads (PT) to central vein (CV). Zone 1 = periportal; zone 2 = mid-acinar; zone 3 = pericentral. The maturational lineage goes from zone 1 to zone 3 with progenitor populations in zone 1 to apoptotic cells in zone 3. Section of liver stained with hematoxylin/eosin. Image is from www.med.huji.ac.ll/mirror/webpath/liver/html.

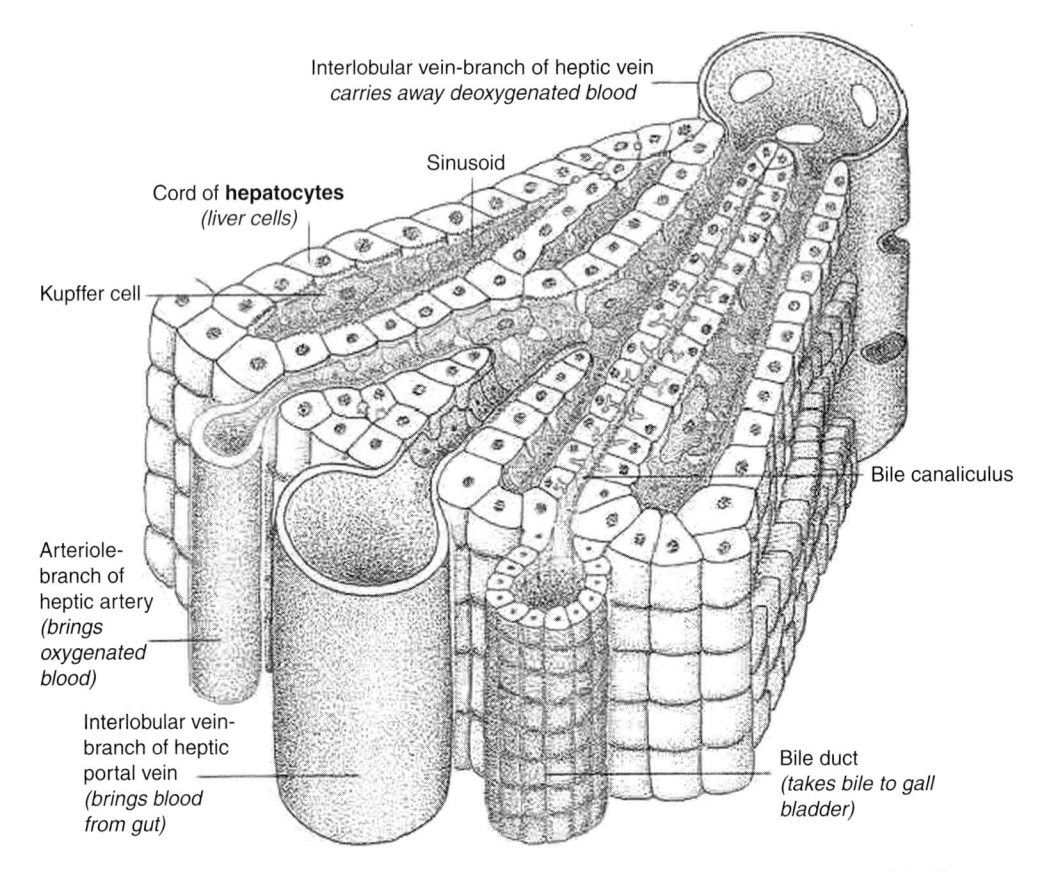

Fig. 2 Schematic of liver acinus. The vascular supply and the cellular components of the liver are defined. Blood flows into the liver at the portal triads (hepatic artery, hepatic vein, bile duct). Blood flow is from the spleen and gut (hepatic vein) and oxygenated blood (hepatic artery) from the lungs. The blood from these two sources mixes in the sinusoids of the liver and then departs the liver via the central vein (connected to the vena cava). The blood flows along cords of parenchymal cells blanketed on either side by sinusoidal endothelial cells that are continuous near the portal triads but fenestrated near the central veins. Image is from http://www.biologymad.com/kidneys/liverlobule.

8. Hepatic Stem Cells and Hepatoblasts: Identification, Isolation, and *Ex Vivo* Maintenance

EpCAM+, N-CAM+, CK19+ ALB±, hedgehog proteins, AFP− → Hepatic stem cells (pluripotent Cells) ⟲ Self renewal

EpCAM+, ICAM1+, CK19+, ALB+, hedgehog proteins ±, AFP+, P450 3A7+ → Hepatoblasts: (bipotent cells) — Probable transit amplifying cells

Committed hepatocyte progenitors ← ALB+, CK19− | ALB−, CK19+ → Committed bile duct progenitors

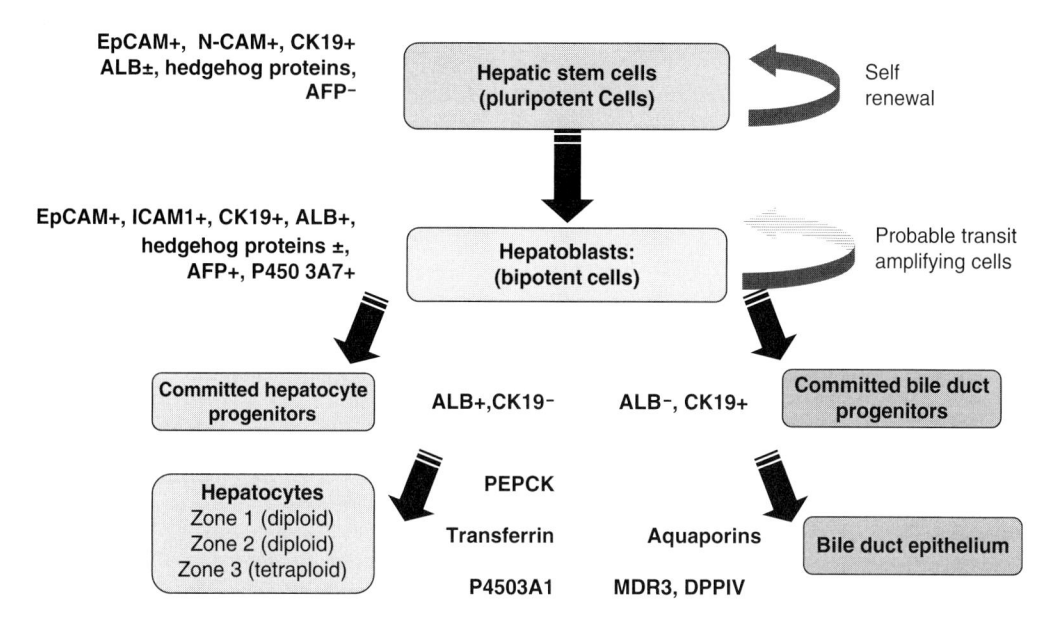

Hepatocytes: Zone 1 (diploid), Zone 2 (diploid), Zone 3 (tetraploid)

PEPCK, Transferrin, P4503A1 | Aquaporins, MDR3, DPPIV → Bile duct epithelium

Fig. 3 Schematic of liver lineage stages. The human liver's maturational lineage begins with hepatic stem cells (HpSCs), pluripotent cells, that give rise to hepatoblasts (HBs), bipotent cells (probable transit amplifying cells of the liver), and then to committed progenitors leading ultimately to the biliary epithelia and hepatocytic lineages. The cell size, ploidy, morphology, gene expression, and growth potential are lineage dependent.

- **Hepatic stem cells (hHpSCs)**
  - **Positive:** N-CAM, claudin 3, indian hedgehog, weak level of albumin; rare cells in culture are CD117+ (unclear if on HpSCs or "mesenchymal companion cells")
  - **Negative:** α-fetoprotein, I-CAM1; all isoforms of P450
  - **Percentage:** ~0.5–0.1.5%
  - **Size:** 7–10 μm
  - **Location:** Ductal (limiting) plate, canals of Hering

- **Hepatoblasts (hHBs)**
  - **Positive:** I-CAM1, α-fetoprotein, albumin; fetal forms of P450s (P450-3A7); indian hedgehog (lower than in HpSCs)
  - **Negative:** N-CAM, claudin 3, CD117 (cKit)
  - **Percentage:** ~80% in fetal livers, <0.01% in adult livers; higher in diseased livers
  - **Size:** 10–12 μm
  - **Location:** most of parenchyma in fetal livers; tethered to canals of Hering in adults; nodules in diseased livers

Shared by both stages:
Positive: CK 8, 18, and 19; EpCAM, CD133/1, E-cadherin
Negative: hemopoietic antigens (e.g. CD34, CD38, CD45, CD14, glycophorin A) and mesenchymal antigens (CD 146, KDR, desmin, α-smooth muscle actin, von willebrand factor)

Fig. 4 Shared versus distinct antigenic profiles of two stages of pluripotent human hepatic progenitors.

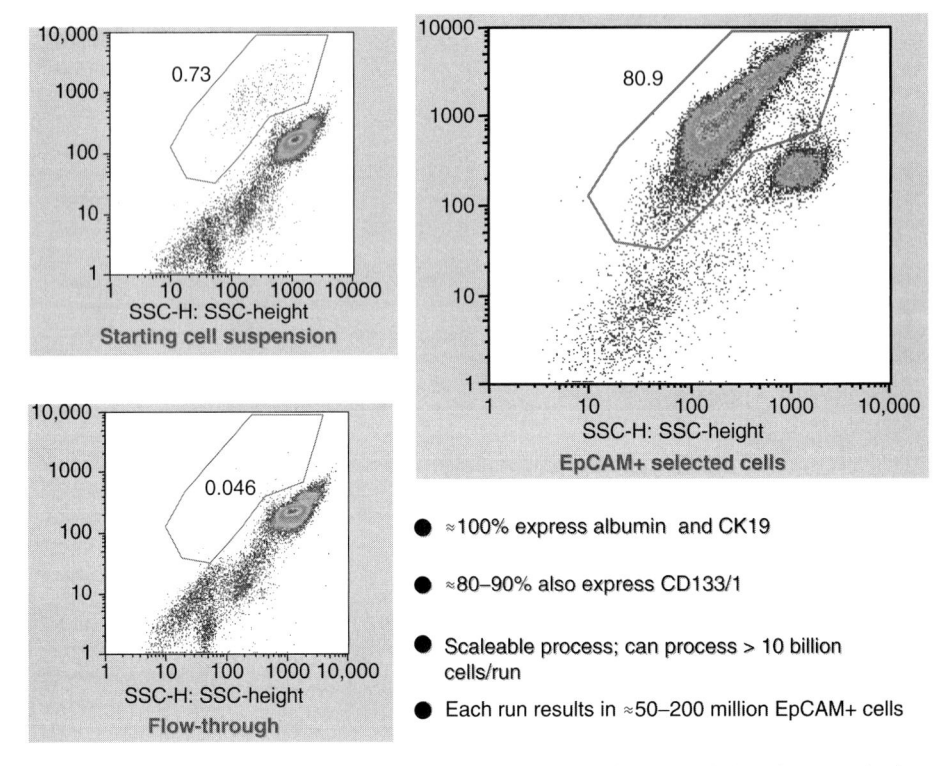

Fig. 5 Enrichment of EpCAM+ cells from adult human liver using magnetic bead immunoselection. Freshly prepared suspensions of adult human liver cells were labeled with Miltenyi beads coupled to antibodies to epithelial cell adhesion molecule (EpCAM). The starting cell suspension had 0.73% EpCAM+ cells. The labeled cells were put through a column that was magnetized so that the EpCAM+ cells were retained in the column. Then the magnetic field was removed, and the EpCAM+ cells flushed through. Just one pass through the column resulted in an enrichment of EpCAM+ cells to 80.9%. The flow through had only 0.046% EpCAM+ cells. The EpCAM+ cells co-expressed albumin and CK19 and more than 90% also expressed CD133/1. Each 10 billion adult liver cells contains at least 50–200 million EpCAM+ cells. A single adult liver can yield ~500 billion cells. Each EpCAM+ cell can undergo clonogenic expansion with division rates of ~1 division/24 h under optimal conditions. From Schmelzer *et al.* (2007). (See Plate no. 12 in the Color Plate Section.)

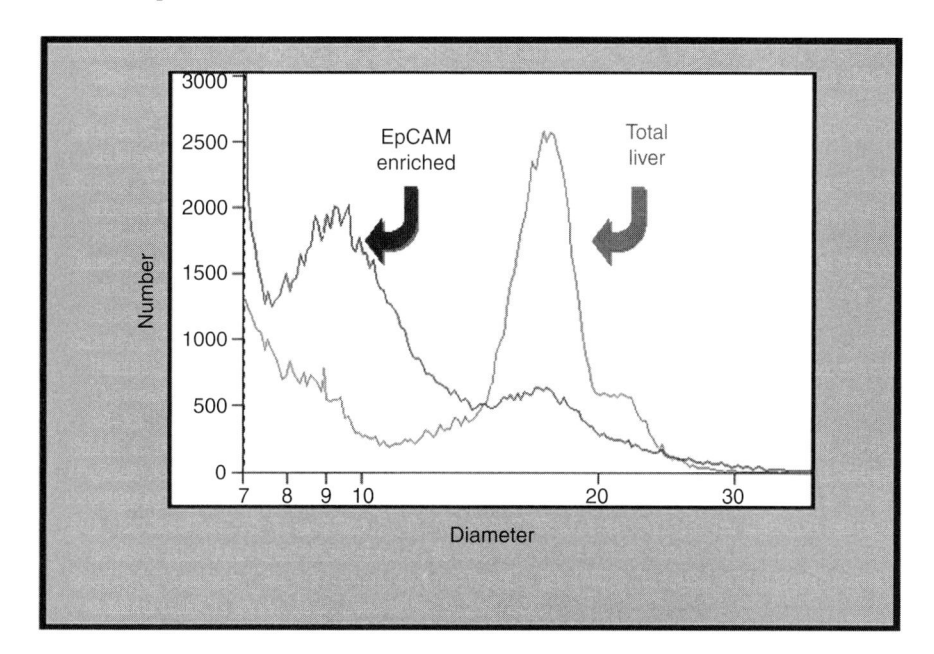

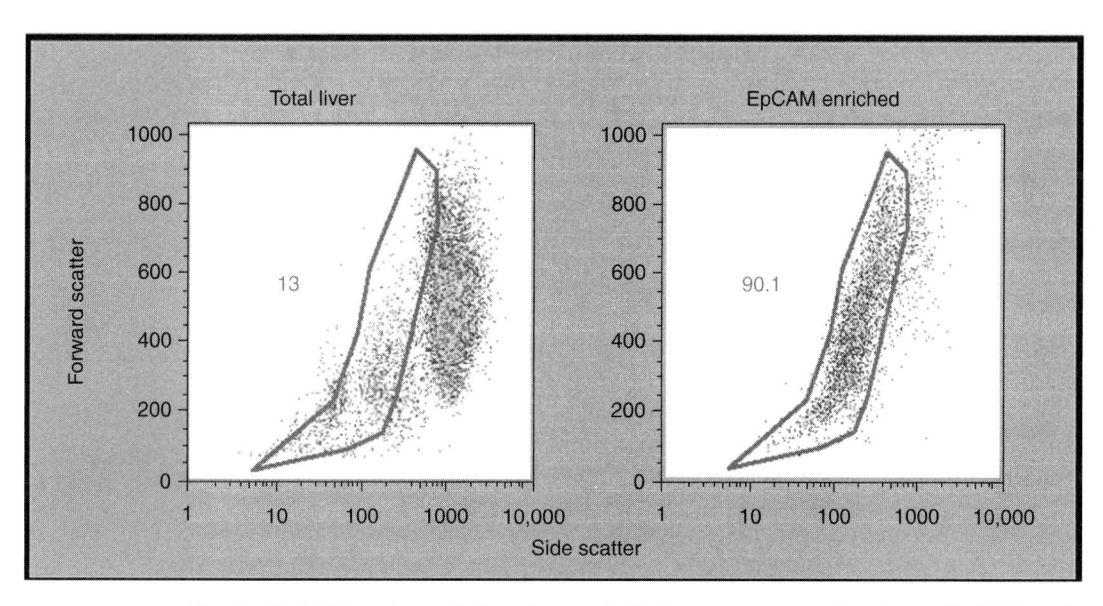

Fig. 6 EpCAM+ cells are distinct phenotypically from mature parenchymal cells. EpCAM+ cells in pediatric and adult livers are HpSCs with a size of 9–10 μm in diameter and with low side scatter (granularity). By contrast mature liver parenchymal cells are 18–25 μm and with high side scatter (Schmelzer et al., 2007). The gene expression profiles of EpCAM+ cells are also distinct from that of mature parenchyma (Schmelzer et al., 2006). (See Plate no. 13 in the Color Plate Section.)

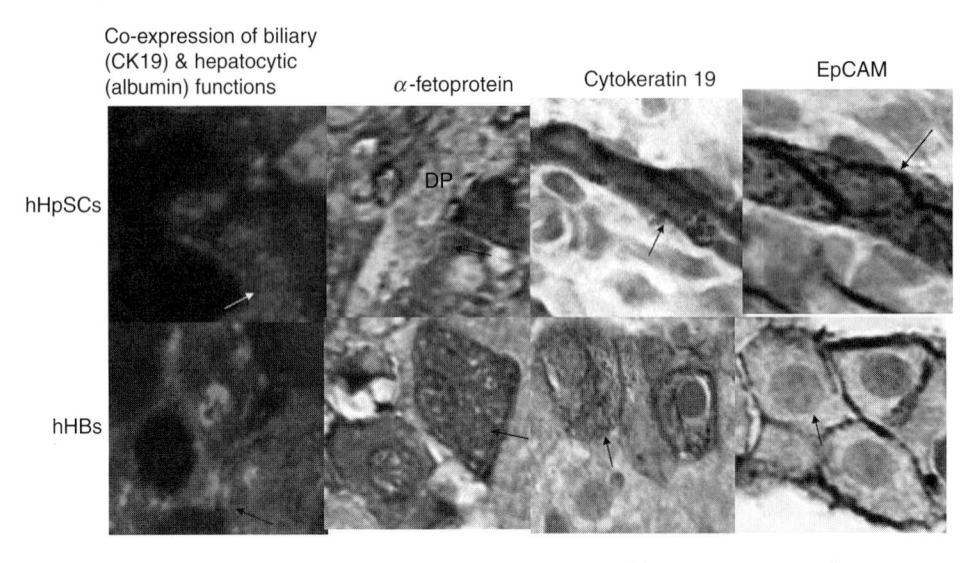

Fig. 7 Human hepatic stem cells (hHpSCs) versus human hepatoblasts (hHBs). Human livers have two pluripotent cell populations: hepatic stem cells (hHpSCs), that are multipotent, and hepatoblasts (hHBs), that are bipotent. hHpSCS are found in the ductal plates of fetal and neonatal livers and in the Canals of Hering of pediatric and adult livers. hHPSCs are ~7-9 μm; hHBs are ~10-12 μm in diameter. They both express albumin and cytokeratin (CK) 19 but can be distinguished by the pattern of expression of EpCAM and CK 19 and by the fact that HpSCs express NCAM and claudin 3 but not AFP; HBs express ICAM-1 and AFP. [From Schmelzer et al., 2007]. (See Plate no. 14 in the Color Plate Section.)

A. Common Requirements

- Nutrient-rich basal media
  - e.g. RPMI 1640
- Lipids
  - High density lipoprotein
  - Mixture of free fatty acids
    - Bound to carrier molecule such as albumin
- Trace elements
  - Zinc, selenium
- Insulin

B. Lineage Dependent Requirements

- **Stem cells and committed Progenitors**
  - Transferrin/Fe
  - Embryonic/fetal matrix substrata (e.g. hyaluronans, type IV collagen)
  - Avoid Cu (drives differentiation)
  - Avoid epidermal growth factor (EGF) -- restricts cells to hepatocytic lineage
  - Paracrine signals from angioblasts/endothelial cells (some identified)

- **Diploid and Polyploid adult cells**
  - Mature matrix substrata (e.g. type I collagen, heparin proteoglycans)
    - Chemistry of the matrix is distinct for growth vs differentiation)
  - Epidermal growth factor
  - No need for transferrin/Fe (since they make it)

Fig. 8 (A) Common requirements for culture of all lineage stages of liver cells. (B) Some of the Lineage-dependent requirements.

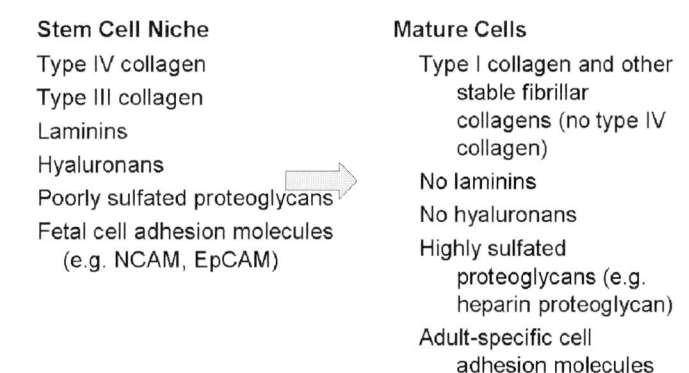

Stem Cell Niche
Type IV collagen
Type III collagen
Laminins
Hyaluronans
Poorly sulfated proteoglycans
Fetal cell adhesion molecules
  (e.g. NCAM, EpCAM)

Mature Cells
Type I collagen and other stable fibrillar collagens (no type IV collagen)
No laminins
No hyaluronans
Highly sulfated proteoglycans (e.g. heparin proteoglycan)
Adult-specific cell adhesion molecules

Fig. 9 Maturational changes in liver matrix chemistry. The matrix chemistry changes gradient fashion from the young cells to the old cells.

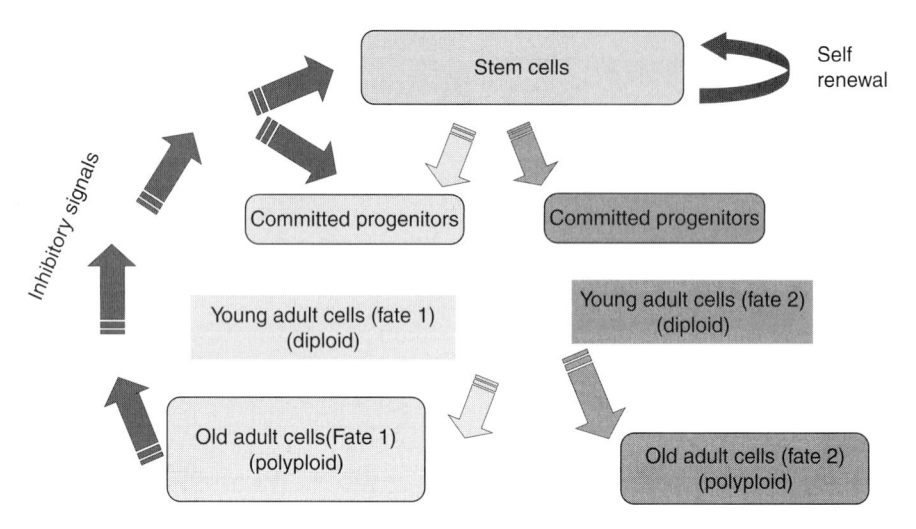

Fig. 10 Feedback loop. *Ex Vivo:* stem cell/progenitors do not grow when: 1) co-cultured with mature liver cells (polyploid liver cells) or 2) cultured with conditioned medium from mature liver cells (polyploid liver cells). *In Vivo:* stem cell/progenitors do not grow *in vivo* unless there is selective loss of mature cells in zone 3 (pericentral zone)–"cellular vacuum" Feedback loop signals–released either into blood of bile; not yet been identified.

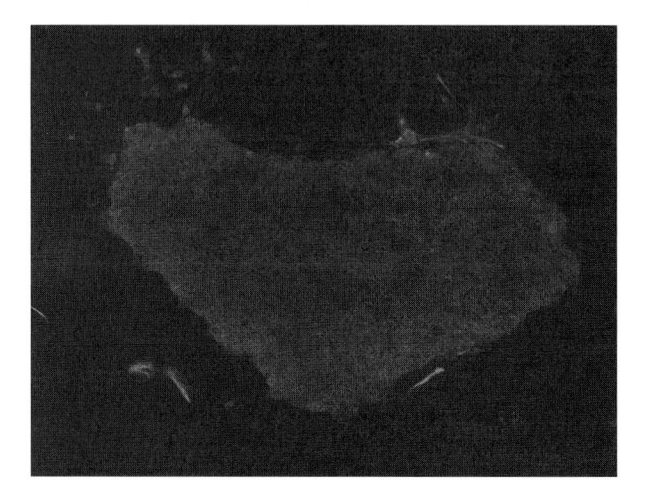

Fig. 11 An hHpSC colony from fetal human liver. Culture of hHpSCs from human fetal liver on tissue culture plastic and in Kubota's Medium. The colony of hHpSCs has been stained for EpCAM (green) and DAPI (blue). Desmin+ cells (red) are hepatic stellate cells that provide critical paracrine signaling (matrix and soluble signals) for the hHpSCs. (See Plate no. 15 in the Color Plate Section.)

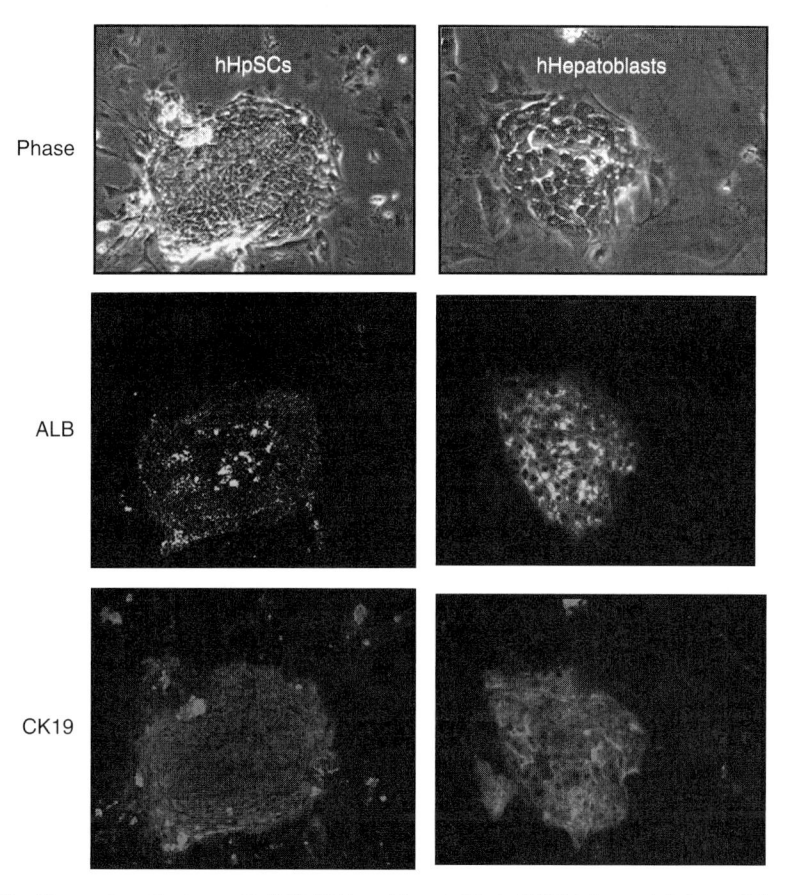

Fig. 12 Human hepatic stem cells (hHpSCs) and hepatoblasts (hHBs) from adult liver. Colonies of hHpSCs (left panels) versus hHBs (right panels) in culture on embryonic stromal feeder, STO cells, and in Kubota's Medium. The top set are phase micrographs; the middle panels are colonies stained for albumin; the bottom panels are ones stained for cytokeratin 19. (Figures from Schmelzer et al., 2007.) (See Plate no. 16 in the Color Plate Section.)

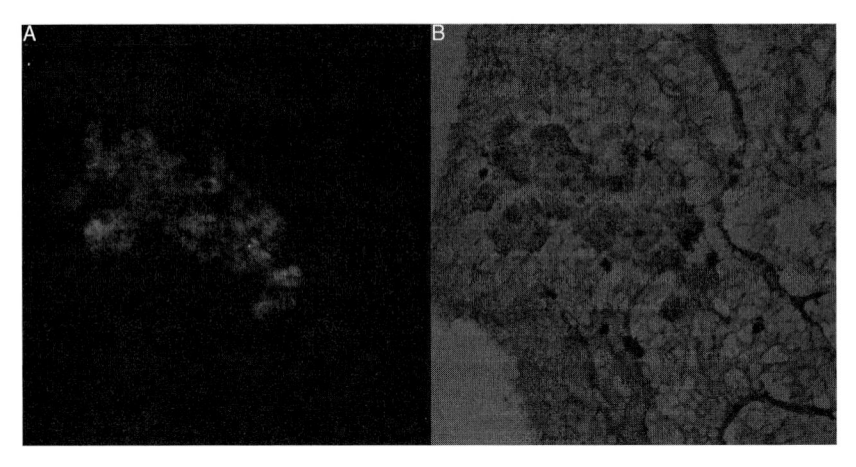

Fig. 13 The hHpSCs embedded in hyaluronan hydrogels. Hepatic progenitors in hyaluronan hydrogels crosslinked by disulfide bridges. 20× magnification of HpSCs in hydrogels were taken by the Olympus FV500 Laser Scanning Microscope. Panel A identifies the HpSCs after 7 days of culture in a hyaluronan hydrogel. Image is of autoflourescence of the cells. Panel B. Digital Image Composite (DIC) shows the cells and the HA hydrogel. The hydrogel reagents were obtained from Glycosan Biosciences (Utah). (From Turner et al., 2007.) (See Plate no. 17 in the Color Plate Section.)

DNA synthesis   Cytokinesis   Hyperplastic growth

- **Hepatic stem cells**
  - Division rates of up to ~1 per day
  - Can be subcultured repeatedly
  - A single cell can generate a colony of >10,000 daughter cells in 2–3 weeks

- **Hepatoblasts**
  - Division rates of ~1 per 2–3 days
  - Can be subcultured but not as easily as the stem cells
  - A single cell can generate a colony of 4000–5000 daughter cells in ~3 weeks

- **Diploid adult hepatocytes ("small hepatocytes")**
  - A single cell can form a colony of ~130 daughter cells in ~3 weeks
  - Limited ability to be subcultured

                                                                                  Hypertrophic growth

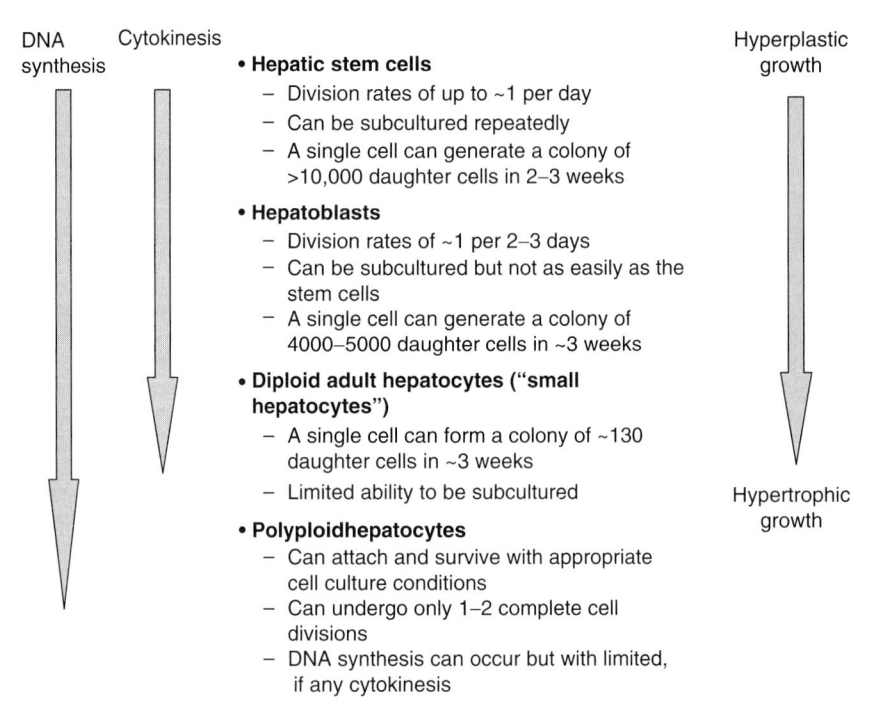

- **Polyploidhepatocytes**
  - Can attach and survive with appropriate cell culture conditions
  - Can undergo only 1–2 complete cell divisions
  - DNA synthesis can occur but with limited, if any cytokinesis

Fig. 14 Regenerative capacity of liver: combination of hyperplastic + hypertrophic growth.

	Glossary/nomenclature/abbreviations
Canals of Hering	Ductular structures around the portal triads of the liver acinus and found to be the stem cell niche in pediatric and adult livers. Assumed to be derived from the ductal plates found in fetal and neonatal livers.
Clonogenic expansion	Cells that can expand from a single cell and that can be repeatedly passaged at single cell seeding densities. Only the pluripotent progenitors (and possibly the unipotent, committed progenitors) are able to undergo clonogenic expansion.
Colony formation	Cells that can form a colony of cells when seeded at low densities. Diploid subpopulations, both progenitors and adult diploid cells, are able to form colonies of cells, but the adult diploid cells are limited in the numbers of divisions and are not able to undergo repeated passaging.
Committed progenitors	Unipotent progenitors capable of maturing into only one adult fate.
Determined stem cells	Pluripotent cells that can develop into some, but not all adult cell types and have self-renewed capacity.
Ductal Plate (also called limiting plate)	A band of cells encircling the portal triads in the acinus of fetal and neonatal livers and separating the connective tissue associated with the portal triads from the parenchyma. Recently found to be the stem cell niche (Schmelzer et al., 2007).
Embryonic stem (ES) cell	Pluripotent cells derived from pre- or post-implantation embryos and that can be maintained in their undifferentiated (unspecialized) state *ex vivo* under specific conditions; they can mature into all the cell types of all three germ layers (ectoderm, mesoderm, endoderm) but cannot give rise to extraembryonic cell types (e.g. amnions, placenta).
Oval cells	Small cells (~10 um in diameter) with oval-shaped nuclei and related to the stem cells and committed progenitors in the liver. They are located near the portal triads and expand in the livers of animals exposed to oncogenic insults. The insults result in cells that are partially or completely transformed. The term is used often as a synonym for the liver's stem cells and progenitors. However, they are distinguishable phenotypically and in their growth regulatory requirements from their normal progenitor counterparts (Oh et al., 2002).
Multipotent stem cells	This term has come into vogue only in the last few years to define the determined stem cells found in postnatal tissues as a way of distinguishing them from pluripotent embryonic stem cells.
Pluripotent cells	There have been political efforts in the last few years to restrict the use of this term to embryonic stem cells. However, the term was defined originally by embryologists and is still used today to mean cells that can self-replicate and that can give rise to daughter cells of more than one mature cell type. This applies to embryonic stem cells and to determined stem cells.
Progenitors or precursors	Broad terms encompassing both stem cells and committed progenitors.
Stem cells	Pluripotent cells that are capable of clonogenic expansion and self-replication (i.e. capable of producing daughter cells identical to the parent).
Totipotent stem cells	Cells capable of producing all cell types from all embryonic germ layers (ectoderm, mesoderm, and endoderm) plus the extraembryonic tissues such as amnions and placenta. Only zygotes and the cells of the first 3 divisions qualify as totipotent stem cells.

(continues)

Appendix 1 (*continued*)

Glossary/nomenclature/abbreviations	

The cellular subpopulations described below comprise lineage stages of epithelia and mesenchymal cells that are cell–cell partners with critical paracrine signaling: [following the list of the subpopulations are lists of markers identifying these subpopulations].

A small letter is used to indicate the species (m, murine; r, rat; h, human) and is coupled with an abbreviation for the cellular subpopulation

Hepatic stem cells (HpSCs)	Lineage stage 1 of the parenchymal cell lineage: In humans, they express EpCAM, NCAM, low levels of albumin, claudin 3, cytokeratins 8, 18, and 19, E-cadherin, and hedgehog proteins but not AFP, ICAM-1, or P450s. (This list is not complete: see Schmelzer *et al.*, 2006a).
Hepatoblasts (HBs)	Lineage stage 2 of the parenchymal cell lineage: Markers overlap with those of HpSCs but are distinctive from them in their expression of ICAM-1, higher levels of albumin, fetal forms of P450s (e.g. P450A7) and α-fetoprotein.
Hepatocytic committed progenitors	Lineage stage 3: Unipotent progenitors that give rise to hepatocytes.
Biliary committed progenitors	Lineage stage 3: Unipotent progenitors that give rise to biliary epithelia.
Angioblasts	Progenitors that give rise to endothelia. There is some evidence (Kubota *et al.*, 2007), albeit not yet conclusive, that they also give rise to hepatic stellate cells.
Hepatic stellate cells (HpSTCs)	Also called Ito cells. Found in the sinusoidal space, between the sinusoids and hepatocytes and represent 5–8% of the total number of liver cells. Their signature features are vitamin A and α-smooth muscle actin. They have been shown to produce many of the signals (matrix and soluble ones) regulating hepatic progenitors (Kubota *et al.*, 2007).
Endothelia	Cells derived from angioblasts and forming the cells lining blood vessels.
Stroma	The term is generic for multiple mesenchymal cell populations that in culture are bipolar and spindly in shape and provide critical paracrine signaling with epithelia.

Markers used to identify the subpopulations:

Parenchymal cells	ALB	Albumin
	AFP	α-Fetoprotein (note, different isoforms of AFP exist, some of them species specific).
	CK	Cytokeratin
	CK 8 and 18	Cytokeratins 8 and 18, ones typifying epithelia
	CK19	A cytokeratin unique to biliary epithelia
	EpCAM	Epithelial cell adhesion molecule. Found on HpSCs and HBs.
	NCAM	Neural cell adhesion molecule. Found on HpSCs.
	ICAM 1	Intercellular cell adhesion molecule. Found on HBs.
	CD133/1	Prominin found on various progenitors including HpSCs, HBs, and endothelia.
Angioblasts/endothelia	KDR (VEGFr2)	flt1; member of VEGF receptor (VEGFR) family
	CD31	Platelet/endothelial cell adhesion molecule or PECAM
	VWF	Von Willebrand's factor; a glycoprotein present in blood and produced by endothelium.
	CD117	c-kit, the receptor for stem cell factor
	CD133/1	Prominin found on HpSCs and on various stem cell populations

(*continues*)

8. Hepatic Stem Cells and Hepatoblasts: Identification, Isolation, and *Ex Vivo* Maintenance

Appendix 1 (*continued*)

	Glossary/nomenclature/abbreviations		
Hepatic stellate cells (HpSTCs)	Desmin	A 52 kDa protein that is a subunit of intermediate filaments found in muscle cells (smooth, cardiac and skeletal muscle cells), myofibroblasts, and HpSTCs.
	ASMA	α-Smooth muscle actin. One of the signature features of HpSTCs. Activation of HpSTCs results in greatly elevated levels of ASMA.
	Vitamin A	Retinoids such as vitamin A are another signature feature of HpSTCs isolated from both fetal and adult tissues.
Shared by angioblasts, endothelia, and hepatic stellate cells	VCAM-1	Vascular cell adhesion molecule-1; CD106. Expressed on endothelial cells stimulated by cytokines; also an endothelial ligand for VLA-4 (very late antigen-1 or α4β1) and for integrin α4β7.
	CD146	Expressed by both activated endothelia and activated HpSTCs. Also called MCAM or MELCAM, A32, MUC18. Functions are unknown.
	β3 integrin	Found on platelets, angioblasts, endothelia, and HpSTCs.

Appendix 2: Details on the Preparation of Buffers/Reagents

1. *10× Leffert's buffer (filtered):*

250 mM HEPES	59.57 g
1.15 M NaCl	69.27 g
50 mM KCl	3.73 g
10 mM KH$_2$PO$_4$	1.36 g
Combine in MQ water, pH to 7.4, and q.s. to 1.0 L	

2. *1 M CaCl$_2$ (autoclave)*

   73.5 g CaCl$_2$

   500 ml MQ H$_2$O

   0.2 M EGTA (filtered):

   38.04 g EGTA

   500 ml MQ H$_2$O

3. *Chelation buffer (A):*

   1× Leffert's/0.5 mM EGTA: 0.625 ml of 0.2 M EGTA stock in final volume of 250 ml

4. *Collagenase buffer (B):*

1× Leffert's/1 mM CaCl$_2$: 0.5 ml of 1 M CaCl$_2$ stock in final volume of 500 ml (digestion buffer). Within an hour of the time of the perfusion, add the collagenase to the buffer; for rat livers, add 70 mg of collagenase/500 ml; for human livers, add ~500–600 mg/l of buffer. The exact amount depends on the biological activity of the collagenase sample and must be adjusted also based on empirical findings with the routine samples being digested.

5. *Buffer C:*

1× Leffert's/2 mM CaCl$_2$. Add 0.5 ml of 1 M CaCl$_2$ stock in final volume of 250 ml. Add 1.5 g BSA at the time of use.

6. *Percoll® buffer:*

After the liver is digested and just prior to fractionation of the cells, mix the culture medium with 90% isotonic Percoll® (Sigma). The ratio of volumes should be 3 parts culture medium to 1 part Percoll (e.g. 30 ml of cells in medium + 10 ml of 90% isotonic Percoll). Adjust the ratio of medium to Percoll depending on the extent of fatty deposits in the liver. The more fat, the lower the Percoll ratio to the culture medium.

7. *Plating medium and culture medium for mature liver cells*

Use RPMI 1640 (GIBCO/BRL) or a 1:1 mixture of Dulbecco's modified Eagle's medium and Ham's F12 (DMEM/F12, GIBCO/BRL) to which is added 20 ng/ml EGF (Collaborative Biomedical Products), 5 µg/ml insulin (Sigma), 10 µg/ml iron-saturated transferrin (Sigma), 4.4×10^{-3} M nicotinamide (Sigma), 0.2% BSA (Sigma), 5×10^{-5} M 2-mercaptoethanol (Sigma), 7.6 µEq free fatty acid mixture (see below for its preparation), 2×10^{-3} M glutamine (GIBCO/BRL), 1×10^{-6} M CuSO$_4$, 3×10^{-8} M Na$_2$SeO$_3$ and antibiotics. For the plating medium add, in addition, 1–2% FBS (Hyclone) and keep the cells in it for only 4–6 h until the cells attach; it will be referred to as a serum-supplemented medium ("SSM") combined with a hormonally defined medium ("HDM") or "SSM/HDM". After the cells have attached, remove the SSM/HDM, rinse gently and switch to a serum-free version that will be called a serum-free, hormonally defined medium, "HDM".

8. *Cell wash used for human fetal liver processing*

- 500 ml RPMI 1640 (GIBCO 11875-093)
- 0.5 g BSA (Sigma # A8806-5G fatty acid free)
- 0.5 ml selenium ($3 \times 10E-5$ M, final $3 \times 10E-8$ M)
- 5.0 ml Antibiotics (GIBCO 15240-062, AAS)

Sterilize by filtering

9. *Kubota's Medium (KM)*
   - 500 ml RPMI 1640 (GIBCO # 11875-093)
   - 0.5 g BSA (Sigma # A8806 fatty acid free)
   - 270 mg Niacinamide (Sigma # N0636)
   - 5 µg/ml Insulin (Sigma # I5500)
   - 10 µg/ml Transferrin/Fe (Ado, bovine) (Sigma # T1283)
   - 5 ml l-Glutamine 200 mM (2 mM, GIBCO # 25030-081)
   - 5.0 ml Antibiotics (GIBCO # 15240-062, AAS)
   - 10E-8 M Hydrocortisone (Sigma # H0888)
   - 1.75 ml β-Mercaptoethanol (5E-5 M, Sigma # M6250)
   - 10E-10 M Selenium (Aldrich # 22,982-7)
   - 10E-10 M Zinc sulfate heptahydrate (10E-10 M, Specpure # JMC156)

Note: Zn is included in the crystallization process used by Sigma, at ~0.5% and the exact amount varies from lot to lot. However, the concentration at which zinc is effective but not toxic is broad. Moreover zinc is an absolute requirement for most cells when in serum-free medium given the large number of enzymes that utilize it as a co-factor. For example, in liver there are more than 30 liver enzymes with zinc as a required cofactor. Therefore, simply add the zinc at 10E-10 M.

   - 38 µl of the free fatty acids mix (see below—Buffer #10 for its preparation)

Note: The free fatty acids are toxic unless they are presented bound to purified, fatty acid free, endotoxin-free serum albumin (e.g. Pentex type V albumin). Albumin (of whatever species desired) is prepared in the basal medium to be used and at a typical concentration of 0.1–0.2%.

   - 10 µg/ml High density lipoprotein from human plasma (Sigma #L8039-solution, Sigma #L1567-lyophilized powder).

Note: HDL can be ignored for media used for plating the cells or for those for cultures to be maintained for only a few days. However, it is essential for cultures to be maintained for a week or longer

   - Sterilize by filtering through low protein-binding filters

10. *Free fatty acid mixture*–Chessebeuf and Padieu (1984)

Preparation of the fatty acid stocks

The free fatty acids are prepared by dissolving each individual free fatty acid in 100% ethanol. Most are readily dissolved in ethanol, though some require heating to achieve fully solubility.

Components are as follows:

Palmitic acid (solid, MW 256.4, Sigma # P0500)

Prepare 1.0 M solution: Weigh out palmitic acid, put into scintillation vial with screw cap. Add appropriate amount of ethanol and close cap tightly. Heat to 50 °C in water bath to dissolve. (256.4 mg/ml 500 mg/1.95 ml 1 g/3.9 ml)

Palmitoleic acid (MW 254.4, Sigma # P9417)
   Prepare 1.0 M solution (254.4 mg/ml 500 mg/1.97 ml 1 g/3.93 ml)
Stearic acid (MW 284.5, Sigma # S4751)
   Prepare 151 mM solution
Oleic acid (MW 282.5, Sigma # O1008)
   Prepare 1.0 M solution (282.5 mg/ml 500 mg/1.77 ml 1 g/3.54 ml)
Linoleic acid (free acid, not sodium salt)
   Prepare 1.0 M solution (280.4 mg/ml 1 g/3.6 ml)
Linolenic acid (MW 278.4, Sigma # L2376)
   Prepare 1.0 M solution (278.4 mg/ml 1 g/3.59 ml)

Palmitoleic, oleic, linoleic (free acid), and linolenic acids are liquids at RT. They are somewhat less dense than water, which can be used as a guide in determining how much alcohol to add to the contents of an ampoule. For example, 1 g of palmitoleic acid will be ~1 ml volume, and should be used to make 3.93 ml of solution. To be safe, ~2.7 ml of ethanol should be added and the total volume checked. Further additions of ethanol can be made to bring the volume to the correct level. These stocks can be stabilized by bubbling through nitrogen in each of them and then storing them at –20 °C.

Preparation of the MIx of Free Fatty Acids:
   Proportions to be used to reach a final concentration of 100 mM for each of the fatty acids as follows:
   31.0 volumes of 1.0 M palmitic acid
   2.8 volumes of 1.0 M palmitoleic acid
   76.9 volumes of 151 mM stearic acid
   13.4 volumes of 1.0 M oleic acid
   35.6 volumes of 1.0 M linolenic acid
   5.6 volumes of 1.0 M linolenic acid
   834.7 volume of alcohol
   The Volume typically used is a microtetter. Thus, the final volume will be 1 ml.

Final solution:
Add 76 ml of the free fatty acid mixture stock per liter of culture medium to achieve a final concentration of 7.6 µeq. The free fatty acids are toxic unless they are presented bound to purified, fatty acid free, endotoxin-free serum albumin (e.g. Pentex type V albumin). Albumin is prepared in the basal medium or PBS to be used and at a typical concentration of 0.1–0.2%.

11. *Digestion buffer (human fetal livers)*
   - 100 ml Cell Wash
   - Collagenase (Sigma # C5138): final 300 U/ml [*ideally this is Liberase*]
   - 30 mg DNAse (Sigma # DN25)
   Sterilize by filtering.

Appendix 3: Commercial Sources of Reagents

Vendors	
Growth factors, hormones, supplements	
Alexis Corp.	Cortex Biochemicals Inc.
American Qualex International Inc.	ICN Biomedicals
Antigenix America Inc	INVITROGEN
Biodesign International is Now Meridian Life Sciences Inc	Novabiochem
	Pepro Tech
BIOTREND Chemikalien	Sigma-Aldrich
Calbiochem	Spectrum Laboratory Products
Chemicon International	TCI America
Clonetics Products	Upstate Biologicals
Collaborative Biomedicals/BECTON DICKINSON	US Biological
Free fatty acids and lipids	
Academy Biomedical Co.	Chemicon International
BIOMOL Research Laboratories, Inc.	ICN Biomedicals
BIOTREND Chemikalien	Sigma-Aldrich
Matrix molecules	
Accurate Chemicals	Calbiochem
Becton Dickinson (includes Collaborative Biomedicals)	CarboMer, Inc.
Amersham Biosciences (microcarrier beads)	Chemicon International
Alexis Corp.	EY Laboratories
BioChemika is SIGMA	PolySciences, Inc.
Biodesign International	Sigma-Aldrich
Biosource International is INVITROGEN	Upstate Biologicals
BIOTREND Chemikalien	Calbiochem
Trace elements	
Alfa Aesar	MV Laoratories, Inc
Chem Services Inc.	Sigma-Aldrich
Crescent Chemicals	Spectrum Laboratory Products
Gallade Chemical, Inc.	Strem Chemicals, Inc.
ICN Biomedicals	
Sulfated proteoglycan and glycosaminoglycans	
Alexis Corp.	ICN Biomedicals
Calbiochem	Seikagaku USA
CarbOMer, Inc.	Sigma-Aldrich (includes Biochemika)
Chemicon International	TCI America
Clonetics	US Biologicals

Appendix 4: Sources of Primary and Secondary Antibodies for Immunoselection and for Characterization of the Cells. (See Below for Suppliers and the Location of the Suppliers)

Markers	Markers (sources)
Hemopoietic markers	Glycophorin A (Caltag #MHGLA04); CD14-FITC (Pharmingen #555397) CD34-FITC (Caltag #CD3458101) CD38-PE (Pharmingen #M030098) CD45-FITC (BD #347463)
Cell surface proteins	EpCAM, 1:800 (Neomarkers; # MS-144-P1ABX) N-CAM (CD56), 1:100 (Novocastra, #NCL-CD56-1B6; also NCAM 16.2-PE (BD # 340363) I-CAM-1 (CD54), 1:200 (Pharmingen; # 664970 or Bender MedSystems; #BMS108) CD54 or I-CAM (BD #347977) c-kit or CD117 (Dako #R7145) CD133-1 and CD133-2, 1:500 (Myltenyi Biotec # 130-080-801 and 130-080-901) Epithelial antigen (Dako # F0860) CD44 (hyaluronan receptor) 1:300 (Molecular Probes/Invitrogen) Claudin 3 (CLDN-3) (Abcam)
Intracellular proteins	Albumin, 1:800 (Sigma, # A6684) α-Fetoprotein (AFP), 1:500 (Sigma, # 8452 or Zymed, #18-0003) CK8/18, 1:1000 (Zymed, #18-0213) CK19, 1:200 (NovoCastra; #NCL-CK19)
Goat anti-mouse isotype-specific antibodies	Alexa Fluor 568 conjugated goat anti-mouse IgG2a, 1:200 (Molecular Probes, A21124) Cy5 conjugated goat anti-mouse IgG1, 1:200 (Southern Biotech; #1070-15)
Conjugated isotype controls	Mouse IgG FITC and PE (BD #'s 349041 and 349043)

Suppliers for antibodies

Abcam (Cambridge, MA)

BD/Pharmingen (San Diego, CA)

BD/Biosciences (San Jose, CA)

Bethyl (Montgomery, TX)

Biodesign (Saco, ME)

Biogenesis—now called AbD Serotec (Raleigh, NC)

Biomeda (Foster City, CA)

Cappel (Solon, OH)

Chemicon (Temecula, CA)

Covance (www.covance.com)

DakPierce (Glostrup, Denmark)
Fitzgerald (Concord, MA)
GE Healthcare/Amersham Biosciences (Piscataway, NJ)
Invitrogen/Biosources, Caltag, Molecular Probes, Zymed (www.invitrogen.com)
Jackson Immunoresearch (West Grove, PA)
Myltenyi Biotec (Auburn, CA)
Novocastra Laboratories (Newcastle upon Tyne, UK)
Pierce (Rockford, IL)
Santa Cruz Biotechnology (Santa Cruz, CA)
Southern Biotech (Birmingham, Al)
US Biological (Swampscott, MA)
Vector (Burlingame, CA)

Acknowledgments

The studies done at the Reid lab were supported by a sponsored research grant from Vesta Therapeutics (Research Triangle Park, Durham, NC), by National Institutes of Health grants (RO1 DK52851, RO1 AA014243, RO1 IP30-DK065933), and by a United States Department of Energy grant (DE-FG02-02ER-63477).

References

Aberle, H., Schwartz, H., and Kemler, R. (1996). Cadherin-catenin complex: Protein interactions and their implications for cadherin function. *J. Cell Biochem.* **61,** 514–523.

Anatskaya, O. V., Vnogradov, A. E., and Kudryavtsev, B. N. (1994). Hepatocyte polyploidy and metabolism/life-history traits: Hypotheses testing. *J. Theor. Biol.* **168,** 191–199.

Anti, M., Marra, G., Rapaccini, G. L., Rumi, C., Bussa, S., Fadda, G., Vecchio, F. M., Valenti, A., Percesepe, A., Pompili, M., Armelao, F., and Gentiloni, N. (1994). DNA ploidy pattern in human chronic liver diseases and hepatic nodular lesions. Flow cytometric analysis on echo-guided needle liver biopsy. *Cancer* **73,** 281–288.

Austin, T. W., Solar, G. P., Ziegler, F. C., Liem, L., and Matthews, W. (1997). A role for the Wnt gene family in hematopoiesis: Expansion of multilineage progenitor cells. *Blood* **89,** 3624–3635.

Baumann, U., Crosby, H. A., Ramani, P., Kelly, D. A., and Strain, A. (1999a). Expression of the stem cell factor receptor c-kit in normal and diseased pediatric liver: Identification of a human hepatic progenitor cell? *Hepatology* **30,** 112–117.

Baumann, U., Crosby, H. A., Ramani, P., Kelly, D. A., and Strain, A. J. (1999b). Expression of the stem cell factor receptor c-kit in normal and diseased pediatric liver: Identification of a human hepatic progenitor cell? *Hepatology* **30,** 112–117.

Beardsley, T. (1999). Stem cells come of age [news]. *Sci. Am.* **281,** 30–31.

Bernfield, M., Kokenyesi, R., Kato, M., Hinkes, M. T., Spring, J., Gallo, R. L., and Lose, E. J. (1992). Biology of the syndecans: A family of transmembrane heparan sulfate proteoglycans. *Annu. Rev. Cell Biol.* **8,** 365–393.

Berry, M. N., and Friend, D. J. (1969). High-yield preparation of isolated rat liver parenchymal cells: A biochemical and fine structure study. *J. Cell Biol.* **43,** 506–520.

Bhatia, S., Yarmush, M., and Toner, M. (1996). Controlling cell interactions by micropatterning in co-cultures: Hepatocytes and 3T3 fibroblasts. *J. Biomed. Mater. Res.* **32,** 1–11.

Bodnar, A. G., Ouellette, M., Frolkis, M., Holt, S. E., Chiu, C. P., Morin, G. B., Harley, C. B., Shay, J. W., Lichtsteiner, S., and Wright, W. E. (1998). Extension of life-span by introduction of telomerase into normal human cells [see comments]. *Science* **279,** 349–352.

Brill, S., Holst, P., Sigal, S., Zvibel, I., Fiorino, A., Ochs, A., Somasundaran, U., and Reid, L. M. (1993). Hepatic progenitor populations in embryonic, neonatal, and adult liver. *Proc. Soc. Exp. Biol. Med.* **204,** 261–269.

Brill, S., Holst, P. A., Zvibel, I., Fiorino, A., Sigal, S. H., Somasundaran, U., and Reid, L. M. (1994). Extracellular matrix regulation of growth and gene expression in liver cell lineages and hepatomas. *In* "Liver Biology and Pathobiology" (I. M. Arias, *et al.*, eds.), 3rd edn., pp. 869–897. Raven Press, New York.

Brinster, R. L. (1974). The effect of cells transferred into the mouse blastocyst on subsequent development. *J. Exp. Med.* **140,** 1049–1056.

Brodskii, V., Tsirekidze, N. N., Kogan, M. E., Delone, G. V., and Aref'eva, A. M. (1983). Measurement of the absolute cell count in the heart and liver. The quantitative preservation of proteins and DNA in isolated cells. *Tsitologiia* **25,** 260–265.

Brodsky, W. Y., Arefyeva, A. M., and Uryvaeva, I. V. (1980). Mitotic polyploidization of mouse heart myocytes during the first postnatal week. *Cell Tissue Res.* **210,** 133–144.

Brodsky, W. Y., and Uryvaeva, I. V. (1977). Cell polyploidy: Its relation to tissue growth and function. [Review] [247 refs.]. *Int. Rev. Cytol.* **50,** 275–332.

Bruder, S. P., Fink, D. J., and Caplan, A. I. (1994). Mesenchymal stem cells in bone development, bone repair, and skeletal regeneration therapy. *J. Cell Biochem.* **56,** 283–294.

Burger, H. J., Gebhardt, R., Mayer, C., and Mecke, D. (1989). Different capacities for aminoacid transport in periportal and perivenous hepatocytes isolated by digitonin/collagenase perfusion. *Hepatology* **9,** 22–28.

Burke, Z., and Oliver, G. (2002). Prox1 is an early specific marker for the developing liver and pancreas in the mammalian foregut endoderm. *Mech. Dev.* **118,** 147–155.

Cantz, T., Zuckerman, D. M., Burda, M. R., Dandri, M., Goricke, B., Thalhammer, S., Heckl, W. M., Manns, M. P., Petersen, J., and Ott, M. (2003). Quantitative gene expression analysis reveals transition of fetal liver progenitor cells to mature hepatocytes after transplantation in uPA/RAG-2 mice. *Am. J. Pathol.* **162,** 37–45.

Caplan, A. I. (1994). The mesengenic process. *Clin. Plast. Surg.* **21,** 429–435.

Carriere, R. (1967). Polyploid cell reproduction in normal adult rat liver. *Exp. Cell Res.* **46,** 533–540.

Cattaneo, E., and McKay, R. (1991). Identifying and manipulating neuronal stem cells. *Trends Neurosci.* **14,** 338–340.

Chen, U. (1992). Careful maintenance of undifferentiated mouse embryonic stemcells is necessary for their capacity to differentiate to hematopoietic lineages *in vitro*. *Curr. Top. Microbiol. Immunol.* **177,** 3–12.

Cheng, N., Yao, H., and Reid, L. (2007). Pluripotent hepatic stem cells and maturational lineage biology. *In* "Principles of Regenerative Medicine" (A. Attala, and R. Lanza, eds.), Elsevier Press, San Diego. Cal. pp. 344–384, 2007.

Chessebeuf, M., and Padieu, P. (1984). Rat liver epithelial cell cultures in a serum-free medium: Primary cultures and derived cell lines expressing differentiated functions. *In Vitro* **20,** 780–795.

Chow, I., and Poo, M. M. (1982). Redistribution of cell surface receptors induced by cell to cell contact. *J. Cell Biol.* **95,** 510–518.

Corbeil, D., Roper, K., Hannah, M. J., Hellwig, A., and Huttner, W. B. (1999). Selective location of the polytopic membrane protein prominin in microvilli of epithelial cells—A combination of apical sorting and retention in plasma membrane protrusion. *J. Cell Sci.* **112,** 1023–1033.

Cowan, C. A., Klimanskaya, I., McMahon, J., Atienza, J., Witmyer, J., Zucker, J. P., Wang, S., Morton, C. C., McMahon, A. P., Powers, D., and Melton, D. A. (2004). Derivation of embryonic stem-cell lines from human blastocysts. *N. Engl. J .Med.* **350,** 1353–1356.

Crissman, H. A., and Hirons, G. T. (1994). Staining of DNA in live and fixed cells. *In* "Methods in Cell Biology," pp. 195–209. Academic Press, New York.

Crosbie, O. M., Reynolds, M., McEntee, G., Traynor, O., Hegarty, J. E., and O'Farrelly, C. (1999). In vitro evidence for the presence of hematopoietic stem cells in the adult human liver. *Hepatology* **29**, 1193–1198.

Crosby, H. A., Kelly, D. A., and Strain, A. J. (2001). Human hepatic stem-like cells isolated using c-kit or CD34 can differentiate into biliary epithelium. *Gastroenterology* **120**, 534–544.

Dan, Y. Y., Riehle, K. J., Lazaro, C., Teoh, N., Haque, J., Campbell, J. S., and Fausto, N. (2006). Isolation of multipotent progenitor cells from human fetal liver capable of differentiating into liver and mesenchymal lineages. *Proc. Natl. Acad. Sci. USA* **103**, 9912–9917.

Deans, R. J., and Moseley, A. B. (2000). Mesenchymal stem cells: Biology and potential uses. *Exp. Hematol.* **28**, 875–884.

Dilworth, C., Hamliton, G. A., George, E., and Timbrell, J. A. (2000). The use of liver spheroids as an *in vitro* model for studying induction of the stress response as a marker of chemical toxicity. *Toxicol. In Vitro* **14**, 169–176.

Dunsford, H., and Sell, S. (1989). Production of monoclonal antibodies to preneoplastic liver cell populations induced by chemical carcinogens in rats and to transplantable Morris hepatomas. *Cancer Res.* **49**, 65–77.

Eckl, P. M., Whitcomb, W. R., Michalopoulos, G., and Jirtle, R. L. (1987). Effects of EGF and calcium on adult parenchymal hepatocyte proliferation. *J. Cell. Physiol.* **132**, 363–366.

Eferl, E., Sibilia, M., Hilberg, F., Fuchsbichler, A., Kufferath, I., Guertl, B., Zenz, R., Wagner, E., and Azatloukal, K. (1999). Functions of c-jun in liver development. *J. Cell Biol.* **145**, 1049–1061.

Eghbali, B., Kessler, J. A., Reid, L. M., Roy, C., and Spray, D. C. (1991). Involvement of gap junctions in tumorigenesis: Transfection of tumor cells with connexin 32 cDNA retards growth *in vivo*. *Proc. Natl. Acad. Sci. USA* **88**, 10701–10705.

Ek, S., Ringden, O., Markling, L., and Westgren, M. (1993). Cryopreservation of fetal stem cells. *Bone Marrow Transplant.* **11**, 123.

Enat, R., Jefferson, D. M., Ruiz-Opazo, N., Gatmaitan, Z., Leinwand, L. A., and Reid, L. M. (1984). Hepatocyte proliferation *in vitro*: Its dependence on the use of serum-free hormonally defined medium and substrata of extracellular matrix. *Proc. Natl. Acad. Sci. USA* **81**, 1411–1415.

Engelhardt, N. V., Factor, V. M., Medvinsky, A. L., Baranov, V. N., Lazareva, M. N., and Poltoranina, V. S. (1993). Common antigen of oval and biliary epithelial cells (A6) is a differentiation marker of epithelial and erythroid cell lineages in early development of the mouse. *Differentiation* **55**, 19–26.

Epstein, C. J., and Gatens, E. A. (1967). Nuclear ploidy in mammalian parenchymal liver cells. *Nature* **214**, 1050–1051.

Fiegel, H. C., Kluth, J., Lioznov, M. V., Holzhuter, S., Fehse, B., Zander, A. R., and Kluth, D. (2003). Hepatic lineages isolated from developing rat liver show different ways of maturation. *Biochem. Biophys. Res. Commun.* **305**, 46–53.

Fiegel, H. C., Park, J. J., Lioznov, M. V., Martin, A., Jaeschke-Melli, S., Kaufmann, P. M., Fehse, B., Zander, A. R., and Kluth, D. (2003). Characterization of cell types during rat liver development. *Hepatology* **37**, 148–154.

Fleming, R. J. (1998). Structural conservation of Notch receptors and ligands. *Semin. Cell Dev. Biol.* **9**, 599–607.

Fraser, J. R. E., and Laurent, T. C. (1989). Turnover and metabolism of hyaluronan. *In* "The Biology of Hyaluronans" (D. Evered, and J. Whelan, eds.), CIBA Foundation Symposium No. 143. pp. 41–59. Wiley and Sons, Hoboken, N.J.

Fraser, J. R. E., Laurent, T. C., and Laurent, U. B. G. (1997). Hyaluronan: Its nature, distribution, functions and turnover. *J. Int. Med.* **242**, 27–33.

Freshney, R. I. (2000). "Culture of Animal Cells." A. John Wiley and Sons, Inc., New York.

Fuchs, E., and Byrne, C. (1994). The epidermis: Rising to the surface. *Curr. Opin. Genet. Dev.* **4**, 725–736.

Fuchs, E., and Raghavan, S. (2002). Getting under the skin of epidermal morphogenesis. *Nat. Rev. Genet.* **3**, 199–209.

Fujimoto, I., Bruses, J. L., and Rutishauser, U. (2001). Regulation of cell adhesion by polysialic acid. Effects on cadherin, immunoglobulin cell adhesion molecule, and integrin function and independence from neural cell adhesion molecule binding or signaling activity. *J. Biol. Chem.* **276**, 31745–31751. Epub 2001 Jun 25.

Fujio, K., Evarts, R. P., Hu, Z., Marsden, E. R., and Thorgeirsson, S. S. (1994). Expression of stem cellfactor and its receptor, c-kit, during liver regeneration from putative stem cells in adult rat. *Lab. Invest.* **70**, 511–516.

Fujita, M., Spray, D. C., Choi, H., Saez, J., Jefferson, D. M., Hertzberg, E., Rosenberg, L. C., and Reid, L. M. (1986). Extracellular matrix regulation of cell-cell communication and tissue-specific gene expression in primary liver cultures. *Prog. Clin. Biol. Res.* **226**, 333–360.

Furthmayr, H. (1993). Basement membrane collagen: Structure, assembly, and biosynthesis. *In* "Extracellular Matrix: Chemistry, Biology, and Pathobiology with Emphasis on the Liver" (M. Zern, and L. M. Reid, eds.), pp. 149–185. Marcel Dekker, New York.

Gaasbeek Janzen, J. W., Gebhardt, R., ten Voorde, G. H., Lamers, W. H., Charles, R., and Moorman, A. F. (1987). Heterogeneous distribution of glutamine synthetase during rat liver development. *J. Histochem. Cytochem.* **35**, 49–54.

Gage, F. H. (1994). Neuronal stem cells: Their characterization and utilization. *Neurobiol. Aging* **15** (supplement 2), S191.

Garrod, D. R., and Nicol, A. (1981). Cell behaviour and molecular mechanisms of cell-cell adhesion. *Biol. Rev.* **56**, 199–242.

Gebhardt, R. (1988). Different proliferative activity *in vitro* of periportal and perivenous hepatocytes. *Scand. J. Gastroenterol. Suppl.* **151**, 8–18.

Gebhardt, R., Wegner, H., and Alber, J. (1996). Perifusion of co-cultured hepatocytes: Optimization of studies on drug metabolism and cytotoxicity *in vitro*. *Cell Biol. Toxicol.* **12**, 57–68.

Gerlyng, P., Abyholm, A., Grotmol, T., Erikstein, B., Huitfeldt, H. S., Stokke, T., and Seglen, P. O. (1993). Binucleation and polyploidization patterns in developmental and regenerative rat liver growth. *Cell Prolif.* **26**, 557–565.

Germain, L., Blouin, M. J., and Marceau, N. (1988). Biliary epithelial and hepatocytic cell lineage relationships in embryonic rat liver as determined by the differential expression of cytokeratins, alpha-fetoprotein, albumin, and cell surface-exposed components. *Cancer Res.* **48**, 4909–4918.

Gill, J. F., Chambers, L. L., Baurley, J. L., Ellis, B. B., Cavanaugh, T. J., Fetterhov, T. J., and Dwulet, F. E. (1995). Safety testing of Liberase, a purified enzyme blend for human islet isolation. *Transplant. Proc.* **27**, 3276–3277.

Gittes, G. K., Galante, P. E., Hanahan, D., Rutter, W. J., and Debase, H. T. (1996). Lineage-specific morphogenesis in the developing pancreas: Role of mesenchymal factors. *Development* **122**, 439–447.

Gluckman, E., Broxmeyer, H. A., Auerbach, A. D., Freidman, H. S., Douglas, G. W., Devergie, A., Esperou, H., Thierry, D., Socie, G., Lehn, P., Cooper, S., English, D., *et al.* (1989). Hematopoietic reconstitution in a patient with Fanconi's anemia by means of umbilical cord blood from an HLA-identical sibling. *N. Engl. J. Med.* **321**, 1174–1178.

Gressner, A. M., and Vasel, A. (1985). Proteochondroitin sulfate is the main proteoglycan synthesized in fetal hepatocytes. *Proc. Soc. Exp. Biol. Med.* **180**, 334–339.

Grisham, J. W. (1997). Hepatocyte lineages: Of clones, streams, patches, and nodules in the liver. *Hepatology* **25**, 250–252.

Gumucio, J. J. (1989). Hepatocyte heterogeneity. *Hepatology* **9**(1), 154–160.

Guyomard, C., Rialland, L., Fremond, B., Chesne, C., and Guillouzo, A. (1996). Influence of alginate gel entrapment and cryopreservation on survival and xenobiotic metabolism capacity of rat hepatocytes. *Toxicol. Appl. Pharmacol.* **141**, 349–356.

Hamilton, G., Westmoreland, C., and George, E. (1996). Liver spheroids as a long-term model for liver toxicity *in vitro*. *Hum. Exp. Toxicol.* **15**, 153.

Hamilton, G. A., Jolley, S. L., Gilbert, D., Coon, D. J., Barros, S., and LeCluyse, E. L. (2001). Regulation of cell morphology and cytochrome P450 expression in human hepatocytes by extracellular matrix and cell-cell interactions. *Cell Tissue Res.* **306,** 85–99.

Harley, C. B., and Villeponteau, B. (1995). Telomeres and telomerase in aging and cancer. *Curr. Opin. Genet. Dev.* **5,** 249–255.

Hentsch, B., Lyons, I., Li, R., Hartley, L., Lints, T. J., Adams, J. M., and Harvey, R. P. (1996). Hlx homebox gene is essential for an inductive tissue interaction that drives expansion of embryonic liver and gut. *Genes Dev.* **10,** 70–79.

Hernandez, J., Zarnegar, R., and Michalopoulos, G. K. (1992). Characterization of the effects of humanplacental HGF on rat hepatocytes. *J. Cell. Physiol.* **150,** 116–121.

Higgins, P. J., Kessler, G. K., Nisselbaum, J. S., and Melamed, M. R. (1985). Characterization and cell cycle kinetics of hepatocyte populations isolated from adult liver tissue by a nonenzymatic procedure. *J. Histochem. Cytochem.* **33,** 672–676.

Hilberg, F., Aguzzi, A., Howells, N., and Wagner, E. F. (1993). c-jun is essential for normal mouse development and hepatogenesis. *Nature* **365,** 179–181.

Hixson, D. C., Faris, R. A., and Thompson, N. L. (1990). An antigenic portrait of the liver during carcinogenesis. *Pathobiology* **58,** 65–77.

Hoffmann, B., and Paul, D. (1990). Basic fibroblast growth factor and transforming growth factor alpha are hepatotrophic mitogens in vitro. *J. Cell. Physiol.* **142,** 149–154.

Ito, S., Tateno, C., Tuda, M., and Yoshitake, A. (1996). Immunohistochemical demonstration of the gap junctional protein connexin 32 and proliferating cell nuclear antigen in glutathione S-transferase placental form—Negative lesions of rat liver induced by diethylnitrosamine and clofibrate. *Toxicol. Pathol.* **24,** 690–695.

Jefferson, D. M., Clayton, D. F., Darnell, J. E., Jr., and Reid, L. M. (1984). Posttranscriptional modulation of gene expression in cultured rat hepatocytes. *Mol. Cell. Biol.* **4,** 1929–1934.

Jensen, C. H., Jauho, E. I., Santoni-Rugiu, E., Holmskov, U., Teisner, B., Tygstrup, N., and Bisgaard, H. C. (2004). Transit-amplifying ductular (oval) cells and their hepatocytic progeny are characterized by a novel and distinctive expression of delta-like protein/preadipocyte factor 1/fetal antigen 1. *Am. J. Pathol.* **164,** 1347–1359.

Joseph, B., Bhargava, K., Malhi, H., Schilsky, M. L., Jain, D., Palestro, C., and Gupta, S. (2003). Sestamibi is a substrate for MDR1 and MDR2 P-glycoprotein genes. *Eur. J. Nucl. Med. Mol. Imaging* **30,** 1024–1031.

Jung, J., Zheng, M., Goldfarb, M., and Zaret, K. S. (1999). Initiation of mammalian liver development from endoderm by fibroblast growth factors. *Science* **284,** 1998–2003.

Kallunki, P., and Tryggvason, K. (1992). Human basement membrane heparan sulfate proteoglycan core protein: A 467-kD protein containing multiple domains resembling elements of the low density lipoprotein receptor, laminin, neural cell adhesion molecules, and epidermal growth factor. *J. Cell Biol.* **116,** 559–571.

Kearney, J. B., Kappas, N. C., Ellerstrom, C., DiPaola, F. W., and Bautch, V. L. (2004). The VEGF receptor flt-1 (VEGFR-1) is a positive modulator of vascular sprout formatin and branching morphogenesis. *Blood* **103,** 4379–4380.

Keng, V. W., Yagi, H., Ikawa, M., Nagano, T., Myint, Z., Yamada, K., Tanaka, T., Sato, A., Muramatsu, I., Okabe, M., Sato, M., and Noguchi, T. (2000). Homeobox gene Hex is essential for onset of mouse embryonic liver development and differentiation of the monocyte lineage. *Biochem. Biophys. Res. Commun.* **276,** 1155–1161.

Knudson, W., and Knudson, C. B. (1991). Assembly of a chondrocyte-like pericellular matrix on nonchondrogenic cells. *J. Cell Sci.* **99,** 227–235.

Kobayashi, Y., Okamoto, A., and Nishinari, K. (1994). Viscoelasticity of hyaluronic acid different molecular weights. *Biorheology* **31,** 235–244.

Koebe, H. G., Dahnhardt, C., Muller-Hocker, J., Wagner, H., and Schildberg, F. W. (1996). Cryopreservation of porcine hepatocyte cultures. *Cryobiology* **33,** 127–141.

Koebe, H. G., Dunn, J. C., Toner, M., Sterling, L. M., Hubel, A., Cravalho, E. G., Yarmush, M. L., and Tompkins, R. G. (1990). A new approach to the cryopreservation of hepatocytes in a sandwich culture configuration. *Cryobiology* **27**, 576–584.

Koide, N., Sakaguchi, K., Koide, Y., Asano, K., Kawaguchi, M., Matsushima, H., Takenami, T., Shinji, T., Mori, M., and Tsuji, T. (1990). Formation of multicellular spheroids composed of adult rat hepatocytes in dishes with positively charged surfaces and under other nonadherent environments. *Exp. Cell Res.* **186**, 227–235.

Koide, N., Shinji, T., Tanabe, T., Asano, K., Kawaguchi, M., Sakaguchi, K., Koide, Y., Mori, M., and Tsuji, T. (1989). Continued high albumin production bymulticellular spheroids of adult rat hepatocytes formed in the presence of liver-derived proteoglycans. *Biochem. Biophys. Res. Commun.* **161**, 385–391.

Kubota, H., and Reid, L. M. (2000). Clonogenic hepatoblasts, common precursors for hepatocytic and biliary lineages, are lacking classical major histocompatibility complex class I antigen. *Proc. Natl. Acad. Sci. USA* **97**, 12132–12137.

Kubota, H., Storms, R. W., and Reid, L. M. (2002). Variant forms of alpha-fetoprotein transcripts expressed in human hematopoietic progenitors. Implications for their developmental potential towards endoderm. *J. Biol. Chem.* **277**, 27629–27635.

Kubota, H., Yao, H., and Reid, L. M. (2007). Identification and characterization of vitamin A-storing cells in fetal liver. *Stem Cells* **25**, 2339–2349, 2007.

Kudryavtsev, B. N., Kudryavtseva, M.V, Sakuta, G. A., and Stein, G. I. (1993). Human hepatocyte polyploidization kinetics in the course of life cycle. *Virchows Arch., B, Cell Pathol. Incl. Mol. Pathol.* **64**, 387–393.

Lagasse, E. (2007). Liver engrafting cells, assays, and uses thereof. U.S. Patent #10177178. Filed June 21, 2002.

Landry, J., Bernjer, D., Oullet, C., Gayette, R., and Marceau, N. (1985). Spheroidal aggregate culture of rat liver cells: Histotypic reorganization, biomatrix deposition and maintenance of functional activities. *J. Cell Biol.* **101**, 914–923.

Landry, J., and Freyer, J. P. (1984). Regulatory mechanisms in spheroidal aggregates of normal and cancerous cells. *Recent Results Cancer Res.* **95**, 50–66.

Laurent, T. C., and Fraser, J. R. E. (1986). The properties and turnover of hyaluronan. *In* "The Biology of Hyaluronans". Ciba Foundation Symposium No. 124, pp. 9–29. Wiley and Sons, Hoboken, N.J.

Laurent, T. C., and Fraser, J. R. E. (1991). Catabolism of hyaluronan. *In* "Degradation of Bioactive Substances: Physiology and Pathophysiology" (H. Henriksen, ed.), pp. 249–265. CRC Press, Boca Raton.

LeCluyse, E. L. (2001). Human hepatocyte culture systems for the *in vitro* evaluation of cytochrome P450 expression and regulation. *Eur. J. Pharm. Sci.* **13**, 343–368.

LeCluyse, E. L., Bullock, P. L., and Parkinson, A. (1996). Strategies for restoration and maintenance of normal hepatic structure and function in long-term cultures of rat hepatocytes. *Adv. Drug. Del. Rev.* **22**, 133–186.

Lecouter, J., Lin, R., and Ferrara, N. (2004). EG-VEGF: A novel mediator of endocrine-specific angiogenesis, endothelial phenotype, and function. *Ann. N. Y. Acad. Sci.* **1014**, 50–57.

Lemaigre, F., and Zaret, K. S. (2004). Liver development update: New embryo models, cell lineage control, and morphogenesis. *Curr. Opin. Genet. Dev.* **14**, 582–590.

Levenberg, S., Khademhosseini, A., and Langer, R. (2004). Embryonic stem cells in tissue engineering. *In* "Handbook of Stem Cells" (R. Lanza, H. Blau, D. Melton, M. Moore, E. D. Thomas, C. Verfailie, I. Weissman, and M. West, eds.), Vol. 1, pp. 737–746. Elsevier, New York.

Lewis, J. (1998). Notch signalling and the control of cell fate choices in vertebrates. *Semin. Cell Dev. Biol.* **9**, 583–589.

Li, A. P., Colburn, S. M., and Beck, D. J. (1992a). A simplified method for the culturing of primary adult rat and human hepatocytes as multicellular spheroids. *In vitro Cell Dev. Biol.* **28A**, 673–677.

Li, A. P., Roque, M. A., Beck, D. J., and Kaminski, D. L. (1992b). Isolation and culturing of hepatoyctes from human livers. *J. Tissue Cult. Methods* **14**, 139–146.

Lin, Y., Weisdorf, D. J., Solovey, A., and Hebbel, R. P. (2000). Origins of circulating endothelial cells and endothelial outgrowth from blood. *J. Clin. Invest.* **105,** 71–77.

Liu, H., Di Cunto, F., Imarisio, S., and Reid, L. M. (2003). Citron kinase is a cell cycle-dependent, nuclear protein required for G2/M transition of hepatocytes. *J. Biol. Chem.* **278,** 2541–2548.

Liu, Z., and Chang, T. M. (2000). Effects of bone marrow cells on hepatocytes: When co-cultured or co-encapsulated together. *Artif. Cells Blood Substit. Immobil. Biotechnol.* **28,** 365–374.

Macdonald, J., Grifin, J., Kubota, H., Griffth, L., Fair, J., and Reid, L. M. (1999). Bioartificial livers. *In* "Cell Encapsulation Technology and Therapeutics" (W. M. Kuhtreiber *et al.*, eds.), pp. 252–286. Birkhauser, Boston.

Macdonald, J. M., Xu, A., Kubota, H., Lecluyse, E., Hamilton, G., Liu, H., Rong, Y., Moss, N., Lodestro, C., Luntz, T., Wolfe, S. P., and Reid, L. M. (2002). Liver cell culture and lineage biology. *In* "Methods of Tissue Engineering" (A. Atala, and R. P. Lanza, eds.), pp. 151–202. Academic Press, London.

Maher, J. J., and Bissell, D. M. (1993). Cell-matrix interactions in liver. *Semin. Cell Biol.* **4,** 189–201.

Mahoney, D. J., Alpin, R. T., Calabro, A., Hascall, V. C., and Day, A. J. (2001). Novel methods for the preparation and characterization of hyaluronan oligosaccharides of defined length. *Glycobiology* **11,** 1025–1033.

Malik, R., Selden, C., and Hodgson, H. (2002). The role of non-parenchymal cells in liver growth. *Semin. Cell Dev. Biol.* **13,** 425–431.

Maltsev, V. A., Rohwedel, J., Hescheler, J., and Wobus, A. M. (1993). Embryonic stem cells differentiate *in vitro* into cardiomyocytes representing sinusnodal, atrial and ventricular cell types. *Mech. Dev.* **44,** 41–50.

Martin, G. R. (1981). Isolation of a pluripotent cell line from early mouse embryos cultured in medium conditioned by teratocarcinoma stem cells. *Proc. Natl. Acad. Sci. USA* **78,** 7634–7638.

Martin, G. R., and Evans, M. J. (1975). Differentiation of clonal lines of teratocarcinoma cells: Formation of embryoid bodies *in vitro*. *Proc. Natl. Acad. Sci. USA* **72,** 1441–1445.

Martinez Barbera, J. P., Clements, M., Thomas, P., Rodriguez, T., Meloy, D., Kioussis, D., and Beddington, R. S. (2000). The homeobox gene Hex is required in definitive endodermal tissues for normal forebrain, liver and thyroid formation. *Dev. Suppl.* **127,** 2433–2445.

Martinez-Hernandez, A., and Amenta, P. S. (1993a). The hepatic extracellular matrix. I. Components and distribution in normal liver [editorial]. *Virchows Arch. A Pathol. Anat. Histopathol.* **423,** 1–11.

Martinez-Hernandez, A., and Amenta, P. S. (1993b). The hepatic extracellular matrix. II. Ontogenesis, regeneration and cirrhosis. *Virchows Arch. A Pathol. Anat. Histopathol.* **423,** 77–84.

Martinez-Hernandez, A., and Amenta, P. S. (1993c). "Morphology, Localization, and Origin of the Hepatic Extracellular Matrix." Marcel Dekker, New York.

Martinez-Hernandez, A., and Amenta, P. S. (1995). The extracellular matrix in hepatic regeneration. *FASEB J.* **140,** 4–1410.

Matsusaka, S., Tsujimura, T., Toyosaka, A., Nakasho, K., Sugihara, A., Okamoto, E., Uematsu, K., and Terada, N. (1999). Role of c-kit receptor tyrosine kinase in development of oval cells in the rat 2-acetylaminofluorene/partial hepatectomy model. *Hepatology* **29,** 670–676.

Matsushita, T., Ijima, H., Koide, N., and Funatsu, K. (1991). High albumin production by multicellular spheroids of adult rat hepatocytes formed in the pores of polyurethane foam. *Appl. Microbiol. Biotechnol.* **36,** 324–326.

Matturri, L., Campiglio, G. L., Lavezzi, A. M., Riberti, C., Cavalca, D., and Azzolini, A. (1991). Cell kinetics and DNA content (ploidy) of human skin under expansion. *Eur. J. Basic Appl. Histochem.* **35,** 73–79.

McClelland, R., and Coger, R. (2000). Use of micropathways to improve oxygen transport in a hepatic system. *J. Biomech. Eng.* **122,** 268–273.

McClelland, R., Denis, R., Reid, L. M., and Macdonald, J. M. (2005). Tissue Engineering. *In* "Introduction to Biomedical Engineering" (John Enderle, Susan Blanchard, and Joseph Bronzino, eds.), Elsevier Academic Press, San Diego, Cal. Chapter 7, pp. 313–400.

McClelland, R. E., and Reid, L. M. (2007). Bioartificial livers. *In* "Principles of Regenerative Medicine" (A. Attala, and R. Lanza, eds.), 3rd edition, Elsevier Academic Press, San Diego, Cal. pp. 928–945.

McClelland, R., Wauthier, E., Uronis, J., and Reid, L. (2008). Differential effects of extracellular matrix from periportal versus pericentral liver on human heptic stem cells. *Tissue Engineering.* **14**(1), 59–70.

McClelland, R. E., Wauthier, E., Schmelzer, E., and Reid, L. (2007). *Ex vivo* conditions for self replication of human hepatic stem cells. Tissue Engineering (in press).

McKeehan, W. L., Barnes, D., Reid, L., Stanbridge, E., Murakami, H., and Sato, G. H. (1990). Frontiers in mammalian cell culture. *In vitro Cell. Dev. Biol.* **26**, 9–23.

Mendiola, M. M., Peters, T., Young, E. W., and Zoloth-Dorfman, L. (1999). Research with human embryonic stem cells: Ethical considerations. *Hastings Cent. Rep.* **29**, 31–36.

Minguet, S., Cortegano, I., Gonzalo, P., Martinez-Marin, J., Andres, B., Salas, C., Melero, D., Gaspar, M., and Marcos, M. (2003). A population of c-kit((low), CD45-, TER119- hepatic cell progenitors of 11 day postcoitus mouse embryliver reconstitutes cell-depleted liver organoids. *J. Clin. Invest.* **112**, 1152–1163.

Mitaka, T., Kojima, T., Mizuguchi, T., and Mochizuki, Y. (1995). Growth and maturation of small hepatocytes isolated from adult rat liver. *Biochem. Biophys. Res. Commun.* **214**, 310–317.

Mitaka, T., Mikami, M., Sattler, G. L., Pitot, H. C., and Mochizuki, Y. (1992). Small cell colonies appear in the primary culture of adult rat hepatocytes in the presence of nicotinamide and epidermal growth factor. *Hepatology* **16**, 440–447.

Mitaka, T., Mizuguchi, T., Sato, F., Mochizuki, C., and Mochizuki, Y. (1998). Growth and maturation of small hepatocytes. [Review] [44 refs.]. *J. Gastroenterol. Hepatol.* **13**, S70–S77.

Mitaka, T., Norioka, K., Nakamura, T., and Mochizuki, Y. (1993). Effects of mitogens and co-mitogens on the formation of small-cell colonies in primary cultures of rat hepatocytes. *J. Cell. Physiol.* **157**, 461–468.

Mizuguchi, T., Hui, T., Palm, K., Sugiyama, N., Mitaka, T., Demetriou, A. A., and Rozga, J. (2001). Enhanced proliferation and differentiation of rat hepatocytes cultured with bone marrow stromal cells. *J. Cell. Physiol.* **189**, 106–119.

Monaghan, A. P., Kaestner, K. H., Grau, E., and Schutz, G. (1993). Postimplantation expression patterns indicate a role for the mouse forkhead/HNF-3 alpha, beta and gamma genes in determination of the definitive endoderm, chordamesoderm and neuroectoderm. *Development* **119**, 567–578.

Morrison, S. J., Prowse, K. R., Ho, P., and Weissman, I. L. (1996). Telomerase activity in hematopoietic cells is associated with self-renewal potential. *Immunity* **5**, 207–216.

Moscono, A. (1961). Rotation-mediated hostgenic aggregation of dissociated cells. *Exp. Cell Res.* **22**, 455–475.

Mossin, L., Blankson, H., Huitfeldt, H., and Seglen, P. O. (1994). Ploidy-dependent growth and binucleation in cultured rat hepatocytes. *Exp. Cell Res.* **214**, 551–560.

Muschel, R., Khoury, G., and Reid, L. M. (1986). Regulation of insulin mRNA abundance and adenylation: Dependence on hormones and matrix substrata. *Mol. Cell. Biol.* **6**, 337–341.

Nimni, M. E. (1993). Fibrillar collagens: Their biosynthesis, molecular structure, and mode of assembly. *In* "Extracellular Matrix: Chemistry, Biology, and Pathobiology with Emphasis on the Liver" (M. Zern, and L. M. Reid, eds.), pp. 121–148. Marcel Dekker, New York.

Nitou, M., Sugiyama, Y., Ishikawa, K., and Shiojiri, N. (2002). Purification of fetal mouse hepatoblasts by magnetic beads coated with monoclonal anti-E-cadherin antibodies and their *in vitro* culture. *Exp. Cell Res.* **279**, 330–343.

Nozaki, T., Masutani, M., Watanabe, M., Ochiya, T., Hasegawa, F., Nakagama, H., Suzuki, H., and Sugimura, T. (1999). Syncytiotrophoblastic giant cells in teratocarcinoma-like tumors derived from Parp-disrupted mouse embryonic stem cells. *Proc. Natl. Acad. Sci. USA* **96**, 13345–13350.

Ogawa, K., Ochoa, E. R., Borenstein, J., Tanaka, K., and Vacanti, J. P. (2004). The generation of functionally differentiated, three-dimensional hepatic tissue from two-dimensional sheets of progenitor small hepatocytes and nonparenchymal cells. *Transplantation* **77**, 1783–1789.

Oh, S. H., Hatch, H. M., and Petersen, B. E. (2002). Hepatic oval 'stem' cell in liver regeneration. *Semin. Cell Dev. Biol.* **13**, 405–409.

Olack, B. J., Swanson, C. J., Howard, T. K., and Mohanakumar, T. (1999). Improved method for the isolation and purification of human islets of Langerhans using Liberase enzyme blend. *Hum. Immunol.* **60,** 1303–1309.

Omori, M., Evarts, R. P., Omori, N., Hu, Z., Marsden, E. R., and Thorgeirsson, S. S. (1997). Expression of alpha-fetoprotein and stem cell factor/c-kit system in bile duct ligated young rats. *Hepatology* **25,** 1115–1122.

Oren, R., Dabeva, M., Petkov, P., Hurston, E., Laconi, E., and Shafritz, D. (1999). Restoration of serum albumin levels in nagase analbuminemic rats by hepatocyte transplantation. *Hepatology* **29,** 75–81.

O'Shea, K. S. (1999). Embryonic stem cell models of development. *Anat. Rec.* **257,** 32–41.

Overturf, A. D. M., Finegold, M., and Grompe, M. (1999). The repopulation potential of hepatocyte populations differing in size and prior mitotic expansion. *Am. J Pathol.* **155,** 2135–2143.

Passegue, E., Jochum, W., Behrens, A., Ricci, R., and Wagner, E. (2002). Jun B can substitute for jun in mouse development and cell proliferation. *Nat. Genet.* **30,** 158–166.

Paterson, D. J., Green, J. R., Jefferies, W. A., Puklavec, M., and Williams, A. F. (1987). The MRC OX-44 antigen marks a functionally relevant subset among rat thymocytes. *J. Exp. Med.* **165,** 1–13.

Pedersen, R. A. (1999). Embryonic stem cells for medicine. *Sci. Am.* **280,** 68–73.

Petersen, B. E., Grossbard, B., Hatch, H., Pi, L., Deng, J., and Scott, E. W. (2003). Mouse A6-positive hepatic oval cells alsexpress several hematopoietic stem cell markers. *Hepatology* **37,** 632–640.

Petzer, A. L., Hogge, D. E., Landsdorp, P. M., Reid, D. S., and Eaves, C. J. (1996). Self-renewal of primitive human hematopoietic cells (long-term-culture-initiating cells) *in vitro* and their expansion in defined medium. *Proc. Natl. Acad. Sci. USA* **93,** 1470–1474.

Plescia, C., Rogler, C., and Rogler, L. (2001). Genomic expression analysis implicates Wnt signaling pathway and extracellular matrix alterations in hepatic specification and differentiation of murine hepatic stem cells. *Differentiation* **68,** 254–269.

Potten, C. S., Clarke, R. B., Wilson, J., and Renehan, A. G. (Eds.) (2006). Tissue stem cells Taylor and Francis Group, New York.

Reid, L. M. (1990). Stem cell biology, hormone/matrix synergies and liver differentiation. *Curr. Opin. Cell Biol.* **2,** 121–130.

Reid, L. M., Agelli, M., and Ochs, A. (1994). Method of expanding hepatic precursor cells U.S. Patent 5,576,207. Date of application: August 7, 1991.

Reid, L. M., Fiorino, A. S., Sigal, S. H., Brill, S., and Holst, P. A. (1992). Extracellular matrix gradients in the space of Disse: Relevance to liver biology. *Hepatology* **15,** 1198–1203.

Reid, L. M., and Jefferson, D. M. (1984). Culturing hepatocytes and other differentiated cells. *Hepatology* **4,** 548–559.

Reid, L. M., and LeCluyse, E. (2000). Liver tissue source U.S. Patent Application# 09/764359. Publication # 2002/0039786.

Reid, L. M., and Luntz, T. L. (1997). *Ex vivo* maintenance of differentiated mammalian cells. In "Basic Cell Culture Protocols" (J. W. Pollard, and J. M. Walker, eds.), pp. 31–57. Humana Press, Totowa, NJ.

Reid, M. Z. A. L. (1993). "Extracellular Matrix Chemistry and Biology." Academic Press, New York.

Reinholt, F. P., Hultenby, K., Tibell, A., Korsgren, O., and Groth, C. G. (1998). Survival of fetal porcine pancreatic islet tissue transplanted to a diabetic patient. *Xenotransplantation* **5,** 222–225.

Resnick, J. L., Bixler, L. S., Cheng, L., and Donovan, P. J. (1992). Long-term proliferation of mouse primordial germ cells in culture [see comments]. *Nature* **359,** 550–551.

Reubinov, B. E., Pera, M. F., Fong, C. Y., Trounson, A., and Bongso, A. (2000). Embryonic stem cell lines from human blastocysts: Somatic differentiation *in vitro*. *Nat. Biotechnol.* **18,** 399–404.

Roberts, D. M., Kearney, J. B., Johnson, J. H., Rosenberg, M. P., Kumar, R., and Bautch, V. L. (2004). The EGF receptor flt-1 (EGFR-1) modulates flk-1 (EGFR-2) signaling during blood vessel formation. *Am. J. Pathol.* **164,** 1531–1535.

Rollini, P., and Fournier, R. E. (1999). The HNF-4/HNF-1alpha transactivation cascade regulates geneactivity and chromatin structure of the human serine protease inhibitor gene cluster at 14q32.1. *Proc Natl. Acad. Sci. USA* **96,** 10308–10313.

Ros, J. E., Libbrecht, L., Geuken, M., Jansen, P. L., and Roskams, T. A. (2003). High expression of MDR1, MRP1, and MRP3 in the hepatic progenitor cell compartment and hepatocytes in severe human liver disease. *J. Pathol.* **200,** 553–560.

Rosenberg, E., Spray, D. C., and Reid, L. M. (1992). Transcriptional and posttranscriptional control of connexin mRNAs in periportal and pericentral rat hepatocytes. *Eur. J. Cell Biol.* **59,** 21–26.

Rubinstein, P., Rosenfield, R. E., Adamson, J. W., and Stevens, C. E. (1993). Stored placental blood for unrelated bone marrow reconstitution. *Blood* **81,** 1679–1690.

Runge, D., Runge, D. M., Bowen, W. C., Locker, J., and Michalopoulos, G. K. (1997). Matrix induced re-differentiation of cultured rat hepatocytes and changes of CCAAT/enhancer binding proteins. *Biol. Chem.* **378,** 873–881.

Runge, D., Runge, D. M., Drenning, S. D., Bowen, W. C., Jr., Grandis, J. R., and Michalopoulos, G. K. (1998). Growth and differentiation of rat hepatocytes: Changes in transcription factors HNF-3, HNF-4, STAT-3, and STAT-5. *Biochem. Biophys. Res. Commun.* **250,** 762–768.

Ruoslahti, E., and Yamaguchi, Y. (1991). Proteoglycans as modulators of growth factor activities. *Cell* **64,** 867–869.

Saeter, G., Lee, C. Z., Schwarze, P. E., Ous, S., Chen, D. S., Sung, J. L., and Seglen, P. O. (1988). Changes in ploidy distributions in human liver carcinogenesis. *J. Natl. Cancer Inst.* **80,** 1480–1485.

Santella, L. (1998). The role of calcium in the cell cycle: Facts and hypotheses. *Biochem. Biophys. Res. Commun.* **244,** 317–324.

Sato, G. H. (1984). Hormonally defined media and long-term marrow culture: General principles. *Kroc Foundation Ser.* **18,** 133–137.

Schmelzer, E., and Reid, L. M. (2008). EpCAM expression in normal, non-pathological tissues. *(Review) Frontiers in Biosciences* **13,** 3096–3100. 2008.

Schmelzer, E., Wauthier, E., and Reid, L. M. (2006). The phenotypes of pluripotent human hepatic progenitors. *Stem Cells* **24,** 1852–1858.

*Schmelzer, E., *Zhang, L., *Bruce, A., Wauthier, E., Yao, H., Moss, NG., Melhem, A., Tallheden, T., McClelland, R., Tallheden, T., Cheng, N., Kulik, M., *et al.* (2007). Human Hepatic Stem Cells from Fetal and Postnatal Donors. *Journal of Experimental Medicine* **204**(8), 1973–1987. [*co-equal first authors].

Schmelzer, E., Zhang, L., Melhem, A., Yao, H., Turner, W., McClelland, R., Wauthier, E., Furth, M., Gerber, D., Gupta, S., Kulik, M., Sherwood, S., *et al.* (2006). Hepatic stem cells and the Liver's Maturational Lineages: Implications for liver biology, gene expression and cell therapies. *In* "Tissue Stem Cells" (C. S. Potten, R. B. Clarke, J. Wilson, and A. G. Renehan, eds.), pp. 161–214. Taylor and Francis Group, New York.

Schuldiner, M., Yanuka, O., Itskovitz-Eldor, J., Melton, D. A., and Benvenisty, N. (2000). Effects of eight growth factors on the differentiation of cells derived from human embryonic stem cells. *Proc. Natl. Acad. Sci. USA* **97,** 11307–11312.

Seglen, P. O. (1976). Preparation of isolated rat liver cells. *Methods Cell Biol.* **13,** 29–83.

Seglen, P. O., Schwarze, P. E., and Saeter, G. (1986). Changes in cellular ploidy and autophagic responsiveness during rat liver carcinogenesis. *Toxicol. Pathol.* **14,** 342–348.

Severin, E., Meier, E. M., and Willers, R. (1984). Flow cytometric analysis of mouse hepatocyte ploidy. I. Preparative and mathematical protocol. *Cell Tissue Res.* **238,** 643–647.

Shah, C., and Barnett, S. (1992). Hyaluronic acid gels. *In* "ACS Symposium Series 480. Polyelectrolyte Gels: Properties, Preparation, and Applications" (H. RS, and R. Prud'homme, eds.), pp. 116–130, Chapter 7. American Chemical Society, Washington, DC.

Shiojiri, N. (1994). Transient expression of bile-duct-specific cytokeratin in fetal mouse hepatocytes. *Cell Tissue Res.* **278,** 117–123.

Shu, X. Z., Liu, Y., Luo, Y., Roberts, M. C., and Prestwich, G. D. (2002). Disulfide cross-linked hyaluronan hydrogels. *Biomacromolecules.* **3,** 1304–1311.

Shumakov, V. I., Bliumkin, V. N., Ignatenko, S. N., Skaletskii, N. N., and Kauricheva, N. I. (1983). Transplantation of cultures of human fetal pancreatic islet cells to diabetes mellitus patients. *Klin. Med.* **61,** 46–51.

Sicklick, J. K., Li, Y. X., Melhem, A., Schmelzer, E., Zdanowicz, M., Huang, J., Caballero, M., Fair, J. H., Ludlow, J. W., McClelland, R. E., *Reid, L. M., and *Diehl, A. M. (2006). Hedgehog signaling maintains resident hepatic progenitors throughout life. *Am. J. Physiol. Gastrointest. Liver Physiol.* **290,** G859–G870. [*co-equal senior authors].

Sigal, S., Brill, S., Fiorino, A., and Reid, L. M. (1993). The liver as a stem cell and lineage system. In "Extracellular Matrix: Chemistry, Biology, and Pathobiology with Emphasis on the Liver" (M. A. Zern, and L. M. Reid, eds.), pp. 507–538. Marcel Dekker, Inc., New York.

Sigal, S. H., Brill, S., Fiorino, A. S., and Reid, L. M. (1992). The liver as a stem cell and lineage system. *Am. J. Physiol.* **263,** G139–G148.

Sigal, S. H., Brill, S., Reid, L. M., Zvibel, I., Gupta, S., Hixson, D., Faris, R., and Holst, P. A. (1994). Characterization and enrichment of fetal rat hepatoblasts by immunoadsorption ("panning") and fluorescence-activated cell sorting. *Hepatology* **19,** 999–1006.

Sigal, S. H., Gupta, S., Gebhard, D. F., Jr., Holst, P., Neufeld, D., and Reid, L. M. (1995). Evidence for a terminal differentiation process in the rat liver. *Differentiation* **59,** 35–42.

Sigal, S. H., Rajvanshi, P., Gorla, G. R., Sokhi, R. P., Saxena, R., Febhard, D. R., Jr., Reid, L. M., and Gupta, S. (1999). Partial hepatectomy-induced polyploidy attenuates hepatocyte replication and activates cell aging events. *Am. J. Physiol.* **276,** 1260–1272.

Smith, A. G., Heath, J. K., Donaldson, D. D., Wong, G. G., Moreau, J., Stahl, M., and Rogers, D. (1988). Inhibition of pluripotential embryonic stem cell differentiation by purified polypeptides. *Nature* **336,** 688–690.

Sosa-Pineda, B., Wigle, J. T., and Oliver, G. (2000). Hepatocyte migration during liver development requires Prox1. *Nat. Genet.* **25,** 254–255.

Spray, D. C., Fujita, M., Saez, J. C., Choi, H., Watanabe, T., Hertzberg, E., Rosenberg, L. C., and Reid, L. M. (1987). Proteoglycans and glycosaminoglycans induce gap junction synthesis and function in primary liver cultures. *J. Cell Biol.* **105,** 541–551.

Stamatoglou, S. C., and Hughes, R. C. (1994). Cell adhesion molecules in liver function and pattern formation. *FASEB J.* **8,** 420–427.

Stein, G. I., and Kudryavtsev, B. N. (1992). A method for investigating hepatocyte polyploidization kinetics during postnatal development in mammals. *J. Theor. Biol.* **156,** 349–363.

Stow, J. L., Kjellen, L., Unger, E., Hook, M., and Farquhar, M. G. (1985). Heparan sulfate proteoglycans are concentrated on the sinusoidal plasmalemmal domain and in intracellular organelles of hepatocytes. *J. Cell Biol.* **100,** 975–980.

Strom, S. C., Pisarov, L. A., Dorko, K., Thompson, M. T., Schuetz, J. D., and Schuetz, E. G. (1996). Use of human hepatocytes to study CYP450 gene induction. *Methods Enzymol.* **272,** 388–401.

Susick, R., Moss, N., Kubota, H., LeCluyse, E., Hamilton, G., Luntz, T., Ludlow, J., Fair, J., Greber, D., Bergstrand, K., White, J., Bruce, A., *et al.* (2002). Hepatic progenitors and strategies for liver cell therapies. *Ann. N. Y. Acad. Sci.* **39,** 8–419.

Suskind, D. L., and Muench, M. O. (2004). Searching for common stem cells of the hepatic and hematopoietic systems in the human fetal liver: CD34þ cytokeratin 7/8þ cells express markers for stellate cells. *J. Hepatol.* **40,** 261–268.

Suzuki, A., Iwama, A., Miyashita, H., Nakauchi, H., and Taniguchi, H. (2003). Role for growth factors and extracellular matrix in controlling differentiation of prospectively isolated hepatic stem cells. *Development* **130,** 2513–2524.

Tanimizu, N., Nishikawa, M., Saito, H., Tsujimura, T., and Miyajima, A. (2003). Isolation of hepatoblasts based on the expression of Dlk/Pref-1. *J. Cell Sci.* **116,** 1775–1786.

Tanimizu, N., Saito, H., Mostov, K., and Miyajima, A. (2004). Long-term culture of hepatic progenitors derived from mouse Dlkþ hepatoblasts. *J. Cell Sci.* **117,** 6425–6434.

Tateno, C., and Yoshizato, K. (1999). Growth potential and differentiation capacity of adult rat hepatocytes *in vitro*. *Wound Repair Regen.* **7,** 36–44.

Taub, M., and Sato, G. (1980). Growth of functional primary cultures of kidney epithelial cells in defined medium. *J. Cell. Physiol.* **105,** 369–378.

Temple, S., and Alvarez-Buylla, A. (1999). Stem cells in the adult mammalian central nervous system. *Curr. Opin. Neurobiol.* **9,** 135–141.

Terada, T., Ashida, K., Kitamura, Y., Matsunaga, Y., Takashima, K., Kato, M., and Ohta, T. (1998). Expression of epithelial-cadherin, alpha-catenin and beta-catenin during human intrahepatic bile duct development: A possible role in bile duct morphogenesis. *J. Hepatol.* **28,** 263–269.

Theise, N. D., Saxena, R., Portmann, B. C., Thung, S. N., Yee, H., Chiriboga, L., Kumar, A., and Crawford, J. M. (1999). The canals of Hering and hepatic stem cells in humans. *Hepatology* **30,** 1425–1433.

Thorgeirsson, S. S., Factor, V. M., and Grisham, J. W. (2004). Early activation and expansion of hepatic stem cells. *In* "Stem Cells" (R. Lanza, H. Blau, D. Melton, M. Moore, E. D. Thomas, C. Verfailie, I. Weissman, and M. West, eds.), Vol. 2, pp. 497–512. Elsevier, London.

Tillotson, L. G., Lodestro, C., Hocker, M., Weidenmann, B., Newcomer, C. E., and Reid, L. M. (2001). Isolation, maintenance, and characterization of human pancreatic islet tumor cells expressing vasoactive intestinal peptide. *Pancreas* **22,** 91–98.

Timchenko, N. A., Wilde, M., and Darlington, G. J. (1999). C/EBPalpha regulates formation of S-phase-specific E2F-p107 complexes in livers of newborn mice. *Mol. Cell. Biol.* **19,** 2936–2945.

Tomizawa, M., Garfield, S., Factor, V. M., and Xanthopoulos, K. (1998). Hepatocytes deficient in CCAAT/enhancer binding protein alpha (C/EBPalpha) exhibit both hepatocyte and biliary epithelial cell character. *Biochem. Biophys. Res. Commun.* **249,** 1–5.

Traber, P. G., Chianale, J., and Gumucio, J. J. (1988). Physiologic significance and regulation of hepatocellular heterogeneity. *Gastroenterology* **95,** 1130–1143.

Turley, E. A. (1989). The role of a cell-associated hyaluronan-binding protein in fibroblast behaviour. *In* "The Biology of Hyaluronan" (D. Evered, and J. Whelan, eds.), CIBA Symposium No. 143. pp. 121–137. John Wiley and Sons, Hoboken, N.J.

Turner, W. S., Schmelzer, E., McClelland, R. E., Chen, W., Wauthier, E., and Reid, L. M. (2006). Human hepatoblast phenotype maintained by hyaluronan hydrogels. *Journal of Biomedical Biomaterials Research B. Applied Biomaterials* **82**(1), 156–168.

Turner, W. S., Seagle, C., Galanko, J., Favorov, O., Prestwich, D. G., Macdonald, J., and Reid, L. M. (2008). Metabolomic footprinting of human hepatic stem cells and hepatoblasts: Applications developed with engineered 3-D hyaluronan hydrogel scaffolds. *Stem Cells* In press, 2008.

Van den Hov, M. J., Vermeulen, J. L., De Boer, P. A., Lamers, W. H., and Moorman, A. F. (1994). Developmental changes in the expression of the liver-enriched transcription factors LF-B1, C/EBP, DBP and LAP/LIP in relation to the expression of albumin, alpha-fetoprotein, carbamoylphosphate synthase and lactase mRNA. *Histochem. J.* **26,** 20–31.

Van Eyken, P., Sciot, R., Callea, F., van der Steen, K., Moerman, P., and Desmet, V. J. (1988). The development of the intrahepatic bile ducts in man: A keratin-immunohistochemical study. *Hepatology* **8,** 1586–1595.

Vongchan, P., Warda, M., Toyoda, H., Toida, T., Marks, R. M., and Linhardt, R. J. (2005). Structural characterization of human liver heparan sulfate. *Biochim. Biophys. Acta* **1721,** 1–8.

Walker, L., Lynch, M., Silverman, S., Fraser, J., Boulter, J., Weinmaster, G., and Gasson, J. C. (1999). The Notch Jagged pathway inhibits proliferation of human hematopoietic progenitors *in vitro. Stem Cells* **17,** 162–171.

Wang, N. D., Finegold, M. J., Bradley, A., Ou, C. N., Abdelsayed, S. V., Wilde, M. D., Taylor, L. R., Wilson, D. R., and Darlington, G. J. (1995). Impaired energy homeostasis in C/EBP alpha knockout mice. *Science* **269,** 1108–1112.

Watanabe, T., and Tanaka, Y. (1982). Age-related alterations in the size of human hepatocytes. A study of mononuclear and binucleate cells. *Virchows Arch. B Cell Pathol. Incl. Mol. Pathol.* **39,** 9–20.

Watts, P., and Grant, M. H. (1996). Cryopreservation of rat hepatocyte monolayer cultures. *Hum. Exp. Toxicol.* **15,** 30–37.

Weigmann, A., Corbeil, D., Hellwig, A., and Huttner, W. B. (1997). Prominin, a novel microvillispecific polytopic membrane protein of the apical surface of epithelial cells, is targeted to plasmalemmal protrusions of non-epithelial cells. *Proc. Natl. Acad. Sci. USA* **94,** 12425–12430.

Westgren, M., Ek, S., Bui, T. H., Hagenfeldt, L., Markling, L., Pschera, H., Seiger, A., Sundstrom, E., and Ringden, O. (1994). Establishment of a tissue bank for fetal stem cell transplantation. *Acta Obstet. Gynecol. Scand.* **73,** 385–388.

Wu, F. J., Friend, J. R., Hsiao, C. C., Zilliox, M. J., Cerra, F. B., and Hu, W. (1996). EYcient assembly of rat hepatocyte spheroids for tissue engineering applications. *Biotechol. Bioeng.* **50,** 404–415.

Wu, F. J., Friend, J. R., Lazar, A., Mann, H. J., Remmel, R. P., Cerra, F. B., and Hu, W. S. (1996). Hollow fiber bioartificial liver utilizing collagen-entrapped porcine hepatocyte spheroids. *Biotechnol. Bioeng.* **52,** 34–44.

Xu, A., Luntz, T., Macdonald, J., Kubota, H., Hsu, E., London, R., and Reid, L. M. (2000). Liver stem cells and lineage biology. *In* "Principles of Tissue Engineering", 2nd edition. (R. Lanza, R. Langer, and J. Vacanti, eds.) Academic Press, San Diego, CA. pp. 559–597.

Xu, R. H., Chen, X., Li, D. S., Li, R., Addicks, G. C., Glennon, C., Zwaka, T. P., and Thomson, J. A. (2002). BMP4 initiates human embryonic stem cell differentiation to trophoblast. *Nat. Biotechnol.* **20,** 1261–1264.

Yagi, K., Tsuda, K., Serada, M., Yamada, C., Kondoh, A., and Miura, Y. (1993). Rapid formation of multicellular spheroids of adult rat hepatocytes by rotation culture and their immobilization within calcium alginate. *Artifcial Organs* **17,** 929–934.

Yamada, K., Kamihira, M., Hamamoto, R., and Iijima, S. (1998). EYcient induction of hepatocyte spheroids in a suspension culture using a water-soluble synthetic polymer as an artificial matrix. *J. Biochem.* **123,** 1017–1023.

Yin, M., Bradford, B. U., Wheeler, M. D., Uesugi, T., Froh, M., Goyert, S. M., and Thurman, R. G. (2001). Reduced early alcohol-induced liver injury in cd14-deficient mice. *J. Immunol.* **166,** 4737–4742.

Young, H. E., Mancini, M. L., Wright, R. P., Smith, J. C., Black, A. C., Jr., Reagan, C. R., and Lucas, P. A. (1995). Mesenchymal stem cells reside within the connective tissues of many organs. *Dev. Dyn.* **202,** 137–144.

Zaret, K. (1999). Developmental competence of the gut endoderm: Genetic potentiation by GATA and HNF3/fork head proteins. *Dev. Biol. (Orlando)* **209,** 1–10.

Zaret, K. S. (2002). Regulatory phases of early liver development: Paradigms of organogenesis. *Nat. Rev. Genet.* **3,** 499–512.

Zvibel, I., Brill, S., and Reid, L. M. (1995). Insulin-like growth factor II regulation of gene expression in rat and human hepatomas. *J. Cell. Physiol.* **162,** 36–43.

Zvibel, I., Halay, E., and Reid, L. M. (1991). Heparin and hormonal regulation of mRNA synthesis and abundance of autocrine growth factors: Relevance to clonal growth of tumors. *Mol. Cell Biol.* **11,** 108–116.

CHAPTER 9

Culture of Pluripotent Neural Epithelial Progenitor Cells from E9 Rat Embryo

Ronghao Li and Jennie P. Mather

Research and Development
Raven Biotechnologies, Inc.
South San Francisco, California 94080

    Abstract
I. Introduction
II. Rationale
III. Methods
    A. Dissection of E9 Neural Plate
    B. Preparation of Esc CoCulture and Conditioned Medium
    C. Primary Culture of E9 Neural Plate
    D. Long-Term Culture
    E. Establishment and Serial Passage of NEP Cell Line
    F. Characterization of NEP by Immunohistochemistry
    G. Differentiation of NEP Cells
    H. Differentiation of NEP Cells by Intracranial Implantation
IV. Materials
    A. Instrument and Equipment
    B. Reagents
V. Discussion
VI. Summary
    References

Abstract

The mammalian central nervous system is developmentally derived from neuroepithelial cells in the neural plate. These neuroepithelial cells grow and differentiate in response to signals from their surrounding environment. Many of those signals have been well characterized and others remain to be discovered.

In cell culture, a conditioned medium, a feeder cell layer, or a tissue extract has been used as supplement in addition to those factors well characterized for maintaining the multipotent status of neural progenitor cells. To date, there have been many types of neural progenitor cells established in culture from various stages of development and from different regions of the nervous system of various species. This chapter will provide a brief introduction to those cultures and a detailed method for culturing rat neural epithelial cells at embryonic stage E9 and characterizing them *in vitro* and *in vivo*.

I. Introduction

Since the introduction of serum-free culture for mammalian cells (Mather and Sato, 1979a,b), a number of neural progenitor or stem cell cultures have been established in the presence of EGF and/or FGF. An EGF-dependent mouse SFME was established as a glial precursor cell line from embryonic brain (Loo *et al.*, 1987); EGF-responsive multipotent neural stem cell cultures were established from mouse embryonic (Reynolds and Weiss, 1996) and adult striatum (Reynolds and Weiss, 1992) in neurospheres; and FGF-responsive self-renewal progenitor/ stem cells have been isolated from embryonic brain (Gage *et al.*, 1995; Kilpatrick and Bartlett, 1995, Mckay RDG, 1997), adult brain, mouse brain (Gritti *et al.*, 1996), and adult rat hippocampus (Palmer *et al.*, 1997). More recently, the isolation and culture of human bipotent and neural progenitor cells have been reported (Carpenter *et al.*, 1999; Sah *et al.*, 1997; Svendsen *et al.*, 1998) in the presence of both EGF and FGF. However, neither EGF nor FGF, alone or in combination with serum, support long-term survival of neuroepithelial cells cultured from earlier stages of development (e.g., E9 rat neural plate). The initiation of neuroepithelial cell culture from E9 neural plate was supported by coculture with or conditioned medium from fetal Schwann cell cultures, similar to the requirement of ES cells for fibroblastic cell feeder layer (Li *et al.*, 1996a).

While a few of the murine neural progenitor or stem cell lines were capable of indefinite growth in serum-free cultures (Li *et al.*, 1996a; Loo *et al.*, 1987), many types of neural progenitor cells cultured *in vitro* tend to differentiate spontaneously. Therefore, long-term culture of neuroepithelial cells has relied on viral or oncogene transformation, which, it was hoped, might halt development at specific stages of differentiation. This approach has resulted in the establishment of several immortalized, or conditionally immortalized, neural precursor cell lines (Bartlett *et al.*, 1988; Frederiksen *et al.*, 1988; Li *et al.*, 2000; Pietsch *et al.*, 1994; Snyder *et al.*, 1992).

However, transformation frequently results in alteration of the neural precursor cell properties and instability of the differentiated phenotype of the cells. For example, SV40 large T-antigen transfection of precursor cells prevented the acquisition of a cerebellar granule cell phenotype (Gao and Hatten, 1994). A v-myc transfected multipotent mouse cerebellar-derived cell line can spontaneously alternate between neuronal and glial phenotypes (Snyder *et al.*, 1992). The alteration caused by immortalizing genes was minimized when regulated promoters were used to drive the expression of the immortalizing genes (Li *et al.*, 2000).

II. Rationale

The maintenance of the undifferentiated state or the self-renewal capability of progenitor/stem cells depends on the microenvironment where stem cells reside. Thus, the removal of certain environmental factors might be expected to result in cell differentiation. With the progression of development, such privilege sites retract until, in adult life, stem cells can be found only in a few small regions in the brain, for example, the subgranular zone of the hippocampal dentate gyrus and the subventricular zone of the lateral ventricle. Recreating such an environment *in vitro* with a mixture of growth factors, matrix protein, tissue extracts, conditioned medium, and feeder cell layers is the key to success in growing progenitor/stem cells. Sometimes, maintaining appropriate cell–cell contact has proven to be crucial to the growth of certain undifferentiated cell types (Li *et al.*, 1997; Li, 1996a). In addition to growth factors, fibronectin or laminin is often used as a substrate for monolayer cultures and bovine pituitary extracts or chicken embryo extracts are often used as growth supplements. With this approach, cell lines have been established that grow indefinitely in culture and maintain the desired stage of differentiation (Li *et al.*, 1997; Li *et al.*, 1996a; Loo *et al.*, 1987).

By the same principle, an appropriate and distinct environment must be provided to cause the differentiation of those stem/progenitor cells. A cell line that tends to differentiate into one lineage of cells in one condition may differentiate into another lineage of cells in another. The simplicity of *in vitro* conditions helps in directing the cell to differentiate into specific pathways while more complex *in vivo* conditions induce more diversified cell differentiation. We describe below methods for obtaining primary cultures from rat E9 (0 somite) neural plates and the isolation and establishment of an early neuroepithelial precursor cell line that can be induced to differentiate into a variety of neural cell types *in vitro* and *in vivo*.

III. Methods

A. Dissection of E9 Neural Plate

In early embryonic development, stem/progenitor cells are induced to differentiate by their neighboring cells. It is important to separate the stem progenitor cells from their adjacent tissues by microdissection before putting them in culture. We use animals at the 0-somite stage of development (rat day 9). The dissection procedure is as follows.

1. Obtain timed pregnant Sprague-Dawley rats.
2. Keep animals in a controlled temperature and humidity environment, with a light period of 0600–1800 h and ad libitum food.
3. Animals are sacrificed by CO_2 asphyxiation on day 9 of pregnancy (plug day was designated as day 0) between 11 am and 2 pm.
4. The uterus is removed and embryos collected by dissection under a dissecting microscope using fine surgical scissors and forceps.
5. E9 rat embryos (0-somite) are selected under the dissecting microscope and dissected using fine needles to remove the extraembryonic membrane. The caudal portion is removed by cutting through the middle of Hensen's node.
6. Incubate the embryos with collagenase and dispase (Boehringer Mannheim, for 10 min at 0 °C) (Fig. 1).
7. Wash the embryos by transfer from plate to plate with 1% BSA.
8. Remove the mesoderm and endoderm with fine needles.

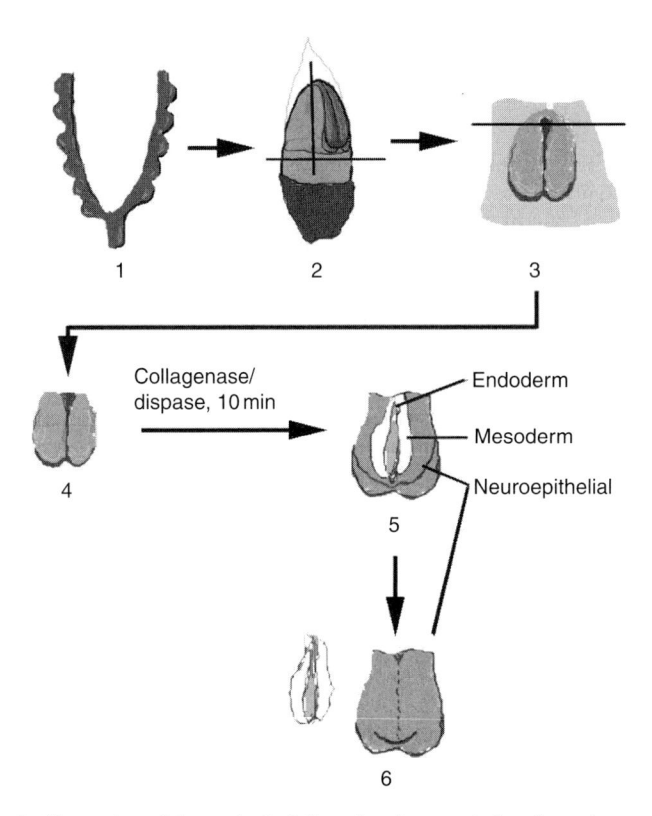

Fig. 1 An illustration of the method of dissecting the neural plate from the rat embryos.

9. The neural plates are then dispersed into small aggregates (preferentially 20–50 cells in each aggregate) by gentle pipetting with a P 200 pipette.
10. Complete dispersion must be avoided.
11. The dispersed cells are plated in 96-well microtiter plates precoated with laminin.
12. Cells from 20 neural plates are divided into 96 wells.

B. Preparation of Esc CoCulture and Conditioned Medium

For coculture, a confluent culture of rat fetal Schwann cells (ESC), cells carried as described by Li *et al.* (1996b), is treated with collagenase/dispase to remove cells from the plate and washed with serum-free medium by gentle centrifugation. Plate cells at 4×10^3 cells per well in a 96-well plate no later than the day before the neural plate cells are to be added. For transwell culture, grow ESC in tissue culture inserts for 24-well plates until near confluence, and plate neural plate cells underneath the inserted wells in the 24-well plates. If conditioned medium is desired, follow the directions below for preparation.

1. Grow ESC cells in tissue culture plates in the serum-free supplements described in Li *et al.* (1996a,b).
2. When the cells are about 50% confluent, remove the spent medium and replace with 10 ml hormone-supplemented serum-free medium.
3. Allow cells to incubate for 48 h.
4. Remove medium from the plates, combine, and filter through 0.1 μ sterilization filter unit to remove any Schwann cells or cell debris.
5. Concentrate the medium 10-fold using a Centricon (Millipore) concentrator of 10,000 MW cutoff.
6. Store concentrated medium at 6 °C in a sterile polypropylene tube. The medium may be used for up to 7 days.

C. Primary Culture of E9 Neural Plate

The dissected tissue can be cocultured with the ESC cells as described below. The ESC and neuroepithelial cells can be easily distinguished phenotypically as shown in Fig. 2. The neuroepithelial cell colonies will grow rapidly in these conditions (Fig. 3 and Fig. 4).

1. Cells prepared as described above are cultured in 96-well laminin-coated plates.
2. Add serum-free medium supplemented with bovine pituitary extract (BPE, optimal concentration for growth stimulation determined for each batch prepared as described in Roberts *et al.*, 1990 or see chapter 10, this volume),

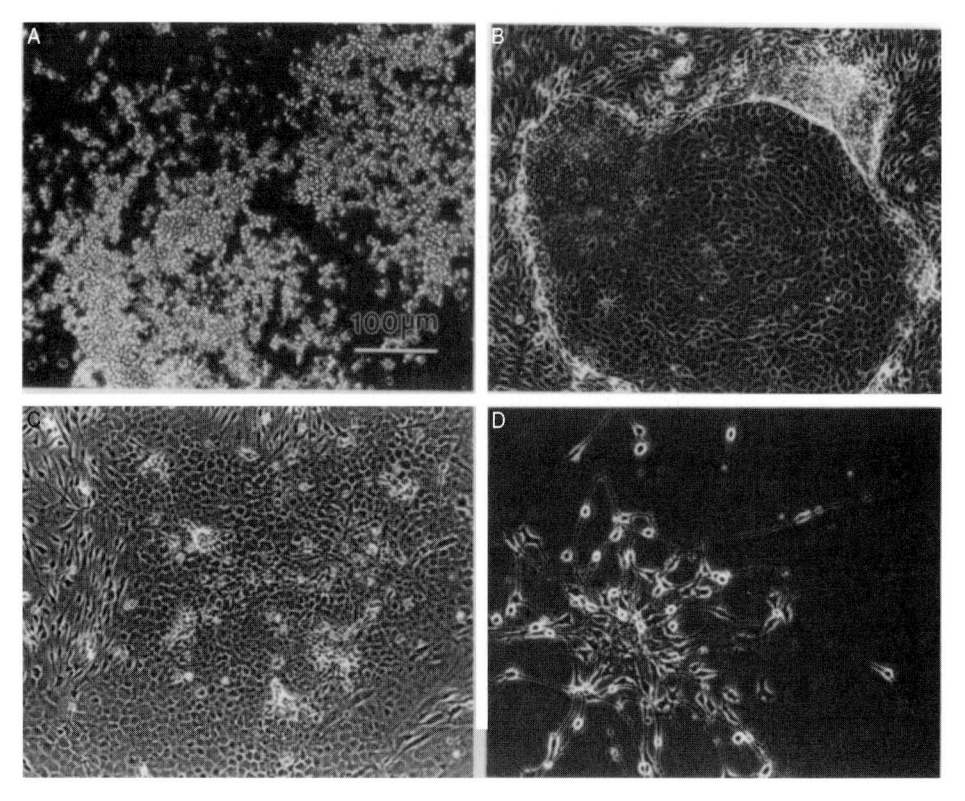

Fig. 2 Primary cocultures of neural plate tissue and ESC showing the growth of a neuroepithelial cell colony.

and usually at 3–8 μl/ml; insulin at 10 μg/ml; forskolin at 5 μM; and additionally either coculture with the rat embryonic Schwann cell line or supplement with 10% 10-fold concentrated ESCCM.
3. In these conditions, some precursor cells survive and proliferate to form large colonies of compacted monolayer epithelial cells containing some differentiated neurons bearing long processes.
4. To maintain mixed primary cultures, change medium every 3–4 days.

D. Long-Term Culture

These cultures can be subcultured either as cocultures or using conditioned medium. To maintain an appropriate ratio of ESC to primary cells, the ESC cells should be removed by a brief treatment and discarded, and then the primary cells taken to clumps of cells and remixed with fresh ESC grown separately. Alternatively, the entire culture can be passed with heregulin omitted from the supplements and the ESC discarded at passage. The timing of the passages, the

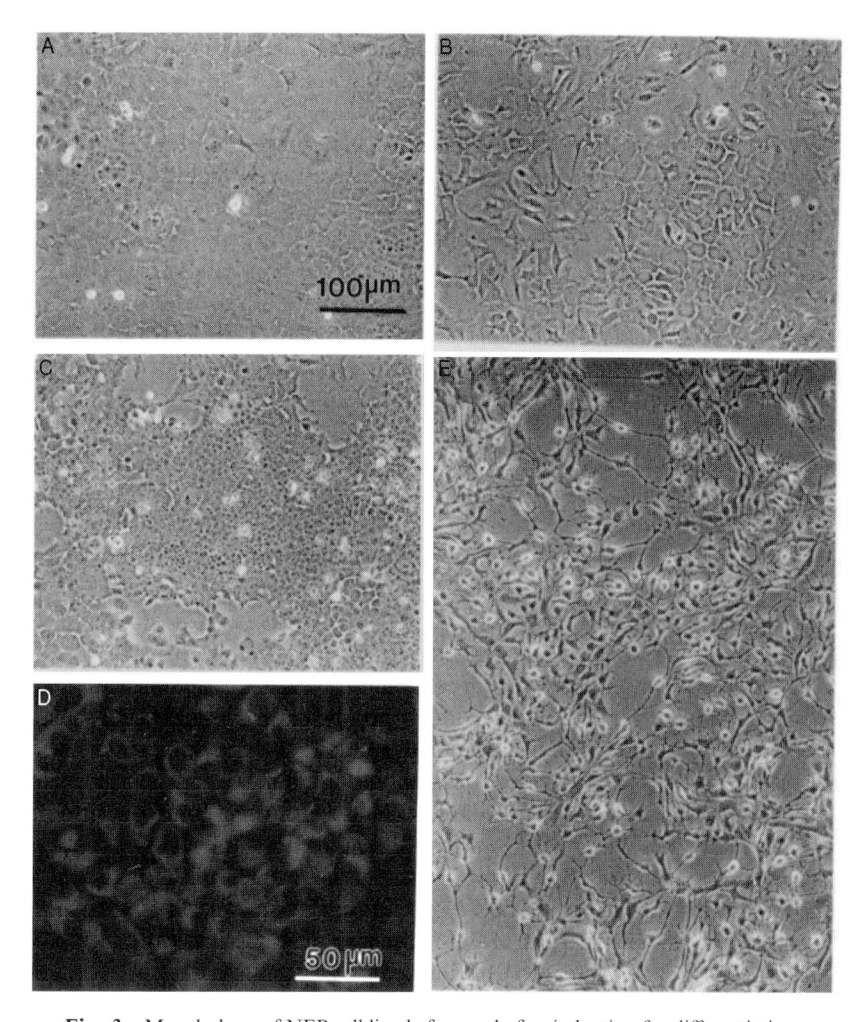

Fig. 3   Morphology of NEP cell line before and after induction for differentiation.

density of the cells at subculture, the substrate, and the culture supplements will determine the mixture of cells present in the secondary and subsequent passages.

1. Colonies of epithelial cells formed in the primary culture can be passaged in F12/DMEM supplemented with 7F and ESCCM:

    insulin, 10 µg/ml
    transferrin, 10 mg/ml
    progesterone, 3×10^{-9} M
    alpha-tocopherol, 5 µg/ml
    BPE (optimal concentration batch-dependent), 3 µl/ml
    forskolin, 5 mM

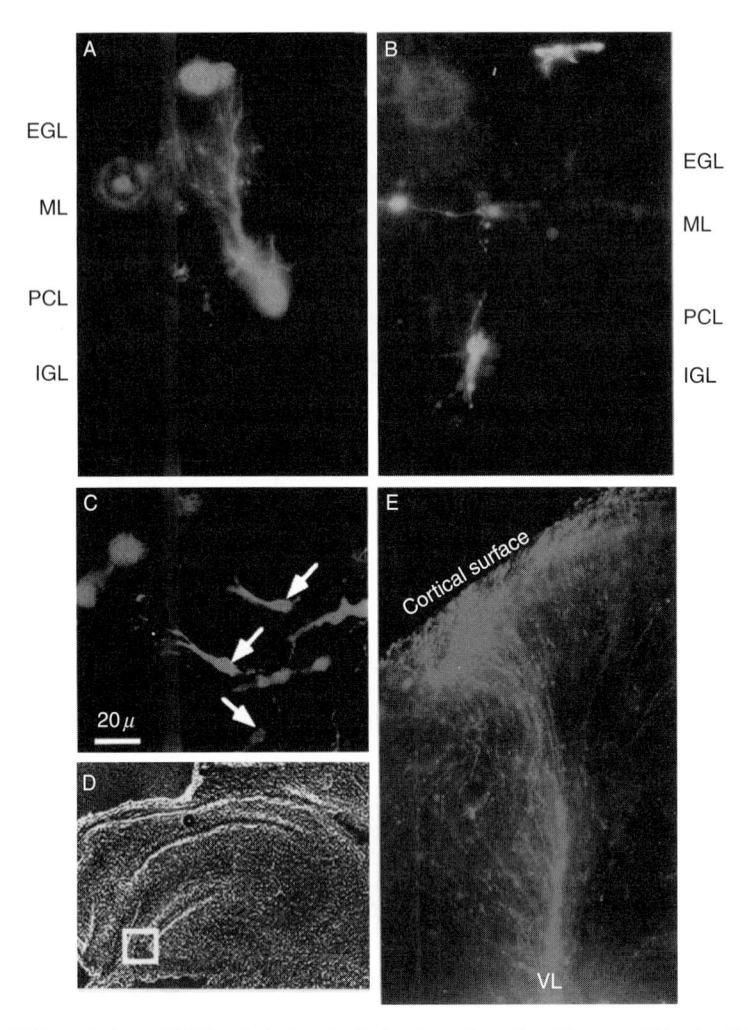

Fig. 4 Differentiation of NEP cells in intrabrain implantation. (See Plate no. 18 in the Color Plate Section.)

recombinant human heregulin (HRG-Beta1 177–244), 10 nM
ESCCM 10% of 10× concentrated

2. Remove cells from the substrate by incubating with 0.2% collagenase/dispase at 37 °C. The neuroepithelial cells will detach as a loose sheet maintaining intercellular adhesion.
3. Cells are washed free of enzymes by centrifugation on a layer of 3% BSA.
4. Plate cells onto laminin-coated 24-well plates.
5. Add F12/DMEM supplemented with 7F.

6. Cells can be subcultured by the same procedure for 5–6 passages.
7. During this period, the cultures will contain two major cell types: compact epithelial cells and bipolar cells resembling Schwann cells or radial glia. Other cell populations persist through the first few subcultures.

E. Establishment and Serial Passage of NEP Cell Line

Starting from the mixed long-term cultures obtained above, the epithelial cells are subsequently removed from the culture by differential enzymatic digestion with collagenase/dispase. The bipolar cells lift off the plate first and are discarded. Removing heregulin from the culture medium, and allowing the epithelial cells to grow at a higher density, also favors the growth of the epithelial component of the coculture. The cells are then carried in 6F medium (7F medium above, omitting heregulin) on laminin-coated plates and passaged every week with collagenase dispase, using a 1:4 split. If the cells are maintained at a high density, the ESCCM can be omitted from the cultures at this point.

The epithelial cell line isolated in our laboratory using this protocol-designated NEP has been maintained in continuous culture for more than 3 years. The cultures can be caused to differentiate in response to forskolin and bFGF. Within 48 h of changing the cells to a forskolin-containing medium (as described below in the III.G), more than 90% of the cells are postmitotic and express neuronal markers (Li *et al.*, 1996a).

F. Characterization of NEP by Immunohistochemistry

The NEP cell line is a nestin-positive neural progenitor cell. Nestin is an intermediate cytoskeletal protein that can be stained by immunofluorescence (see fig. 3D). Monoclonal antinestin antibody is now available from Chemicon (Catalog number mab353). To prepare for immunofluorescence, NEP cells are grown on a glass chamber slide and fixed by absolute alcohol prechilled at −20 °C for 10 min and followed by air dry for 1 h. The fixed slides are then incubated sequentially with 10% normal goat serum for 30 min and with antinestin antibody (5 µg/ml) for 1 h at 37 °C; rinsed 3× with PBS, and then with anti-mouse IgG-FITC (10 µg/ml) for 1 h; and mounted with Vectashield (Vector Labs) mounting medium. The slide is examined under fluorescence microscope. Positively stained cells will show green fluorescence in cytoskeletal network in the cytoplasm.

G. Differentiation of NEP Cells

Neuronal differentiation is induced when NEP cells are treated with forskolin and FGF-II. Cells are plated in 6F medium at a 1:8 split in laminin-coated 24-well plates with F12/DMEM supplemented with 6F for 48 h and then cultured with fresh medium containing the following:

insulin, 10 µg/ml
transferrin, 10 µg/nl

alpha-tocopherol, 10 μg/ml

bFGF (recombinant human bFGF, Gibco BRL), 30 ng/ml

forskolin (Calbiochem), 5 μM

The extent of cell differentiation can be assessed by immunofluorescence staining of the cultures with various markers that are common in characterized neural cell types as listed in Table I. Further functional analysis of differentiated neurons could be performed by electrophysiology with patch clamp techniques (Li et al., 2000).

H. Differentiation of NEP Cells by Intracranial Implantation

The differentiation potential of NEP cells is further examined by *in vivo* grafting in neonatal rat brain. The method was adapted from Gao and Hatten (1994). Prior to implantation, NEP cells are labeled with fluorescent tracer PKH-26 (Sigma) at 5 μM concentrations for 7 min at room temperature according to the protocol provided by the supplier. Labeled cells are incubated with serum-free medium containing 100 ng/ml FGF-II and 5 μM forskolin at 37 °C for 4 hours. This predifferentiation step improves engraftment in the cortex. The cells are counted and a suspension of partially dissociated cells is prepared at a concentration of 5×10^7 cells/ml. The cell suspension is drawn into a Hamilton syringe.

To prepare for surgical implantation, neonatal rat pups are rendered unconscious by chilling the animals on an ice water bath and placed in a Stoelting Stereotaxic device fitted with a neonatal rat adapter and a vertical holder for a Hamilton syringe. The skin overlying the forebrain is rinsed with 70% ethanol, a small incision is made in the skin, and the Hamilton syringe needle is lowered gently through the incision to a position of lateral ventricular zone. The injection site can be evaluated by injecting blue solution instead of cell suspension and visualizing under a light microscope after sectioning. For each injection, inject about 2 μl of the above-labeled cell suspension. Then the syringe is removed, the skull is rinsed with a solution of penicillin-streptomycin (0.25%), and the skin is

Table I
Common Markers for Immunochemical Characterization of CNS Neural Cells

Cell type	Markers
Neural progenitor	Nestin
Neurons	βIII-tubulin, neurofilament 68 (NF68), neurofilament 160 (NF160), neurofilament 200 (NF200), microtubule-associated protein 2a/b (MAP2a/b), Tau, synaptophysin, glutamate, GABA, choline acetyl transferase
Astrocyte	Glial fibrillary acidic protein (GFAP)
Oligodendrocyte	Gal C, CNPase

replaced and sealed with Vetbond (Henry Schein, Inc.). The animals are then warmed to 35.5 °C and returned to the litter. After survival of 1–5 days, animals are anesthesized with ketamine prior to perfusion with 4% paraformaldehyde in 0.1 M phosphate buffer (pH 7.4). The brain hemispheres are removed and post-fixed in the same fixative. The fixed tissues are rinsed with PBS and embedded in 3% agar gel. Serial sections (200 μ thick) are cut with a vibratome and labeled cells are visualized under epifluorescence microscope. Morphology of labeled cells is recorded by fluorescence photography.

IV. Materials

A. Instrument and Equipment

Microdissection needles (black), needle holders, forceps, and scissors were obtained from Fine Scientific Tools, Inc. Tissue culture laboratory was equipped with Biosafety hoods (Forma Scientific), Allegra 60 Centrifuge (Beckman), inverted microscope (Nikon), dissecting microscope (Zeiss), and CO_2 incubators (Sanyo). Stoelting Stereotaxic device was purchased from Harvard Apparatus.

B. Reagents

Recombinant human heregulin (rHRGB 177–244) is available from R&D Systems. A stock solution (1000×) was prepared in PBS and stored in small aliquots at −80 °C. Bovine pituitary extract (also available from Gibco BRL) is prepared as described in Roberts *et al.* (1990). Collagenase/dispase (Roche Molecular Chemicals), insulin (recombinant human, Novo Nordisk), aprotinin (Roche Molecular Chemicals), and forskolin (Calbiochem) were purchased from the indicated sources. Laminin, FGFII, transferrin, and F12/DME medium powder were obtained from Gibco BRL (Grand Island, NY); progesterone and alpha-tocopherol from Sigma.

V. Discussion

It is important that the animals be sacrificed between 11 am and 2 pm, on day 9 (0 somite) of development, if the cell cultures described here are to be obtained. All embryos should be dissected from the uterus and then kept in medium on ice until further dissection. If the embryos are taken at an earlier or later stage, the characteristics of the culture will differ and the neuroepithelial cell type described will not be apparent in the cultures. Collagenase/dispase is recommended for both dissection and subsequent passaging of the cell cultures, because cells require cell–cell contact for survival both in primary culture and in subsequent passage. The dispersion of NEP cells into single cells must be avoided at all times, otherwise the separated NEP cells will die. The survival of totally dispersed cells will indicate a phenotypic change of the culture. NEP cells after differentiation as described above will survive single-cell dispersion. Again, in order to maintain the cell–cell contact throughout the culture process, a 1:4 split of a confluent or near confluent plate is the correct density. If the cells are not confluent after 4 days, refeed with the hormones and let the plate grow to confluency. It is important to split the cells on the recommended schedule and split ratio if the neuroepithelial cells are to be favored (initially) and then maintained in the culture.

The ESC, fetal rat Schwann, cell line can support the growth of these primaries in coculture, transwell culture, or by the addition of concentrated medium. Thus, there seems to be a secreted factor responsible for the survival activity. An adult Schwann cell line, a melanoma cell line (M2R) (Mather and Sato, 1979b), and a pre-astroglial cell line SFME (Loo *et al.*, 1987) were all tried as sources of conditioned medium. They will not support the growth of the NEP-like cells in the culture. More than 25 hormones and growth factors have been tested as a replacement for the conditioned medium without success (Li, 1996a). In the absence of the ESC cell line one can prepare primary cultures of E14 embryonic Schwann cells as described (Li, 1996b) and use conditioned medium from these cells. These cells have a strong preference for laminin over other matrix proteins such as collagen, polylysine, and fibronectin as a substrate.

The dependence of NEP cells on intercellular contact may reflect the *in vivo* situation. Such neural progenitor cells live most likely in clusters and could be nourished from nearby glial cell produced factors. The differentiation of these cells will progress along with the migration away from those cell clusters. Such phenomenon is common in neural differentiation during development and is consistent with the observations on adult neural stem cells.

VI. Summary

With the support of a Schwann cell feeder layer or the conditioned medium from Schwann cell culture, primary cultures can be obtained from the dissected neural plates of the 0-somite day 9 rat embryo in hormone-supplemented serum free medium and ESCCM. By differential enzymatic treatment, removal of heregulin from the medium supplement and subculture in serum-free medium with hormone supplements, a neuroepithelial cell type can be selected from the mixed cultures that is nestin positive and undergoes continuous division in culture. Maintenance of this cell line requires the maintenance of cell–cell contact, a relatively high cell density, and prevention of exposure of the cells to trypsin or serum. These cells retain the ability to differentiate to a nonmitotic cell type having the properties of neurons on exposure to bFGF and forskolin *in vitro* and are multipotent in differentiating into a variety of neuronal and glial cells when implanted into neonatal rat brain.

Acknowledgments

We thank Dr. W-Q Gao and Penny Roberts for discussion and encouragement during the course of this work.

References

Bartlett, P. F., Reid, H. H., Bailey, K. A., and Bernard, O. (1988). Immortalization of mouse neural precursor cells by the c-myc oncogene. *Proc. Natl. Acad. Sci. USA* **85**, Erratum in: *Proc Natl Acad Sci USA*. **86**, 1103.

Carpenter, M. K., Cui, X., Hu, Z. Y., Jackson, J., Sherman, S., Seiger, A., and Wahlberg, L. U. (1999). *In vitro* expansion of a multipotent population of human neural progenitor cells. *Exp. Neurol.* **158**, 265–278.

Frederiksen, K., Jat, P. S., Valtz, N., Levy, D., and McKay, R. (1988). Immortalization of precursor cells from the mammalian CNS. *Neuron* **1**, 439–448.

Gage, F. H., Ray, J., and Fisher, L. J. (1995). Isolation, characterization, and use of stem cells from the CNS. *Rev. Neurosci.* **18**, 159–192.

Gao, W. Q., and Hatten, M. E. (1994). Immortalizing oncogenes subvert the establishment of granule cell identity in developing cerebellum. *Development* **120**, 1059–1070.

Gritti, A., Parati, E. A., Cova, L., Frolichsthal, P., Galli, R., Wanke, E., Faravelli, L., Morassutti, D. J., Roisen, F., Nickel, D. D., and Vescovi, A. L. (1996). Multipotential stem cells from the adult mouse brain proliferate and self-renew in response to basic fibroblast growth factor. *J. Neurosci.* **16**, 1091–1100.

Kilpatrick, T. J., and Bartlett, P. F. (1995). Cloned precursors from the mouse cerebrum require FGF-2, whereas restricted precursors are stimulated with either FGF-2 or EGF. *J. Neurosci.* **15**, 3653–3661.

Li, R., Gao, W. Q., Moore, A., and Mather, J. P. (1996a). Multiple factors control the proliferation and differentiation of rat early embryonic (day 9) neuroepithelial cells. *Endocr. J.* **5**, 205–217.

Li, R., Phillips, D. M., Moore, A., and Mather, J. P. (1997). Follicle-stimulating hormone induces terminal differentiation in a predifferentiated rat granulosa cell line (ROG). *Endocrinology* **138**, 2648–2657.

Li, R., Sliwkowski, M., Lo, J., and Mather, J. P. (1996b). Establishment of Schwann cell lines from normal adult and embryonic rat dorsal root ganglia. *J. Neurosci. Methods* **67**, 57–69.

Li, R., Thode, S., Zhou, J., Richard, N., Pardinas, J., Rao, M. S., and Sah, D. W. (2000). Motoneuron differentiation of immortalized human spinal cord cell lines. *J. Neurosci. Res.* **59**, 342–352.

Loo, D. T., Fuquay, C. L., and Barnes, D. W. (1987). Extended culture of mouse embryo cells without senescence: Inhibition by serum. *Science* **236**, 200–202.

Mather, J. P., and Sato, G. H. (1979a). The growth of mouse melanoma cells in supplemented, serum-free medium. *Exp. Cell Res.* **120**, 191–200.

Mather, J. P., and Sato, G. H. (1979b). The use of hormone-supplemented serum-free media in primary cultures. *Exp. Cell Res.* **124**, 215–221.

McKay, R. (1997). Stem cells in the central nervous system. *Science* **276**, 66–71.

Palmer, T. D., Takahashi, J., and Gage, F. H. (1997). The adult rat hippocampus contains primordial neural stem cells. *Mol. Cell Neurosci.* **8**, 389–404.

Pietsch, T., Scharmann, T., Fonatsch, C., Schmidt, D., Ockler, R., Freihoff, D., Albrecht, S., Wiestler, O. D., Zeltzer, P., and Riehm, H. (1994). Characterization of five new cell lines derived from human primitive neuroectodermal tumors of the central nervous system. *Cancer Res.* **54**, 3278–3287.

Reynolds, B. A., and Weiss, S. (1992). Generation of neurons and astrocytes from isolated cells of the adult mammalian central nervous system. *Science* **255**, 1707–1710.

Reynolds, B. A., and Weiss, S. (1996). Clonal and population analyses demonstrate that an EGF-responsive mammalian embryonic CNS precursor is a stem cell. *Development* **175**, 1–13.

Roberts, P. E., Phillips, D. M., and Mather, J. P. (1990). A novel epithelial cell from neonatal rat lung: Isolation and differentiated phenotype. *Am. J. Physiol.* **259,** L415–L4125.

Sah, D. W., Ray, J., and Gage, F. H. (1997). Bipotent progenitor cell lines from the human CNS. *Nat. Biotechnol.* **15,** 574–580.

Snyder, E. Y., Deitcher, D. L., Walsh, C., Arnold-Aldea, S., Hartwieg, E. A., and Cepko, C. L. (1992). Multipotent neural cell lines can engraft and participate in development of mouse cerebellum. *Cell* **68,** 33–51.

Svendsen, C. N., ter Borg, M. G., Armstrong, R. J., Rosser, A. E., Chandran, S., and Ostenfeld, M. A. (1998). A new method for the rapid and long-term growth of human neural precursor cells. *J. Neurosci. Methods* **85,** 141–152.

CHAPTER 10

Primary and Multipassage Culture of Human Fetal Kidney Epithelial Progenitor Cells

Deryk Loo, Claude Beltejar, Jeff Hooley, and Xiaolin Xu

Raven Biotechnologies Inc.
South San Francisco, California 94080

Abstract
I. Introduction
II. Rationale
III. Methods
    A. Isolation and Primary Culture
    B. Subculturing—First Passage
    C. Subculturing—Subsequent Passages
    D. Cryopreservation and Subsequent Thawing and Reculturing
IV. Materials
    A. Equipment
    B. Reagents
V. Discussion
VI. Summary
    References

Abstract

Homogeneous, well-characterized cultures of kidney cells representative of defined cellular phenotypes comprising the developing and adult kidney provide important tools to investigate kidney biology. Further, the development of defined media for these culture systems provides opportunities to investigate the role of nutrients, hormones, and matrix components, as well as exogenous insults, in renal development, function, and toxicity. The current explosion in stem cell research has fueled an expanded effort to develop techniques to isolate and culture kidney progenitor and stem cells, which have the potential to treat various forms of renal disease. In this chapter, we outline methods to initiate and propagate long-term cultures of highly homogeneous fetal kidney epithelial progenitor cells. By utilizing a low calcium-containing serum-free culture medium together with a set of defined hormones and extracellular matrix, kidney epithelial progenitor cells can be cultured for more than 60 population doublings without loss of growth potential or phenotypic signs of differentiation. The cultures appear to represent early kidney epithelial progenitors based on cellular marker expression. The cells express the mRNA encoding the embryonic kidney mesenchyme/epithelial marker PAX-2, the stem cell protein CD133, the kidney embryonic progenitor protein CD24, as well as CD29 and CD44. The cells are negative for E-cadherin when grown under low calcium conditions (<0.05 mM); however, E-cadherin expression is induced when cells are cultured under normal calcium conditions (1.2 mM), suggesting that differentiation of the kidney epithelial progenitor culture can be modulated in part by altering the calcium concentration of the medium.

I. Introduction

Through the pioneering work of Taub, Sato and colleagues, a serum-free hormone-supplemented medium was developed that supported the culture of the Madin-Darby canine kidney epithelial cell line (Taub *et al.*, 1979), a cell line initially established using conventional, serum-containing medium. Soon after, serum-free, hormone-supplemented medium was shown to support the primary culture of canine kidney epithelial cells (Jefferson *et al.*, 1980; Taub *et al.*, 1979) and primary cultures of baby mouse kidney cells (Taub and Livingston, 1981; Taub and Sato, 1980). The development of a media formulation devoid of serum was paramount not only because it opened up the possibility of examining the impact of individual nutrients and hormones to cultures but also because serum promoted contamination and overgrowth by unwanted fibroblasts, and inhibited cell growth and promoted differentiation of many cell types, including nontransformed fetal cells (Loo *et al.*, 1987, 1994). Studies by Detrisac *et al.* (1984) extended the early discoveries of Taub, Sato and colleagues by showing that a serum-free, hormonally defined culture medium enabled the selection and extended passage of human kidney epithelial cells specifically of proximal tubule origin. Over the subsequent two decades, culture conditions have been refined by the efforts of many laboratories to enable the selection and short-term propagation of cells representing the major defined cellular phenotypes comprising the mature kidney. Together, this body of work has provided a framework for the exploration of the nutrient and hormone requirements of kidney epithelial cultures and the identification of factors that play a role in the regulation of epithelial cell function and differentiation, and have provided researchers with *in vitro* models to study renal development, function, and toxicity.

With the current heightened interest and advances in stem cell biology, work has begun to explore the development of highly refined separation techniques and culture conditions that allow for the isolation, propagation, and characterization of putative kidney progenitor or stem cells from both fetal and adult tissues (Bussolati *et al.*, 2005; Gupta *et al.*, 2006; Sagrinati *et al.*, 2006). Several reports have highlighted the potential importance of this area of research with respect to regenerative medicine (Al-Awqati and Oliver, 2002; Dekel *et al.*, 2003; Rookmaaker *et al.*, 2004; Zerbini *et al.*, 2006).

The goal of this chapter is to provide investigators with the methodology to generate primary and long-term cultures of homogenous kidney epithelial progenitor cell cultures and provide a jumping-off point for the use of kidney progenitor cell cultures as a tool to study kidney cell biology and development. Implicit in this goal is advancing our understanding of the nutrient and hormone requirements of the cultures, the role these factors play in selecting for (and against) specific subpopulations of cells, and the role of these factors in determining the differentiation fate of the cultures. It is our aim to develop culture conditions, considering both the growth factor requirements and the nutrient and salt requirements of the cultures, that allow for long-term passage of homogeneous kidney epithelial progenitor cells that will provide an accessible tool by which to investigate renal cell biology and toxicity.

II. Rationale

Our approach to the development of kidney epithelial progenitor cultures has been to consider both the nutrient requirements of the culture (salts, vitamins, trace elements, etc.) as well as the growth factor and extracellular matrix requirements of the culture, and develop balanced media formulations that allow for long-term, serial passagable cultures. Our initial studies using commercially available F12/DMEM medium, together with a mixture of hormones and extracellular matrix, indicated that cultures of highly enriched primary epithelial cultures could be grown for several passages before the cultures lost growth potential, likely due to terminal differentiation. This suggested to us that the nutrient, salt, and hormone balance was suboptimal for the maintenance and continuous culture of a kidney epithelial progenitor cell type. Subsequently, through a systematic analysis of the nutrient, salt, and hormone requirements of the kidney cultures, we developed a fructose-based culture medium (termed I/3F) that enabled the propagation of a homogeneous population of kidney epithelial progenitor cells that could be maintained for at least 8 passages without exhibiting a differentiated phenotype. The basis for the I/3F medium formulation is published in United States Patent US 20050101011 (Tsao, 2005), and provides recommended ranges of nutrient, salt, and hormone concentrations for the culture. This serum-free, hormone-supplemented culture medium satisfied the goal of selecting for kidney epithelial progenitor cells (and preventing fibroblast overgrowth); however, the growth of the cultures slowed over time, resulting in diminishing numbers of cells at later passages. On the basis of the reports implicating a role for extracellular calcium in regulating differentiation of human keratinocytes (Peehl and Ham, 1980; Tsao et al., 1982), mouse epidermal cells (Hennings et al., 1980), and human mammary epithelial cells (Soule and McGrath, 1986), we reasoned that a low calcium formulation of the I/3F medium may extend the growth potential of human kidney progenitor cells without initiating differentiation. By switching to the low calcium-containing I/3F medium (<0.05 mM), following culture initiation and one or two passages in normal calcium-containing I/3F medium (1.2 mM), we were able to culture a homogeneous population of kidney epithelial progenitor cells through more than 30 passages (>60 population doublings). We were also successful in maintaining cultures of the kidney epithelial progenitor cells in low calcium-containing F12/DMEM for more than 7 passages—this culture maintained a similar morphology as the cultures grown in low calcium I/3F medium and had a similar growth rate. Longer-term cultures in low calcium-containing F12/DMEM are likely possible, but were not attempted. Representative growth curves of the fetal kidney epithelial progenitor cultures under low calcium and normal calcium culture conditions are shown in Fig. 1. Photomicrographs of fetal kidney epithelial progenitor

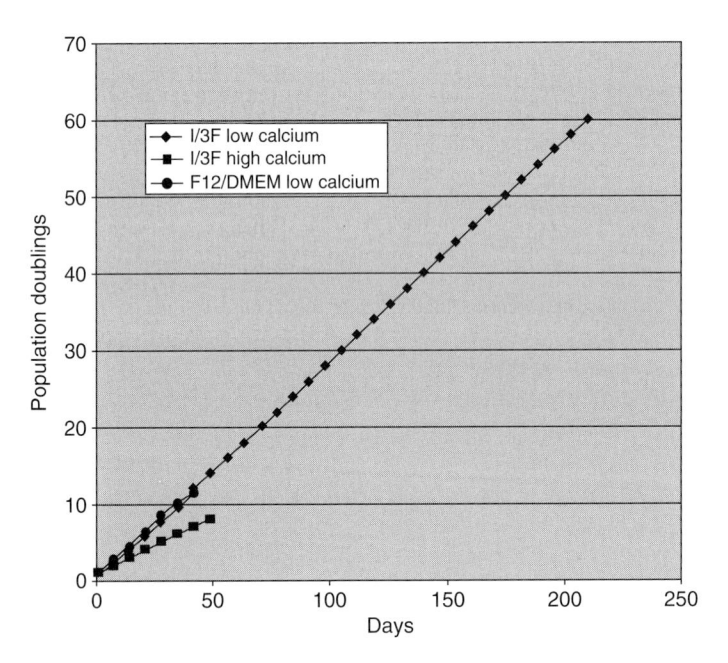

Fig. 1 Representative growth curves of cultures of fetal kidney epithelial progenitor cultures initiated and grown for the first passage under normal calcium conditions, then cultured in either low calcium-containing I/3F medium (diamonds), normal calcium-containing I/3F medium (squares), or low calcium-containing F12/DMEM medium (circles), for the remaining passages shown. Cultures were subcultured weekly and population doublings were calculated based on the culture split ratios.

cultures at various passages, under normal calcium and low calcium culture conditions, are shown in Figs. 2 and 3.

Herein, we outline the methods our laboratory has used to initiate and propagate cultures of kidney epithelial progenitor cells using the media formulations described above. As is always the case, the growth medium we have developed is but one possible formulation, and there remains ample opportunity for the reader to refine the media formulations and culture conditons outlined in this chapter to meet specific needs.

III. Methods

A. Isolation and Primary Culture

1. Procure human fetal kidneys and ship to the lab in sterile culture medium on wet ice. Our laboratory has been successful in culturing fetal kidneys ranging in gestational age from 10 to 18 weeks. Tissue should be cultured within 24 h following isolation.

2. Immediately upon arrival, transfer kidneys to wash medium (cold PBS containing penicillin/streptomycin and gentamicin). Carefully remove outer membranes with sterile forceps and wash the kidneys in 70% ethanol, then rinse twice in wash medium.

3. Place kidneys in a dry 10 cm culture dish and mince the kidneys into ∼1 mm cubes using sterile curved surgical scissors.

4. Add 5 ml of culture medium to the culture dish, transfer the tissue pieces into a 15 ml centrifuge tube, then centrifuge at ∼160 × g (1000 rpm using a Beckman GH-3.6 rotor) for 5 min. Resuspend the tissue pieces in serum-free culture medium containing insulin (10 µg/ml), transferrin (10 µg/ml), epidermal growth factor (EGF; 20 ng/ml), growth hormone (10 ng/ml), pig pituitary extract (0.2%), gentamicin (100 µg/ml), penicillin/streptomycin (1×), and collagenase/dispase (0.1%). Incubate at 4 °C overnight.

5. The following day, centrifuge the digested tissue pieces at ∼160 × g for 5 min and then wash twice with culture medium.

6. Resuspend the pelleted tissue pieces in 10 ml of culture medium containing insulin (10 µg/ml), transferrin (10 µg/ml), EGF (20 ng/ml), growth hormone (10 ng/ml), and pig pituitary extract (0.2%) and plate into a fibronectin-precoated 10 cm culture dish. Cultures are routinely initiated by plating each fetal kidney in 1–3 10 cm dishes.

Under these culture conditions, the human fetal kidney cells will attach to the substrate-coated plates and grow to form a confluent monolayer of bright, tightly packed cells over 5–10 days, depending on the size of the kidney and the condition of the tissue.

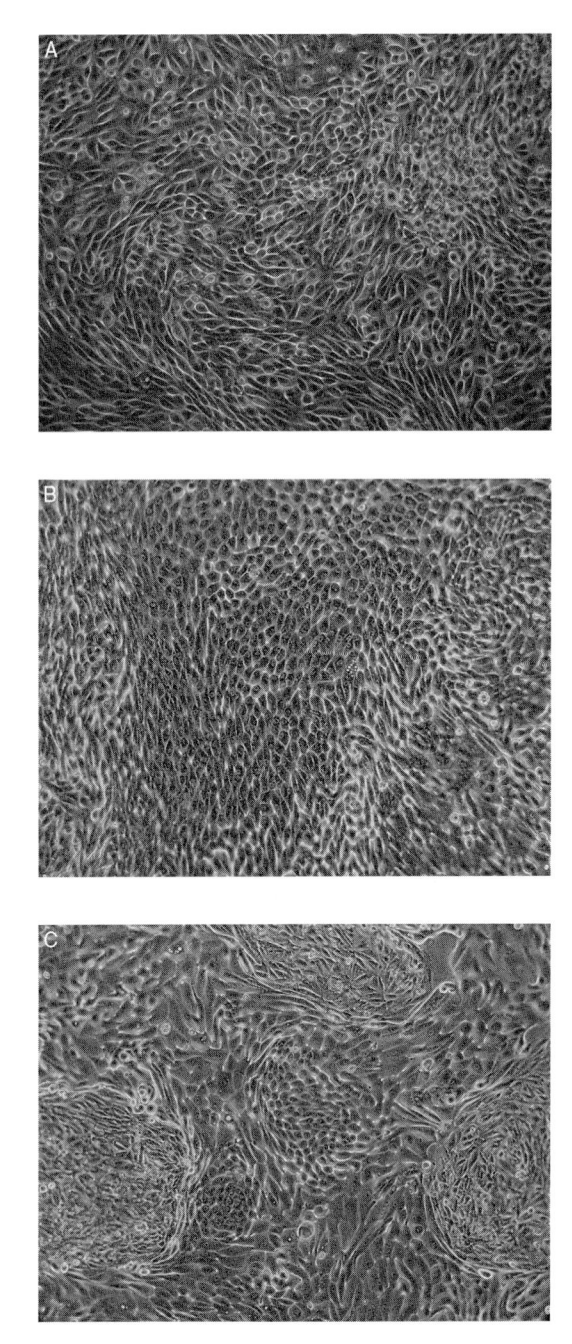

Fig. 2 (*continues*).

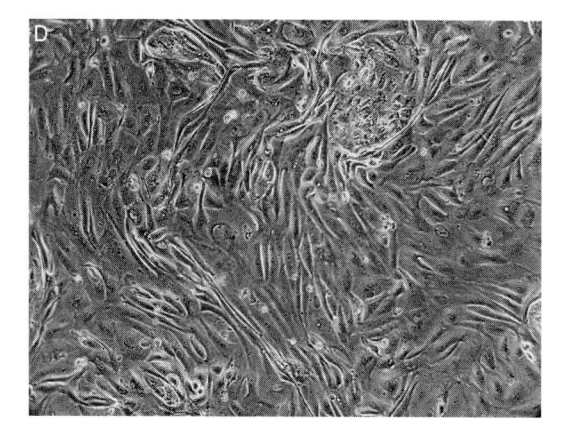

Fig. 2 Representative photomicrographs of human fetal kidney progenitor cultures grown in normal calcium-containing I/3F medium. (A) Confluent primary culture, (B) Confluent culture at passage 4, (C) Confluent culture at passage 6 with areas of the culture exhibiting a differentiated phenotype, (D) Confluent culture at passage 7 with the majority of cells exhibiting a differentiated phenotype. The culture at passage 7 showed a marked increase in doubling time and was not capable of further subculturing.

7. Replace the culture medium every 2–4 days by carefully aspirating ∼75% of the culture medium from the dish and replacing with fresh culture medium containing insulin (10 μg/ml), transferrin (10 μg/ml), EGF (20 ng/ml), growth hormone (10 ng/ml), and pig pituitary extract (0.2%). Floating, nonadherent cells and tissue fragments may be replated into a fresh fibronectin-precoated culture dish.

B. Subculturing—First Passage

1. When the primary culture reaches 70–80% confluence (typically 4–8 days), rinse the cell monolayer with fresh culture medium, add 2 ml trypsin/EDTA solution, and incubate at 37 °C for 2–4 min—until the majority of the cells round up and detach from the culture surface. Avoid extended exposure to the trypsin/EDTA solution as this can greatly reduce cell viability. Detachment of the cells from the culture surface may be aided by gently pipetting the trypsin/EDTA solution across the cell monolayer if necessary.

2. After the cells have become detached, add 2 ml of soybean trypsin inhibitor solution to the plate and transfer the cell suspension to a 15 ml conical centrifuge tube. Pellet the cell suspension by centrifugation at ∼160 × g for 5 min at room temperature. Carefully aspirate the supernatant and resuspend the cells in a small volume (1–2 ml) of culture medium and determine the cell concentration using standard methods, if desired.

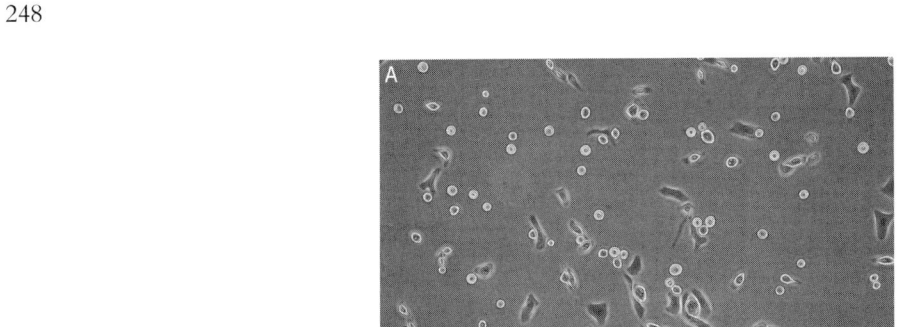

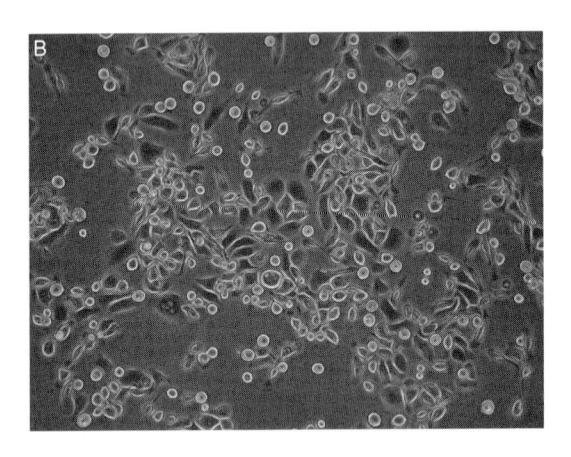

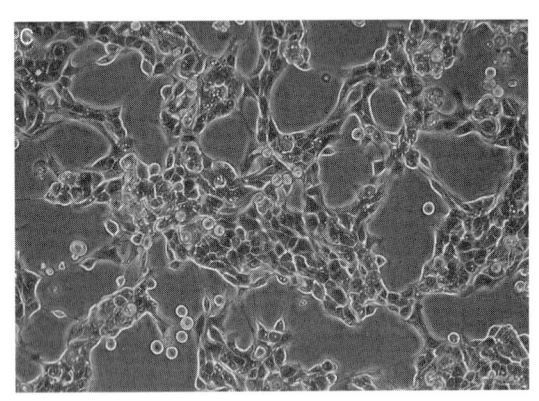

Fig. 3 Representative photomicrographs of human fetal kidney progenitor cultures continuously grown and passaged in low calcium-containing I/3F medium (from passage 2 onward). (A) Low density culture at passage 9 growing in low calcium I/3F medium. Note that the subpopulation of round, bright cells is strongly adhered to the fibronectin matrix, but has not yet spread out on the matrix. (B) Higher

3. Resuspend the desired volume of cells in culture medium containing insulin (10 µg/ml), transferrin (10 µg/ml), EGF (20 ng/ml), growth hormone (10 ng/ml), and pig pituitary extract (0.2%) and replate into fibronectin-precoated 10 cm culture dishes. Additionally, conditioned medium from high-density kidney epithelial cultures may be added when subculturing to reduce the recovery time of the culture and improve viability. A split ratio of 1:2–1:4 is recommended when subculturing the cells into a fresh 10 cm culture dish. Split ratios greater than 1:4 may be used; however, this may extend the recovery time and increase the doubling time of the culture.

4. Replace the culture medium every 2–4 days, as described previously, until the culture reaches near confluence.

C. Subculturing—Subsequent Passages

Cultures maintained in normal calcium-containing medium (1.2 mM) can be maintained for at least eight passages at a split ratio of 1:2–1:4. However, in our experience, the doubling time of the cultures noticeably slowed over the course of the eight passages and lead to reduced cell density and/or a reduced split ratio over subsequent passages.

Switching the culture, following passage 1 or 2, into low calcium-containing medium (<0.05 mM) enables the culture to maintain growth potential for more than 30 passages at a split ratio of 1:4. When near confluent, subculture as described in Section III B, except replace the normal calcium-containing culture medium with low calcium-containing culture medium. Similarly, conditioned low calcium-containing culture medium from high-density kidney epithelial cultures may be added when subculturing to reduce the recovery time of the culture and improve viability. A split ratio of 1:4 is recommended when subculturing the near confluent culture into a fresh 10 cm culture dish under low calcium-containing culture medium conditions. Replace the culture medium every 2–4 days, as described previously, until the culture reaches near confluence.

D. Cryopreservation and Subsequent Thawing and Reculturing

1. When the culture reaches near confluence (70–80% confluent), rinse the cell monolayer with fresh culture medium, add 2 ml trypsin/EDTA solution, and incubate at 37 °C for 2–4 min—until the majority of the cells round up and detach from the culture surface.

---

density culture at passage 30 growing in low calcium I/3F medium. Again, round, bright cells are strongly attached to the fibronectin matrix. (C) Sister culture at passage 30 that had been growing in low calcium I/3F medium, and was then switched into normal calcium-containing I/3F medium for 24 h. Note the change in morphology to a more spread phenotype with tighter cell–cell interactions. This culture has begun to express E-cadherin protein as determined by flow cytometry.

2. After the cells have become detached, add 2 ml of soybean trypsin inhibitor solution to the plate and transfer the cell suspension to a 15 ml conical centrifuge tube. Pellet the cell suspension by centrifugation at ~160 × g for 5 min at room temperature. Carefully aspirate the supernatant and resuspend the cells in freezing medium (90% serum-free culture medium/10% DMSO) at 1×10^6 to 5×10^6 cells/ml and transfer to 1.5 ml sterile cryogenic vials. Place vials into a Nalgene cryogenic freezing container, or sandwich the vials between two styrofoam 15 ml culture tube racks, and place at $-80\,°C$ overnight. The following day, transfer vials to a liquid nitrogen dewar.

3. To reculture a frozen vial of cells, remove the vial from liquid nitrogen and thaw quickly by swirling in a 37 °C water bath. Transfer the cell suspension to a 15 ml conical centrifuge tube, add 5 ml of fresh culture medium, and pellet the cells by centrifugation at ~160 × g for 5 min at room temperature. Resuspend the cell pellet in fresh culture medium containing insulin (10 μg/ml), transferrin (10 μg/ml), EGF (20 ng/ml), growth hormone (10 ng/ml), and pig pituitary extract (0.2%) and plate into a fibronectin-precoated 10 cm culture dish as described above for the initial subculturing.

IV. Materials

A. Equipment

1. For Preparation of Porcine Pituitary Extract:

> Heavy duty glass blender
> High-speed refrigerated centrifuge
> JA-20 centrifuge rotor or equivalent
> 50 ml Oak Ridge high-speed polycarbonate centrifuge tubes
> Stir plate in 4 °C cold room or refrigerator

2. For Cell Culture:

> 5% CO_2 incubator at 37 °C
> Low-speed table top centrifuge
> 10 cm tissue culture dishes
> 15 and 50 ml conical polypropylene centrifuge tubes
> 1.5 ml cryogenic freezing vials
> Surgical scissors and forceps

B. Reagents

A list of commonly used reagents, supplier information, and working concentrations is presented in Table I.

Table I

List of Commonly used Reagents, Supplier Information and Culture Supplements used to Culture Fetal Kidney Epithelial Progenitor Cells

Reagent	Formulation	Working Concentration	Vendor	Catalog Number
Human Recombinant Insulin	10 mg/ml in 0.01 N HCl	10 μg/ml	Sigma	H-9892
Human Transferrin	10 mg/ml in PBS	10 μg/ml	Sigma	T-1147
Human Recombinant EGF	10 μg/ml in PBS	20 ng/ml	Invitrogen	13247–051
Growth Hormone (Human pituitary)	10 μg/ml in 10 mM sodium bicarbonate	10 ng/ml	Calbiochem	869008
Collagenase/Dispase	10% in F12/DMEM	neat	Roche	10 270 172
Trypsin/EDTA	0.05%	neat	Invitrogen	25300
Soybean Trypsin Inhibitor	1 mg/ml in PBS	neat	Sigma	T-9003
Penicillin/Streptomycin	100×	1×	Sigma	P-0781
Gentamycin	100 mg/ml in water	100 μg/ml	Sigma	G-1264

1. Basal Medium

F12/DMEM, a 1:1 mixture of Ham's F12 and high-glucose Dulbecco's modified Eagle medium (DMEM), supplemented with 15 mM HEPES, pH 7.4, 1.2 g/L sodium bicarbonate, and 2 mM glutamine, can be obtained as a powder or liquid from Invitrogen (Carlsbad, CA). Custom formulated low-calcium or calcium-free F12/DMEM and fructose-based media can also be obtained from Invitrogen. Guidelines for the recipe for fructose-based I/3F medium is described in United States Patent US 20050101011 (Tsao, 2005). Calcium/magnesium-free phosphate-buffered saline (PBS) can be obtained from Invitrogen.

2. Growth Substrate

Historically, exogenously supplied serum has served to provide the extracellular matrix components necessary for nontransformed cells to attach and spread on tissue culture plastic surfaces. However, under serum-free culture conditions, a source of extracellular matrix must be provided. The kidney epithelial cultures will adhere strongly to fibronectin-coated plastic. To coat tissue culture dishes, add fibronectin (0.1% stock solution; F1141, Sigma) to the dish at a concentration of 10 μg/ml in culture medium and incubate for 1–24 h at 37 °C in a level incubator. Typically, dishes are precoated in one-half to one-third the volume used for culturing the cells (i.e., 3 ml/10 cm dish). Following incubation, remove the fibronectin solution and wash the dish once with culture medium. The vessel is now ready for cell plating. Avoid allowing the precoated dish to dry out.

3. Growth Factors

Table I lists the growth factors and working concentrations that are used to support the growth of the kidney epithelial progenitor cell cultures. Growth factors readily adhere to glass and plastic; therefore, they should be added directly to the culture medium in the culture dish. If many dishes of cells are being prepared simultaneously, the required volume of culture medium for all of the dishes can be aliquoted to a polypropylene tube and the growth factors added to the culture medium-containing tube immediately prior to dispensing into the culture dishes.

4. Porcine Pituitary Extract

Porcine pituitary extract is prepared using the following protocol. Tissue can be obtained from Pel-Freez Biologicals (Rogers, AR).

1. To a chilled glass blender add 250 ml of 150 mM NaCl at 4 °C.
2. Add 100 g of porcine pituitary tissue
3. Blend by pulsing until mixed thoroughly, then puree for 2 min.
4. Transfer puree to a beaker and mix on a stir plate for 90 min at 4 °C.
5. Transfer mixture to 50 ml Oak Ridge high-speed centrifuge tubes. Centrifuge for 40 min at $\sim 8000 \times g$ (10,000 rpm in a Beckman JA-20 rotor) at 4 °C.
6. Carefully remove supernatant, pool, then aliquot into fresh 50 ml Oak Ridge high-speed centrifuge tubes and centrifuge for an additional 40 min at $\sim 8000 \times g$ at 4 °C.
7. Repeat centrifugation, if necessary, until supernatant appears clear.
8. Carefully decant supernatant and filter sequentially through 0.8, 0.45, and 0.2 μm low protein-binding tissue culture grade filters. Poor or very slow filtration indicates the need for additional centrifugation.
9. Dispense 0.2 μm filter sterile porcine pituitary extract into sterile cryotubes. Store aliquots at −80 °C. Typical batches of porcine pituitary extract have a protein concentration of ~10 mg/ml.

The aliquots will remain stable for at least 1 year at −80 °C. It is recommended that the extract be stored in small aliquots (1–5 ml), if convenient, to avoid repeated freeze/thaw cycles. The working aliquot should be stored at 4 °C and used within 1 month. A small amount of reddish precipitate may form over time at 4 °C—formation of a precipitate does not compromise the extract; however, it is recommended that care be taken not to introduce the precipitate material into the cultures.

5. Antibiotics

Contamination is one of the main causes of failure in primary cultures; therefore, it is recommended to include antibiotics in the primary and early passage cultures to control for possible bacterial contamination. The most commonly used antibiotics include penicillin, streptomycin, and gentamicin. Penicillin/streptomycin is available as a 100× mixture (Cat# P-0781; Sigma, St. Louis, MO). Gentamicin is prepared as a 100 mg/ml solution in water (Cat# G-1264; Sigma) and used at a final concentration of 100 µg/ml.

6. Dissociation Reagents

a. Collagenase/Dispase

Prepare collagenase/dispase as a 10% solution in culture medium (10 g/100 ml culture medium). After the collagenase/dispase has dissolved, sterile filter through a 0.2 µm filter and store in aliquots at −20 °C. Collagenase/dispase remains stable at −20 °C for at least 6 months. Store thawed aliquots at 4 °C and use within 1 month. A brownish precipitate may form—this does not significantly alter the enzymatic activity of the collagenase/dispase; however, care should be taken to avoid introduction of the precipitate into the cultures.

b. Trypsin/EDTA and Soybean Trypsin Inhibitor

Trypsin/EDTA is available as a 0.05% solution from Invitrogen and is used neat. Soybean trypsin inhibitor is available in powder form from Sigma. Prepare a 1 mg/ml solution of soybean trypsin inhibitor in PBS and sterilize through a 0.2 µm filter. Store aliquots at −20 °C. Soybean trypsin inhibitor remains stable for more than 1 year at −20 °C. Store thawed aliquots at 4 °C and use within 2 months.

7. Conditioned Medium

Kidney epithelial culture conditioned medium is a rich source of autocrine growth-promoting and maintenance factors and may be used to reduce the stress of passaging, thereby improving recovery time and viability of the culture. Culture medium collected from high-density kidney epithelial cultures prior to passaging is an ideal source of conditioned medium. Collect culture medium that has been exposed to the culture for 2–4 days, centrifuge at ∼160 × g to remove cell debris, then sterile filter through a 0.2 µm filter. Store conditioned medium at 4 °C and use within 1–2 weeks of harvest.

V. Discussion

This chapter outlined methods for the isolation and extended culture of fetal kidney epithelial progenitor cells for greater than 60 population doublings. The fetal kidney progenitor cells express the mRNA encoding the embryonic kidney mesenchymal/epithelial transcription factor PAX-2 (Dressler and Douglass, 1992), the stem cell surface antigen protein CD133 (Miraglia et al., 1997) which is expressed in embryonic kidney (Corbeil et al., 2000), the kidney embryonic progenitor marker protein CD24 (Challen et al., 2004), as well as CD29 and CD44 proteins. The cells do not express the hematopoietic cell marker proteins CD34 or CD45 (data not shown).

One of the critical factors that enabled long-term propagation of the fetal kidney progenitor cells was the use of a low calcium-containing culture medium. Cultures of fetal kidney cells initiated and continuously cultured in normal calcium-containing medium lost growth potential and appeared phenotypically to undergo differentiation within 10–15 population doublings. Cultures of fetal kidney progenitor cells grown under low calcium conditions maintained a high plating efficiency, retained an epithelial-like morphology, and continued to grow with a doubling time of 24–48 h. Consistent with a possible role of extracellular calcium in differentiation of the culture, we observed that when mature cultures of fetal kidney progenitor cells growing under low calcium conditions were transferred to normal calcium conditions, the cultures grew slower, developed a more flattened morphology with tighter cell–cell contact, and were induced to express E-cadherin within 24 h following transfer into normal calcium conditions (data not shown).

We are continuing to refine the culture conditions for propagation of the kidney epithelial progenitor cells and beginning to examine their differentiation potential *in vivo*. Our hope is that this chapter will aid investigators in the isolation and generation of kidney progenitor cell lines, and encourage investigators to extend these methods to the further development and characterization of kidney progenitor cells.

VI. Summary

This chapter provides methods for the initiation and long-term propagation of highly homogeneous kidney epithelial progenitor cells from human fetal kidney tissue. By utilizing an optimized low calcium-containing serum-free culture medium, in concert with a defined set of hormones and extracellular matrix, fetal kidney epithelial progenitor cells can be cultured for more than 60 population doublings without loss of growth potential or evidence of differentiation. The cultures display cellular markers described for early kidney epithelial progenitor cells, including PAX-2, CD133, CD24, CD29, and CD44, and do not express the hematopoietic cell markers CD34 or CD45. Consistent with a possible role of calcium in differentiation of the culture, cells transferred to normal calcium conditions grew slower, adopted a flattened morphology with tighter cell–cell contact, and expressed E-cadherin.

Acknowledgments

The authors thank Penelope Roberts, Mary Tsao, and Jennie P. Mather for helpful discussions, and Aimee Mann.

References

Al-Awqati, Q., and Oliver, J. A. (2002). Stem cells in the kidney. *Kidney Int.* **61,** 387–395.
Bussolati, B., Bruno, S., Grange, C., Buttiglieri, S., Deregibus, M. C., Cantino, D., and Camussi, G. (2005). Isolation of renal progenitor cells from adult human kidney. *Am. J. Pathol.* **166,** 545–555.

Challen, G. A., Martinez, G., Davis, M. J., Taylor, D. F., Crowe, M., Teasdale, R. D., Grimmond, S. M., and Little, M. H. (2004). Identifying the molecular phenotype of renal progenitor cells. *J. Am. Soc. Nephrol.* **15,** 2344–2357.

Corbeil, D., Roper, K., Hellwig, A., Tavian, M., Miraglia, S., Watt, S. M., Simmons, P. J., Peault, B., Buck, D. W., and Huttner, W. B. (2000). The human AC133 hematopoietic stem cell antigen is also expressed in epithelial cells and targeted to plasma membrane protrusions. *J. Biol. Chem.* **275,** 5512–5520.

Dekel, B., Burakova, T., Arditti, F. D., Reich-Zeliger, S., Milstein, O., Aviel-Ronen, S., Rechavi, G., Friedman, N., Kaminski, N., Passwell, J. H., and Reisner, Y. (2003). Human and porcine early kidney precursors as a new source for transplantation. *Nat. Med.* **9,** 53–60.

Detrisac, C. J., Sens, M. A., Garvin, A. J., Spicer, S. S., and Sens, D. A. (1984). Tissue culture of human kidney epithelial cells of proximal tubule origin. *Kidney Int.* **25,** 383–390.

Dressler, G. R., and Douglass, E. C. (1992). PAX-2 is a DNA-binding protein expressed in embryonic kidney and Wilms tumor. *Proc. Natl. Acad. Sci. USA* **89,** 1179–1183.

Gupta, S., Verfaillie, C., Chmielewski, D., Kren, S., Eidman, K., Connaire, J., Heremans, Y., Lund, T., Blackstad, M., Jiang, Y., Luttun, A., and Rosenberg, M. E. (2006). Isolation and characterization of kidney-derived stem cells. *J. Am. Soc. Nephrol.* **17,** 3028–3040.

Hennings, H., Michael, D., Cheng, C., Steinert, P., Holbrook, K., and Yispa, S. H. (1980). Calcium regulation of growth and differentiation of mouse epidermal cells in culture. *Cell* **19,** 245–254.

Jefferson, D. M., Cobb, M. H., Gennaro, J. F., Jr., and Scott, W. N. (1980). Transporting renal epithelium: Culture in hormonally defined serum-free medium. *Science* **210,** 912–914.

Loo, D. T., Althoen, M. C., and Cotman, C. W. (1994). Down regulation of nestin by TGF-beta or serum in SFME cells accompanies differentiation into astrocytes. *Neuroreport* **5,** 1585–1588.

Loo, D. T., Fuquay, J. I., Rawson, C. L., and Barnes, D. W. (1987). Extended culture of mouse embryo cells without senescence: Inhibition by serum. *Science* **236,** 200–202.

Miraglia, S., Godfrey, W., Yin, A. H., Atkins, K., Warnke, R., Holden, J. T., Bray, R. A., Waller, E. K., and Buck, D. W. (1997). A novel five-transmembrane heatopoietic stem cell antigen: Isolation, characterization, and molecular cloning. *Blood* **90,** 5013–5021.

Peehl, D. M., and Ham, R. G. (1980). Clonal growth of human keratinocytes with small amounts of dialyzed serum. *In Vitro* **16,** 526–540.

Rookmaaker, M. B., Verhaar, M. C., van Zonneveld, A. J., and Rabelink, T. J. (2004). Progenitor cells in the kidney: Biological and therapeutic perspectives. *Kidney Int.* **66,** 518–522.

Sagrinati, C., Netti, G. S., Mazzinghi, B., Lazzeri, E., Liotta, F., Frosali, F., Ronconi, E., Meini, C., Gacci, M., Squecco, R., Carini, M., Gesualdo, L., *et al.* (2006). Isolation and characterization of multipotent progenitor cells from the Bowman's capsule of adult human kidneys. *J. Am. Soc. Nephrol.* **17,** 2443–2456.

Soule, H. D., and McGrath, C. M. (1986). A simplified method for passage and long-term growth of human mammary epithelial cells. *In Vitro Cell. Dev. Biol.* **22,** 6–12.

Taub, M., Chuman, L., Saier, M. H., Jr., and Sato, G. (1979). Growth of Madin-Darby canine kidney epithelial cell (MDCK) line in hormone-supplemented, serum-free medium. *Proc. Natl. Acad. Sci. USA* **76,** 3338–3342.

Taub, M., and Livingston, D. (1981). The development of serum-free hormone-supplemented media for primary kidney cultures and their use in examining renal function. *In* "Annals New York Acad. Sci." (W. N. Scott, and D. B. P. Goodman, eds.), pp. 406–421. New York Acad. Sci., New York.

Taub, M., and Sato, G. (1980). Growth of functional primary cultures of kidney epithelial cells in defined medium. *J. Cell Physiol.* **105,** 369–378.

Tsao, M. C. (2005). Inventor; Raven biotechnologies, inc., assignee. Cell culture media United States patent US 20050101011. 2005 May 12.

Tsao, M. C., Walthall, B. J., and Ham, R. G. (1982). Clonal growth of normal human epidermal keratinocytes in a defined medium. *J. Cell. Physiol.* **110,** 219–229.

Zerbini, G., Piemonti, L., Maestroni, A., Dell'Antonio, G., and Bianchi, G. (2006). Stem cells and the kidney: A new therapeutic tool? *J. Am. Soc. Nephrol.* **17,** S123–S126.

CHAPTER 11

Isolation of Human Mesenchymal Stem Cells from Bone and Adipose Tissue

Susan H. Bernacki, Michelle E. Wall,[1] and Elizabeth G. Loboa

Joint Department of Biomedical Engineering at
NC State University and UNC
Chapel Hill, Burlington Laboratories
Raleigh, NC 27695

[1] Flexcell International corporation,
Hillsborough, NC 27278

    I. Introduction
   II. Isolation of hMSCs from Trabecular Bone
       A. Bone Samples
       B. Protocol
  III. Isolation of hMSCs from Adipose Tissue
       A. Adipose Tissue Samples
       B. Protocol
  IV. Growth, Subculture, and Cryopreservation of Undifferentiated hMSCs
       A. Feeding
       B. Subculture
       C. Cryopreservation
       D. Thawing Cryopreserved Cells
   V. Surface Marker Characterization of hMSCs
       A. Micromass Cell Cultures
       B. Fixation and Immunostaining
  VI. hMSC Differentiation Assays for Osteogenic, Adipogenic, and Chondrogenic Pathways
       A. Osteogenic and Adipogenic Differentiation
       B. Chondrogenic Differentiation in Attached Micromass Culture
       C. Media Formulations
 VII. Summary
     References

I. Introduction

Stem cells from adult tissues are attractive materials for cell therapy, gene therapy, and tissue engineering. These cells generally have restricted lineage potential when compared to embryonic stem cells, and this may be advantageous from the standpoint of controlling cell growth and differentiation in certain therapeutic applications (reviewed in Barrilleaux *et al.*, 2006; Barry and Murphy, 2004; Haynesworth *et al.*, 1998).

In 1961, bone marrow was shown to contain hematopoietic progenitor cells (Till and McCulloch, 1961). Beginning in the 1970s, numerous investigators demonstrated that bone marrow also contained cells with fibroblastic morphology that could differentiate into bone, cartilage, fat, and muscle (reviewed in Prockop, 1997). These cells have been variously designated as marrow stromal cells or mesenchymal stem cells, and ambiguously abbreviated as "MSCs." Pittenger *et al.* (1999) demonstrated that individual cells from the bone marrow stromal population possessed multilineage potential.

Since the identification of MSCs in bone marrow, cells with similar multilineage potential have been isolated from other tissues, including trabecular bone (Noth *et al.*, 2002; Sottile *et al.*, 2002) and adipose tissue (Lee *et al.*, 2004; Zuk *et al.*, 2001). The presence of MSCs in adipose tissue has generated special interest because harvesting fat tissue is generally less traumatic to the donor than harvesting bone marrow, and greater quantities may be available. Adipose-derived MSCs are also called adipose-derived stem cells (ADSCs) and adipose-derived adult stromal or stem cells (ADAS cells).

MSCs from different sources have similar multilineage potential; however, recent evidence suggests they may differ in behavior depending on the culture conditions. MSCs derived from adipose tissue deposit histologically detectable quantities of mineral, lipid, and cartilage matrix in monolayer culture when maintained in appropriate media (Zuk *et al.*, 2002). In a mouse model, adipose- and bone marrow–derived MSCs were equally effective in healing skeletal defects (Cowan *et al.*, 2004); adipose-derived MSCs were also able to heal bone defects in a rat model (Yoon *et al.*, 2007). However, under stimulation by transforming growth factor-beta (TGFβ), bone marrow–derived MSCs expressed more cartilage extracellular matrix proteins and exhibited a more mature chondrogenic phenotype than adipose-derived MSCs (Afizah *et al.*, in press; Mehlhorn *et al.*, 2006).

Bone marrow–derived MSCs initially were separated from hematopoietic progenitor cells based on their substantial affinity for tissue culture plastic (Friedenstein *et al.*, 1976). More recent isolation protocols often include separation of bone marrow cells on a Percoll® or Ficoll® density gradient prior to plating (see Colter *et al.*, 2000; Yoo *et al.*, 1998), the use of magnetic beads with selective antibodies (Aslan *et al.*, 2006), or fractionation by fluorescence-activated cell sorting (FACS; Buhring *et al.*, in press; Simmons and Torok-Storb, 1991). It has been suggested that MSCs from bone marrow are similar regardless of the method of isolation (Lodie *et al.*, 2002).

Here, we describe general protocols for the isolation of human MSCs (hMSCs) from trabecular bone and adipose tissues using adhesion-based methods. These protocols employ standard experimental techniques and can be readily performed in laboratories with the following basic equipment items: biosafety cabinet (sterile hood), humidified CO_2 incubator, centrifuge (maximum required speed $500 \times g$ for bone, $5000 \times g$ for adipose tissue), inverted microscope, and autoclave. If cells are to be cryopreserved, an ultralow temperature freezer ($-86\,^\circ$C) and liquid nitrogen cryostorage unit are also required. Where a supplier is specified, this indicates that the particular reagent has been used successfully by the authors. Equivalent reagents from other sources may be equally suitable. Knowledge of standard tissue culture technique and general laboratory practice is assumed. A flowchart summarizing the process is shown in Fig. 1.

Biohazard note: All human tissues, and the cells derived from them, are considered potential sources of bloodborne pathogens, including hepatitis B, hepatitis C, and HIV, and should be handled using Universal Precautions (Centers for Disease Control, 1988). Consult with the institutional biosafety or environmental health safety office regarding appropriate training for personnel, and handling and disposal of waste. If tissue samples are to be obtained from a hospital or clinic and transported to a laboratory for processing, check local and institutional regulations regarding transport of medical specimens. Secondary containment (e.g., a sealed bag and/or a cooler) and biohazard labeling are generally required.

Note on Institutional Review Board (IRB) approval: Tissues for isolation of MSCs are commonly obtained from patients undergoing therapeutic or elective surgical procedures. These tissues would normally be discarded but can often be recovered for research purposes. Policies regarding use of excess tissue for research vary depending on the institution. The IRB offices of both the hospital and the research laboratory, if at a different institution, should be consulted prior to obtaining surgical specimens.

II. Isolation of hMSCs from Trabecular Bone

MSCs from trabecular bone were initially obtained from explant cultures of bone fragments from the femoral heads of patients undergoing hip surgery. Small pieces of bone were placed in culture, either directly (Sottile *et al.*, 2002) or following collagenase digestion to remove adherent material (Noth *et al.*, 2002). After 2–4 weeks, adherent cells were harvested. The authors compared explant cultures of digested and undigested bone fragments, and also of the cells released by digestion, and found that the best yields of hMSCs were obtained by culturing the released cells.

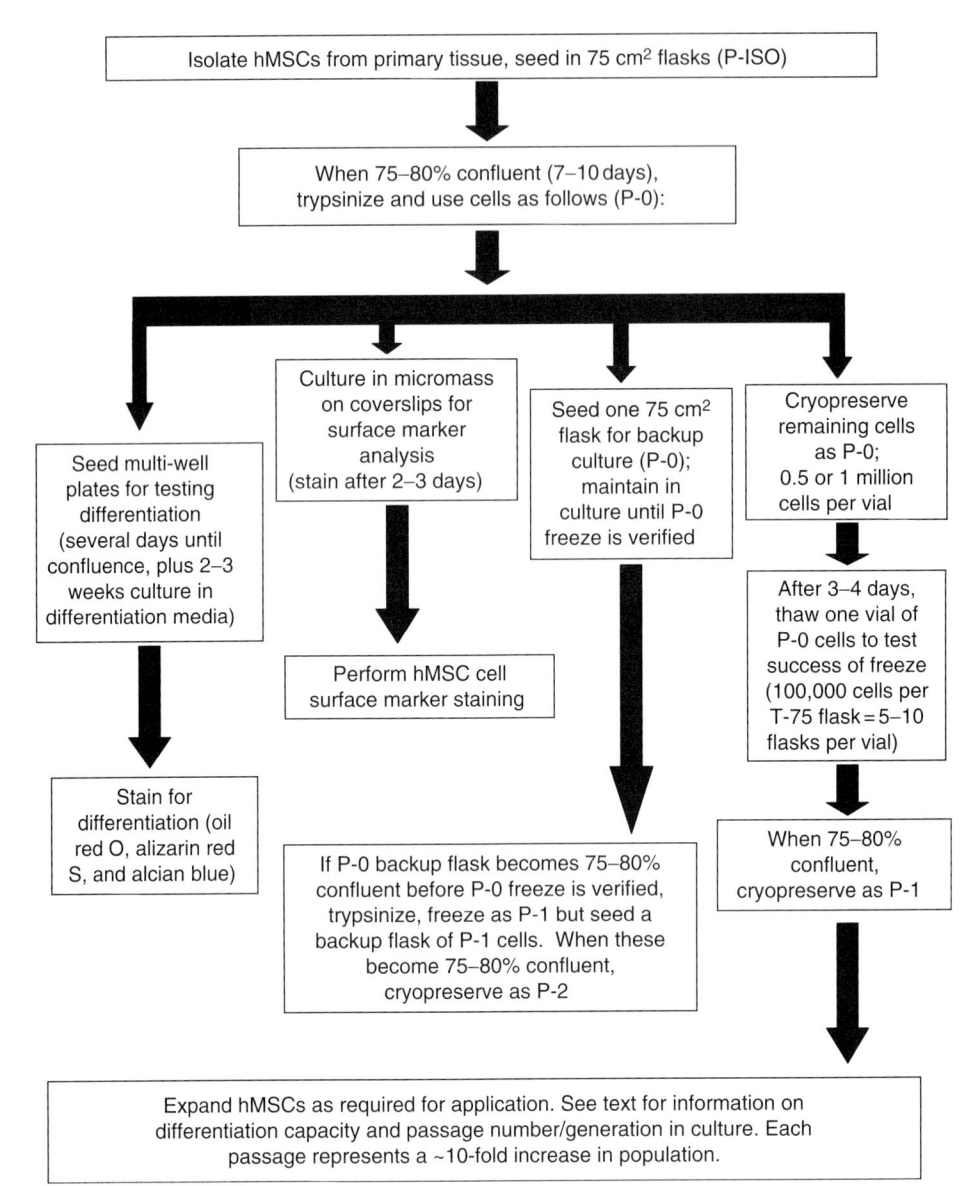

```
┌─────────────────────────────────────────────────────────────┐
│ Isolate hMSCs from primary tissue, seed in 75 cm² flasks (P-ISO) │
└─────────────────────────────────────────────────────────────┘
                              ↓
┌─────────────────────────────────────────────────────────────┐
│        When 75–80% confluent (7–10 days),                   │
│        trypsinize and use cells as follows (P-0):           │
└─────────────────────────────────────────────────────────────┘
```

Fig. 1 Flowchart illustrating sequence of isolation, differentiation assessment, surface marker analysis, expansion, and cryopreservation of human mesenchymal stem cells isolated from human adipose tissue or bone. Approximate time from tissue isolation to completion of differentiation assessment is 4 weeks.

Branches from P-0:

- Seed multi-well plates for testing differentiation (several days until confluence, plus 2–3 weeks culture in differentiation media) → Stain for differentiation (oil red O, alizarin red S, and alcian blue)
- Culture in micromass on coverslips for surface marker analysis (stain after 2–3 days) → Perform hMSC cell surface marker staining
- Seed one 75 cm² flask for backup culture (P-0); maintain in culture until P-0 freeze is verified → If P-0 backup flask becomes 75–80% confluent before P-0 freeze is verified, trypsinize, freeze as P-1 but seed a backup flask of P-1 cells. When these become 75–80% confluent, cryopreserve as P-2
- Cryopreserve remaining cells as P-0; 0.5 or 1 million cells per vial → After 3–4 days, thaw one vial of P-0 cells to test success of freeze (100,000 cells per T-75 flask = 5–10 flasks per vial) → When 75–80% confluent, cryopreserve as P-1

Expand hMSCs as required for application. See text for information on differentiation capacity and passage number/generation in culture. Each passage represents a ~10-fold increase in population.

A. Bone Samples

MSCs have been isolated from various types of bone (see Barrilleaux *et al.*, 2006; Barry and Murphy, 2004), and the protocol described below should be adaptable to most trabecular bone tissue. The particular type of bone to be used for a given application will depend on the research focus and on the tissues available. Bone from the femoral head and tibia may be most readily obtainable due to the frequency of hip- and knee-replacement procedures.

The sample should contain as much trabecular bone and as little cortical bone as possible. Tissue samples from surgical reaming procedures are excellent starting material, because they are already well fragmented, and are generally predominantly trabecular bone with minimal cortical bone. Some surgical procedures generate quantities of small cortical bone fragments which are very sharp. These samples require special handling to avoid reducing the yield of viable MSCs (see below).

Good communication with the surgical team regarding sample specifications is important, and it may be possible to have the bone fragmented into appropriate sized pieces (2–3 mm) at the time the sample is obtained, as tools for this purpose may be available in the operating room. It may be necessary to obtain special tools to fragment the pieces in the laboratory if the sample consists of large pieces of bone.

After harvest, the bone sample should be placed in a sterile specimen container for transport to the laboratory. Samples that will not be processed immediately should be stored on ice. The authors have isolated viable MSCs from bone samples stored on ice in the refrigerator for up to 4 days; however, results are variable, and samples should be processed as soon as possible.

MSC yield varies substantially depending on the type of sample. In the authors' laboratory, a typical preparation of MSCs from 10 g of tibial reaming tissue yields ~10 million cells if harvested 7–10 days following isolation.

B. Protocol

1. Materials and Supplies

    Note: All supplies, tools, and solutions must be sterile.

    - Phosphate-buffered saline (PBS)
    - hMSC complete growth medium (see below)
    - 100× antibiotic stock solution (10,000 units/ml penicillin, 10,000 μg/ml streptomycin)
    - Collagenase solution (type XI-S; Sigma C9697), 3 mg/ml in alpha-modified minimum essential medium (aMEM) (without serum) with 1× antibiotics, filter sterilized. Solution may be prepared in advance and frozen at −20 °C in aliquots
    - Centrifuge tubes, pipets, sterile tissue culture plasticware
    - Forceps, scalpels, and other tools for handling tissue
    - Cell strainers (BD Falcon™, 100 μm mesh size)
    - Rotisserie-style rotator

2. Procedure

Note: All activities that involve exposed tissue must be performed in a biosafety cabinet. Maintain sterility of the preparation throughout the procedure. Handle all human tissues and cells using Universal Precautions and dispose of all biohazardous waste as per institutional requirements.

1. Rinse bone fragments several times in PBS with 1× antibiotics to remove excess blood. Aspirate PBS.

2. If bone is not already broken into 2–3 mm fragments, transfer bone sample to sterile Petri dish. Remove excess adipose tissue and cartilage with forceps and scalpel. Using appropriate tools, dice or break bone fragments into 2–3 mm pieces. Do not attempt to cut cortical bone with a scalpel. If excess blood is present, rinse using PBS with 1× antibiotics.

3. Place bone fragments into sterile, conical centrifuge tubes and add 2.5 ml collagenase solution per 1–3 g of bone fragments. Combined volume should not exceed one half the total tube volume (e.g., no more than 7.5 ml in a 15-ml tube). Incubate on rotator (rotisserie mode) at 37 °C for 3 h. *Note*: Small, sharp fragments of cortical bone in the sample may reduce yield of viable cells. If large numbers of these fragments are present, do not use rotisserie rotation, rather place the tubes horizontally on a rocker and agitate very slowly for 3 h, inverting the tubes periodically by hand to ensure thorough mixing.

4. Following collagenase digestion, add an equal volume of hMSC complete growth medium (includes 10% serum) to the digest to neutralize the collagenase.

5. Allow bone fragments to settle, and filter supernatant through a 100-μm cell strainer.

6. Centrifuge filtered solution at $500 \times g$ for 5 min to pellet cells, and carefully aspirate supernatant.

7. Resuspend pelleted cells in hMSC complete growth medium, and transfer to tissue culture flasks containing complete growth medium. For each 1–2 g of bone fragments, use one 75-cm^2 flask. In Fig. 1, these cells are "isolation passage" or P-ISO.

8. Place flasks in a 37 °C humidified incubator with 5% CO_2, and allow cells to attach. After 24 h, wash flasks twice with sterile PBS to remove nonadherent cells, then add fresh complete growth medium. Additional washing with PBS may be necessary if large numbers of red blood cells are still present.

9. Change medium every 3–4 days. When cells become 75–80% confluent (about 7–10 days), harvest using standard trypsinization method. Harvested cells may be cryopreserved or seeded at ~1500 cells per cm^2 for continued expansion (P-0, Fig. 1). At this point, cultures may also be established for surface marker characterization, and to test differentiation capacity (see Sections V and VI below).

III. Isolation of hMSCs from Adipose Tissue

hMSCs were initially isolated from lipoaspirate (Zuk *et al.*, 2001) and subsequently from intact adipose tissue (Lee *et al.*, 2004). Recent evidence suggests that lipoaspirate may yield more cells with greater differentiation potential than resected adipose tissue (Vermette *et al.*, in press). However, tissue harvested by either method produces satisfactory results. hMSCs do not contain lipid droplets and are therefore more dense than adipocytes, which comprise the bulk of adipose tissue. An initial centrifugation step to separate the adipocytes from other cell types is commonly used. Denser cells from the pellet are then seeded on plastic for an additional separation step based on the adhesion of hMSCs to plastic. Most protocols for isolation of MSCs from adipose tissue do not employ a gradient purification step, which is commonly used for hMSCs from bone marrow.

A. Adipose Tissue Samples

Excess adipose tissue (fat) is often available in the form of resected tissue from elective plastic surgery procedures (lipectomy), for example, abdominoplasty or breast reduction. Lipoaspirate from liposuction procedures is also a convenient source of tissue for isolation of hMSCs. The fat tissue in lipoaspirate is already significantly dissociated, so processing is simplified. As discussed above for bone samples, communication with the surgical team is important regarding the type and quantity of tissue required. The tissue should be placed in a sterile specimen container at the time of harvest. Tissue should be stored at ambient temperature and processed within 8 h. Refrigeration should be avoided if possible because the lipid hardens at low temperatures. Tissue may be stored overnight at 4 °C; however, it should equilibrate at ambient temperature prior to processing.

Fifty grams of adipose tissue or 50 ml of lipoaspirate can be conveniently processed using eight 15-ml sterile, disposable conical centrifuge tubes. Cell yield varies; however, the authors generally obtain 10–20 million cells from 50 g of tissue processed as below, and cultured ~7–10 days until 75–80% confluent.

B. Protocol

1. Materials and Supplies

Note: All supplies, tools, and solutions must be sterile.

- PBS
- hMSC complete growth medium (see below)
- 100× antibiotic stock solution (10,000 units/ml penicillin, 10,000 μg/ml streptomycin)

- Collagenase type I (code CLS-1, Worthington Biochemical Corporation, Lakewood, NJ) 0.075% in α-MEM (without serum) plus 1× antibiotics, sterile filtered. Solution may be prepared in advance and frozen at −20 °C in aliquots
- 160 mM NH_4Cl in deionized H_2O, sterile filtered. Solution may be prepared in advance and frozen at −20 °C in aliquots
- Centrifuge tubes, pipets, and sterile tissue culture plasticware
- Forceps, scalpels, and other tools for handling tissue
- Cell strainers (BD Falcon™, 100 μm mesh size)
- Rotisserie-style rotator

2. Procedure

Note: All activities that involve exposed tissue must be performed in a biosafety cabinet. Maintain sterility of the preparation throughout the procedure. Handle all human tissues and cells using Universal Precautions and dispose of all biohazardous waste as per institutional requirements.

1. If sample is liposuction fluid (lipoaspirate), skip steps (1) and (2) and start at step (3). If sample is resected fat tissue, rinse with an equal volume of sterile PBS with 1× antibiotics to remove excess blood. Repeat until most of the blood has been washed off.

2. In a sterile Petri dish, mince adipose tissue into small (4–5 mm) pieces with surgical scissors and/or scalpel. Remove any attached skin and as much connective tissue as possible.

3. If using 15-ml tubes for the preparation, dispense ∼4–5 ml adipose tissue into each tube together with 4 ml of 0.075% type I collagenase solution prepared as above. Minced tissue and lipoaspirate can be drawn up into a 10-ml sterile plastic pipet with the tip broken off. The procedure can be scaled up to use larger tubes if desired.

4. Incubate adipose tissue-collagenase suspension on a rotisserie-style rotator at 37 °C for 30 min.

5. After 30-min incubation, add an equal volume (4 ml or more) of hMSC complete growth medium (with serum) to the digest to neutralize the collagenase. Total volume in a 15-ml conical tube will be ∼12–14 ml.

6. Centrifuge at $5000 \times g$ for 10 min to pellet the MSC-rich dense cell fraction. Adipocytes and fat will be apparent as a yellow oily layer at the top of the tube.

7. Decant supernatant (the oily layer and collagenase solution). A transfer pipet or other sterile tool may be used to dislodge the oily layer. Be careful not to pour off or disturb the cell pellet.

8. Resuspend each pellet in 2 ml of 160 mM NH_4Cl and incubate at room temperature (RT) for 10 min to lyse red blood cells. Samples may be pooled for convenient handling. Add more NH_4Cl if necessary to achieve lysis of most of the blood cells. Unlysed red blood cells appear refractive and "donut" shaped by phase microscopy.

9. Transfer samples to new centrifuge tubes and centrifuge at $1200 \times g$ for 10 min to pellet the MSC-rich dense cell fraction.

10. Remove supernatant. Resuspend pelleted cells in hMSC complete growth medium.

11. Filter cell suspension through a 100-µm cell strainer.

12. Seed suspension into culture flasks containing the appropriate amount of medium. We have found that cells isolated from ~2.5 g of fat will become 75–80% confluent in ~7–10 days if seeded in a 75-cm^2 flask (e.g., a 50-g tissue sample would be processed in eight 15-ml tubes, and the resulting cells seeded into sixteen 75-cm^2 flasks). This corresponds to the isolation passage (P-ISO) in Fig. 1.

13. Incubate flasks at 37 °C in a humidified incubator with 5% CO_2 for 24 h. Wash flasks twice with PBS to remove nonadherent cells and add fresh growth media.

14. Feed cells every 3–4 days. When cells reach 70–80% confluency (about 7–10 days), harvest using standard trypsinization method. Harvested cells may be cryopreserved or seeded at ~1500 cells per cm^2 for continued expansion (P-0, Fig. 1). At this point, cultures may also be established for surface marker characterization, and to test differentiation capacity (see Sections V and VI below).

IV. Growth, Subculture, and Cryopreservation of Undifferentiated hMSCs

hMSCs are typically cultured in a basic medium containing essential nutrients plus 10% fetal bovine serum (FBS) using standard tissue culture practices. hMSCs from various sources, including bone marrow and adipose tissue, have been found to maintain an undifferentiated phenotype and various degrees of multilineage differentiation potential up to 30 generations (population doublings) or more in culture (Bruder *et al.*, 1997; Digirolamo *et al.*, 1999; Zuk *et al.*, 2002). In our studies (Wall *et al.*, in press), adipose-derived hMSCs maintained osteogenic and adipogenic differentiation potential through 30 generations; however, changes in proliferation rate and other properties were observed 20–25 generations following isolation. hMSC preparations should be evaluated for the properties of interest (e.g., differentiation potential; see below) before use in specific applications.

hMSCs tolerate cryopreservation well, and generally recover without significant loss of viability. The cryopreservation medium described below is of standard composition. Other formulations and commercial preparations may be equally suitable.

A. Feeding

Medium should be changed every 3–4 days. A 75-cm^2 flask seeded with 100,000 cells will reach 75–80% confluence in 7–10 days (doubling time 2–3 days; see Wall *et al.*, in press).

1. Warm hMSC complete growth medium to 37 °C.
2. In a sterile hood, aspirate medium from flask. Replace with warmed medium and return flask to incubator.

B. Subculture

Subculture flasks when cells reach 75–80% confluence. The effect of contact between undifferentiated hMSCs is not well understood; however, if the cells are to be maintained in an undifferentiated state, they should be subcultured before complete confluence is attained to avoid any potential effects of contact inhibition.

1. Wash cell monolayers twice with sterile PBS. Add sufficient trypsin solution to cover the monolayer (0.05% trypsin with EDTA, available from tissue culture suppliers).
2. Incubate at ambient temperature or at 37 °C until cells are released from the flask (no more than 5 min). Flask may be gently tapped to assist release.
3. Add a volume of complete growth medium equal to the volume of trypsin to the flask in order to inactivate the enzyme. Transfer cell suspension to centrifuge tube, and centrifuge at 500 × g for 5 min.
4. Aspirate supernatant and resuspend pellet in a few milliliters of growth medium.
5. Count cells using a hemacytometer and reseed in flasks with prewarmed growth medium at ∼100,000 cells per 75-cm^2 flask (∼1500 cells per cm^2).
6. Place flasks in incubator, and allow cells to attach for 12 h without disturbance.

C. Cryopreservation

1. Prepare cryopreservation medium as follows. Combine 35 ml MEM (or other base medium) with 10 ml human serum albumin (25% solution, SeraCare Life Sciences, Inc., Oceanside, CA). Add 5 ml DMSO, mix and filter sterilize. Cryopreservation medium may be stored in aliquots at −20 °C.
2. Trypsinize cells as described above.
3. Count cells in trypsin/medium suspension, centrifuge 5 min at 500 × g.
4. Resuspend cells in cryopreservation medium at 500,000–1,000,000 cells per ml, and aliquot into cryotubes (0.5–1.0 ml per tube). The concentration of cells and the number frozen in each tube can be adjusted based on eventual application.

5. Place tubes in cryogenic controlled rate freezing container (Nalgene "Mr. Frosty") or between styrofoam 15-ml centrifuge tube racks placed face to face, and placed in an ultralow temperature freezer (−86 °C) overnight.

6. Transfer tubes to liquid nitrogen cryostorage unit. *Caution*: Tubes stored in liquid phase present an increased explosion hazard. Tubes should be maintained in the vapor phase of the cryostorage unit (see manufacturer's instructions).

7. After 3–4 days, thaw one tube of cells to verify success of freezing process.

D. Thawing Cryopreserved Cells

Caution: Tubes stored in liquid nitrogen present an explosion hazard. Use appropriate personal protection when handling frozen vials.

1. Prepare appropriate number of flasks with prewarmed complete growth medium and place in incubator until required. Dispense 5 ml of medium into a sterile 15-ml centrifuge tube.
2. Remove tube of cells from cryostorage unit and place in sterile hood. Briefly loosen cap to release any pressure.
3. Retighten cap and thaw vial in 37 °C water bath until ice is almost melted.
4. Return vial to sterile hood and transfer contents of vial to 5 ml medium in centrifuge tube. Mix gently.
5. Centrifuge for 5 min at $500 \times g$.
6. Aspirate supernatant and resuspend in 5–10 ml of medium. If desired, assess viability using trypan blue according to manufacturer's instructions.
7. Seed cells in flasks with prewarmed medium at desired density. Return flasks to incubator, and monitor cell growth as appropriate.

V. Surface Marker Characterization of hMSCs

Cell surface proteins may characterize particular cell types or lineages. In some cases, the function of a specific surface protein, and its role in the biology of the cell type, is known. However, often the function of the protein has not been determined, but the protein has been shown to be associated with a certain type of cell and can serve as a marker. Exclusive, diagnostic surface markers for hMSCs have not been identified; however, several surface markers have been found to be commonly associated with hMSCs, including STRO-1, CD105 (endoglin), CD166 (activated leukocyte cell adhesion molecule, ALCAM) (reviewed in Barry and Murphy, 2004; Gronthos *et al.*, 2001), and more recently CD271 (low affinity nerve growth factor receptor, LNGFR; Buhring *et al.*, in press; Quirici *et al.*, 2002). Surface marker antigens can be used to characterize the cells in a specific preparation, and monitor their differentiation. Surface markers that are uniquely positive for a different cell type, for example, the hematopoietic surface markers CD45 and CD34, can be used to look for contamination of MSC preparations with other cell types. Surface markers have also been used for positive and negative immunoselection of MSC cell populations (Buhring *et al.*, in press; Simmons and Torok-Storb, 1991). The method below describes a standard indirect immunofluorescent staining technique modified to examine surface characteristics of both recently isolated and cultured hMSCs.

A. Micromass Cell Cultures

Small cultures of hMSCs grown on plain glass coverslips provide a convenient format for immunostaining.

1. Glass coverslips (22 mm × 22 mm, thickness no. 1) can be sterilized by separating them between layers of gauze in a glass Petri dish or other shallow, lidded container and autoclaving.
2. Place one coverslip in each well of a six-well multi-well culture dish, or in individual 35-mm Petri dishes.
3. Trypsinize cells as described above, centrifuge, and resuspend at a concentration of 3000 cells per 10 μl in complete growth medium.
4. Using a micropipet, pipet 2–4× 10 μl spots on each coverslip (see Fig. 2). The multiple spots will act as a replicates. All spots on one coverslip will be labeled with the same antibody.
5. Cover culture dish, place in humidified 37 °C incubator and allow cells to adhere for 45–60 min.
6. Gently pipet 2–3 ml complete growth medium into each well, flooding the coverslips.
7. Grow cells for a minimum of 2–3 days in complete growth medium to establish the micromass cultures before staining.

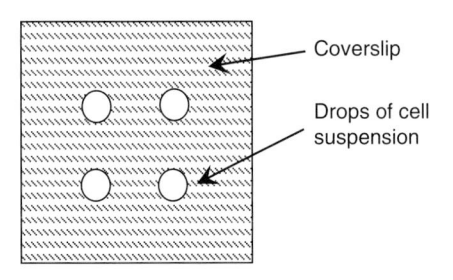

Fig. 2 Arrangement of 10-μl drops of cell suspension on coverslip for micromass culture; two to four drops per coverslip are recommended.

B. Fixation and Immunostaining

Many immunostaining protocols exist, and either fluorescent or nonfluorescent staining methods may be used. This protocol describes a basic indirect immunofluorescent staining method that has been used successfully to visualize surface antigens in hMSCs. Fluorescent dyes should be protected from light throughout the protocol.

1. Rinse coverslips in culture dishes twice with PBS.

2. Fix cells for 30 min at ambient temperature with 10% buffered formalin (contains 3.7% formaldehyde), using ~1–2 ml formalin per well. Formalin fixation is recommended because it generally preserves the immunogenicity of surface markers; however, other fixatives may also be appropriate. *Note*: All procedures involving formaldehyde should be performed in a location separate from the cell culture area, and with tools and glassware reserved for histological protocols. Formaldehyde is highly toxic to cells and can persist on surfaces, tools, and glassware even after washing. Dispose of formaldehyde waste according to institutional policies.

3. Wash coverslips twice with PBS. If cells cannot be stained immediately, the coverslips may be stored in PBS with 0.02% azide for up to 2 weeks at 4 °C. Following storage, wash twice with PBS before staining.

4. Permeabilize cell membranes and block nonspecific binding by incubating fixed micromass cultures in blocking buffer (0.2% Triton X-100/0.5% bovine serum albumin in PBS) for 40 min, using ~1 ml per well. (*Note*: this solution can be stored at 4 °C; use within 60 days.)

5. Dilute antibodies in blocking buffer according to manufacturer's recommendations. Appropriate dilutions may need to be empirically determined. Final concentrations may be in the range of 10–20 µg/ml for unlabelled primary mouse monoclonal antibodies and 4 µg/ml for fluorescently labeled polyclonal secondary antibodies. Fifty microliters each of the primary and secondary antibody dilutions will be required for each coverslip.

6. Pipet 50 µl of diluted primary antibody onto a piece of Parafilm®. Remove one coverslip from the blocking buffer, drain briefly, and invert on top of the drop of antibody solution. Repeat for other coverslips. Incubate for 2 h, then return coverslips to culture dishes and wash three times for 5 min in PBS.

7. Repeat using a new piece of Parafilm® and 50 µl of diluted secondary antibody for each coverslip. Incubate for 1 h, protected from light, then wash three times for 5 min in PBS.

8. *Optional*: DAPI (4′-6-diamidino-2-phenylindole, dihydrochloride) and/or fluorescently labeled phalloidin can be used to label nuclei and actin filaments, respectively, to aid in visualizing the cells. Three-color images showing cells with stained nuclei, cytoskeleton, and surface markers can be obtained with appropriate fluors.

DAPI staining is most easily accomplished by using a mounting medium containing DAPI, for example, Prolong Gold® with DAPI (Invitrogen Corporation, Carlsbad, CA). Phalloidin can be included with the secondary antibody at 5 units/ml [e.g., Alexa Fluor® 594-labeled phalloidin (Invitrogen) with an Alexa Fluor® 488-labeled secondary antibody for red/green staining]. DAPI can also be added to the secondary antibody solution at 300 nM.

9. After the final PBS wash, drain coverslips and mount on glass slides, inverted in 1–2 drops Prolong Gold®, with or without DAPI. Dry several hours or overnight before viewing.

10. *Note*: For a negative background control, a primary antibody of the same isotype as the specific antibodies, but with no known binding on the target cells, should be used. Cat. No. M5284 from Sigma-Aldrich (St. Louis, MO) is a suitable isotype control for mouse $IgG_1\kappa$ primary antibodies.

VI. hMSC Differentiation Assays for Osteogenic, Adipogenic, and Chondrogenic Pathways

Preparations of hMSCs from different individuals, or even cells from the same individual, but prepared by different methods, may differ substantially in multilineage potential (Afizah *et al.*, in press). The multilineage potential of the cells can be assessed by culturing the hMSCs in osteogenic, adipogenic, and chondrogenic media. Many formulations for differentiation media have been developed (Colter *et al.*, 2001; Halleux *et al.*, 2001; Sekiya *et al.*, 2001, 2002a; Sottile *et al.*, 2002; Zuk *et al.*, 2002), but generally have common factors. The media compositions suggested below are derived from published formulations. The base medium contains essential nutrients and is supplemented with serum for growth in the undifferentiated state. For osteogenic differentiation, ascorbate, β-glycerolphosphate, and dexamethasone or vitamin D_3 are typically added to the growth medium. Adipogenic differentiation is generally induced by supplementing growth medium with insulin, isobutylmethylxanthine (IBMX), and a peroxisome proliferator-activated receptor-gamma (PPARγ) agonist such as indomethacin, plus other factors. For chondrogenic differentiation, a low-serum or serum-free formulation that includes TGFβ is typically used. TGFβ1, 2, and 3 have been used successfully, although some studies indicate that TGFβ3 is more effective than TGFβ1 (Barry *et al.*, 2001; Miyanishi *et al.*, 2006; Zuk *et al.*, 2002). hMSCs in monolayer culture will undergo osteogenic and adipogenic differentiation. Pellet or micromass cultures (Sekiya *et al.*, 2002b; Zuk *et al.*, 2002) are recommended for chondrogenic differentiation. These culture conditions may be appropriate for an initial assessment of multilineage potential of a given preparation of cells; however, for experimental systems involving three dimensional scaffolds, or agarose or collagen gels, for example, additional tests using the experimental growth conditions are recommended.

Materials and supplies

- 6-well and 12-well culture dishes, tissue culture treated
- Complete growth, osteogenic, adipogenic, and chondrogenic media (see formulations below)

- PBS
- Buffered formalin, 3.6% formaldehyde
- 99% isopropyl alcohol
- 60% isopropyl alcohol
- Oil red O (Fisher Scientific, Suwanee, GA, BP112-10)
- Alizarin red S (Fisher Scientific AC40048-0250)
- Harris-modified hematoxylin with acetic acid (Fisher Scientific SH26-500D)
- Alcian Blue 8GX (Sigma-Aldrich A3157-10g)
- 0.1N HCl

A. Osteogenic and Adipogenic Differentiation

Suggested media formulations are given below. A six-well plate provides a convenient format to compare hMSCs cultured in complete growth medium, osteogenic medium, and adipogenic medium, and stained for mineralized matrix (calcium) using alizarin red and lipid droplets using oil red O. See Wall (2007b) and Zuk *et al.* (2002) for images of stained cultures.

1. Staining Solutions

- *Alizarin red S solution*: Dissolve 2 g alizarin red S in 100 ml deionized water; filter. Adjust pH to 4.1–4.3 with 1N ammonium hydroxide (dye can be rinsed from pH meter electrode with water). The solution is stable at RT.
- *Oil red O stock solution*: Dissolve 300 mg oil red O in 100 ml 99% isopropyl alcohol. The stock solution is stable at RT.
- *Oil red O working stain solution*: Mix stock solution 3:2 with deionized water (e.g., 6 ml stock, 4 ml water). Filter and let stand for 10 min. Do not agitate solution prior to use, or precipitate may be stirred up. Use within 2 h.

2. Protocol

1. Seed one 6-well plate with hMSCs at a density of 50,000 cells per well. Culture in complete growth medium (3 ml per well) until 100% confluent, replacing medium every 3–4 days.

2. When cells are fully confluent (usually 3–7 days), aspirate growth medium and add osteogenic medium to two wells, adipogenic medium to two wells, and complete growth medium to the remaining two wells. Culture for 2–3 weeks, replacing media every 3–4 days.

3. Aspirate culture media and wash each well 2× with PBS. *Caution*: confluent monolayers and calcium deposits are easily dislodged.

4. Fix for 15–30 min in buffered formalin, then rinse 2× with deionized water.

5. Label three wells (one of each type of medium) for oil red O staining and three for alizarin red staining.

6. For oil red O staining, remove water and add 2–3 ml 60% isopropyl alcohol to each well (leave dH$_2$O in alizarin red wells). Gently but thoroughly mix the alcohol in each well to ensure water is dispersed. Incubate at RT for 2–5 min.

7. Remove liquid from all wells and add 1–2 ml appropriate stain, 2% alizarin red S or oil red O working solution. Stain for 3 min.

8. Wash very gently with water until excess stain is removed, about 3–4 washes. A dark reddish precipitate may form when water is added to the oil red O wells. This is normal, and the precipitate is easily distinguished from stained lipid vacuoles under the microscope.

9. Macroscopic calcium deposits may be visible on the bottom and sides of the culture well. View stained plates as soon as possible. Lipid droplet morphology is best viewed while plates are still wet; however, mineral deposits may dissolve in residual wash water. Under phase microscopy, alizarin red stained calcium deposits appear as irregular red-orange crystals. Oil red O stained lipid droplets appear as cherry red spheres within individual cells. A dark reddish precipitate, which is not specific for lipid droplets, may be present in wells stained with oil red O.

10. If desired, cells may be counterstained with Harris hematoxylin (commercially available in a ready to use form). Hematoxylin stains nuclei a pale bluish-purple, and the overall cell morphology is outlined faintly in the same color.

B. Chondrogenic Differentiation in Attached Micromass Culture

The following protocol, adapted from Zuk *et al.* (2002), is similar to pellet culture methods referenced above, except that the micromass of cells usually remains attached to the plastic culture surface, which simplifies feeding and staining.

1. Staining Solutions

- *Alcian Blue solution*: Dissolve 1 g alcian blue 8GX in 100 ml deionized water and filter. The solution is stable at ambient temperature. Re-filter before use if solution has been stored.
- *Destaining solution*: 0.1N HCl in deionized water.

2. Protocol

1. Trypsinize a flask of hMSCs and prepare a cell suspension with a concentration of 8 million cells per ml.

2. Pipet 2–3× 10 μl droplets (80,000 cells per droplet) into the bottom of a Petri dish or well of a multi-well plate. Twelve-well multi-well plates are recommended to conserve medium.
3. Incubate for 2 h at 37 °C in humidified incubator. Do not allow drops to dry out. Sterile water added to unused wells of a multi-well plate will help maintain humidity.
4. Gently flood wells with chondrogenic medium or complete growth medium for control. Culture 2–3 weeks, replacing medium every 3–4 days. If micromass cultures detach, continue culture and staining but use care to avoid aspirating the micromasses.
5. To stain with alcian blue, wash cultures twice in PBS.
6. Fix 15 min in buffered formalin.
7. Wash 3× in PBS, 10 min each wash.
8. Stain 30 min in alcian blue solution. Overnight staining may be required for cultures with large proportions of compact cartilage.
9. Destain with repeated washes in 0.1N HCl until excess stain is removed. Stained cultures may be stored in 0.1N HCl for several days. Cartilage will appear bright blue.
10. *Optional*: Formalin-fixed micromass cultures may be imbedded in paraffin for sectioning and staining using standard histological procedures.

C. Media Formulations

These media were developed based on numerous published studies, and have been used successfully over the course of 3 years work in the Cell Mechanics Laboratory at North Carolina State University (http:www.bme.ncsu.edu\labs\cml). The media have been used for hMSCs derived from bone marrow, adipose tissue, and trabecular bone. For the convenience of investigators new to the field of stem cell biology, suppliers and suggested stock concentrations for some media components are provided; however, other suppliers and different stocks may be equally suitable. Cell culture–certified reagents should be used whenever possible. Commercial formulations of media optimized for growth and differentiation of hMSCs are also available. *Note on serum*: the ability of each lot of serum to support differentiation of hMSCs should be verified.

1. Complete Growth Medium

Components with suggested stock concentrations:

- α-MEM, 1× liquid, alpha modified; with L-glutamine, no ribonucleosides, no deoxyribonucleosides; *Invitrogen Product Number: 12561–056; 500 ml*

- FBS; *Atlanta Biologicals Premium Select: S11550, lot selected* (Atlanta Biologicals, Lawrenceville, GA)
- L-Glutamine 200 mM solution (29.2 mg/ml); 100×; *Mediatech/Cellgro Product Number: 25-005-CI*
- 10,000 I.U. penicillin/10,000 μg/ml streptomycin antibiotic solution (pen/strep); 100×; *Mediatech/Cellgro Product Number: 30-002-CI*

Formulation (stock concentration volume for 500 ml of media):

- α-MEM (440 ml)
- 10% FBS (50 ml)
- 100 I.U. penicillin/100 μg/ml streptomycin (5 ml)
- 2.0 mM L-glutamine (5 ml)

2. Osteogenic Differentiation Medium

Components (in addition to growth medium components) with suggested stock concentrations:

- Ascorbic acid, stock 50 mM in α-MEM; *Sigma-Aldrich #A4403-100 mg*
- Dexamethasone, stock 0.1 mM; dissolve 1 mg in 1 ml 100% EtOH, add 24.5 ml α-MEM; *Sigma-Aldrich #D8893-1 mg*
- β-Glycerolphosphate, 1 M in α-MEM; glycerol 2-phosphate disodium salt hydrate; *Sigma-Aldrich #G9891-10 g*

Formulation (stock concentration volume for 500 ml of media):

- α-MEM (434 ml)
- 10% FBS (50 ml)
- 50 μM ascorbic acid (0.5 ml)
- 0.1 μM dexamethasone (0.5 ml)
- 10 mM β-glycerolphosphate (5.0 ml)
- 100 I.U. penicillin/100 μg/ml streptomycin (5 ml)
- 2.0 mM L-glutamine (5 ml)

3. Adipogenic Differentiation Medium

Components (in addition to growth medium components) with suggested stock concentrations:

- Dexamethasone, stock 0.1 mM; dissolve 1 mg in 1 ml 100% EtOH, add 24.5 ml α-MEM; *Sigma-Aldrich #D8893-1 mg*

- Insulin, human recombinant; 10 mg/ml in 25 mM HEPES; *Sigma-Aldrich #I9278-5 ml*
- Indomethacin, stock 25 mM in 100% EtOH; *Sigma-Aldrich #I7378-5 g*
- IBMX, stock 0.5 M in DMSO; *Sigma-Aldrich #I5879-1 g*

Formulation (stock concentration volume for 500 ml of media):

- α-MEM (432 ml)
- 10% FBS (50 ml)
- 1.0 μM dexamethasone (5.0 ml)
- 10 μg/ml insulin (0.5 ml)
- 100 μM indomethacin (2.0 ml)
- 500 μM IBMX (0.5 ml)
- 100 I.U. penicillin/100 μg/ml streptomycin (5 ml)
- 2.0 mM L-glutamine (5 ml)

4. Chondrogenic Differentiation Medium (Serum Free)

Components (in addition to growth medium components) with suggested stock concentrations:

- Dulbecco's Modified Eagle's Medium, high glucose, 1× liquid (DMEM-H), without glutamine, with sodium pyruvate; *Mediatech #15-013 CV*
- Dexamethasone, stock 0.1 mM; dissolve 1 mg in 1 ml 100% EtOH, add 24.5 ml α-MEM; *Sigma-Aldrich #D8893-1 mg*
- Ascorbic acid, stock 50 mM in α-MEM; *Sigma-Aldrich #A4403-100 mg*
- Sodium pyruvate, 100 mM in DMEM-H; *Sigma-Aldrich #p5280-25 g*
- L-Proline, 4 mg/ml in DMEM-H; *Sigma-Aldrich #P5607-25 g*
- ITS+ (insulin/transferrin/selenium/linoleic acid) 100×; *Sigma-Aldrich #I2521-5 ml*
- Transforming growth factor β3 (TGFβ3), recombinant human; various suppliers; reconstitute according to manufacturer's instructions

Formulation (stock concentration volume for 500 ml of media):

- DMEM-H (471.5 ml)
- 0.1 μM dexamethasone (0.5 ml)
- 50 μM ascorbic acid (0.5 ml)
- 1 mM sodium pyruvate (5.0 ml)
- 40 μg/ml proline (5.0 ml)
- ITS+ (5.0 ml)
- 10 ng/ml TGFβ3 (see manufacturer's instructions)

- 100 I.U. penicillin/100 µg/ml streptomycin (5 ml)
- 2.0 mM L-glutamine (5 ml)

VII. Summary

The above protocols are intended to provide practical, generally applicable methods for isolating hMSCs from different tissues. They have been used successfully by the authors to prepare and characterize hMSCs from over 50 bone and adipose tissue samples over 3 years. However, the protocols should be considered as guidelines, and a starting point for researchers interested in preparing hMSCs for their own applications. Investigators should not hesitate to modify the protocols to correspond with ongoing research projects, individual laboratory practices, or to use available equipment or supplies. The authors have not attempted to isolate MSCs from nonhuman animals, however, minor modifications, for example, in the composition of media, may allow these protocols to be used successfully with other species.

Acknowledgments

The authors thank Dr. Wolf Losken, M.D.; Dr. John van Aalst, M.D.; Dr. Douglas Dirschl, M.D.; Dr. Ron Graff, Ph.D.; and their colleagues, patients, and staff at UNC Hospitals for providing tissue samples. We also thank the graduate and undergraduate students in the Cell Mechanics Laboratory, including Allison Finger, Ariel Hanson, Carla Haslauer, Jennifer Jassawalla, Wayne Pfeiler, Carolyn Sargent, Ruwan Sumanasinghe, and Catherine Ward, who spent innumerable hours isolating cells and testing protocols. This study was supported by the Ralph E. Powe Junior Faculty Enhancement Award (EGL); North Carolina Biotechnology Center Institutional Development Grant (EGL); NCSU Faculty Research and Professional Development Grants (EGL); and North Carolina Biotechnology Center Multi-Disciplinary Research Grant (EGL).

References

Afizah, H., Yang, Z., Hui, J. H. P., Ouyang, H.-W., and Lee, E.-H. (2007). A comparison between the chondrogenic potential of human bone marrow stem cells (BMSCs) and adipose-derived stem cells (ADSCs) taken from the same donors. *Tissue Eng.* **13**(4), 659–666.

Aslan, H., Zilberman, Y., Kandel, L., Liebergall, M., Oskouian, R. J., Gazit, D., and Gazit, Z. (2006). Osteogenic differentiation of noncultured immunoisolated bone marrow-derived cd105$^+$ cells. *Stem Cells* **24**, 1728–1737.

Barrilleaux, B., Phinney, D. G., Prockop, D. J., and O'Connor, K. C. (2006). Review: *Ex vivo* engineering of living tissues with adult stem cells. *Tissue Eng.* **12**, 3007–3019.

Barry, F., Boynton, R. E., Liu, B., and Murphy, J. M. (2001). Chondrogenic differentiation of mesenchymal stem cells from bone marrow: Differentiation-dependent gene expression of matrix components. *Exp. Cell Res.* **268**, 189–200.

Barry, F. P., and Murphy, J. M. (2004). Mesenchymal stem cells: Clinical applications and biological characterization. *Int. J. Biochem. Cell Biol.* **36**, 568–584.

Bruder, S. P., Jaiswal, N., and Haynesworth, S. E. (1997). Growth kinetics, self-renewal, and the osteogenic potential of purified human mesenchymal stem cells during extensive subcultivation and following cryopreservation. *J. Cell. Biochem.* **64,** 278–294.

Buhring, H.-J., Battula, V. L., Treml, S., Schewe, B., Kanz, L., and Vogel, W. (2007). Novel markers for the prospective isolation of human MSC. *Ann. N. Y. Acad. Sci.* **1106,** 262–271.

Centers for Disease Control (1988). Perspectives in disease prevention and health promotion update: Universal precautions for prevention of transmission of human immunodeficiency virus, hepatitis b virus, and other bloodborne pathogens in health-care settings. *Morb. Mortal. Wkly. Rep.* **37,** 377–388.

Colter, D. C., Class, R., DiGirolamo, C. M., and Prockop, D. J. (2000). Rapid expansion of recycling stem cells in cultures of plastic-adherent cells from human bone marrow. *Proc. Natl. Acad. Sci. USA* **97,** 3213–3218.

Colter, D. C., Sekiya, I., and Prockop, D. J. (2001). Identification of a subpopulation of rapidly self-renewing and multipotential adult stem cells in colonies of human marrow stromal cells. *Proc. Natl. Acad. Sci. USA* **98,** 7841–7845.

Cowan, C. M., Shi, Y. Y., Aalami, O. O., Chou, Y. F., Mari, C., Thomas, R., Quarto, N., Contag, C. H., Wu, B., and Longaker, M. T. (2004). Adipose-derived adult stromal cells heal critical-size mouse calvarial defects. *Nat. Biotechnol.* **22,** 560–567.

Digirolamo, C. M., Stokes, D., Colter, D., Phinney, D. G., Class, R., and Prockop, D. J. (1999). Propagation and senescence of human marrow stromal cells in culture: A simple colony-forming assay identifies samples with the greatest potential to propagate and differentiate. *Br. J. Haematol.* **107,** 275–281.

Friedenstein, A. J., Gorskaja, J. F., and Kulagina, N. N. (1976). Fibroblast precursors in normal and irradiated mouse hematopoietic organs. *Exp. Hematol.* **4,** 267–274.

Gronthos, S., Franklin, D. M., Leddy, H. A., Robey, P. G., Storms, R. W., and Gimble, J. M. (2001). Surface protein characterization of human adipose tissue-derived stromal cells. *J. Cell. Physiol.* **189,** 54–63.

Halleux, C., Sottile, V., Gasser, J. A., and Seuwen, K. (2001). Multi-lineage potential of human mesenchymal stem cells following clonal expansion. *J. Musculoskelet. Neuronal. Interact.* **2,** 71–76.

Haynesworth, S. E., Reuben, D., and Caplan, A. I. (1998). Cell-based tissue engineering therapies: The influence of whole body physiology. *Adv. Drug Deliv. Rev.* **33,** 3–14.

Lee, R. H., Kim, B., Choi, I., Kim, H., Choi, H. S., Suh, K., Bae, Y. C., and Jung, J. S. (2004). Characterization and expression analysis of mesenchymal stem cells from human bone marrow and adipose tissue. *Cell. Physiol. Biochem.* **14,** 311–324.

Lodie, T. A., Blickarz, C. E., Devarakonda, T. J., He, C., Dash, A. B., Clarke, J., Gleneck, K., Shihabuddin, L., and Tubo, R. (2002). Systematic analysis of reportedly distinct populations of multipotent bone marrow-derived stem cells reveals a lack of distinction. *Tissue Eng.* **8,** 739–751.

Mehlhorn, A. T., Niemeyer, P., Kaiser, S., Finkenzeller, G., Stark, G. B., Sudkamp, N. P., and Schmal, H. (2006). Differential expression pattern of extracellular matrix molecules during chondrogenesis of mesenchymal stem cells from bone marrow and adipose tissue. *Tissue Eng.* **12,** 2853–2862.

Miyanishi, K., Trindade, M. C. D., Lindsey, D. P., Beaupre, G. S., Carter, D. R., Goodman, S. B., Schurman, D. J., and Smith, R. L. (2006). Effects of hydrostatic pressure and transforming growth factor-β3 on adult human mesenchymal stem cell chondrogenesis *in vitro*. *Tissue Eng.* **12,** 1419–1428.

Noth, U., Osyczka, A. M., Tuli, R., Hickok, N. J., Danielson, K. G., and Tuan, R. S. (2002). Multi-lineage mesenchymal differentiation potential of human trabecular bone-derived cells. *J. Orthop. Res.* **20,** 1060–1069.

Pittenger, M. F., Mackay, A. M., Beck, S. C., Jaiswal, R. K., Douglas, R., Mosca, J. D., Moorman, M. A., Simonetti, D. W., Craig, S., and Marshak, D. R. (1999). Multilineage potential of adult human mesenchymal stem cells. *Science* **284,** 143–147.

Prockop, D. J. (1997). Marrow stromal cells as stem cells for nonhematopoietic tissues. *Science* **276,** 71–74.

Quirici, N., Soligo, D., Bossolasco, P., Servida, F., Lumini, C., and Deliliers, G. L. (2002). Isolation of bone marrow mesenchymal stem cells by anti-nerve growth factor receptor antibodies. *Exp. Hematol.* **30,** 783–791.

Sekiya, I., Colter, D. C., and Prockop, D. J. (2001). BMP-6 enhances chondrogenesis in a subpopulation of human marrow stromal cells. *Biochem. Biophys. Res. Commun.* **284,** 411–418.

Sekiya, I., Larson, B. L., Smith, J. R., Pochampally, R., Cui, J. G., and Prockop, D. J. (2002a). Expansion of human adult stem cells from bone marrow stroma: Conditions that maximize the yields of early progenitors and evaluate their quality. *Stem Cells* **20,** 530–541.

Sekiya, I., Vuoristo, J. T., Larson, B. L., and Prockop, D. J. (2002b). In vitro cartilage formation by human adult stem cells from bone marrow stroma defines the sequence of cellular and molecular events during chondrogenesis. *Proc. Natl. Acad. Sci. USA* **99,** 4397–4402.

Simmons, P. J., and Torok-Storb, B. (1991). Identification of stromal cell precursors in human bone marrow by a novel monoclonal antibody, STRO-1. *Blood* **78,** 55–62.

Sottile, V., Halleux, C., Bassilana, F., Keller, H., and Seuwen, K. (2002). Stem cell characteristics of human trabecular bone-derived cells. *Bone* **30,** 699–704.

Till, J. E., and McCulloch, C. E. (1961). A direct measurement of the radiation sensitivity of normal mouse bone marrow cells. *Radiat. Res.* **14,** 213–222.

Vermette, M., Trottier, V., Menard, V., Saint-Pierre, L., Roy, A., and Fradette, J. (2007). Production of a new tissue-engineered adipose substitute from human adipose-derived stromal cells. *Biomaterials* **28**(18), 2850–2860.

Wall, M. E., Bernacki, S. H., and Loboa, E. G. (2007). Effects of serial passaging on the adipogenic and osteogenic differentiation potential of adipose-derived human mesenchymal stem cells. *Tissue Eng.* **13**(6), 1291–1298.

Yoo, J. U., Barthel, T. S., Nishimura, K., Solchaga, L., Caplan, A. I., Goldberg, V. M., and Johnstone, B. (1998). The chondrogenic potential of human bone-marrow-derived mesenchymal progenitor cells. *J. Bone Joint Surg. Am.* **80,** 1745–1757.

Yoon, E., Dhar, S., Chun, D. E., Gharibjanian, N. A., and Evans, G. R. (2007). In vivo osteogenic potential of human adipose-derived stem cells/poly lactide-co-glycolic acid constructs for bone regeneration in a rat critical-sized calvarial defect model. *Tissue Eng.* **13,** 619–627.

Zuk, P. A., Zhu, M., Ashjian, P., De Ugarte, D. A., Huang, J. I., Mizuno, H., Alfonso, Z. C., Fraser, J. K., Benhaim, P., and Hedrick, M. H. (2002). Human adipose tissue is a source of multi-potent stem cells. *Mol. Biol. Cell* **13,** 4279–4295.

Zuk, P. A., Zhu, M., Mizuno, H., Huang, J., Futrell, J. W., Katz, A. J., Benhaim, P., Lorenz, H. P., and Hedrick, M. H. (2001). Multilineage cells from human adipose tissue: Implications for cell-based therapies. *Tissue Eng.* **7,** 211–228.

CHAPTER 12

Culture of Mesenchymal Stem/Progenitor Cells in Adhesion-Independent Conditions

Dolores Baksh[*] and John E. Davies[†]

[*]Organogenesis Inc.
Canton, Massachusetts 02021

[†]Faculty of Dentistry
Institute of Biomaterials and Biomedical Engineering
University of Toronto
Ontario, Canada M5S 3G9

I. Introduction
II. Suspension Culture of Bone Marrow-Derived MPCs
    A. Bioreactor Set-up
    B. MPC Sources
    C. Choice of Cytokines
    D. Media Formulation
    E. Passaging Suspension-Cultured Cells
III. Analysis of Suspension Culture
    A. Total Viable Cells
    B. Phenotypic Analysis
    C. Multilineage Differentiation Potential
IV. Mechanism of MPC Expansion in Suspension
    A. The Role of Hematopoietic Cells
    B. Adhesion Receptor Expression
V. Discussion
    References

I. Introduction

One of the major goals of therapeutic strategies utilizing stem cells is the delivery of a sufficient number of functional cells. Mesenchymal stem/progenitor cells (MSCs/MPCs), one of two types of stem cells found in bone marrow, the other being hematopoietic stem cells, have already demonstrated clinical potential both as a source of cells for cell-based therapies (Noort *et al.*, 2002) and as a platform for gene therapy (Dayoub *et al.*, 2003; Shi *et al.*, 2002). Unfortunately, MPCs directly derived from their native sources are limited in their numbers [<1% of adult human bone marrow comprises MSCs (Caplan, 1991)] and, thus, require an *ex vivo* expansion step to generate larger numbers of these cells. The current and most accepted method for the culture of MPCs relies on the adhesion and subsequent culture of these cells on tissue culture dishes. Expansion is achieved by serial passaging of the adherent cell population. However, one of the major limitations of the current approach to the culture of MPCs is the loss of developmental potential during (adherent) culture expansion; that is, the progeny of successive mitotic division fail to maintain their ability to self-renew and form all the connective tissue types specific to the mesenchymal cell lineage (Bianchi *et al.*, 2003; Bruder *et al.*, 1997a; Shi *et al.*, 2002). Our own experience with culturing MPCs in this traditional manner corroborates such reported findings (Baksh *et al.*, 2003), and suggests that cell adhesion and subsequent propagation on a solid surface may play a significant role in compromising the developmental potential of the expanding cell population. The mechanism by which this happens is still unknown. Nevertheless, there is a necessity to develop optimal bioprocess configurations and protocols that result in the generation of sufficient numbers of functional MPCs for these clinical applications.

This chapter, therefore, describes a novel culture paradigm for the cultivation and expansion of MPCs that aims to circumvent some of the limitations which are characteristic of the traditional method for the culture of MPCs. By applying a systematic experimental design regimen, a novel bioprocess approach to the culture of MPCs was developed. The MPCs generated under these conditions maintain their "stemness" potential and have the ability to form functional mesenchymal tissue types including bone, cartilage, fat, and myoblasts. This approach challenges current biological paradigms surrounding the culture of these MPCs but, importantly, offers an alternative strategy to generate functional MPCs for a variety of cell-based therapeutic strategies.

II. Suspension Culture of Bone Marrow-Derived MPCs

A. Bioreactor Set-up

The approach to expansion utilizing suspension cultures is different to the one intuitively practiced, which relies on attachment of cells to a surface, whether it be a 2-dimensional surface like culture plastic (Bruder *et al.*, 1997b; Caplan and Bruder, 1997) or microcarriers maintained in suspension (Botchwey *et al.*, 2001;

Qiu *et al.*, 1999). A stirred suspension culture configuration, similar to that typically used for the culture of hematopoietic stem cells in the absence of a supported adherent layer, is employed (Collins *et al.*, 1998; Sardonini and Wu, 1993; Zandstra *et al.*, 1994). Cells are inoculated at 1×10^6 cells/ml into siliconized (Sigma) stirred suspension spinner flasks (glassware available from Bellco Biotechnology), suspended in growth media. The interior of the glassware should be treated with a siliconizing agent prior to culture. Cells are maintained in the suspension via constant agitation achieved through the use of magnetic stir bars affixed to impellers. The bioreactors are placed on a magnetic stir table housed at 37 °C in humidified atmosphere of 5% CO_2 in air. To ensure that cell survival in suspension is not attributed to aggregate formation, the culture system is configured such that (i) the impeller is placed in the middle of the bioreactor at a 90° angle to the solution surface to maintain axial flow, (ii) a constant mixing speed of 40 rpm is used to maintain the cells in suspension (Coulson, 1999), and (iii) the agitator is positioned three quarters of the way down the vessel to ensure uniform mixing (Geankoplis, 1993) (Fig. 1A). Passing the culture contents at each medium exchange through a 70 µm filter, followed by inspecting both the flow-through and the surface of the filter under high power magnification, is sufficient to confirm the absence of aggregate formation (Fig. 1B–J).

B. MPC Sources

Suspension culture of MPCs is achievable from a variety of animal sources, including mouse, rat, and human bone marrow cells. However, for this chapter, suspension culture of MPCs from human bone marrow specimens will be the focus, although similar techniques do apply to other nonhuman cell sources. Human bone marrow-derived cells (hBMDC) may be obtained from a variety of sources including consenting normal bone marrow donors or bone marrow cells harvested from the femoral heads of consenting patients who are undergoing an orthopedic procedure such as a total hip arthroplasty. If the samples are contaminated with large amounts of red blood cells (RBCs), a standard RBC lysis protocol may be used to lyse the RBCs. A standard density gradient approach is used to isolate mononuclear cells, such as Ficoll-Paque™ (1.077 g/ml, Sigma, St. Louis, MO). The starting cell population is inoculated at 1×10^6 cells/ml into spinner flasks, suspended in growth media.

C. Choice of Cytokines

Suspension culture of hematopoietic stem/progenitors has proven to be successful due to the identification of various growth factors and cytokine combinations that could stimulate HSC expansion in the absence of feeder cell layers (Collins *et al.*, 1998; Sardonini and Wu, 1993; Zandstra *et al.*, 1994). Therefore, in the development of a suspension culture system for the expansion of MPCs, the importance that soluble factors might have on achieving adhesion-independent expansion of MPCs was recognized. To this end, a factorial design approach to

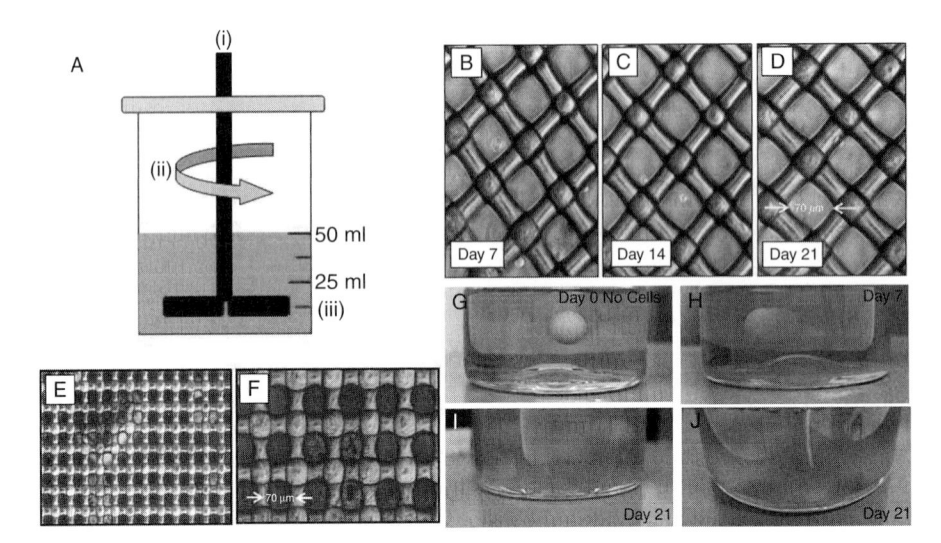

Fig. 1 Bioreactor setup and macroscopic appearance of suspension cultures. (A) The culture system is configured such that (i) the impeller is placed in the middle of the bioreactor at a 90° angle to the solution surface to maintain axial flow, (ii) a constant mixing speed of 40 rpm is used to maintain the cells in suspension, and (iii) the agitator is positioned three quarters of the way down the median level in the vessel to ensure uniform mixing. In this example, 50 ml of media is used and therefore, the agitator is placed at the 12.5 ml mark. (B)–(D) Representative images after suspension cells were passed through a 70 μm filter on days 7, 14, and 21, respectively. There were no detectable aggregates observed. (E) and (F) Representative low- and high-power microscope images, respectively, of aggregates which form after adherent cells were cultured under suspension culture conditions. Aggregates typically measured ∼200 μm. (G) Suspension culture on day 0 with no cells. (H) Suspension cultures inoculated with 1×10^6 cells/ml on day 0 and appearance of culture on day 7. (I) and (J) Suspension culture on day 21. Note the lack of any visible aggregates in the cultures on day 21. (See Plate no. 19 in the Color Plate Section.)

identify the optimum soluble factor and combination(s) for the maintenance and expansion of MPC was employed (Baksh et al., 2005). These experiments were performed in the absence of serum. The analysis revealed that the ability of suspension-derived MPCs to survive and expand in suspension was influenced by the addition of specific combinations of soluble factor(s) (both in the presence and absence of serum). Specifically, factorial design analysis, designed to study the effect of a variety of cytokines (e.g., SCF, IL3, PDGF, and FGF) and their 2^4 possible combinations on MPCs expansion, revealed that SCF + IL3 (in serum-free conditions) was a potent stimulator in the survival and expansion of colony forming unit—fibroblast (CFU-F) and CFU-osteoblast (O) relative to the No Cytokine control condition. SCF and IL3 are not typical cytokines used for the culture of mesenchymal cell types. IL3 (as well as GM-CSF and IL5) and SCF are characteristically involved with the regulation of hematopoiesis, specifically modulating the proliferation, differentiation, and survival of various hematopoietic cell lineages and their precursors (i.e., macrophage, neutrophils, eosinophils, megakaryocytes, and erythroid cells) (de Groot et al., 1998).

D. Media Formulation

The choice of growth media may include MyeloCult™ medium (containing serum) (StemCell Technologies Inc., Vancouver, Canada) or Dulbecco's Minimal Essential Media (DMEM) (Gibco), supplemented with 10% fetal bovine serum. To ensure survival and expansion of cells, cultures require growth factors, at minimum 2 ng/ml of purified recombinant human interleukin 3 (rhIL-3) (StemCell) and 10 ng/ml of recombinant human stem cell factor (rhSCF) (StemCell). If this system is used for experimental research purposes, a 10% antibiotic solution can be added to medium which comprises penicillin G (167 units/ml) (Sigma), gentamicin (50 µg/ml) (Sigma), and amphotericin B (0.3 µg/ml) (Sigma). Depending on the application, suspension cultures may also be performed in the absence of serum. Suspension culture has been tested under serum-free conditions using StemSpan™ medium (StemCell). Under these conditions, the media is supplemented with 20 ng/ml rhIL3 and 100 ng/ml rhSCF, in addition to 10^{-6} M hydrocortisone.

E. Passaging Suspension-Cultured Cells

Passaging of the cells in suspension culture occurs when the cell density is greater than input (i.e., $>1 \times 10^6$ cells/ml). Typically, the first time to passage occurs between 5 and 7 days postinoculation. At this point, one-third the medium, including cells, is replaced with fresh media, either MyeloCult™ medium (already containing serum) or DMEM. If DMEM is used, then 10% fetal bovine serum must be added to the medium. The fresh portion of medium is supplemented with 2 ng/ml of rhIL3 and 10 ng/ml of rhSCF. If serum-free media is used, then 20 ng/ml of rhIL3 and 100 ng/ml of rhSCF is added to the fresh medium. Cell counts, flow cytometric analysis, progenitor assays, etc. may be performed using the cells at each time point or depending on the cell yield, the cells in this fraction may be used to set up new suspension cultures at 1×10^6 cells/ml. Suspension cultures may be maintained for several weeks (i.e., 6 weeks), with one-third the media being replaced with fresh media every 2–3 days.

III. Analysis of Suspension Culture

A successful suspension culture may be determined by evaluating a number of output parameters (see below).

A. Total Viable Cells

As a first screen, the total number of viable cells should be determined using conventional methods such as hemacytometer readings or other more rigorous methods such as viability dyes (propidium iodide or 7 AAD), combined with flow cytometry. The viability of the cells is typically >95%. Suspension cells should also

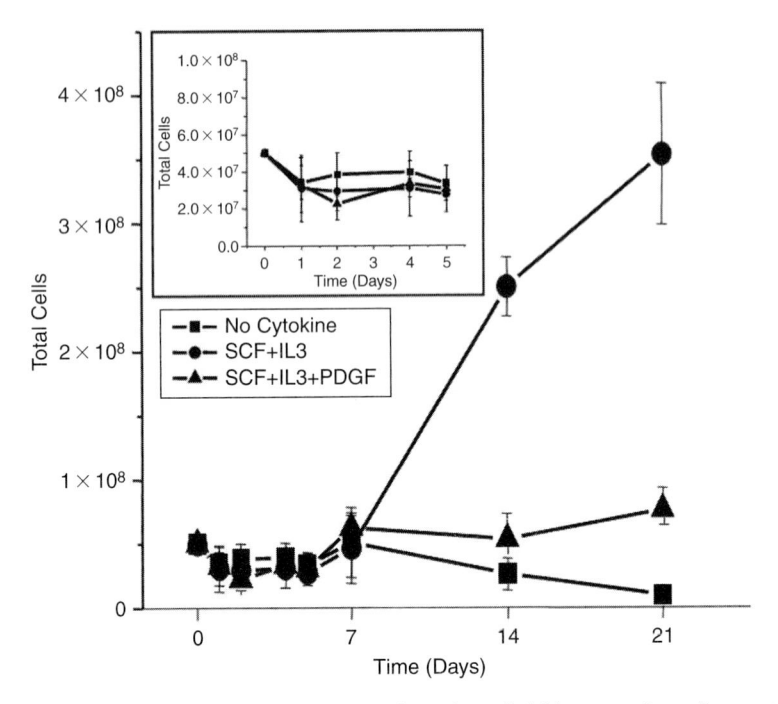

Fig. 2 Cell growth in suspension cultures. Total number of viable suspension cells counted as a function of time. Insert plot shows that there is very little change in cell number within the first 5 days of suspension culture. Each data point represents $n = 3$ experiments.

be examined under a microscope, revealing a distribution in the size of the cells, which may vary from 6 to 9 μm in diameter. Cells should appear as single cells. There is generally no significant change in the cell number within the first 5 days of culture (Fig. 2). In fact, there is typically a decrease in cell number observed from input values within the first few days post setup. After 5 days, there is an approximate two-fold increase in cell number relative to input values. Ideally, a cell count should be performed within the first 24 h of cell inoculation. This value would realistically reflect the starting cell number and should be recorded as the day 0 cell number in cell expansion calculations.

B. Phenotypic Analysis

1. MPCs are Located in the CD45⁻ Suspension Cell Fraction

To date, no single marker has been shown to definitively delineate a MSC population. Historical reports, such as those of Pittenger et al. (1999), Gronthos et al. (1994), and Haynesworth et al. (1992), reveal the lack of consensus on definitive phenotypic markers associated with this stem cell type. Nevertheless, the expression of STRO-1 (Gronthos and Simmons, 1995) and SH2 (Haynesworth *et al.*, 1992) has been strongly associated with an MSC phenotype, and for this reason we used this phenotype to track the relative expression of these surface antigens as a first attempt at identifying a suspension-derived MPC cell population. However, flow cytometric analysis of suspension-derived cells revealed that <1% of the expanding cell population expressed STRO-1 and SH2, and, at times, these markers were undetectable. Our observations suggested that suspension-derived cells, with MPC potential, may not express these surface antigens during suspension culture; however, on plating, this phenotype is detected (Baksh, 2001, unpublished data). To date, only adherent-derived MPCs have been phenotypically characterized and, therefore, there is little known about the phenotypic profile of tissue-derived MSCs, prior to culture manipulation. However, the isolation and recovery of CD45⁻ cells, grown in suspension cultures via FACS, revealed that ~100% of the total CFU-F (colony forming unit-fibroblast) and CFU-O (-osteoblast) could be recovered from an expanding CD45⁻ cell population. The cells in this fraction, giving rise to CFU-F and CFU-O, lacked any detectable level of SH2 and STRO-1 as well as PDGFαRc expression (Baksh, 2001, unpublished data). However, we did identify that the CD45⁻ cells, giving rise to CFU-F and CFU-O, expressed a low level of CD49e (mean fluorescence intensity (MFI) of 295 ± 3 as determined using flow cytometry cell analysis software) throughout suspension culture (Baksh, *et al.*, 2007). Adherent-derived MPCs express high levels of CD49e (MFI in CD49e expression is typically 560 ± 4). Taken together, our results indicate that suspension-derived MPCs may have a distinct phenotype compared to that characterized for adherent-derived MPCs.

2. Phenotypic Screening

Phenotypic analysis of the suspension cells using standard flow cytometric techniques should be performed on a routine basis to track the hematopoietic and nonhematopoietic cell composition. As a first screen, the CD45 cell surface antigen may be used to track both the hematopoietic and nonhematopoietic cell populations based on the expression and lack of expression of CD45, respectively. For the purpose of culturing MPCs in suspension, it is critically important to perform phenotypic analyses at various time points (e.g., at 1, 2, 3 weeks) to ensure that the CD45⁻ cell population is expanding, that is, the number of CD45⁻ cells should increase with time (Fig. 3A), as this is the fraction of cells which contain MPCs. In parallel, a standard CFU-F assay (Castro-Malaspina *et al.*, 1980) should be performed using either the unsorted or CD45⁻ sorted suspension cell populations. The CFU-F assay has been typically used by many in the field as an *in vitro* correlate of MPC potential and, therefore, can also be used to ensure that an MPC population is surviving and expanding under suspension culture conditions. Briefly, aliquots of the starting cell population and cell suspensions at each time point are plated (1×10^4 cells/cm^2) under CFU-F assay conditions. The cells are cultured in basal medium [10% FBS (Gibco) in DMEM (Invitrogen Corp.)] in tissue culture plates at 37 °C with 5% humidified CO$_2$. After 14 days, CFU-F

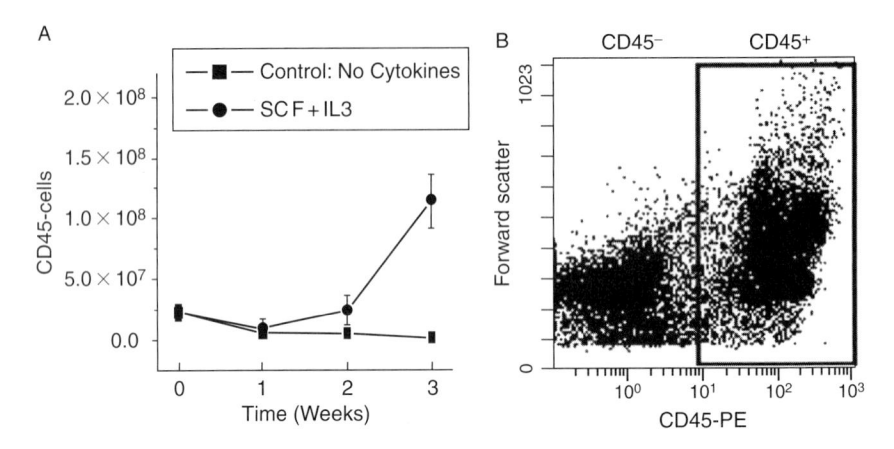

Fig. 3 Tracking and isolating CD45⁻ cells in suspension. (A) Flow cytometry is used to track the CD45 negative and positive cells present in suspension culture as a function of time. Note that the SCF + IL3 group results in the highest yield of CD45⁻ cells relative to control conditions (no cytokines). Each data point represents $n = 3$ experiments. (B) Flow cytometric dot plot revealing CD45 negative and positive cells populations. The gate drawn (□) can be used to isolate these two populations FACS.

cultures are terminated and stained with Giemsa modified solution (Sigma) to visualize the cell nuclei and cytoplasm. Colonies of fibroblasts containing >50 Giemsa positive colonies are counted. If suspension cultures are successful, a significant increase in CFU-F frequency should be detected at later time points relative to day 0 values. In addition to phenotypic cell tracking, suspension cells may also be sorted on the basis of CD45 expression (Fig. 3B) to isolate the fraction of cells (CD45⁻) containing MPCs and may be used to inoculate new suspension cultures with this more homogenous cell population.

C. Multilineage Differentiation Potential

As there is no single definite marker (phenotypic or molecular signature) for MPCs described to date, it is, therefore, important to employ functional assays to readout the development potential of MPCs generated under suspension culture conditions. In particular, the CFU-F and CFU-osteoblast (O) assays should be used to assess the fibroblastic (described above) and osteogenic capacity, respectively, of the cells generated in this novel bioprocess configuration. Although the CFU-F assay is used as *in vitro* correlate of MSC potential, careful considerations should be made in interpreting the significance of detected CFU-F derived from suspension-grown cells since it has been shown that the clones isolated in CFU-F assays are heterogeneous with respect to size (Owen *et al.*, 1987) and developmental potential (Muraglia *et al.*, 2000; Okamoto *et al.*, 2002). Therefore, to definitively assess the mesenchymal developmental potential of suspension-derived cells, considerable effort should be focused on characterizing other mesenchymal tissue derivatives, including the osteogenic, chondrogenic, myogenic, and adipogenic potential of suspension-derived cells at various times points (Fig. 4) (culture details described below).

1. Osteogenesis

An aliquot of suspension cells is removed from bioreactor cultures and plated at 1×10^4 cells/cm^2 and grown in osteogenic growth medium [10 nM dexamethasone (DEX), 5 mM β-glycerophosphate, 50 μg/ml ascorbic acid (AA), and 10 nM 1,25-dihydroxy vitamin D$_3$] (All reagents purchased from Sigma). On day 21, cultures are stained for alkaline phosphatase (ALP) activity (Sigma) and mineralization assessed by von Kossa staining (2% silver nitrate) (Sigma) after 5 weeks (Fig. 4B).

2. Chondrogenesis

Cell suspensions are removed from bioreactor cultures and sorted based on CD45 expression/lack of expression. The CD45$^-$ cells are isolated and grown under monolayer conditions. After 2 weeks, the cell population can then be

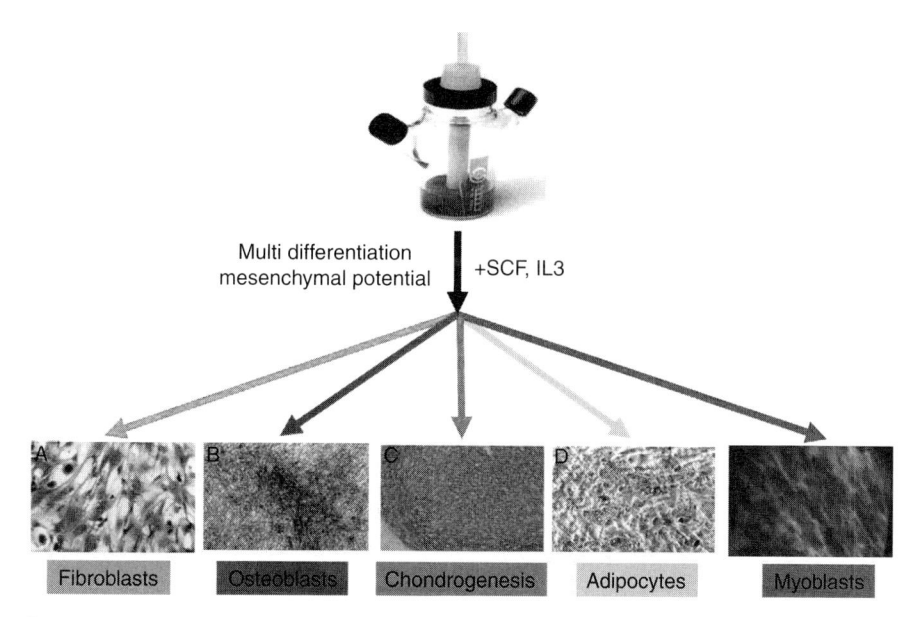

Fig. 4 The multilineage differentiation potential of suspension-grown cells. Bone marrow-derived cells grown under suspension culture conditions and in the presence of SCF and IL3 demonstrate the capacity to differentiate down the fibroblast, osteoblastic, chondrogenic, adipocytic, and myoblastic lineages. (A) Suspension cells grown under CFU-F assay conditions and stained with hematoxylin and eosin. (B) Phase-contrast micrograph of a von Kossa-stained CFU-O assay. (C) Representative Alcian blue stained histological section. Suspension cells were harvested, plated for 2 days, and then grown as high-density pellet cultures and stained with Alcian blue to visualize sGAG accumulation. (D) Phase-contrast micrograph of suspension cells grown under adipogenic conditions. Note the lipid accumulation within the cells. (E) Fluorescence image of desmin-stained cultures of suspension cells grown under myogenic conditions. Day 21 suspension cells were used for all differentiation assays shown. (See Plate no. 20 in the Color Plate Section.)

grown as high-density pellets (2.5×10^5 cells) for 21 days in serum-free medium [DMEM, ITS-premix (BD Biosciences), 50 µg/ml AA, 40 µg/ml L-proline, 100 µg/ml sodium pyruvate, 0.1 µM DEX], with and without 10 ng/ml recombinant human TGF-β3 (R&D Systems). On day 21, pellets are prepared for histology to detect sulfated glycosaminoglycan (sGAG) (Alcian blue staining) (Fig. 4C). sGAG content may be quantified using the Blyscan sGAG assay (Accurate Chemical & Scientific Corp., Wesbury, NY).

3. Adipogenesis

Suspension cells are cultured in monolayer at 1×10^4 cells/cm^2 in the presence of adipogenic supplements (1 µM DEX, 1 µg/ml insulin, and 0.5 mM 3-Isobutyl-1-methylxanthine (IBMX); purchased from Sigma). On day 21, cultures are stained with Oil Red O stain (Sigma) (Fig. 4D) and dye content quantified by isopropanol elution and spectrophotometry.

4. Colony Forming Unit—Myoblast Assay

The CFU-M assay is designed to evaluate the number of bone marrow mesenchymal progenitors capable of giving rise to a colony of myoblasts. Briefly, cell suspensions are prepared at a concentration of 1×10^4 cells/cm^2 and plated in 35 mm tissue culture dishes. The cells are maintained in MCDB 120 medium (provided by Dr. J.P. Tremblay, Centre de Recherche en Génétique Humanie, Sainte-Foy, QC, Canada) completed with 15% fetal bovine serum (FBS) for 2 weeks (myoblast proliferation medium, MPM). At 2 weeks, the serum level in the basal medium (MPM) is dropped to 2% (myoblast differentiation medium) and the cultures are terminated after 7 days. The cultures are refed three times a week with appropriate culture medium. At termination, cultures are rinsed three times with phosphate buffered saline (PBS), fixed in cold methanol ($-20\ °C$) for 5 min, and then incubated with 2% FBS in PBS for 20 min. Cultures are incubated with purified mouse anti-desmin (dilution 1:40 in PBS, BD Pharmingen) for 1 h. After washing three times with PBS, the cultures are incubated with the secondary antibody [Alexa Fluor 488 goat anti-mouse (dilution 1:00, Molecular Probes)] for 1 h. The cultures are washed with PBS and resuspended in 0.1M sodium cacodylate buffer (pH 7.4). Positive desmin colonies are imaged and enumerated at 4× using an inverted microscope (Fig. 4E). The immunostained cultures are rinsed twice in PBS and then stained with Geimsa (Sigma) to enumerate the total number of colonies present in each culture dish.

IV. Mechanism of MPC Expansion in Suspension

A. The Role of Hematopoietic Cells

With the exception of blood cell precursors, the current culture methods of culturing normal animal cells (including mesenchymal cells) involve anchorage of these cell types to solid surfaces in order to divide and proliferate in culture (Folkman and Moscona, 1978)—this phenomenon is known as anchorage dependence of cell division (O'Neill et al., 1986). However, the results of our work demonstrate a potentially novel property associated with the MPC lineage that allows these cells to expand in suspension as single cells. Given the current opinions on how these cells are typically cultured, we realize the necessity to propose a mechanism to explain how these cells may be cultured under adhesion-independent conditions.

The early of work Dexter et al. in the 1970s documented that bone marrow stromal cells play an integral role in hematopoiesis (Dexter et al., 1977; Dexter and Lajtha, 1974). Others since Dexter and colleagues have determined that specific growth factors secreted by marrow fibroblasts contribute to a suitable microenvironment desirable for hematopoietic stem/progenitor cell growth (Kim and Broxmeyer, 1998). What is less studied, however, is the role that hematopoietic cells play on the survival and growth of MPCs. To answer this question, studies were performed whereby bone marrow-derived cells were grown in stirred suspension cultures in the presence and absence of hematopoietic cells (characterized by the expression of CD45) (Baksh et al., 2005). From these studies, we reported that the expansion of CFU-F and CFU-O was significantly greater than that obtained in the absence of $CD45^+$ cells. These results suggest that MPC expansion may be influenced by the presence of hematopoietic cell types. Specifically, $CD45^+$ cells may be involved in the recruitment of MPCs through the release of soluble factors that then target MPCs in suspension, resulting in signal transduction events leading to proliferation, such as that described in adherent culture systems (Oprea et al., 2003; Wang et al., 1990). This work demonstrates, for the first time, the stimulatory role that hematopoietic cells may have on the survival and growth of mesenchymal cells (Fig. 5). Furthermore, analysis of the CFU-F and CFU-O frequencies obtained from unsorted and $CD45^-$ sorted cell fractions provides support that there may exist at least two types of MPCs, one requiring the presence of $CD45^+$ cells and the other type which does not require $CD45^+$ cells to achieve expansion in suspension. Taken together, these results suggest that there may exist a unique MPC type with multilineage potential, which does not require $CD45^+$ cells for their survival and expansion in suspension.

B. Adhesion Receptor Expression

Cell adhesion involves complex orchestrated events, which include the formation of focal contacts (mediated by cell adhesion receptors), for the attachment of extracellular matrix molecules (e.g., fibronectin). These focal contacts relay signals from the extracellular matrix to the cytoskeleton via tyrosine kinases that ultimately transduce signals which have been shown to regulate growth, morphology, movement, and differentiation of the cell (Aplin et al., 1999). In our studies, we tracked the expression of the adhesion receptor subunit CD49e [membrane-expressed α_5 subunit which heterodimerizes with the β_1-integrin subunit to form a functional adhesion receptor complex (Aplin et al., 1999)] on adherent-derived cells (expressing $CD45^-$) and observed an increase in their CD49e expression

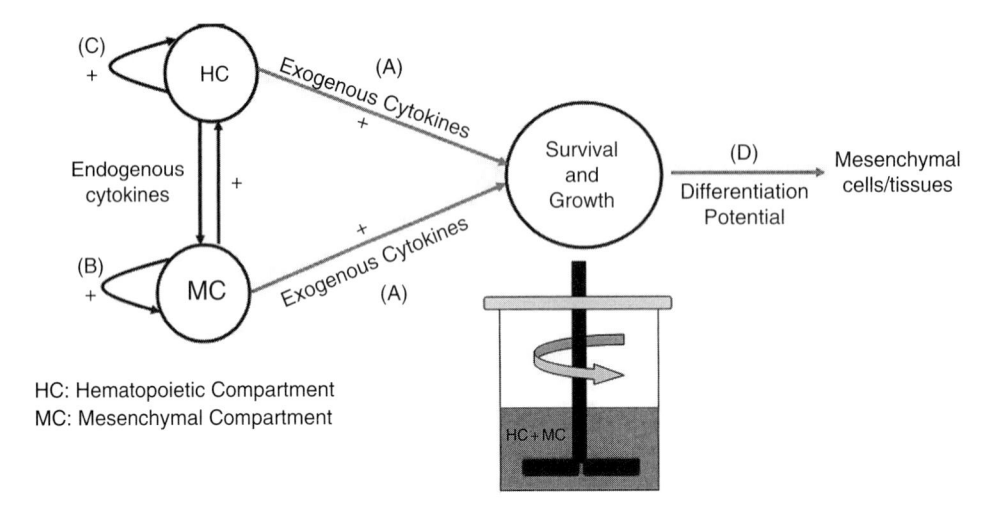

Fig. 5 Schematic representation of suspension culture compartments leading to the survival and growth of hematopoietic and mesenchymal cells. (A) Exogenously added cytokines act on both HC and MC, leading to survival and/or growth stimuli. (B) MC cells secrete cytokines that act on target cells (i.e., hematopoietic stem/progenitor cells) and are also self-regulated in an autocrine fashion by these cytokines (Majumdar et al., 1998). (C) HC cells secrete cytokines that act on target cells (i.e., mesenchymal stem/progenitor cells) and are also regulated by them in an autocrine fashion. (D) MC and HC cytokine secretion and action happens concomitantly.

(>80%), as a function of time in culture (Baksh et al., 2005), suggesting that perhaps the cycling of adherent cultured cells is preferentially regulated by integrin-mediated interactions with their substrate. However, when the same cells were placed in suspension culture, they were incapable of further proliferation. These findings suggest that once cells contact a surface and upregulate their cell surface expression of adhesion receptors, they may no longer be capable of cell proliferation in suspension. Furthermore, any given substrate that a cell contacts will generally be covered with a thin layer (10–15 nm) comprising mixtures of absorbed proteins, soluble factors (from serum), and extracellular matrix (secreted by the cell itself); therefore, the combinatorial effect of each interaction with the cell may also contribute to a variety of responses including proliferation and differentiation. Therefore, we can hypothesize that cell adhesion and subsequent growth on a solid substrate, not only predisposes the cell to participate in adhesion-mediated growth (almost exclusively) but also provides an early differentiation cue (due to direct contact with other cells and ECM), leading to the eventual loss of long-term self-renewing capacity. Aspects of this hypothesis are supported by the work of Franceschi et al. (2003) who demonstrated that binding of type I collagen to membrane-expressed $\alpha_2\beta_1$ integrins on osteoblast progenitors results in the activation of the MAP kinase pathway, leading to the phosphorylation of Cbfa1/Runx2, resulting in the switch to differentiation (Franceschi et al., 2003), with the concomitant downregulation of osteogenic cell proliferation.

We have also shown in our own work that cells cultured in CFU-F assays not only express extracellular matrix-specific genes (i.e., type I collagen) but also secrete their own matrix (perhaps indicative of cell differentiation) throughout culture expansion, suggesting that integrin-mediated interactions with extracellular matrix play a role in controlling and coordinating multiple signals involved in not only cell proliferation but also differentiation.

In our studies we determined that an expanding $CD45^-$ suspension cell population maintained a constant and low expression level of CD49e throughout suspension culture. However, this population ($CD45^-CD49e^+$ suspension-derived cells isolated via FACS) demonstrated an approximately three-fold increase in mean fluorescence intensity in CD49e expression on plating in CFU-F assays within 24 h. These results suggest that $\alpha_5\beta_1$-integrin-mediated interactions (via cell–cell or cell–ECM interactions) do not contribute to, nor are necessary for, cell cycle progression of suspension-grown MPC since we observed an expansion of MPC assayed by their ability to generate increased numbers of CFU-F and CFU-O as a function of time. In hematopoietic stem cell (HSC) culture systems, it has been demonstrated that β_1-dependent adhesion is not necessary for $CD34^+$ cell growth (Verfaillie and Catanzaro, 1996); in fact contact with stroma inhibits their cell proliferation (Hurley *et al.*, 1995). Given these observations, perhaps adult human bone marrow comprises MPC that have the capacity (and/or property, i.e., low $\alpha_5\beta_1$-integrin expression) for adherent-independent proliferation, similar to that capable of HSC.

V. Discussion

This chapter describes the development and utilization of a novel bioprocess approach to the culture expansion of MPCs. Through suspension culture, we demonstrated that MPCs express low levels of adhesion receptor(s), specifically CD49e, which, we hypothesize, attributes to their ability to survive and proliferate in suspension. This finding suggests that the expansion of anchorage-capable stem cells may no longer be limited by surface area constraints. In our design strategy, we determined that the addition of soluble factors (e.g., SCF and IL3) and initiating suspension cultures with a suitable innoculum density (1×10^6 cells/ml), as well as establishing a suitable feeding protocol [refeeds every 2–3 days, with one-third cell and medium removal if the cells density exceeds $>1 \times 10^6$ cells/ml], is critical in attaining >10-fold expansion of putative MPCs over a 21-day culture period. While the system used to culture MPCs in suspension is sufficient for small-scale benchtop scientific work, the development and optimization of a sophisticated large-scale cell expansion system (>1 L) would be necessary for achieving suitable numbers of functional cells. As bioprocess design is an iterative process, future studies should continue to identify other types of interactions and their dynamic affects on culture outputs. However, to render stem cell bioprocess design efficient, faster, and more predictive, readout measures and/or assays are required. New and innovative technologies, such as microarrays, may provide a realistic approach to generate efficient and near to real-time measurements essential to bioprocess optimization.

References

Aplin, A. E., Howe, A. K., and Juliano, R. L. (1999). Cell adhesion molecules, signal transduction, and cell growth. *Curr. Opin. Cell Biol.* **11**, 737–744.

Baksh, D., Davies, J. E., and Zandstra, P. W. (2003). Adult human bone marrow-derived mesenchymal progenitor cells are capable of adhesion-independent survival and expansion. *Exp. Hematol.* **31**, 723–732.

Baksh, D., Davies, J. E., and Zandstra, P. W. (2005). Soluble factor crosstalk between human bone marrow-derived hematopoietic and mesenchymal cells enhances *in vitro* CFU-F and CFU-O growth and reveals heterogeneity in the mesenchymal progenitor cell compartment. *Blood* **9**, 3012–3019.

Baksh, D., Zandstra, P. W., and Davies, J. E. (2007). A non-contact suspension culture approach to the culture of osteogenic cells derived from a CD49elow subpopulation of human bone marrow-derived cells. *Biotechnol Bioeng.* **6**, 1195–1208.

Bianchi, G., Banfi, A., Mastrogiacomo, M., Notaro, R., Luzzatto, L., Cancedda, R., and Quarto, R. (2003). Ex vivo enrichment of mesenchymal cell progenitors by fibroblast growth factor 2. *Exp. Cell Res.* **287**, 98–105.

Botchwey, E. A., Pollack, S. R., Levine, E. M., and Laurencin, C. T. (2001). Bone tissue engineering in a rotating bioreactor using a microcarrier matrix system. *J. Biomed. Mater. Res.* **55**, 242–253.

Bruder, S. P., Jaiswal, N., and Haynesworth, S. E. (1997a). Growth kinetics, self-renewal, and the osteogenic potential of purified human mesenchymal stem cells during extensive subcultivation and following cryopreservation. *J. Cell Biochem.* **64**, 278–294.

Bruder, S. P., Jaiswal, N., and Haynesworth, S. E. (1997b). Growth kinetics, self-renewal, and the osteogenic potential of purified human mesenchymal stem cells during extensive subcultivation and following cryopreservation. *J. Cell Biochem.* **64**, 278–294.

Caplan, A. I. (1991). Mesenchymal stem cells. *J. Orthop. Res.* **9**, 641–650.

Caplan, A. I., and Bruder, S. P. (1997). Cell and molecular engineering of bone regeneration. *In* "Principles of Tissue Engineering" (R. Lanza, R. Langer, and J. Vacanti, eds.), pp. 603–617. Academic Press, San Diego.

Castro-Malaspina, H., Gay, R. E., Resnick, G., Kapoor, N., Meyers, P., Chiarieri, D., McKenzie, S., Broxmeyer, H. E., and Moore, M. A. (1980). Characterization of human bone marrow fibroblast colony-forming cells (CFU-F) and their progeny. *Blood* **56**, 289–301.

Collins, P. C., Miller, W. M., and Papoutsakis, E. T. (1998). Stirred culture of peripheral and cord blood hematopoietic cells offers advantages over traditional static systems for clinically relevant applications. *Biotechnol. Bioeng.* **59**, 534–543.

Coulson, J. M. R. (1999)."Chemical Engineering." Vol. 1, Butters Worth Heinemann, Boston.

Dayoub, H., Dumont, R. J., Li, J. Z., Dumont, A. S., Hankins, G. R., Kallmes, D. F., and Helm, G. A. (2003). Human mesenchymal stem cells transduced with recombinant bone morphogenetic protein-9 adenovirus promote osteogenesis in rodents. *Tissue Eng.* **9**, 347–356.

de Groot, R. P., Coffer, P. J., and Koenderman, L. (1998). Regulation of proliferation, differentiation and survival by the IL-3/IL-5/GM-CSF receptor family. *Cell. Signal.* **10**, 619–628.

Dexter, T. M., and Lajtha, L. G. (1974). Proliferation of haemopoietic stem cells *in vitro*. *Br. J. Haematol.* **28**, 525–530.

Dexter, T. M., Wright, E. G., Krizsa, F., and Lajtha, L. G. (1977). Regulation of haemopoietic stem cell proliferation in long term bone marrow cultures. *Biomedicine* **27**, 344–349.

Folkman, J., and Moscona, A. (1978). Role of cell shape in growth control. *Nature* **273**, 345–349.

Franceschi, R. T., Xiao, G., Jiang, D., Gopalakrishnan, R., Yang, S., and Reith, E. (2003). Multiple signaling pathways converge on the Cbfa1/Runx2 transcription factor to regulate osteoblast differentiation. *Connect. Tissue Res.* **44**(Suppl 1), 109–116.

Geankoplis, J. (1993). "Transport Processes and Unit Operations." Prentice Hall PTR, New Jersey.

Gronthos, S., Graves, S. E., Ohta, S., and Simmons, P. J. (1994). The STRO-1+ fraction of adult human bone marrow contains the osteogenic precursors. *Blood* **84**, 4164–4173.

Gronthos, S., and Simmons, P. J. (1995). The growth factor requirements of STRO-1-positive human bone marrow stromal precursors under serum-deprived conditions *in vitro*. *Blood* **85**, 929–940.

Haynesworth, S.E, Baber, M.A, and Caplan, A.I (1992). Cell surface antigens on human marrow-derived mesenchymal cells are detected by monoclonal antibodies. *Bone* **13**, 69–80.

Hurley, R. W., McCarthy, J. B., and Verfaillie, C. M. (1995). Direct adhesion to bone marrow stroma via fibronectin receptors inhibits hematopoietic progenitor proliferation. *J. Clin. Invest.* **96**, 511–519.

Kim, C. H., and Broxmeyer, H. E. (1998). *In vitro* behavior of hematopoietic progenitor cells under the influence of chemoattractants: Stromal cell-derived factor-1, steel factor, and the bone marrow environment. *Blood* **91**, 100–110.

Majumdar, M. K., Thiede, M. A., Mosca, J. D., Moormen, M., and Gerson, S. L. (1998). Phenotypic and functional comparison of cultures of marrow-derived mesenchymal stem cells (MSCs) and stromal cells. *J. Cell Physiol.,* **1**, 57–66.

Muraglia, A., Cancedda, R., and Quarto, R. (2000). Clonal mesenchymal progenitors from human bone marrow differentiate *in vitro* according to a hierarchical model. *J. Cell Sci.* **113**(Pt 7), 1161–1166.

Noort, W., Kruisselbrink, A., in't, A. P., Kruger, M., van Bezooijen, R., de Paus, R., Heemskerk, M., Lowik, C., Falkenburg, J., Willemze, R., and Fibbe, W. (2002). Mesenchymal stem cells promote engraftment of human umbilical cord blood-derived CD34(+) cells in NOD/SCID mice. *Exp. Hematol.* **30**, 870.

O'Neill, C., Jordan, P., and Ireland, G. (1986). Evidence for two distinct mechanisms of anchorage stimulation in freshly explanted and 3T3 Swiss mouse fibroblasts. *Cell* **44**, 489–496.

Okamoto, T., Aoyama, T., Nakayama, T., Nakamata, T., Hosaka, T., Nishijo, K., Nakamura, T., Kiyono, T., and Toguchida, J. (2002). Clonal heterogeneity in differentiation potential of immortalized human mesenchymal stem cells. *Biochem. Biophys. Res. Commun.* **295**, 354–361.

Oprea, W. E., Karp, J. M., Hosseini, M. M., and Davies, J. E. (2003). Effect of platelet releasate on bone cell migration and recruitment *in vitro*. *J. Craniofac. Surg.* **14**, 292–300.

Owen, M. E., Cave, J., and Joyner, C. J. (1987). Clonal analysis *in vitro* of osteogenic differentiation of marrow CFU-F. *J. Cell Sci.* **87**(Pt 5), 731–738.

Pittenger, M. F., Mackay, A. M., Beck, S. C., Jaiswal, R. K., Douglas, R., Mosca, J. D., Moorman, M. A., Simonetti, D. W., Craig, S., and Marshak, D. R. (1999). Multilineage potential of adult human mesenchymal stem cells. *Science* **284**, 143–147.

Qiu, Q. Q., Ducheyne, P., and Ayyaswamy, P. S. (1999). Fabrication, characterization and evaluation of bioceramic hollow microspheres used as microcarriers for 3-D bone tissue formation in rotating bioreactors. *Biomaterials* **20**, 989–1001.

Sardonini, C. A., and Wu, Y. J. (1993). Expansion and differentiation of human hematopoietic cells from static cultures through small-scale bioreactors. *Biotechnol. Prog.* **9**, 131–137.

Shi, S., Gronthos, S., Chen, S., Reddi, A., Counter, C. M., Robey, P. G., and Wang, C. Y. (2002). Bone formation by human postnatal bone marrow stromal stem cells is enhanced by telomerase expression. *Nat. Biotechnol.* **20**, 587–591.

Verfaillie, C. M., and Catanzaro, P. (1996). Direct contact with stroma inhibits proliferation of human long-term culture initiating cells. *Leukemia* **10**, 498–504.

Wang, Q. R., Yan, Z. J., and Wolf, N. S. (1990). Dissecting the hematopoietic microenvironment. VI. The effects of several growth factors on the *in vitro* growth of murine bone marrow CFU- F. *Exp. Hematol.* **18**, 341–347.

Zandstra, P. W., Eaves, C. J., and Piret, J. M. (1994). Expansion of hematopoietic progenitor cell populations in stirred suspension bioreactors of normal human bone marrow cells. *Biotechnol. Bioeng.* **12**, 909–914.

CHAPTER 13

Purification and Long-Term Culture of Multipotent Progenitor Cells Affiliated with the Walls of Human Blood Vessels: Myoendothelial Cells and Pericytes

Mihaela Crisan,[†,‡,§] Bridget Deasy,[*,‡] Manuela Gavina,[‡,¶] Bo Zheng,[‡] Johnny Huard,[‡,¶] Lorenza Lazzari,[#] and Bruno Péault[†,‡,**]

[*]Departments of Orthopaedic Surgery and Bio-Engineering
University of Pittsburgh
Pennsylvania 15213

[†]Department of Pediatrics
Children's Hospital of Pittsburgh of UPMC
Pittsburgh, Pennsylvania 15213

[‡]Stem Cell Research Center
Children's Hospital of Pittsburgh of UPMC
Pittsburgh, Pennsylvania 15213

[§]Hillman Cancer Center
University of Pittsburgh Cancer Institute
Pittsburgh, Pennsylvania 15213

[¶]Department of Orthopaedics
University of Pittsburgh School of Medicine
Pittsburgh, Pennsylvania 15213

[#]Fondazione Ospedale Maggiore Policlinico
Department of Regenerative Medicine
Milan, Italy

[**]McGowan Institute for Regenerative Medicine
Pittsburgh, Pennsylvania 15219

Abstract
I. Introduction
II. Materials and Methods: Cell Analysis and Purification by Flow Cytometry
    A. Tissue Procurement
    B. Myoendothelial Cell Characterization in Fetal and Adult Skeletal Muscle
    C. Pericyte Characterization from Fetal and Adult Tissues
    D. Cell Sorting
III. Materials and Methods: Long-Term Culture of Human Blood Vessel-Associated Progenitor Cells
    A. Myoendothelial Cell Culture
    B. Pericyte Culture
    C. Clonal Growth of Cells
IV. Materials and Methods: Proliferation Kinetics of Human Myoendothelial Cells and Pericytes
    A. Long-Term Cell Culture Potential
    B. Use of Time-Lapsed Microscopy in Clonal Analysis
V. Materials and Methods: Genotypic and Phenotypic Analyses of Long-Term Cultured Cells
    A. Flow Cytometry Analysis
    B. Immunocytochemistry on Cultured Pericytes
    C. Malignant Transformation Analysis
VI. Conclusion
    References

Abstract

We have identified with molecular markers and purified by flow cytometry two populations of cells that are developmentally and anatomically related to blood vessel walls in human tissues: myoendothelial cells, found in skeletal muscle and coexpressing markers of endothelial and myogenic cells, and pericytes—aka mural cells—which surround endothelial cells in capillaries and microvessels. Purified myoendothelial cells and pericytes exhibit multilineage developmental potential and differentiate, in culture and *in vivo*, into skeletal myofibers, bone, cartilage, and adipocytes. Myoendothelial cells and pericytes can be cultured on the long term with sustained marker expression and differentiation potential and clonal populations thereof have been derived. Yet, these blood vessel wall-derived progenitors exhibit no tendency to malignant transformation upon extended culture. Our results suggest that multipotent progenitor cells, such as *mesenchymal stem cells*, previously isolated retrospectively from diverse cultured adult tissues are derived from a subset of perivascular cells. We present in this chapter the main strategies and tactics used to purify, culture on the long term, and phenotypically characterize these novel multipotent cells.

Single color stained cells are used as compensation controls. Endothelial cells are sorted as CD45⁻, CD56⁻, CD146⁻, and CD34⁺. Pericytes are sorted as CD45⁻, CD56⁻, CD146⁺, and CD34⁻. Myoendothelial cells are sorted as CD45⁻, CD56⁺, CD146⁺, CD34⁺, and CD144⁺.

III. Materials and Methods: Long-Term Culture of Human Blood Vessel-Associated Progenitor Cells

Long-term culturing of stem cell candidates can serve two scientific questions. First, it can help to answer the questions of whether the cells have the stem cell property of long-term self-renewal. Second, it aids in establishing clinical expansion rates, for example, how long will it take to obtain a defined number of cells from a specific number of founder cells? Or, in a translational research perspective, how many cells can we expect after, for example, 3 weeks of culturing? Finally, growth rates can be used to characterize the different types of cells, even distinguishing stem cell candidates from more committed stem cells.

A unique property of human myoendothelial and perivascular cells is that these cells can be cultured on the long term with no significant loss of developmental potential.

A. Myoendothelial Cell Culture

FACS-sorted myoendothelial cells are seeded in collagen-coated 96-well plates, at a density of 500 cells per well, in proliferation medium (DMEM, 20% FCS, 5% HS, 2% chick embryo extract) at 37 °C in a 5% CO_2 atmosphere. At 60% confluence, cells are detached with trypsin/EDTA and replated after washing at densities ranging from $1.0 \times 10^3/cm^2$ to $2.5 \times 10^3/cm^2$.

B. Pericyte Culture

Wells from a 48-well culture plate are coated with gelatin 0.2% for 10 min at 4 °C. 2×10^4 sorted pericytes are cultured per well at 37 °C and 5% CO_2 in endothelial growth medium (EGM2, Cambrex). EGM2 is changed first after 7 days, then every 4 days until confluence. Cells are detached by replacing culture medium by trypsin 1×-EDTA for 8–10 min at 37 °C. Cells are harvested, washed, and the content of one well (1 cm²) is replated into one uncoated well from a 12-well culture plate (each well = 3.6 cm²), in DMEM high glucose, 20% FCS, 1% PS until confluence (passage 1). The procedure is repeated and cells are replated from one well into two uncoated wells from a 6-well culture dish (each well = 9.6 cm²) (passage 2). The procedure is repeated and cells from both wells are replated into one uncoated T-25 culture flask (passage 3). The procedure is repeated and cells from one T-25 culture flask are replated into one uncoated T-75 culture flask (passage 4). The procedure is repeated and cells from one T-75 culture flask are replated into three uncoated T-75 culture flasks (passage 5). From then on until 16–17 weeks, cells can be passaged 1:7, using the same protocol. From the 17th week of culture, cells are passaged 1:3. The same medium is used throughout the culture.

C. Clonal Growth of Cells

Individual clones of CD146$^+$,CD34$^-$,CD45$^-$ pericytes have been obtained from human fetal muscle. Sorted cells were first plated in 96-well plates coated with 0.2% gelatin (Sigma), at a density of 10 cells/cm^2. After reaching 80% confluence, cells were resuspended by gentle pipetting and cloned by limiting dilution. Briefly, cells were diluted with culture medium 10-fold, from 1000 to 10 cells/ml. one cell/100 µl of medium was seeded per well in 96-well plates. One day after cell seeding, each well was examined microscopically. Only wells with single cells were cultured, whereas empty wells or wells containing more than one cell were deleted. Clones were passaged from 96-well plates into 24-well plates coated with 0.2% gelatin, at a density of 6 × 10^4 cells/well. The cloning efficiency was 18–20%, ~1.8 cell in 10. Alternatively, single myoendothelial cells were sorted as described above and deposited in individual wells of 96-well plates in culture medium, using the automatic cell deposition unit (ACDU) device equipping the FACSAria flow cytometer.

Several media were tested in order to obtain the best proliferation of cloned pericytes. DMEM/Ham-F12 (1:1) supplemented with 10% FCS, 50 ng/ml hepatocyte growth factor (HGF), and a defined hormone and salt mixture composed of insulin (25 µg/ml), apo-tranferrin (100 µg/ml), progesterone (20 ηM), putrescine (60 µM), uridine (10 µM), and sodium selenite (30 ηM); or a proliferative medium (PM), composed of RPMI 1640/Ham-F12 (1:1)/human endothelial (Gibco) (1:1), enriched with 5% HS and 5% FCS. This growth medium was supplemented with 50 ng/ml HGF, 100 ng/ml insulin growth factor (IGF-1), and a defined hormone and salt mixture composed of insulin (25 mg/ml), apo-tranferrin (100 µg/ml), progesterone (20 ηM), putrescine (60 µM), uridine (10 µM), and sodium selenite (30 ηM), and PM supplemented with the same growth factors and a different serum concentration, 2.5% HS, and 2.5% FCS. Culture medium was changed every 2 or 3 days, and cells passaged when they achieved 80% confluence. In the presence of DMEM/Ham-F12, all clones proliferated slowly and underwent senescence after ~28 days. In contrast, in the presence of RPMI/Human Endo (PM), the clones underwent more than 23 population doublings (PDs) at a high proliferation rate, with a doubling time of 36 h. At both early and late passages, cells maintained a diploid karyotype and exhibited robust telomerase activity, as assessed by using a telomerase repeat amplification protocol (TRAP). Moreover, the flow cytometry characterization of clonally expanded cells (see paragraph V below), which was confirmed by RT–PCR analysis, showed maintenance of the CD146 surface antigens after 30 days of culture. Expression of α-smooth muscle actin (αSMA), platelet-derived growth factor receptor β (PDGFRβ), and NG2 was also sustained whereas desmin and CK7 were only faintly expressed.

IV. Materials and Methods: Proliferation Kinetics of Human Myoendothelial Cells and Pericytes

A. Long-Term Cell Culture Potential

Standard assays for extended *in vitro* culture of stem cell candidates involve serial replating before the cells reach confluency and may initiate cell differentiation pathways. The long-term culture of endothelial cells (CD56$^-$, CD34$^+$, CD144$^+$) and myoendothelial cells (CD56+, CD34+, CD144+) was examined. The growth rates of these cells were compared to those of the more committed CD56+, CD34$^-$, CD144$^-$ myogenic cells. In this type of continuous culture assay, cells are plated at an initial seeding density and then replated after a given amount of time, to the initial seeding density (Deasy *et al.*, 2005). The cell counts and cellular dilution factor were recorded and used to calculate the expansion potential, number of PDs, and the doubling time. These calculations are based on the standard exponential growth equation; $N_t = N_0 e^{kt}$, where N_0 is the number of cells at time 0, and N_t the number of cells after time t. Cells were plated at an initial seeding density of 600 cells/cm^2. After 2–3 days of growth (confluency < 50%), cells were trypsinized, counted using a hemocytometer, and replated at 600 cells/cm^2. This continued for a 2-week period, and cell growth was examined in three types of media. Kinetics were followed to examine PD time and to predict expansion yield. It was first established that the growth rate was dependent on the culture medium. In comparison to DMEM, 20% serum, and skeletal muscle basal medium, the growth rates were higher in EGM2. Further, the myoendothelial cells demonstrated significantly faster growth than did the endothelial cells; the endothelial cell population required nearly 40 h before undergoing a doubling in size, while the PD time for the myoendothelial population was ~15 h per doubling. For comparison, Tagliafico *et al.* (2004) examined a population of mesoangioblasts (which share similarities with myoendothelial cells) and showed a doubling rate of ~24 h, while Vilquin *et al.* (2005) showed that human CD56$^+$ myoblasts had a PD time of ~30 h per doubling. Differences in PD time may arise because of the passage at which the cells are examined. For example, we previously reported that the PD time of muscle-derived stem cells decreased as cells were expanded (Deasy *et al.*, 2005). Cells which are freshly isolated can be expected to contain more nondividing cells (apoptotic or quiescent) and hence the doubling time is slower. In addition, variations among investigators may arise because of the timepoints which are included in the calculation (PDT = time/log$_2$ (N_t/N_0). Finally, we reported that myoendothelial cells could be expanded to at least 22 PDs (Zheng *et al.*, 2007), and since we have observed several populations of human muscle stem cells which can be expanded to more than 50 PDs.

B. Use of Time-Lapsed Microscopy in Clonal Analysis

Time-lapse microscopic imaging of cells has been in use for more than 50 years, although only more recently has advanced robotics and software allowed for extended and high-throughput cell culture assays (Curl *et al.*, 2004). We obtain live cell imaging (LCI), as a sequence of jpg image files, by use of a custom-built enclosed humidified environment which is maintained at 37 °C and 5% CO_2 (Deasy *et al.*, 2003). A multiwell plate (up to 400 wells) can be used in the combinatorial experimental design. Frequency of image acquisition (phase, brightfield, or fluorescence) is user defined and assay specific. A robotic controller returns the stage to defined x,y coordinates at set intervals. Stacks of jpg images are subsequently processed for object (cell) data and object track data and then the data is queried in a database. Unique software can measure parameters such as morphology, motility, colony formation, cytokinesis, or ultrastructure. The bioimaging system will provide real-time, nondestructive kinetic and morphological analysis of phenotypic changes in living cells.

In the study of muscle-derived myoendothelial cells, we have used this unique technology to visually confirm clonality of the myoendothelial cells. Green fluorescent protein (GFP)-transfected myoendothelial cells were single cell sorted by FACS, as described above, to a 384-well plate and analyzed by LCI. Using a 4× objective, at least 97% of the well is within the viewfield, and in this way we can feel confident that colonies are initiated with one single cell. In addition, the video record can confirm that no additional cells are present, and that no new cells move into the region. Finally, for further confirmation, we also acquired time-lapsed green florescent images to verify the presence of a single colony-initiating cell. In this analysis, we acquired images every 10 min, for a period of 20 days.

V. Materials and Methods: Genotypic and Phenotypic Analyses of Long-Term Cultured Cells

In an effort to test the long-term self-renewal potential and identify what the limits are for expansion of populations of blood vessel-derived stem cells, *in vitro* phenotypic alterations, as well as signs of aging and transformation, in myoendothelial cells and pericytes extensively expanded under normal culture conditions were examined.

A. Flow Cytometry Analysis

For flow cytometry analysis, cells are detached with 0.05% trypsin and 1 mM EDTA for 5–10 min and washed with PBS. Before staining, nonspecific binding is blocked with PBS, 2% normal serum for 5 min and then cells are incubated for 20 min at room temperature (RT) in the dark with the following cell-specific monoclonal antibodies: CD13-PE, CD34-PE, CD44-FITC, CD45-PECy7, CD56 PE,

CD73-PE (SH3), CD90-PE (Thy-1), CD105-FITC (SH2; Endoglin), CD106-PE (VCAM-1), CD31-PE, CD133-APC, CD146-PE, a-SMA-FITC, HLA-ABC, and HLA-DR. Isotype-matched immunoglobulins: IgG1-PE/-FITC, IgG1-PECy7, IgG1-APC are used as negative controls. Cells are permeabilized with 0.1% Triton X-100 when necessary. After staining the cells are washed once with PBS-containing 0.1% BSA. At least 50,000 events are acquired on a flow cytometry. We have used both the Cytomics FC500 (Beckman Coulter, Fullerton, CA, http//www.beckmancoulter.com), followed by analysis using the CXP-analysis software, or the FACSCalibur (BD Biosciences, San Jose, CA, http://www.bdbiosciences.com).

B. Immunocytochemistry on Cultured Pericytes

One thousand pericytes per well are seeded into 48-well plates for 24 h in DMEM high glucose, 20% FCS, 1% PS. The next day, adherent cells are fixed in cold methanol/acetone (1:1) or 4% paraformaldehyde (for Pax 7 staining) for 10 min at 4 °C. Cells are then washed three times in PBS and incubated with the blocking buffer solution (PBS, 5% goat serum) for 1 h at RT. Cells are incubated with primary antibodies overnight at 4 °C in blocking buffer solution. After three rinses in PBS, cells are incubated with the appropriate secondary antibody: biotinylated goat antimouse antibody, then streptavidin Cy3 for monoclonal antibodies, and Alexa 594-conjugated donkey antirabbit antibody for rabbit antisera, all used at 1:500 dilution and incubated during 1 h at RT in the dark. After rinsing, the second primary antibody (anti-αSMA–FITC) is added for an additional hour at RT for 1 h in the dark. 4'-6-diamidino-2-phenylindole (DAPI) is used (1:2000 in PBS, 5 min at RT) to stain nuclei. After three final washes, coverslips are mounted on slides using glycerol/PBS (1:1) and analyzed under a fluorescence microscope. The following monoclonal uncoupled mouse antihuman antibodies were used in this study: CD146, CD56, CD31, CD34, CD90, CD49b, CD44, NG2, slow and fast myosin heavy chain (MHC), desmin, CD45, αSMA, all at a 1:100 dilution. The uncoupled mouse antimouse Pax7 antibody was used at 1:20. Other antibodies used are the FITC–anti-α SMA monoclonal antibody and the uncoupled rabbit antihuman PDGFRβ (1:100). Negative controls do not receive a primary antibody and are incubated in PBS, 5% goat serum overnight at 4 °C, then with corresponding secondary antibodies and finally with streptavidin Cy3 and DAPI. For detection of intracellular antigens: MHC, desmin, α SMA and Pax7, all solutions (rinsing and incubation) are supplemented with 0.1% Triton X-100.

C. Malignant Transformation Analysis

For transformation analysis, growth in soft agar was examined by plating 2000 cells per 9.6 cm^2 well in 0.35% low melting point agar as a top layer. The bottom layer consisted of 0.7% agar in 20% serum medium. Colonies were allowed to grow for 21 days at 37 °C, at which time the colonies were scored for both number and size using the Northern Eclipse imaging software. DNA content was analyzed by flow cytometry by fixing cells in cold 70% ethanol for 2 h and then resuspending cells in Ipecal and propidium iodide solution. DNA content was examined using flow cytometry and ModFitLT (v2.0). Three replicate experiments were performed and averaged. Differences among the timepoints or doubling levels were compared for the 2N, 4N, and greater than 4N peaks using one-way ANOVA.

VI. Conclusion

Besides multipotent stem cells abundant in the early embryo, and tissue-restricted progenitors present in all organs during the whole life, the existence, within developed tissues derived from the three germ layers, of cells capable of multilineage differentiation has been indirectly documented by independent investigators. Seminal to the isolation of these cells has been their ability to survive and multiply in primary cultures of the tissues of origin, which allowed the enrichment and, ultimately, the growth of homogenous populations of these progenitors. We have identified and purified from human tissues two cell types that are closely associated with blood vessel walls. Myoendothelial cells are present in fetal and postnatal skeletal muscles and are typified by coexpression of markers of endothelial and myogenic cells. Pericytes encircle capillaries and microvessels in the whole organism. We have, for the first time, purified these two cell types to homogeneity and, at first, demonstrated in culture and *in vivo* that these cells are endowed with robust myogenic potential (Crisan *et al.*, in revision Zheng *et al.*, 2007). Further experiments revealed that these cells can give rise to other mesodermal derivatives and their ability to differentiate into bone, cartilage, and fat cells has been documented. While the description of these developmental studies is beyond the scope of the present review, we have summarized above the purification of these cells by flow cytometry, using unique combinations of cell surface markers. We stress the fact that whereas myoendothelial are, by essence, restricted to skeletal muscle, pericytes are encountered in all organs. We have characterized pericytes in bone marrow, lung, skeletal muscle, pancreas, brain, myocardium, placenta, skin, and other organs. Of particular interest in the perspective of a therapeutic utilization of these cells, we have also purified pericytes from human adipose tissue harvested for cosmetic purposes, providing the proof of principle that these multipotent progenitors can be harvested from a dispensable tissue, using a simple and painless procedure, for eventual autologous transplantation (Crisan *et al.*, in revision). Most striking has been the ability of both purified myoendothelial cells and pericytes to grow extensively in culture, including in monoclonal conditions (Zheng *et al.*, 2007), as described in the preceding sections. Culture conditions for these cells are simple; use classic media and exclude the addition of any defined growth factor. Similarly, all categories of adult multipotent progenitor cells identified in the past years, such as MSC, MAPC, MDSC, or fat derived progenitors, are characterized by an outstanding ability to proliferate on the long term *in vitro*. This further supports our hypothesis that these elusive, retrospectively characterized adult multipotent cells are derived from perivascular cells. In agreement, we have observed using the strategies described herein that long-term cultured pericytes express all known markers of MSC such as CD9, CD73, CD90, CD105, and *frizzled 9* (Crisan *et al.*, in revision). Most important, we have observed that cultured myoendothelial cells and pericytes retain their potential for multilineage differentiation. Yet, neither tendency to undergo malignant transformation nor any karyotypic alterations were observed (Zheng *et al.*, 2007).

In 1961, Hayflick first reported that normal human fibroblasts undergo a limited number of divisions *in vitro* (Hayflick and Moorhead, 1961), and it was concluded that the maximum number of (PDs) was ~50. Since then many investigators have similarly reported a finite number of PDs for euploid cells in culture. On the other hand, the current thinking is that stem cells possess a unique proliferation potential that is several times greater than this number or possibly an unlimited replicative potential (Rubin, 2002; Reyes *et al.*, 2001). Indeed long-term proliferation potential is often used to characterize stem cells. Pluripotency and normal karyotype have been reported for human ES cells expanded to 250 PDs (Amit *et al.*, 2000) or 130 PDs (Xu *et al.*, 2001). Non-ES cells, such as human hematopoietic stem cells, appear to have a slower doubling rate though have been expanded 2×10^6 –fold over a 6-month period corresponding to 20 PDs (8) and ~1400-fold over a 3-month period (10 PDs) (Gilmore *et al.*, 2000; Piacibello *et al.*, 1997). MSC and MAPC, which we assume to be derived from perivascular cells, have been expanded 10^9-fold or to 30 PDs (Colter *et al.*, 2000) and to 120 PDs (Jiang *et al.*, 2003), respectively. Our results on myoendothelial cells and pericytes appear to be in agreement with these numbers.

Extensively expanded stem cells may become targets of transformation. On another level, there is an association between stem cells and cancer cells which may exist even before long-term expansion. The self-renewal pathway, which is a defining characteristic of stem cells, may share some signals associated with oncogenesis. Recently, the notch, sonic hedgehog, and wnt signaling pathways have been implicated in self-renewal of hematopoietic stem cells and interestingly, it has been suggested that hematopoietic stem cells are more likely than committed progenitors to undergo transformation (reviewed in Reya *et al.*, 2001). All these results being taken together, the use of expanded stem cells demands examination into the quality of the cells. In this respect, our observations that long-term cultured blood vessel wall derived progenitors do not undergo malignant transformation are encouraging.

Acknowledgments

We are indebted to Alison Logar for sharing her expertise in flow cytometry, David Humiston for critically reading the manuscript, and Roseanne Perry for secretarial assistance.

References

Amit, M., Carpenter, M. K., Inokuma, M. S., Chiu, C. P., Harris, C. P., Waknitz, M. A., Itskovitz-Eldor, J., and Thomson, J. A. (2000). Clonally derived human embryonic stem cell lines maintain pluripotency and proliferative potential for prolonged periods of culture. *Dev. Biol.* **227**(2), 271–278.

Colter, D. C., Class, R., DiGirolamo, C. M., and Prockop, D. J. (2000). Rapid expansion of recycling stem cells in cultures of plastic-adherent cells from human bone marrow. *Proc. Natl. Acad. Sci. USA* **97**(7), 3213–3218.

Curl, C. L., Harris, T., Harris, P. J., Allman, B. E., Bellair, C. J., Stewart, A. G., and Delbridge, L. M. (2004). Quantitative phase microscopy: A new tool for measurement of cell culture growth and confluency *in situ*. *Pflugers Arch.* **448**(4), 462–468.

Deasy, B. M., Gharaibeh, B. M., Pollett, J. B., Jones, M. M., Lucas, M. A., Kanda, Y., and Huard, J. (2005). Long-term self-renewal of postnatal muscle-derived stem cells. *Mol. Biol. Cell* **16**(7), 3323–3333.

Deasy, B. M., Jankowski, R. J., Payne, T. R., Cao, B., Goff, J. P., Greenberger, J. S., and Huard, J. (2003). Modeling stem cell population growth: Incorporating terms for proliferative heterogeneity. *Stem Cells* **21**(5), 536–545.

Gilmore, G. L., DePasquale, D. K., Lister, J., and Shadduck, R. K. (2000). *Ex vivo* expansion of human umbilical cord blood and peripheral blood CD34(+) hematopoietic stem cells. *Exp. Hematol.* **28**(11), 1297–1305.

Hayflick, L., and Moorhead, P. S. (1961). The serial cultivation of human diploid cell strains. *Exp. Cell Res.* **25**, 585–621.

Jiang, Y., Henderson, D., Blackstad, M., Chen, A., Miller, R. F., and Verfaillie, C. M. (2003). Neuroectodermal differentiation from mouse multipotent adult progenitor cells. *Proc. Natl. Acad. Sci. USA* **100**(Suppl. 1), 11854–11860.

Jiang, Y., Jahagirdar, B. N., Reinhardt, R. L., Schwartz, R. E., Keene, G. D., Ortiz-Gonzales, X. R., Reyes, M., Lenvik, T., Lund, T., Blackstad, M., Du, J., Aldrich, S., *et al*. (2002). Pluripotency of mesenchymal stem cells derived from adult marrow. *Nature* **418**, 41–49.

Péault, B., Rudnicki, M., Torrente, Y., Cossu, G., Tremblay, J. P., Partridge, T., Gussoni, E., Kunkel, L. M., and Huard, J. (2007). Stem and progenitor cells in skeletal muscle development, maintenance, and therapy. *Mol. Ther.* **15**(5), 867–877.

Piacibello, W., Sanavio, F., Garetto, L., Severino, A., Bergandi, D., Ferrario, J., Fagioli, F., Berger, M., and Aglietta, M. (1997). Extensive amplification and self-renewal of human primitive hematopoietic stem cells from cord blood. *Blood* **89**(8), 2644–2653.

Pittenger, M. F., MacKay, A. M., Beck, S. C., Jaiswal, R. K., Douglas, R., Mosca, J. D., Moorman, M. A., Simonetti, D. W., Craig, S., and Marshak, D. R. (1999). Multilineage potential of adult human mesenchymal stem cells. *Science* **284**, 143–147.

Qu-Petersen, Z., Deasy, B. M., Jankowski, R., Ikezawa, M., Cummins, J., Pruchnic, R., Cao, B., Mytinger, J., Gates, C., Wernig, A., and Huard, J. (2002). Identification of a novel population of muscle stem cells in mice: Potential for muscle regeneration. *J. Cell Biol.* **157**(5), 851–864.

Reya, T., Morrison, S. J., Clarke, M. F., and Weissman, I. L. (2001). Stem cells, cancer, and cancer stem cells. *Nature* **414**(6859), 105–111.

Reyes, M., Lund, T., Lenvik, T., Aguiar, D., Koodie, L., and Verfaillie, C. M. (2001). Purification and *ex vivo* expansion of postnatal human marrow mesodermal progenitor cells. *Blood* **98**(9), 2615–2625.

Rubin, H. (2002). The disparity between human cell senescence *in vitro* and lifelong replication *in vivo*. *Nat. Biotechnol.* **20**(7), 675–681.

Spangrude, G. J., Heimfeld, S., and Weissman, I. L. (1988). Purification and characterization of mouse hematopoietic stem cells. *Science* **241**(4861), 58–62.

Tagliafico, E., Brunelli, S., Bergamaschi, A., De Angelis, L., Scardigli, R., Galli, D., Battini, R., Bianco, P., Ferrari, S., Cossu, G., and Ferrari, S. (2004). TGFbeta/BMP activate the smooth muscle/bone differentiation programs in mesoangioblasts. *J. Cell Sci.* **117**(Pt. 19), 4377–4388.

Vilquin, J. T., Marolleau, J. P., Sacconi, S., Garcin, I., Lacassagne, M. N., Robert, I., Ternaux, B., Bouazza, B., Larghero, J., and Desnuelle, C. (2005). Normal growth and regenerating ability of myoblasts from unaffected muscles of facioscapulohumeral muscular dystrophy patients. *Gene Ther.* **12**(22), 1651–1662.

Xu, C., Inokuma, M. S., Denham, J., Golds, K., Kundu, P., Gold, J. D., and Carpenter, M. K. (2001). Feeder-free growth of undifferentiated human embryonic stem cells. *Nat. Biotechnol.* **19**(10), 971–974.

Zheng, B., Cao, B., Crisan, M., Sun, B., Li, G., Logar, A., Yap, S., Pollett, J. B., Drowley, L., Cassino, T., Gharaibeh, B., Deasy, B. M., *et al.* (2007). Prospective identification of myogenic endothelial cells in human skeletal muscle. *Nat. Biotechnol.* **25**(9), 1025–1034.

Zuk, P. A., Zhu, M., Mizuno, H., Huang, J., Futrell, J. W., Katz, A. J., Benhaim, P., Lorenz, H. P., and Hedrick, M. H. (2001). Multilineage cells from human adipose tissue: implications for cell-based therapies. *Tissue Eng.* **7**, 211–228.

CHAPTER 14

Isolation and Culture of Colon Cancer Stem Cells

Patrizia Cammareri,* Ylenia Lombardo,[†] Maria Giovanna Francipane,[†] Sebastino Bonventre,[†] Matilde Todaro,[†] and Giorgio Stassi[†]

*Department of GENURTO
University of Palermo
129–90127 Palermo
Italy

[†]Department of Surgical and Oncological Sciences
University of Palermo
5–90127 Palermo
Italy

    Abstract
 I. Introduction
    A. Colon Cancer Stem Cells
 II. Identification of CSCs in Colon Cancer Tissue
 III. Isolation and Propagation of Colon CSCs
    A. Tissue Digestion
    B. Culture of Colon CSCs
    C. MACS Technology
    D. FACS Technology
 IV. Discussion
    References

Abstract

Cancer stem cells (CSCs) resemble normal stem cells in several ways. Both cell types are self-renewing and when they divide, one of the daughter cells differentiates while the other retains stem cell properties, including the ability to divide in the same way again. CSCs have been demonstrated to exist in several solid tumors, including colon carcinoma; these cells are able to initiate and sustain tumor growth.

There are essentially three different methods to isolate CSCs: establishment culture, the MACS (magnetic cell sorting) technology, and the FACS (fluorescence-activated cell sorting) technology.

I. Introduction

The first experimental evidence of the existence of cancer stem cells (CSCs) dates to the late 1990s. Bonnet and Dick isolated a subpopulation of leukemic cells that expressed a specific surface marker CD34, but lacked the CD38 marker. The $CD34^+/CD38^-$ subpopulation was capable of initiating leukemia in nonobese diabetic, severe-combined immunodeficient (NOD/SCID) mice, contrary to both $CD34^+/CD38^+$ and $CD34^-$ fractions (Bonnet and Dick, 1997).

The CSCs are defined by their stem cell-like properties. Like normal adult stem cells, CSCs have the ability to perpetuate themselves through self-renewal and to differentiate into the different cell types of a tumor. Stem cells can accomplish these two tasks throughout asymmetric division by which each stem cell divides to generate one daughter with a stem cell fate and one daughter that differentiates (Clevers, 2005). CSCs have also been demonstrated to exist in breast and brain tumors (Al-Hajj et al., 2003; Singh et al., 2003); moreover, recent works extend this principle to other neoplasia such as prostate, melanoma, pancreatic and colon carcinomas, and melanoma. (Grichnik et al., 2006; O'Brien et al., 2007; Olempska et al., 2007; Ricci-Vitiani et al., 2007; Richardson et al., 2004).

A. Colon Cancer Stem Cells

The colon is a very structured organ. The intestinal epithelium is folded to form a number of invaginations or crypts (Fig. 1). These crypts increase the surface area occupied by the epithelium, allowing efficient adsorption of nutritional compounds.

Now, it is widely accepted that the colon crypt cell replacement is achieved by multipotent stem cells located at the base of each crypt. Approximately, four to six stem cells divide asymmetrically, move upward, toward the lumen of the colon, and differentiate into the three epithelial cell types: enterocytes, globet cells, and entero-endocrine cells.

Colon stem cells could be identified through the expression of specific markers. Musashi 1 (Msi-1) protein has been identified as the first colon stem cell marker. Msi-1 is an RNA-binding protein initially identified in *Drosophila* where it plays an essential role in the early asymmetric divisions of the sensory organ precursor cells (Nakamura et al., 1994). The mammalian homologue is likely involved in the early asymmetric divisions that give rise to differentiated cells from neural stem cells or progenitor cells (Kaneko et al., 2000). Msi-1 suppresses its mRNA target,

- wash in dH$_2$O;
- heat slides for antigen retrieval in 10 mM sodium citrate in a microwave oven for 1 min at 450 W followed by 5 min at 100 W;
- equilibrate sections at room temperature for at least 30 min;
- after rinsing in dH$_2$O, incubate sections in 3% H$_2$O$_2$ for 5 min to suppress endogenous peroxidase activity;
- wash twice in a saline buffer;
- incubate sections with 10% human serum for 20 min to block unspecific staining;
- eliminate the excess of serum;
- incubate with specific antibodies against CD133/2 (AC141) or against other stemness markers, overnight at 4 °C or 1 h at 37 °C;
- wash twice in a saline buffer;
- incubate with enzyme-labeled specific secondary antibody 1 hour at 37 °C;
- reveal the staining using an appropriate substrate-chromogen solution.
- mount the slides with an appropriate mounting medium.

III. Isolation and Propagation of Colon CSCs

There are essentially three different methods to isolate and propagate colon CSCs from the tissue digest: establishment culture, the MACS (magnetic cell sorting) technology, and the FACS (fluorescence-activated cell sorting) technology. Each of these methods is described next, followed by a discussion of their advantages and disadvantages.

A. Tissue Digestion

To purify colon CSCs, tumor samples have to be dissociated by mechanical and enzymatic methods. Because of its anatomic site, colon tissues are prone to bacterial contamination once cultivated. Therefore, after surgical resection it is opportune to extensively wash the colon sample in phosphate-buffered saline (PBS) or Dulbecco's modified Eagle medium (DMEM) containing antibiotics and antimycotics to remove any contaminating factor. After removing fatty and necrotic parts, tissue has to be cut into small pieces with sterile scissors and then digested with a mix of proteolytic enzymes.

After surgical resection:

- extensively wash the colon sample in PBS or DMEM/F12 containing 500 units/ml penicillin, 500 µg/ml streptomycin, 1.25 µg/ml amphotericin B;
- incubate cancer tissue in the above medium overnight at 4 °C.

The day after:

- remove fatty and necrotic parts;
- cut the tissue into small pieces with sterile scissors;

- digest with 1.5 mg/ml collagenase and 20 µg/ml hyaluronidase with gentle agitation for 1 h at 37 °C in DMEM/F12;
- allow the fragments to sediment and collect the supernatant in a new tube;
- centrifuge at 1200 rpm for 5 min;
- discard the supernatant;
- resuspend the pellet in an appropriate volume ($\sim 10^5$ cells/ml) of specific medium (see below).

B. Culture of Colon CSCs

The digest obtained from mechanical and enzymatic dissociation is heterogeneous inasmuch as it is composed of different cell types. To select colon CSCs, it is best to culture the digest in a specific medium into ultralow adhesion flasks. The stem cell medium contains DMEM/F12 supplemented with several factors that favor stem cells growth, in particular, the addition of basic fibroblast growth factor (bFGF) and epidermal growth factor (EGF) and the lack of fetal bovine serum (FBS) opt for immature tumor cells, while nonmalignant or differentiated cells are negatively selected and die through anoikis.

In these culture conditions, surviving immature tumor cells slowly proliferate, grow as nonadherent clusters, termed tumor spheres. The formation of such spheres, containing about 100 cells, will take at least 1–2 months (Fig. 3).

Unlike freshly isolated primary differentiated colon carcinoma cells that die after 5 days of culture in soft-agar, disaggregated colon spheroids survive and maintain the growth potential under these conditions, confirming that spheroid cultures are anchorage independent (Fig. 4).

Sphere-forming cells can be expanded by mechanical dissociation of spheres, followed by re-plating of single cells and residual small cell aggregates in complete fresh medium.

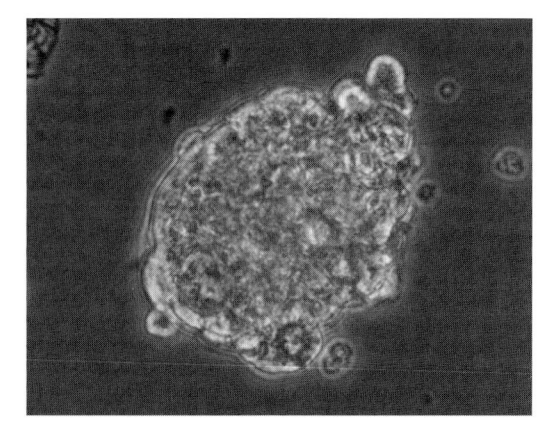

Fig. 3 Phase contrast photo of a colon cancer sphere (40×). (See Plate no. 23 in the Color Plate Section.)

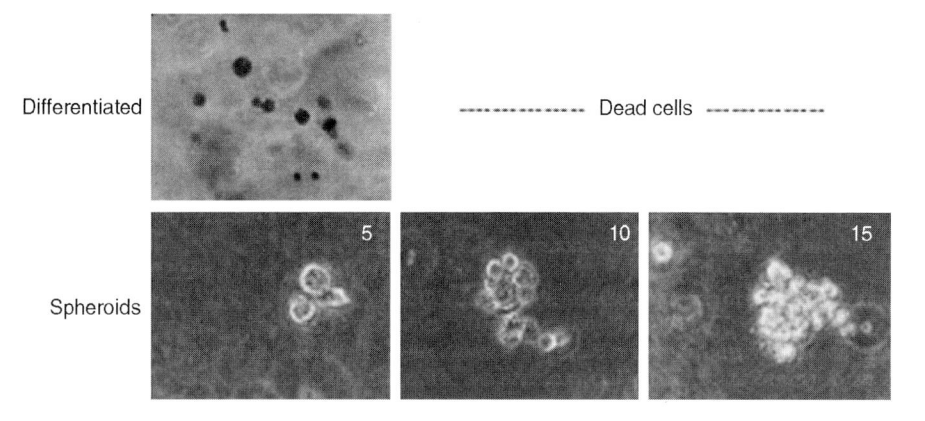

Fig. 4 Morphology of anchorage-independent growth performed in soft-agar up to 15 days cultures from primary colon cancer cells and disaggregated colon cancer sphere cells (40×).

All of the dissociated CD133$^+$ tumor cells show the capacity to form other CD133$^+$ tumor spheres with multiple passages. To characterize CSCs, whole free floating spheres of dissociated cells can be cytospun, fixed in 4% paraformaldehyde and stained for stemness markers. Finally, differentiated cells can be obtained from tumor spheres after growth factors removal and addition of serum or using three-dimensional (3D) culture technique.

1. Culture Medium

   The stem cell medium contains DMEM/F12 supplemented with:
   - 6 mg/ml glucose;
   - 1 mg/ml NaHCO$_3$;
   - 5 mM HEPES;
   - 2 mM L-glutamine;
   - 4 µg/ml Heparin;
   - 4 mg/ml BSA;
   - 10 ng/ml bFGF;
   - 20 ng/ml EGF;
   - 100 µg/ml apotrasferrin;
   - 25 µg/ml insulin;
   - 9.6 µg/ml putrescin;
   - 30 nM sodium selenite anhydrous;
   - 20 nM progesterone.

2. Dissociation

- Centrifuge CSCs at 800 rpm for 5 min;
- discard the supernatant;
- resuspend the pellet in 3 mM ethylenediaminetetraacetic acid (EDTA) plus 0.05 mM dithiothreitol (DTT) in PBS or in 0.05% trypsin plus 0.02% EDTA until the cells are not well dissociated;
- incubate at 37 °C for 5 min;
- centrifuge at 800 rpm for 5 min;
- discard the supernatant;
- resuspend the pellet in the specific medium.

3. Differentiation

For tumor spheres differentiation, dissociated cells have to be cultured in DMEM-High glucose without EGF and bFGF with addition of 10% FBS into a collagen-coated flask. The collagen forms a thin layer of gel that favors cellular adhesion. After one day of culture, floating undifferentiated cells attach to the flask and gradually differentiate into adherent cells with a phenotype similar to that of differentiated epithelial colon cells (Fig. 5). Alternatively, colon CSCs can be cultured in the same medium in Matrigel, which represents a reconstituted 3D culture system (Fig. 6).

To perform a 3D cell culture:

- Mix 30,000 dissociated cells into liquid 50% (vol/vol) Matrigel solution: 10× DMEM-High into Matrigel (growth factor-reduced, BD Biosciences, Bedford, MA) at 1:9 volume ratio (DMEM-Matrigel) at 4 °C;
- put the cell suspension in 6-well plates and incubate at 37 °C for 1 h;
- after the Matrigel solidification, add 1 ml of DMEM-High supplemented with 10% FBS on the top of Matrigel.

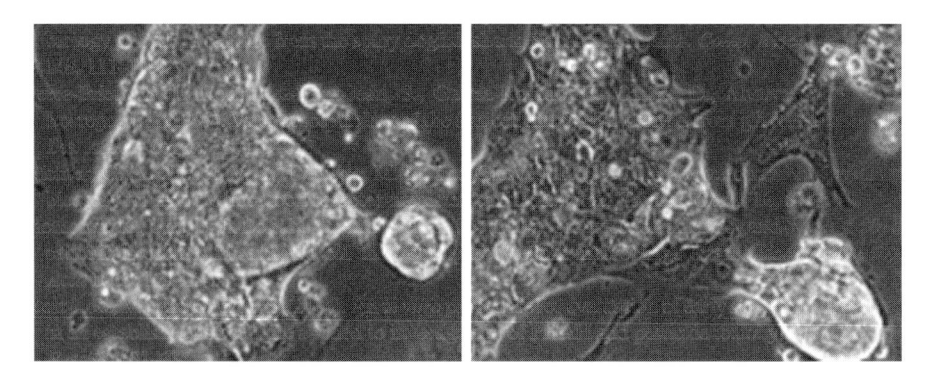

Fig. 5 Gross morphology of differentiated colorectal cancer spheres after 7 days (left) and 14 days (right) of exposure to 10% FBS (40×).

14. Purification of Colon Cancer-Initiating Cells 319

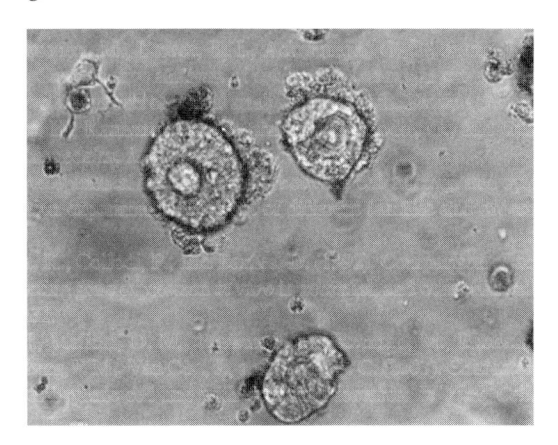

Fig. 6 Inverted phase contrast microscopy of purified colon cancer spheres after 15 days of culture in Matrigel (40×).

During *in vitro* differentiation, CD133$^+$ cells are able to generate colonies organized in crypt-like structures. Of note, cells that are about to differentiate display gradually acquisition of colon epithelia markers, such as CK20, with the concomitant reduction of CD133 stem cell marker expression.

C. MACS Technology

MACS technology is one of the most often used techniques to select cells based on the expression of a specific stemness marker, such as CD133, in a fast and specific way directly after mechanical and enzymatic dissociation of colon carcinoma tissue (www.miltenyibiotec.com).

Using this technology, cells of interest are separated by a positive selection strategy.

For an optimal performance, it is important to obtain a single-cell suspension by dissociation before magnetic separation (see above).

Target cells are specifically labeled with super-paramagnetic MACS MicroBeads conjugated with a highly specific monoclonal antibody anti-CD133/1 (AC133, mouse IgG1, Miltenyi) or other stemness markers to isolate highly pure cells that express the corresponding antigen.

MACS MicroBeads are about one million times smaller in volume than an eukaryotic cell and are biodegradable decomposing when cells are cultured.

After magnetic labeling, cells are passed through a separation column which is placed in a strong permanent magnet. The magnetically labeled cells are retained in the column and separated from the unlabeled cells, which pass through. After removing the column from the magnetic field, the fraction can be eluted with a specific buffer (Fig. 7).

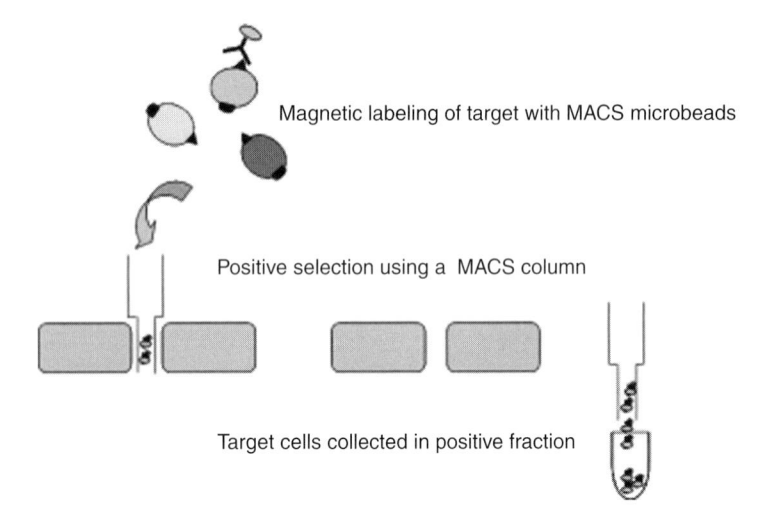

Fig. 7 Scheme of CD133⁺ selection by MACS technology (www.miltenyibiotec.com).

It is possible to control the quality of MACS sorting by flow cytometry analysis using an antibody against the different epitope CD133/2 (AC141, mouse IgG1 or 293C3, mouse IgG2b, Miltenyi) inasmuch as the conjugated microbeads recognize the glycosylated epitope CD133/1 (AC133).

After magnetic sorting, cell viability is assessed by trypan blue exclusion.

The positively selected cells can be immediately used for culturing in the specific stem cell medium or for further downstream applications.

CD133⁺ cells are able to exponentially grow *in vitro* as undifferentiated tumor spheres in serum-free medium for more than one year.

D. FACS Technology

An alternative method to isolate CD133⁺ cell subpopulation is FACS flow cytometric sorting. FACS is a type of flow cytometry capable of sorting cells in suspension into two or more containers with high purities and recoveries. Fluorochromes with different emission wavelengths can be used concurrently, allowing for multiparameter separations.

In a direct immunofluorescence staining, cells are incubated with an antibody anti-CD133/1 (AC133) or other stemness markers directly conjugated to a fluorochrome (e.g., phycoerythrin or fluorescein isothiocyanate); in an indirect immunofluorescence staning, the unlabeled antibody is applied directly to the cells that are then treated with a fluorochrome-conjugated secondary antibody.

Labeled cells are analyzed and sorted in the flow cytometer where the sample is hydrodynamically focused to a tiny stream of single cells (Fig. 8).

These cells can be individually detected and the cellular pool obtained can be cultivated in the specific stem cell medium or used for further analysis.

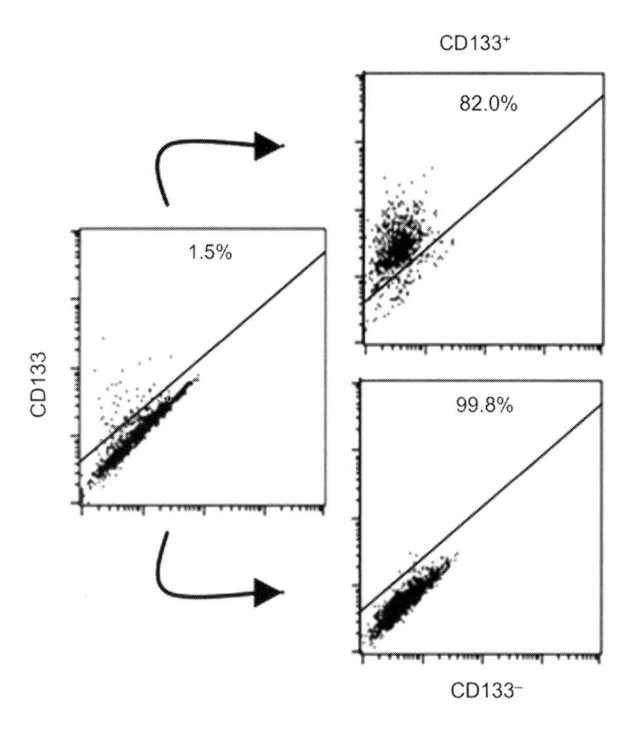

Fig. 8 Flow cytometry profiles of CD133 expression before (left) and after sorting of positive (right, top) and negative (right, bottom) colon cancer cells (Ricci-Vitiani et al., 2007).

The disadvantages of FACS are the low viability of recovered cells, the high cost, and difficulty of using a complex equipment.

1. Direct Staining Protocol

- Harvest and wash the cells, and adjust cell suspension to a concentration of 1×10^6 to 5×10^6 cells/ml in ice cold PBS;
- add labeled antibody against CD133/1 (AC133) or other stemness markers diluted in PBS;
- incubate for at least 30 min at room temperature or overnight at 4 °C;
- wash the cells three times by centrifuging at 400 × g for 5 min and resuspending in 500 µl to 1 ml of ice cold PBS. Keep the cells in the dark on ice or at 4 °C until your scheduled time for analysis;
- for best results, analyze the cells by FACS immediately;
- optional: to verify the purity of the CD133-sorted population, cells should be analyzed by flow cytometry using a CD133/2 (AC141 or 293C3) conjugated antibody (Fig. 8).

2. Indirect Staining Protocol

- Harvest and wash the cells, and determine the total cell number;
- resuspend the cells to $\sim 1 \times 10^7$ to 5×10^7 cells/ml in ice cold PBS;
- add 100 µl of cell suspension to each tube;
- add the antibody against CD133/1 (AC133) or other stemness markers diluted in PBS;
- incubate for at least 30 min at room temperature in the dark or overnight at 4 °C;
- wash the cells three times in PBS by centrifuging at $400 \times g$ for 5 min and resuspending in ice cold PBS;
- dilute the fluorochrome-labeled secondary antibody in PBS;
- incubate for at least 20–30 min at room temperature in the dark;
- wash the cells three times in PBS by centrifuging at $400 \times g$ for 5 min and resuspending in 500 µl to 1 ml ice cold PBS;
- store the cell suspension immediately at 4 °C in the dark;
- for best results, analyze the cells by FACS immediately;
- optional: to verify the purity of the sorted population, cells should be analyzed by flow cytometry using a CD133/2 (AC141 or 293C3) conjugated antibody (Fig. 8).

IV. Discussion

This chapter has provided important tools in the isolation and propagation of colon CSCs. As in myeloid leukemia, breast and brain tumors, a rare population of cells with stemness features, responsible for tumor formation and maintenance, has been recently characterized also in colon carcinoma.

Here, supported in part by our own experience, we have provided a broad view of the main methods for the purification, culture, and propagation of colon CSCs. To this purpose, the number of cells required for each test, the incubation times, the temperature conditions in addition to the precise volume of all the necessary reagents have been indicated.

Briefly, colon digest is cultivated in a specific serum-free medium which contains growth factors and hormones, all compounds that allow the selection of CSCs and the formation of characteristic spheres. Moreover, colon CSCs can be identified and isolated through the CD133 or other markers' expression. Particularly, CD133$^+$ cells can be purified by two different techniques, the MACS and the FACS technologies, as discussed in the text.

The possibility of isolating and expanding CSCs *in vitro* has considerable therapeutic implications. The cells obtained can be characterized at the molecular level and this information can be used to develop new therapeutic strategies more selectively than the existing ones. These latter are based on the use of anticancer agents, a class of cytotoxic compounds able to trigger the intrinsic apoptotic pathway only in dividing cancer cells, often sparing CSCs that seem to have a slow rate of cell turnover apart from overexpressing ABC (ATP-binding cassette) trasporters and antiapoptotic factors.

Recently, it has been demonstrated that colon CSCs refractoriness is mediated by autocrine production of IL-4 and that its blockage sensitizes them to death stimuli through the down-regulation of antiapoptotic proteins (Todaro *et al.*, 2007).

The possibility of isolating and propagating CSCs *in vitro* will allow further investigations of selectively activated signaling transduction pathways in the population of CSCs to better understand the mechanism underlying their survival strategy and define novel approaches in anti-tumor therapy.

Acknowledgments

This work was supported by grants from AIRC to G. Stassi and M. Todaro, Programmi di Ricerca Scientifica di Rilevante Interesse Nazionale (PRIN) 2005 prot. 2005052122 to G. Stassi. Y. Lombardo is a AIRC fellowship recipient. M.G. Francipane is a PhD student in Immunopharmacology at the University of Palermo.

References

Al-Hajj, M., Wicha, M. S., Benito-Hernandez, A., Morrison, S. J., and Clarke, M. F. (2003). Prospective identification of tumorigenic breast cancer cells. *Proc. Natl. Acad. Sci. USA* **100,** 3983–3988.

Barker, N., van Es, J. H., Kuipers, J., Kujala, P., van den Born, M., Cozijnsen, M., Haegebarth, A., Korving, J., Begthel, H., Peters, P. J., and Clevers, H. (2007). Identification of stem cells in small intestine and colon by marker gene Lgr5. *Nature* **449,** 1003–1007.

Bonnet, D., and Dick, J. E. (1997). Human acute myeloid leukemia is organized as a hierarchy that originates from a primitive hematopoietic cell. *Nat. Med.* **3,** 730–737.

Clevers, H. (2005). Stem cells, asymmetric division and cancer. *Nat. Genet.* **37,** 1027–1028.

Dalerba, P., Dylla, S. J., Park, I. K., Liu, R., Wang, X., Cho, R. W., Hoey, T., Gurney, A., Huang, E. H., Simeone, D. M., Shelton, A. A., Parmiani, G., *et al.* (2007). Phenotypic characterization of human colorectal cancer stem cells. *Proc. Natl. Acad. Sci. USA* **104,** 10158–10163.

Grichnik, J. M., Burch, J. A., Schulteis, R. D., Shan, S., Liu, J., Darrow, T. L., Vervaert, C. E., and Seigler, H. F. (2006). Melanoma, a tumor based on a mutant stem cell? *J. Invest. Dermatol.* **126,** 142–153.

Imai, T., Tokunaga, A., Yoshida, T., Hashimoto, M., Mikoshiba, K., Weinmaster, G., Nakafuku, M., and Okano, H. (2001). The neural RNA-binding protein Musashi1 translationally regulates mammalian numb gene expression by interacting with its mRNA. *Mol. Cell. Biol.* **21,** 3888–3900.

Kaneko, Y., Sakakibara, S., Imai, T., Suzuki, A., Nakamura, Y., Sawamoto, K., Ogawa, Y., Toyama, Y., Miyata, T., and Okano, H. (2000). Musashi1: An evolutionarily conserved marker for CNS progenitor cells including neural stem cells. *Dev. Neurosci.* **22,** 139–153.

Nakamura, M., Okano, H., Blendy, J. A., and Montell, C. (1994). Musashi, a neural RNA-binding protein required for *Drosophila* adult external sensory organ development. *Neuron* **13,** 67–81.

Nishimura, S., Wakabayashi, N., Toyoda, K., Kashima, K., and Mitsufuji, S. (2003). Expression of Musashi-1 in human normal colon crypt cells: A possible stem cell marker of human colon epithelium. *Dig. Dis. Sci.* **48,** 1523–1529.

O'Brien, C. A., Pollett, A., Gallinger, S., and Dick, J. E. (2007). A human colon cancer cell capable of initiating tumour growth in immunodeficient mice. *Nature* **445,** 106–110.

Olempska, M., Eisenach, P. A., Ammerpohl, O., Ungefroren, H., Fandrich, F., and Kalthoff, H. (2007). Detection of tumor stem cell markers in pancreatic carcinoma cell lines. *Hepatobiliary Pancreat. Dis. Int.* **6,** 92–97.

Ricci-Vitiani, L., Lombardi, D. G., Pilozzi, E., Biffoni, M., Todaro, M., Peschle, C., and De Maria, R. (2007). Identification and expansion of human colon-cancer-initiating cells. *Nature* **445,** 111–115.

Richardson, G. D., Robson, C. N., Lang, S. H., Neal, D. E., Maitland, N. J., and Collins, A. T. (2004). CD133, a novel marker for human prostatic epithelial stem cells. *J. Cell Sci.* **117,** 3539–3545.

Salven, P., Mustjoki, S., Alitalo, R., Alitalo, K., and Rafii, S. (2003). VEGFR-3 and CD133 identify a population of CD34 + lymphatic/vascular endothelial precursor cells. *Blood* **101,** 168–172.

Singh, S. K., Clarke, I. D., Terasaki, M., Bonn, V. E., Hawkins, C., Squire, J., and Dirks, P. B. (2003). Identification of a cancer stem cell in human brain tumors. *Cancer Res.* **63,** 5821–5828.

Todaro, M., Perez Alea, M., Di Stefano, A., Cammareri, P., Vermeulen, L., Iovino, F., Tripodo, C., Russo, A., Gulotta, G., Medema, J., and Stassi, G. (2007). Colon cancer stem cells dictate tumor growth and resist cell death by production of interleukin-4. *Cell Stem Cell* **1,** 389–402.

Todaro, M., Perez Alea, M., Scopelliti, A., Medema, J. P., and Stassi, G. (2008). IL-4 mediated drug resistance in colon cancer stem cells. *Cell Cycle* **7,** 309–313.

Uchida, N., Buck, D. W., He, D., Reitsma, M. J., Masek, M., Phan, T. V., Tsukamoto, A. S., Gage, F. H., and Weissman, I. L. (2000). Direct isolation of human central nervous system stem cells. *Proc. Natl. Acad. Sci. USA* **97,** 14720–14725.

Yin, A. H., Miraglia, S., Zanjani, E. D., Almeida-Porada, G., Ogawa, M., Leary, A. G., Olweus, J., Kearney, J., and Buck, D. W. (1997). AC133, a novel marker for human hematopoietic stem and progenitor cells. *Blood* **90,** 5002–5012.

CHAPTER 15

Isolation and Establishment of Human Tumor Stem Cells

Penelope E. Roberts

Raven Bio-technologies, Inc.
South San Francisco, California 94080

    Abstract
I. Introduction
II. Rationale
III. Methods
    A. Isolation and Establishment
    B. Tissue Treatment
    C. Preparation of Cells from Xenograft
    D. Cell Surface Phenotype
    E. Medium Selection for the Growth of Cancer Stem Cells
    F. Effect of Serum on Isolation and Establishment and Growth of CSCs
    G. Immunohistochemistry
    H. Renal Capsule Grafting
IV. Results
    A. Tumor-Forming Capability
    B. Expression of Cell Surface Markers
    C. Importance of Growth Conditions in Maintaining the CSCs Phenotype
V. Discussion
VI. Materials
    References

Abstract

Current cancer therapies are based on the ability to inhibit the growth of rapidly dividing cells, the majority of which constitute the tumor. Although for decades, sporadic literature has posited the existence of cancer stem cells (CSCs), only recently has this type of cell been isolated and characterized from solid tumors. Like stem cells from their normal counterpart, CSCs are a rare population that can reconstitute a new tumor with similar composition and phenotype to the tumor of origin. These CSCs represent a small subset of the original tumor, grow indefinitely *in vitro*, and can form tumors in animals from a very few cells. The cells are slow cycling, capable of self-renewal and give rise to daughter cells that are either self-renewing and pluripotent or transit amplifying, and terminally differentiated. Thus far, CSCs have been isolated from only a small number of tumor types. In most instances, the cells are obtained using selection of, and enrichment for, cells with prospectively identified cell surface markers (Al-Hajj M, *et al.*, 2003). This yields a very limited number of cells, and in many cases these cells cannot be cultured. There is a need for a method for isolation, purification, and expansion of stem cells from a greater spectrum of tumors. There is also evidence for "…a link between normal stem cell regulation and the control of cancer stem cells" (NCI Think Tanks in Cancer Biology, Executive Summary of the Tumor Stem Cell and Self-renewal Genes Think Tank1). We present here a strategy for the isolation and establishment of tumor cell lines that represent a minority of cells in the original tumor. They have the ability to grow indefinitely *in vitro*, form tumors in mice from less than 100 cells, and share many of the growth requirements and cell surface antigens of normal tissue stem cells from which they may arise.

I. Introduction

Historically, the use of viable, well-characterized human cell lines, both normal and neoplastic, has proved invaluable in biomedical research. The problem has been that, over time, these cultures have often not reflected the characteristics and phenotype of their tissue of origin. The cell is the smallest living unit of an organism, yet it embodies all the complexity of the entire organism. To quote Max Delbruck, "Any living cell carries with it the experience of a billion years of experimentation by its ancestors." To capture this complexity in a monoculture requires culture conditions that most closely approximate the *in vivo* conditions from which the cultures were derived. While this may sound like belaboring the obvious, it has only been in the last few decades that *in vitro* culture conditions have developed to more accurately reflect the paracrine, hemocrine, and autocrine environment of their tissues of origin and in doing so, to maintain their phenotype *in vitro* (Gerst *et al.*, 1986; Mather and Sato, 1979b). This has been due in large part to the development of media formulations and defined culture conditions, including research in the use of attachment factors, hormones, and other growth and differentiation components (Lang, S.H., *et al.*, 2001). Stem cells, whether normal fetal or normal adult, or cancer stem cells (CSCs), represent a special challenge for media formulation, since the maintenance of a stem-like phenotype is often dependent upon a "niche," with special hormone and growth factors present (Mather, *et al.*, 1984). The control of division and differentiation of cancer and normal stem cells may thus be linked via regulation of their environment. This would suggest that a medium developed for normal stem cell maintenance might share properties with that required for maintenance and growth of a CSC from the same tissue.

Research in cancer biology depends in large part on the ability to study cancer cells *in vitro*. The majority of human tumor cell lines that are in widespread use by researchers today have been established over several decades and have proven to be extremely useful tools. Historically, however, the establishment of a permanent cell line from a given tumor tissue was a rare event, with many tumors failing to grow *in vitro*, and those that have succeeded may not represent the current focus of tumor biology today: the tumor stem cell.

II. Rationale

There is ever increasing evidence that supports the hypothesis that cancers arise from CSCs, and that these CSCs, in fact, arise from the normal tissue progenitor, that is, the somatic stem cell.

In our lab, we have focused over the past 15 years on establishing tissue progenitor cell lines from several species, starting with cells which derive from normal rodent fetal tissue (Stephen JP, *et al.*, 1999). These are arrested at a particular (undifferentiated, or partially differentiated) stage of development (Li *et al.*, 1996a,b, 1997; NCI Think Tanks in Cancer Biology, 2004; Roberts *et al.*, 1990, 1992). More recently, we have established serum-free media that select for the growth and survival of human progenitor (or tissue stem) cells *in vitro* (Li and Mather, 2003; Li *et al.*, 2003; Roberts and Mather, 2002). In addition, we have demonstrated that a defined medium, derived to support the growth of a murine melanoma cell line, for example, is capable of selectively supporting the survival and growth of only the metastatic melanoma cells from a number of whole organs placed in culture (Mather and Sato, 1979b). It is generally true that defined media derived for the growth of one cell type will support the growth of a similar cell type from different organs (Mather and Sato, 1979a) or even from different species.

We have used this approach to derive cancer stem (or progenitor) cells, by exploiting the growth and differentiation requirements for their normal tissue stem or progenitor counterparts, and then determining whether these conditions, with some modification, could select for the CSC.

III. Methods

A. Isolation and Establishment

Following surgical resection, and in accordance with Institutional Review Board-approved guidelines, tumor tissue was aseptically placed in 50 ml conical tubes containing F12/DME plus 100 μg/ml gentamicin, shipped on wet ice, and reached our facility, usually within 24 h, or the same day of excision. All tissue was handled using safety precautions for biologics including the use of a Class II Biological Safety Cabinet for all procedures (U.S. Department of Health and Human Services Public Health Service Centers for Disease Control and Prevention and National Institute of Health (2000)).

B. Tissue Treatment

1. Depending on how easy the tumor was to mince using iris scissors, the tissue was either (a) cut into <1 mm pieces and no enzymatic digestion was necessary (i.e., rectal cancer) or (b) subjected to digestion with a collagenase/dispase solution (50 µl/ml of a 10% w/v stock) containing 10 µl/ml of DNase (2 mg/ml stock in phosphate-buffered saline, PBS) and 200 µg/ml soybean trypsin inhibitor (STI), incubated at 37 °C for 20 minutes-2 h, depending on the tissue, with periodic pipetting to loosen tissue clumps.

2. When small aggregates of cells could be easily observed in suspension (Fig. 1), enzymatic digestion was stopped by initially washing the cell suspension by centrifugation in 10 ml F12/DME for 5 minutes at 800 rpm using a GH3.8 rotor in a Beckman Allegra 6R centrifuge (all subsequent centrifugations were carried out using this rotor) The preferred enzymatic digestion mix for different solid tumor types is shown in Table I.

3. Subsequently, the pellet was resuspended in 4 ml F12/DME and layered over 3 ml of a 5% BSA (Bovine serun albumin) solution.

4. This was then centrifuged for 6 minutes at 1100 rpm, followed by a final wash by centrifugation of the resulting pellet in 10 ml F12/DME for 5 minutes at 800 rpm.

5. This final pellet was then resuspended in serum-free defined growth media and aliquoted into tissue culture dishes precoated with laminin, fibronectin, or collagen.

6. Any tissue remaining in suspension after 48 h was transferred to new precoated dishes, and the medium renewed on the original plate.

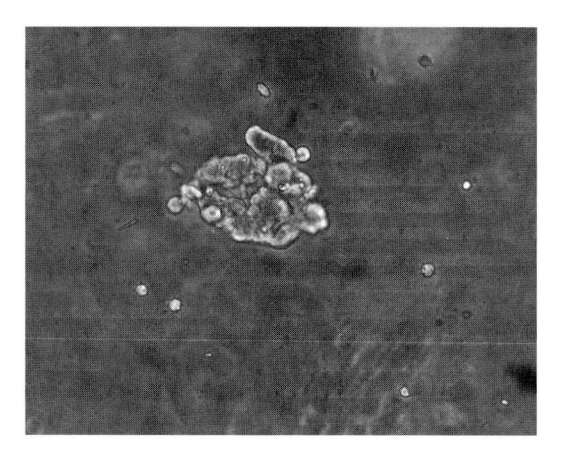

Fig. 1 Small aggregates of cells indicate when enzymatic dissociation should be stopped.

Table I
Parameters for Dissociation/Matrix, and Adherence Properties of Various Tumor Cell Types

Designation	Tumor type	Enzymatic dissociation	Matrix	Adherent/suspension
PRCA629	Prostate	C/D	Laminin	Adherent
CRCA0404	Colon (right ascending)	C/D	Fbn	Adherent/suspension
CRCA1115	Colon	C/D	Fbn	Adherent/suspension
MCL	Mantle cell lymphoma	C/D	Fbn	Adherent
CA130	Lung adenocarcinoma	C/D	Fbn	Adherent
CTL	T-cell lymphoma	N/A	N/A	Suspension
MCC	Merkel carcinoma	C/D-N/A	Fbn/NA[a]	Suspension
RECA0515	Rectal carcinoma	N/A	Fbn/Laminin	Adherent/suspension
RECA1208	Rectal carcinoma	N/A	Fbn/Laminin	Adherent/suspension
PACA9926	Pancreatic ductal carcinoma	Collagnease Hyaluronidase	Fbn	Adherent
BRCA1103	Breast ductal carcinoma	C/D	Fbn	Adherent
BCCA	Basal cell carcinoma	C/D	Fbn	Adherent

[a]MCC parental cells adhere to matrix and require enzymatic dissociation while metastases derived from these cells require no enzymatic dissociation and no matrix as they adhere lightly to the surface and often grow in suspension.

C/D, collagenase-dispase; N/A, nonadherent; Fbn, fibronectin.

7. The medium was changed every 2 days until the majority of cells began to die off, and colonies of small, round "stem-like" cells appeared. This required from 1 to 12 weeks, depending on the tumor type.

8. Cells were observed daily and recorded photographically to capture any morphological change.

9. At the first subculture, the cells were split 1:2 or 1:3, the medium changed every other day and subsequently subcultured every 5 days when cultures were semiconfluent and exhibited logarithmic growth. Initially, it was important to carefully dissociate the cultures, so that cells remained aggregated, as trypsinization into single cells often resulted in loss of the phenotype and eventual scenescence. Eventually most cultures were split 1:3 or 1:5 routinely every 5 days. It is important to note that in some cases initially, the target cell was not clearly observable, particularly in cultures in which the surrounding stroma or mesenchyme appeared to play a role in the *in vitro* outgrowth and eventual proliferation of the target cells. One example of this was the basal cell carcinoma cell line, BCCA, which represented a good illustration of the epithelial mesenchymal "cross-talk" that often occurs during tumor transformation (Fig. 2). The target epithelial cell appeared only after the cells had been in culture for nearly a month. The addition of a glucocorticoid (hydrocortisone, 5×10^{-8} M) to the medium then resulted in inhibition of the stromal component and exponential growth of the epithelial cells. This outgrowth then occurred fairly rapidly, within 10 days from

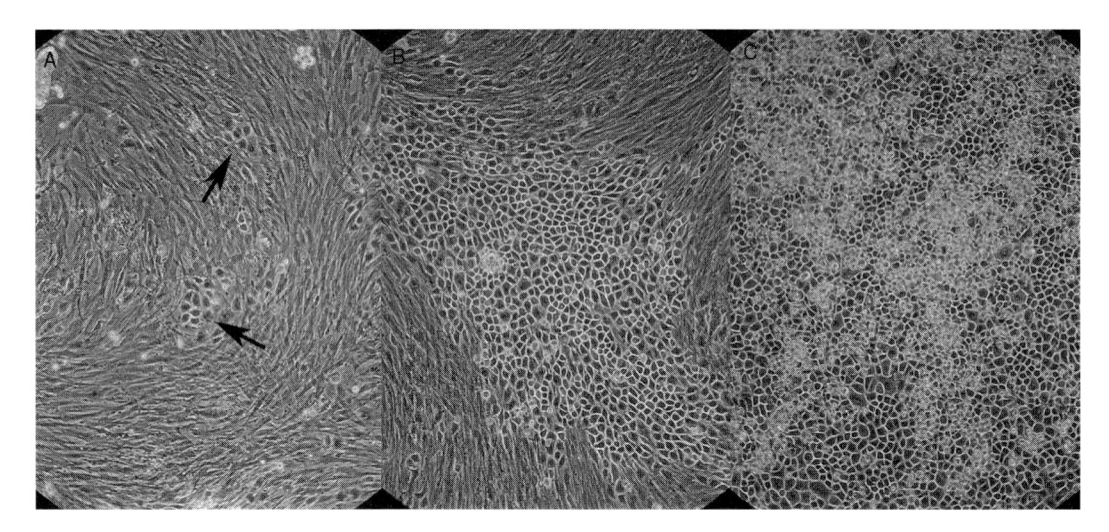

Fig. 2 A–C. Isolation of a basal cell carcinoma BCCA1. A few epithelial cells can be seen in this field initially consisting largely of stroma. Media derived for these epithelial cells supports exponential outgrowth of these cells, whose population eventually outstrips the growth of the surrounding stroma.

the first appearance of the epithelial cells, which could then be removed from the dish, first by a quick trypsinization [1 ml trypsin–EDTA (ethylenediaminetetraacetic acid) for 1–2 min], which easily removed the stromal cells, and then a longer (~5 min) treatment with a 1:20 dilution of 10% collagenase–dispase. Generally, this resulted in a 90% pure population after only the first passage. Alternatively, in the case of a colon carcinoma (CRCA1115), the medium, itself, selected against cells in the heterogeneous population, which did not survive in the culture, while actively supporting exponential growth of the target cell type. In this particular case after one month in culture, the target cell formed loosely attached colonies that could be easily pipetted from the plate, transferred to a new plate, and expanded as a pure population (Fig. 3).

C. Preparation of Cells from Xenograft

1. Tissue from xenografts, grown in the subrenal capsule of immune-compromised mice (nu/nu or severe-combined immunodeficiency, SCID), were harvested aseptically and placed in dishes where they were minced with iris scissors to <1 mm.

2. Tissue pieces were dissociated with 0.5% collagenase-dispase, and placed at 37 °C in 95% humidified incubator for 30′ with repeated pipetting until small aggregates of cells were observed.

3. The enzymatic activity was stopped by passing the cell suspension over 5% BSA and washed by centrifugation.

4. The resulting pellet was resuspended in growth medium, and plated on surfaces precoated with matrix (laminin or fibronectin).

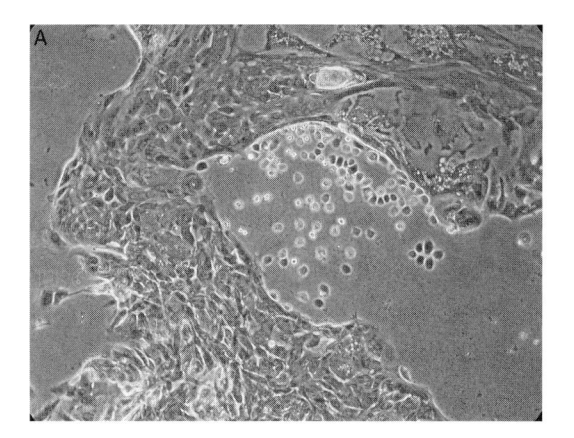

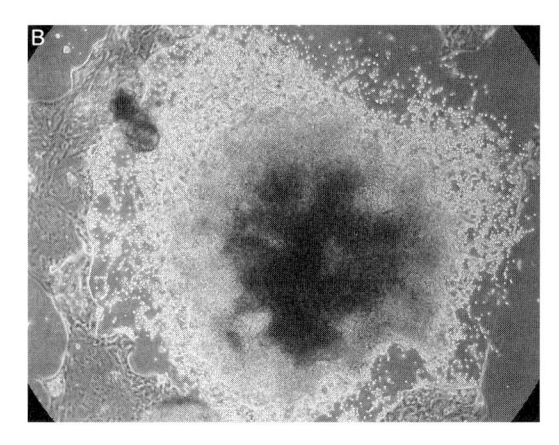

15. Isolation and Establishment of Human Tumor Stem Cells

Fig. 3 Isolation of a colon carcinoma stem cell CRCA1115. Sparse fields of a few cells can be observed among large islands of mixed cell types (42-day culture). Over a period of 5 days these cells grow exponentially while the other cells become vacuolated and die. Time following initial observation of 20 cells: A, 72 h (100×), B, 176 h (40×).

5. The cultures were refed every other day until a sufficient number of cells could be generated for analysis by flow cytometry, usually 4–7 days. Isolating the cell type of interest from xenografts does not require the extended lag phase to remove stromal components. In the appropriate growth medium, the CSC is stimulated to grow exponentially within a week from explant.

D. Cell Surface Phenotype

At the initial subculture, or at a specific passage number, or subsequent to tissue harvest from xenograft models, 10^6 viable cells were used for analysis of marker expression by fluorescence on a log scale using either a FACSCAN (Becton Dickinson) or a Guava PCA-96 (Guava Technologies).

The antibodies used were anti-CD34 (FITC), anti-CD44 (FITC), anti-CD24 (FITC), anti-CD133 (PE-Miltenyi), anti-CD31 (PE), anti-CD45 (PE), and anti-CD141. Unless noted, antibodies were purchased from Becton Dickinson. Dead cells were eliminated by using the viability dyes, PI or 7AAD.

Cells were prepared by removal of the monolayer with collagenase-dispase (50 μl/ml of a 10% wt/vol stock solution prepared in PBS), combined with 200 μg/ml STI. The dissociated cells were resuspended in 4 ml HBSS Hank's Balanced Salt Solution, layered over 3 ml 5% BSA, and centrifuged at 1100 rpm for 6 min using a GH3.8 rotor in a Beckman Allegra 6R centrifuge (The monolayer can also be dissociated with 10 mM EDTA, although the viability can be compromised with this treatment). The resulting pellet was resuspended in HBSS, containing 1% BSA (analysis buffer), and diluted as follows (all dilutions in analysis buffer):

Isotype control: 1×10^5 viable cells/50 μl.

To this volume was added 2 μl 1° antibody. The volume was brought up to 200 μl with analysis buffer and incubated 20 minutes on ice or at 4 °C

Primary antibody: 5×10^4 to 1×10^5 viable cells in 50 μl–100 μl. To this volume was added 2 μl of antibody. The volume was brought up to 200 μl with analysis buffer and incubated 20 minutes on ice or at 4 °C. Cell suspensions were washed by centrifugation after the incubation period by increasing the volume to 250 μl with analysis buffer and centrifuging in a Beckman Allegra 6R centrifuge using a GH3.8 rotor at 1200 rpm for 5 minutes. The supernatant was removed and the wash step was repeated 1×. The resulting pellet was brought up in a 200 μl volume of analysis buffer to which was added 5 μl 7AAD or PI for Live/Dead gating just before acquisition. Cytometric data analysis was performed using FlowJo software (Tree Star, Inc.).

E. Medium Selection for the Growth of Cancer Stem Cells

1. We first tested media derived at Raven for the isolation and establishment of normal tissue progenitor cells to isolate and establish tumor progenitor cells. With little modification of the tissue progenitor medium, all tumor cell lines were established by placing the entire tissue in culture, and allowing the microenvironment to select for a particular cell type, while not supporting, or actively inhibiting, the majority of other cell types in the medium. Ideally, one would wish a completely selective medium (i.e., selects for CSCs and actively counterselects other cell types). However, in practice, if the medium sufficiently selects for differential CSC growth, pure populations can be obtained after a few subcultures. Differential enzymatic digestion at subculture can accelerate this process, as well as manual selection of clones. In Fig. 2, a primary culture of a BCCA appears to consist largely of stroma, with a few small colonies of epithelial stem cells. In this case, while the medium does not actively inhibit stromal outgrowth, it stimulates the stem cells to grow exponentially, soon outstripping the growth of the other cell types. Figure 3 illustrates the concept of a media that supports the growth of a single cell type, a

stem cell of a colon tumor. After placing the tissue in culture, the many different cell types form large islands consisting of normal cells, transit-amplifying cells, stromal cells, and stem cells. Over the subsequent 30 days, only the stem cells are dividing, while other cells in the heterogeneous population become vacuolated and begin to die. In almost all cases, the putative CSCs appeared as clusters of small round cells, phase dark with a large nucleus and dense cytoplasm. The timing of media changes and subculture was critical to a successful outcome: too soon a subculture resulted in a loss of the cell type of interest; too late a subculture resulted in critical media components being exhausted, the cell type of interest becoming quiescent, and the entire culture eventually senescing. Generally, we have found that the cultures do best when subcultured at near-confluency at around 80% confluence. The preferred basal media formulations for several CSC types are listed in Table II.

F. Effect of Serum on Isolation and Establishment and Growth of CSCs

2. Both primary cultures and established cell lines, isolated and propagated in serum-free medium, were tested for their ability to grow in the presence of serum. Figure 4 shows the effect of serum on an established colon CSC line CRCA0404 (Panel A) and on an established prostate CSC line, PRCA629A (Panel B). Both cell lines were grown in medium containing growth factors ±5% fetal bovine serum. At 96 h, the cells in serum-free culture are growing logarithmically and nearing confluence, while cells in the presence of serum have become squamous and quiescent (CRCA0404) or have been unable to attach, spread, and divide (PRCA629A). Moreover, with prostate tumor isolates, primary outgrowth of cells in the presence of 5% serum resulted in fibroblast overgrowth, and even lesser amounts of serum (1–2%) resulted in squamous, terminally differentiated cultures (Fig. 5).

Table II
Tissue Progenitor Cell Lines were the Basis for Determining Media Requirements for Tumor Stem Cell Isolation

Tumor	Tumor-derived cell lines	Medium used
Pancreas	PACA9926[a]	F12/DME (1 mM $CaCl_2$) 14 factors
Lung	CA130, 9979	F12/DME (0.1 mM $CaCl_2$) 6 factors
Skin (Merkel, Melanoma, basal cell carcinoma)	MCC[a] hMEL, BCCA	F12/DME (1 mM $CaCl_2$) 11/9/9 factors
Breast	BRCA1103	F12/DME (1 mM $CaCl_2$) 6 factors
Prostate	PRCA, TDH	F12/DME (0.1 mM $CaCl_2$) 8 factors
Ovary	OVCA2	Opti-MEM (0.1 mM $CaCl_2$) 4 factors
GI Tract	CRCA, RECA[a]	F12/DME (1 mM $CaCl_2$) 11 factors
Immune system	NHL, MCL, CTL[a]	F12/DME (1 mM $CaCl_2$) 8 factors

[a]These cells spontaneously metastasize.

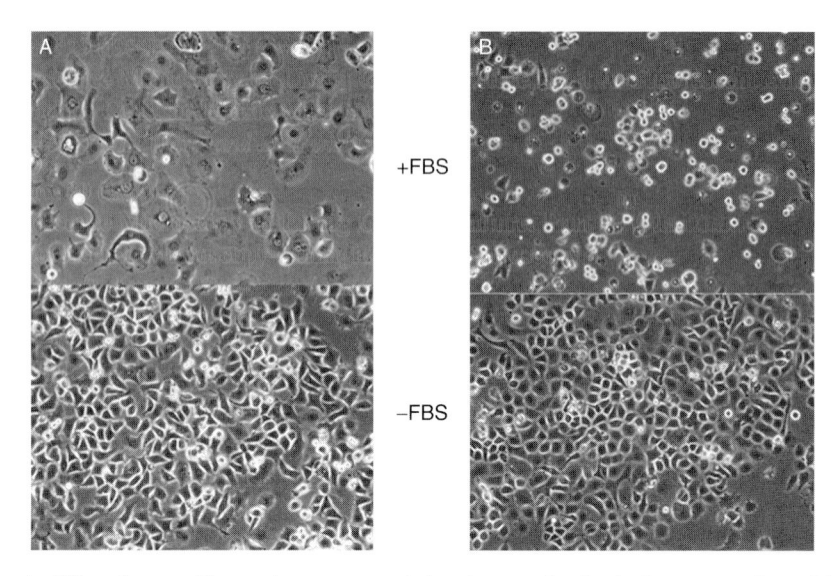

Fig. 4 Effect of serum. These cultures were seeded at the same density, in the same medium with 5% fetal bovine serum (upper panels) or in the absence of fetal bovine serum (lower panels). After 96 h, colon cultures (A) exhibited a squamous morphology and no observable cell division (upper panel), while cultures in the absence of serum were nearing confluence and growing logarithmically. In (B), prostate cultures were never able to grow in the presence of serum (upper panel), while dividing rapidly in serum-free media designed specifically for prostate cell lines.

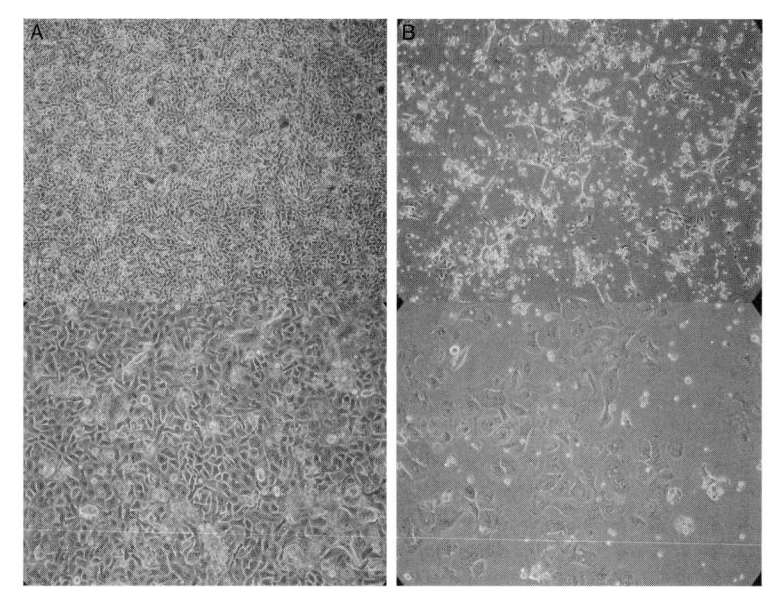

Fig. 5 Primary outgrowth of prostate tumor cells in the presence or absence of serum. Panel A shows morphology of prostate tumor cells grown in serum-free defined media, while panel B shows the effect of serum on this particular cell type (upper panels, 40×, lower panels, 100×).

15. Isolation and Establishment of Human Tumor Stem Cells

3. CSCs, like normal fetal progenitor cells, prefer a biologic substrate to plastic for attached growth. The preferred attachment factors are shown in Table I.

4. Once the medium and attachment factor have been selected, the appropriate hormones and growth factors should be added to the basal medium just before use, or they can be added directly to the culture dish. For preparation and handling of media and additives, see Brinster, 1974, Chapter 8.

The growth factors and other media additions for various CSCs are shown in Table III. Note that reduced calcium is required for prostate CSC selection.

G. Immunohistochemistry

Tissue samples were fixed in 10% phosphate-buffered formalin and embedded in paraffin. Formalin-fixed, paraffin-embedded sections were cut 4 μm thick, mounted on poly-l-lysine-coated slides (Sigma Aldrich, St. Louis, MO), and

Table III
Media Components Used in the Isolation and Growth of Various Tumor Lines

	PACA9926	CA130	PRCA629	MCLY	BCCA	MCC	RECA	CRCA	CTLY	BRCA
Insulin	√	√	√	√	√	√	√	√	√	√
TF	√		√	√	√	√	√	√	√	√
EGF	√	√	√	√	√	√	√	√	√	√
Se	√	√	√	√	√	√	√	√		
Eth	√				√	√	√	√		
pEth	√		√		√	√	√	√		
HC	√	√	√	√			√	√		
Prog	√					√				
FK	√					√				
Aprot	√									
Vit E							√	√		
HRG	√									
T			√							
T_3	√		√			√		√	√	
NGF						√				
PGE1										√
IL2				√					√	
IL4									√	
FLK3									√	
FGF7									√	
GLUC								√		
Ca[a]	√		√ 0.1 mM	√	√	√	√	√	√	√
PPE[b]		√	√		√	√		√		√
BSA[b]	√									

[a]Exogenously added to F12/DME calcium-free medium.
[b]Not required for growth.

TF, transferrin; EGF, epidermal growth factor; Se, selenium; Eth, ethanolamine; pEth, phosphoethanoloamine; HC, hydrocortisone; Prog., progesterone; FK, forskolin; Aprot., aprotinin; HRG, heregulin; T, testosterone; T_3, triiodothyronine; NGF, nerve growth factor; GLUC, glucagon; PPE, porcine pituitary extract.

dried overnight at 37 °C. Sections were then dewaxed in xylene, rehydrated according to standard histopathologic procedures, and stained with H&E.

H. Renal Capsule Grafting

Cells from each tumor line were dissociated from the tissue culture plate as mentioned previously and seeded in 50 µl of rat tail collagen, in separate wells of a 4-well dish (Nunc), containing 500 µl growth media for the particular cell type, and allowed to gel in a 37 °C humidified incubator overnight. The following day, the collagen gels were grafted under the renal capsule of SCID mice (The process of renal capsule grafting is described (Brody, J., et al., 1999) and illustrated in detail at the following website: http://mammary.nih.gov/tools/mousework/Cunha001/index/html). Additional collagen gels were processed for quantitative RT–PCR, using probes specific for human DNA. Grafts were harvested after 7 weeks. After removal, half of the grafts were fixed overnight in 10% phosphate-buffered formalin (Fisher Scientific, Fairlawn, NJ), embedded in paraffin, sectioned, and stained by hematoxylin and eosin using standard procedures. The remaining grafts were processed for quantitative RT–PCR.

IV. Results

A. Tumor-Forming Capability

All cell lines tested were able to form tumors when implanted subcutaneously or grafted under the renal capsule of SCID mice (Table IV). To test whether or not the cell lines established could form tumors from a small number of cells, experiments were conducted in which <20, 100, 500, and 1000 cells, from each of 5 cell

Table IV
Cell Lines Tested for Their Ability to Form Tumors in Mice

Cell line designation	Renal capsule	Subcutaneous
CA130[a]	√	√
PACA9926	√	√
MCLY	√	
CTL	√	
MCC[a]	√	√
CRCA0404[a]	√	
CRCA1115	√	
RECA0515[a]	√	
RECA1208	√	
PRCA629A[a]	√	√
PRCA1004	√	√
BRCA1103	√	

[a]These cell lines form tumors from <100 cells.

lines, were suspended in 50 µl collagen gels, in growth media, and allowed to polymerize overnight. The following morning the gels were placed under the renal capsule of SCID mice ($n = 3$ animals per group). Additional collagen gels containing the same number of cells were prepared under identical conditions for quantitative RT–PCR. After 7 weeks, the animals were euthanized and the presence or absence of tumor tissue was confirmed by visual observation (Fig. 7), histological examination, or by quantitative RT–PCR (data not shown).

B. Expression of Cell Surface Markers

Antibodies directed against CD34, CD44, CD24, and CD133 were used to evaluate the expression of these CSC markers on several cell lines: MCC, PACA9926, CA130, PRCA629A, RECA0515, BRCA1103, MCLY, and CRCA0404. In addition, because there was unanticipated CD34 expression on some of these tumor lines, the vascular endothelial markers, CD31, CD45, and CD141, were evaluated to rule out the possibility that these were endothelial-derived cells. Interestingly, five of the seven lines tested were positive for CD34, including the hematopoietic cell line MCLY, derived from a Mantle Cell lymphoma. Of those five lines, four expressed some level of CD44. Only the BRCA1103 cell line expressed CD24 as well as CD44. None of the CD34 positive cell lines expressed the endothelial markers.

C. Importance of Growth Conditions in Maintaining the CSCs Phenotype

We have shown above that the elimination of serum from the culture media is necessary for the selection and maintenance of CSCs (Figs. 4 and 5). We have also emphasized the importance of media changes and subculturing, using a rigorously adhered to schedule (Fig. 6). The following two examples will illustrate the fact that even such frequently used procedures as media and growth factor

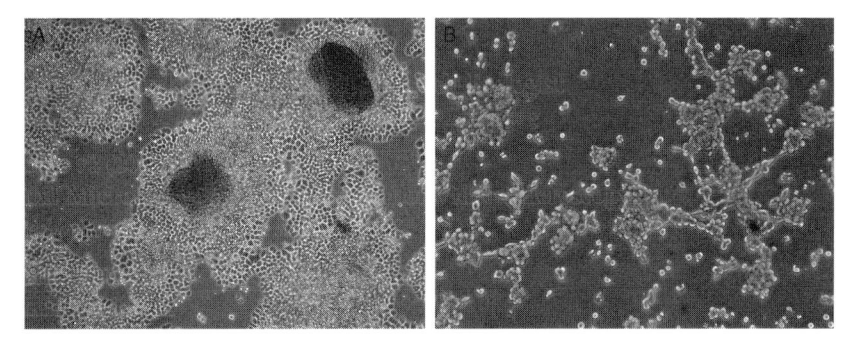

Fig. 6 Effect of media exhaustion in CRCA1115 rectal cancer cell line. This culture did not appear to be overgrown, but growth factors necessary for attachment and survival may have been exhausted. Regular media changes and timely subculture are critical for growth and maintenance of this phenotype.

optimization, for increased growth rate and animal culture passage, can alter the phenotype of the CSC.

Example 1. In an effort to increase the growth rate of the BRCA1103 cell line, several additional growth factors were examined. One of those factors, forskolin, a stimulator of adenylate cyclase, which increases cellular cAMP levels, was shown to stimulate growth of the cell line. A separate culture was carried for four passages in the presence of the usual growth factors and the addition of forskolin. The morphology of these cells differed from the parent line and, in addition, we were able to observe downregulation of CD44 and CD24 expression when these cells were grown in the presence of forskolin (Fig. 8).

Example 2. Several CSC lines which were CD34 positive and CD44 negative *in vitro* expressed high levels of CD44 and no CD34 after animal passage, whether the cells were grown subcutaneously or by implantation under the renal capsule (Fig. 9). Cell lines were established from these xenografts using the same media derived to select from the original tumor.

V. Discussion

We have isolated and established cell lines from a subpopulation of solid tumors of prostate, pancreas, lung, spleen, breast, colon, ovary, and skin (BCCA, melanoma, and Merkel cell carcinoma) as well as a subpopulation of cells from mantle cell lymphoma, non-Hodgkins lymphoma, and cutaneous T-cell lymphoma. These cells are highly tumorigenic, grow indefinitely *in vitro* and, of the cell lines tested thus far, can form tumors from fewer than 200 cells (5 lines have been tested and grow from <20 cells) when implanted under the renal capsule of SCID mice

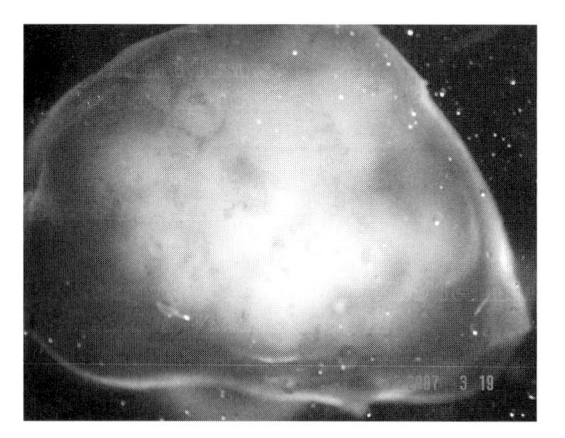

Fig. 7 One hundred cells from the RECA1208 cell line implanted in the kidney capsule resulted in a tumor seen to completely overgrow the kidney after 8 weeks.

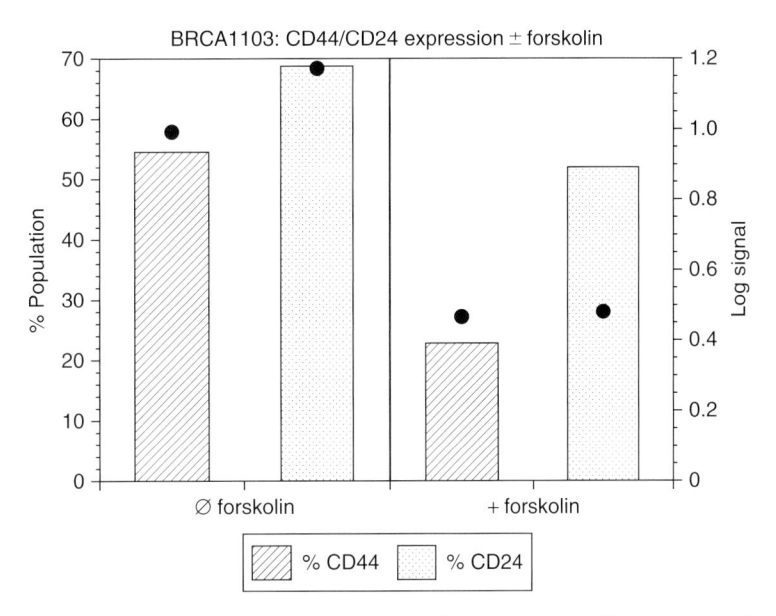

15. Isolation and Establishment of Human Tumor Stem Cells

Fig. 8 The effect of growth factors on expression of cell surface makers. The addition or deletion of growth or differentiation factors can change the marker profile. The breast cancer cell line BRCA1103 requires forskolin for optimal growth *in vitro*, but this results in a significant inhibition of expression of CD44 and to a lesser extent the expression of CD24.

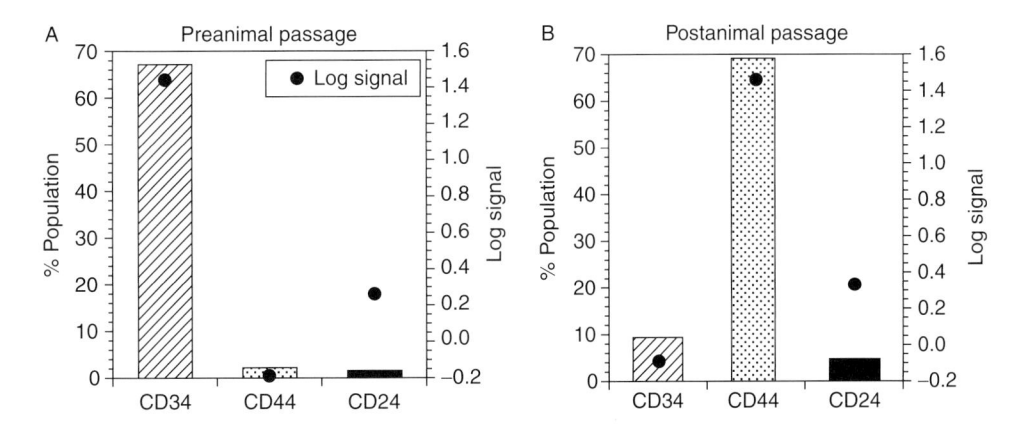

Fig. 9 Animal passage has a dramatic effect on the expression of CD34 and CD44. *In vitro*, the colon cancer cell line CRCA0404 has a high level of expression of CD34, representing nearly 70% of the population (A). There is no discernible CD44 expression. After animal passage, the purified populations of CRCA0404 analyzed by flow cytometry show that CD34 expression is minimal, but CD44 expression is present on the majority of the population (B).

(Fig. 8). Similarity in growth requirements supports the hypothesis that these cells may arise from their normal tissue progenitor (or stem cell) counterpart (Kim CF, et al., 2005). We have shown that deriving a medium for CSCs based upon similar medium and growth conditions derived for normal tissue progenitor cells would preferentially select for that particular cell type in a number of different solid tumors. Tumor stem cells, by definition, have a clonal origin, and can be present at levels <1% of levels of the transit-amplifying daughter cells and/or tumor stroma and vascular elements. As we have shown by the example of the colon cancer line CRCA1115, there are very few cells available to select by most cell sorting technologies. Providing a microenvironment that supports the growth of that single cell type, while not supporting or actively inhibiting the growth of other cells in the population, does not unnecessarily bias the culture in favor of a particular surface marker, as we have shown that those markers can change with changes in culture conditions, including growth and differentiation factors, factors such as low or high calcium concentrations, and matrix. Cells selected and propagated under rigorously defined conditions, which represent a small subset of the original population and which can form tumors from a few cells implanted *in vivo,* can then be used to develop and identify specific markers for the assessment of CSC origin.

VI. Materials

F12/DME medium with and without calcium—Invitrogen

Calcium chloride—Sigma Aldrich

Insulin (human recombinant) Millipore Incelligent™ SG Insulin, Recombinant Human, EP, USP

EGF (epidermal growth factor, human recombinant)—Millpore

Transferrin (human, recombinant)—Sigma Aldrich

Ethanolamine/phosphoethanolamine—Sigma Aldrich

Triiodothyronine (T3)—Sigma Aldrich

Selenous acid 97%—Alfa Aesar

Vitamin E acetate—Alfa Aesar

Nerve growth factor (NGF 7S)—Invitrogen

Hydrocortisone—Sigma Aldrich

Progesterone—Sigma Aldrich

Forskolin—Calbiochem

Fibronectin—Sigma Aldrich

Laminin—Invitrogen

Collagenase type IV—Worthington

Hyaluronidase—Worthington

Collagenase-dispase—Roche Applied Science

Porcine Pituitary Extract—This can be purchased or prepared from frozen pituitaries (Pel Freez biologicals): The addition of 2–10 μl/ml of this extract can frequently increase growth of cells in hormone-supplemented, serum-free medium. *Note:* Such medium is not defined, since it is impossible to know what factor in the extract is the active component and different cell types may respond to different components of the pituitary extract (Mather, *et al.*, 1998).

1. Homogenize 105 g mixed sex porcine pituitaries (Pel Freeze) in 250 ml cold 0.15 M NaCl for 10 min in a blender.
2. Transfer the homogenate to a cold beaker and stir for 90 min at 4 °C.
3. Centrifuge for 40 min (9800 rpm) using a Beckman JA10 rotor at 4 °C.
4. Discard the pellet.
5. At this point, the supernatant can be aliquoted into 50-ml polypropylene tubes and stored at −20 °C.
6. Before use, the supernatants should be filter sterilized through a 0.2-μm low-protein-binding filter.

Note: This can be made easier by centrifuging the thawed supernatants (9800 rpm for 20 min at 4 °C as above) to remove particulate material and then passing it through successive 0.8-μm, 0.4-μm, and then 0.2-μm filters.

7. Aliquot into 5-ml snap-cap tubes and store at −20 °C.
8. Thawed pituitary extract can usually be stored at 4 °C for up to 2 weeks without loss of activity.
9. A dose response of each batch of extract must be done for each cell line. The extract always has a biphasic dose response, with too high a concentration being toxic to many cells, especially in serum-free medium.

References

Al-Hajj, M., Wicha, M. S., Benito-Hernandez, A., Morrison, S. J., and Clarke, M. F. (2003). Prospective identification of tumorigenic breast cancer cells. *PNAS* **100**(7), 3983–3988.

Brody, J., Young, P., and Cunha, G. R. (1999). http://mammary.nih.gov/tools/mousework/Cunha001/index/html.

Gerst, J. E., Sole, J., Mather, J. P., and Salomon, Y. (1986). Regulation of adenylate cyclase by beta-melanotropin in the M2R melanoma cell line. *Mol. Cell Endocrinol.* **46**(2), 137–147.

Kim, C. F., Jackson, E. L., Woolfenden, A. E., Lawrence, S., Babar, I., Vogel, S., Crowley, D., Bronson, R. T., and Jacks, T. (2005). Identification of bronchioalveolar stem cells in normal lung and lung cancer. *Cell* **121**(6), 823–835.

Lang, S. H., Sharrard, R. M., Stark, M., Villette, J. M., and Maitland, N. J. (2001). Prostate epithelial cell lines form spheroids with evidence of glandular differentiation in three-dimensional Matrigel cultures. *Br. J. Cancer* **85**, 590–599.

Li, R., and Mather, J. P. (2003). Human mullerian duct-derived epithelial cells and methods of isolation and uses thereof. U.S. Patent 20030040110.

Li, R. H., Sliwkowski, M. X., Lo, J., and Mather, J. P. (1996a). Establishment of Schwann cell lines from normal adult and embryonic rat dorsal root ganglia. *J. Neurosci. Methods* **67**, 57–69.

Li, R., Gao, W. Q., and Mather, J. P. (1996b). Multiple factors control the proliferation and differentiation of rat early embryonic (day 9) neuroepithelial cells. *Endocrine* **5,** 205–217.

Li, R., Bald, L. N., and Mather, J. P. (2003). Human ovarian mesothelial cells and methods of isolation and uses thereof. U.S. Patent 20030207449.

Li, R., Phillips, D. M., Moore, A., and Mather, J. P. (1997). Follicle-stimulating hormone induces terminal differentiation in a predifferentiated rat granulosa cell line (ROG). *Endocrinology* **138,** 2648–2657.

Mather, J. P. (ed.), (1984). "Mammalian Cell Culture. The Use of Serum-Free Hormone-Supplemented Media" pp. 284. Plenum Press, New York.

Mather, J. P., and Roberts, P. E. (1998). *In* "Introduction to Cell and Tissue Culture, Theory and Technique. Introductory Cell and Molecular Biology Techniques." Plenum Press, New York.

Mather, J. P., and Sato, G. H. (1979a). The use of hormone-supplemented serum-free media in primary cultures. *Exp. Cell Res.* **124**(1), 215–221.

Mather, J. P., and Sato, G. H. (1979b). The growth of mouse melanoma cells in hormone-supplemented, serum-free medium. *Exp Cell Res.* **120**(1), 191–200.

NCI Think Tanks in Cancer Biology (2004). Executive Summary of the Tumor Stem Cell & Self-Renewal Genes Think Tank. http://www.cancer.gov/think-tanks-cancer-biology/page4.

Roberts, P. E., and Mather, J. P. (2002). Human pancreatic epithelial progenitor cells and methods of isolation and use thereof. U.S. Patent 20020192816.

Roberts, P. E., Phillips, D. M., and Mather, J. P. (1990). A novel epithelial cell from neonatal rat lung: Isolation and differentiated phenotype. *Am. J. Physiol.* **3,** (Lung Cell and Mol Physiol) L415–L425.

Roberts, P. E., Chichester, C., Plopper, C. G., Lakritz, J., Phillips, D. M., and Mather, J. P. (1992). Characterization of an airway epithelial cell from neonatal rat. *In* "Animal Cell Technology: Basic and Applied Aspects" (H. Murakami, S. Shivahata, and H. Tachibana, eds.), pp. 335–341. Kluwer Academic Publishers.

Stephan, J. P., Roberts, P. E., Bald, L., Lee, J., Gu, Q., Devaux, B., and Mather, J. P. (1999). Selective cloning of cell surface proteins involved in organ development: Epithelial glycoprotein is involved in normal epithelial differentiation. *Endocrinology* **140**(12), 5841–5854.

U.S. Department of Health and Human Services Public Health Service Centers for Disease Control and Prevention *and* National Institutes of Health. (2000). Primary Containment for Biohazards: Selection, Installation and Use of Biological Safety Cabinets. 2nd edn., September 2000.

CHAPTER 16

Stem Cells from Cartilaginous and Bony Fish

David W. Barnes,[*] Angela Parton,[*] Mitsuru Tomana,[*] Jae-Ho Hwang,[*] Anne Czechanski,[*] Lanchun Fan,[†] and Paul Collodi[†]

[*]Mount Desert Island Biological Laboratory
Salisbury Cove, Maine

[†]Department of Animal Sciences
Purdue University
West Lafayette Indiana

I. Introduction to Models and Uses
   A. Concepts and Approaches to Derivation of Fish Cell and Tissue Culture Systems
   B. Teleost and Chondricthyian Cells *In Vitro*
II. Embryonal Stem Cells
   A. *In Vitro* Methods
   B. Concepts and Procedures for Transgene Integration, Expression, and Homologous Recombination
III. Tissue-Specific Stem Cells
   A. Stem Cell Lines and the Methodological Understanding of Cell Culture Development
   B. Applications to Biology and Medicine
IV. Outlook and Future Contributions
   References

I. Introduction to Models and Uses

Historically, cell culture approaches applied to models of development, differentiation, mutagenesis, virology, and other fields have contributed to a variety of major discoveries (Barnes and Sato, 2000). Today, cell culture increasingly is applied to complement the techniques and hypotheses of molecular biology. Despite the major advances using cell cultures from homeothermic organisms, progress with poikilotherms, such as the common fish model species, has been much more limited. Early cultures generally did not extend beyond a few species of Cypriniformes (e.g., carp, goldfish), Salmoniformes (trout, salmon), and Siluriformes (catfish); all of these are fundamentally freshwater fish (Borenfreund and Puerner, 1984; Lopez-Doriga *et al.*, 2001; Lorenzen *et al.*, 1999; Mcallister, 1997). Only one marine fish cell line was established during this early period: The GF cell line developed by Clem *et al.* (1961) from fin of the grunt *Haemulon sciurus*. Some of the most widely used cell lines from this period are EPC (carp epithelial cell), BG/F-2 (bluegill), CHSE-214 (Chinook salmon embryo), RTH-149 (rainbow trout hepatoma), and BB (brown bullhead catfish) (Babich *et al.*, 1990; Kamer *et al.*, 2003; Martin-Alguacil *et al.*, 1991; Ruiz-Leal and George, 2004).

Until recently, the continuously proliferating fish cell lines generally available were often generic in biological properties, regardless of the tissue of origin, and not well suited for sophisticated biological and molecular questions. They were, however, excellent substrates for studies of fish pathogens, including viruses, bacteria, and parasitic eukaryotes, and several of the cell lines initially were developed for this purpose. Representatives from each of the major virus families have been propagated and studied in cell culture (Barnes *et al.*, 2006; Bernard and Bremont, 1995; Caipang *et al.*, 2003; Moredock *et al.*, 2003; Mork *et al.*, 2004; Wolf, 1988). Fish and fish cell cultures also traditionally have been useful models in toxicology and as environmental sentinels, and some of the earliest-established fish cell lines also were used for these purposes (Ackerman *et al.*, 2002; Babich *et al.*, 1990; Barnes and Collodi, 2005; Bols *et al.*, 2003).

In addition, *Xiphophorus* species and cross-species hybrids have contributed critical concepts in cancer genetics and *in vitro* mutagenesis research (Moredock *et al.*, 2003), and cell cultures from these animals have improved the understanding of mechanisms of these processes. Goldfish and carp cell lines also are used for studies of tissue regeneration and carcinogenesis (Hasegawa *et al.*, 1997; Smith *et al.*, 2002). In the last few years, application of new techniques for development of *in vitro* systems has led to definition of culture conditions that allow the continuous proliferation of a wider variety of cell types. This work has been driven by the realization that fish provide excellent models for stem cell studies and comparative approaches for understanding the nature of evolution and embryological development.

A. Concepts and Approaches to Derivation of Fish Cell and Tissue Culture Systems

Conventional mammalian cell culture methodology using serum as the major biologically derived supplement introduces a variety of unknown components. Some of these may act on the cultured cells to promote proliferation and/or differentiation; others may act negatively on these processes. The newer approaches described below afford some alternatives that may mitigate these effects and help to develop cell lines that represent interesting *in vitro* models. The critical concept is to approach cell culture with knowledge of the physiology of the cell type to be cultured (Barnes and Sato, 2000). The advent of recombinant DNA technology has allowed access to sufficient quantities of growth-promoting peptides and other critical molecules affecting cellular growth or differentiation to carry out systematic cell culture studies of this type, and has greatly increased the potential for development of new culture systems.

B. Teleost and Chondricthyian Cells *In Vitro*

The first fish cell lines were established in basic nutrient media formulations developed for mammalian cells and using mammalian serum as a supplement. As time progressed, new cell lines from a variety of species began to be reported. Examples are cultures from the Asian striped snakehead, koi, grass carp, flounder, grouper, seabream, and Atlantic salmon (Chi *et al.*, 1999; Hart *et al.*, 1996; Kamer *et al.*, 2003; Lu *et al.*, 1990; Perez-Prieto *et al.*, 1999; Shimizu *et al.*, 2003; Syed *et al.*, 2003; Wergeland and Jakobsen, 2001; Zhang *et al.*, 2003). Further advancement in this field relies to some degree on media more precisely tailored to fish in general, and often directed to specific genera or species.

In these approaches, addition of purified growth factors, fish embryo extract, or other medium supplements may be necessary (Collodi and Barnes, 1990). Using these methods, a number of embryonal stem cell-like and tissue-specific stem cell lines have been established from zebrafish (*Danio rerio*) embryo and adult tissues (Collodi *et al.*, 1994; Miranda *et al.*, 1993; Sun *et al.*, 1995). The approach is also adaptable to other model freshwater species: *Orzias* or medaka, *Xiphophorus* sp., or swordtail/platyfish and *Fundulus* sp. or killifish (Fig. 1) (Barnes *et al.*, 2006; Hong *et al.*, 2004a; Wakamatsu *et al.*, 1994). Further adaptation of culture media has allowed the development of cell lines from some model marine species, including pufferfish genera (*Fugu* sp. and *Tetraodon nigroviridis*) (Bradford *et al.*, 1997; Grutzner *et al.*, 1999). Like zebrafish, the entire genome of these species has been sequenced. Modified minimally, these cell culture concepts also have been applied successfully to the establishment of other marine species, such as the moray eel, *Gymnothorax prasinus* or *obesis* (Buck *et al.*, 2001) (Fig. 1).

As an example, culture of pufferfish cells requires the use of LDF, a basal nutrient medium originally developed for salmonid and zebrafish cells (Barnes

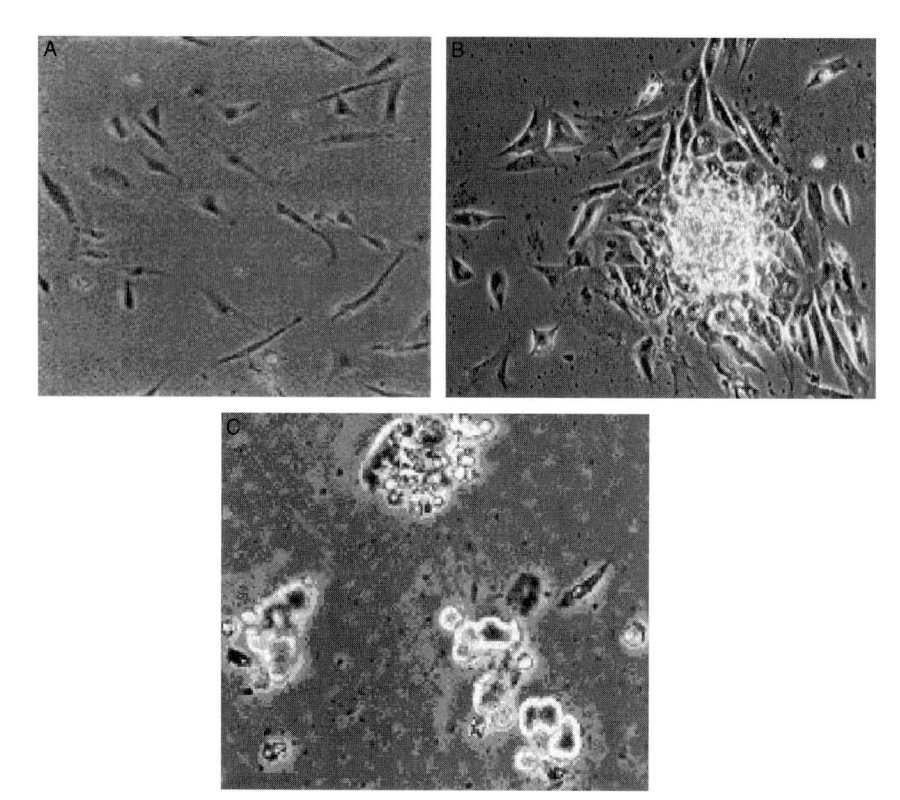

Fig. 1 Cell cultures from tissues of freshwater and marine species using fish species-specific medium formulations. (A) Culture from skin tumor of green moray eel (Buck *et al.*, 2001). (B) Culture from medaka (*Oryzias latipes*) gill. (C) Culture from killifish (*Fundulus heteroclitus*) spleen.

and Collodi, 2005). LDF is 50% Leibovitz's L-15, 35% Dulbecco's modified Eagles, and 15% Ham's F-12 media, with 0.18 mg/ml sodium bicarbonate and 15 mM HEPES (4-(2-hydroxyethyl)-1-piperazineethanesulfonic acid) buffer (pH 7.2). LDF medium contains a bicarbonate concentration that allows culture at ambient carbon dioxide levels. Pufferfish cell lines are further supplemented with the following:

Epidermal growth factor (EGF) (50 ng/ml),
Fetal calf serum (5%),
Fibroblast growth factor (FGF) (10 ng/ml),
Fish embryo extract from zebrafish or trout (25 mg protein/ml) (Collodi and Barnes, 1990),
Fish serum (e.g., trout serum, heat inactivated) (0.25%),
Insulin (1 mg/ml),
2-Mercaptoethanol (55 μM).

The *Fugu* cell lines express telomerase, an indicator of indefinite growth potential (Bradford *et al.*, 1997), and recently pufferfish telomerase has been molecularly cloned from one of these lines. At least one pufferfish cell line, developed from *Fugu* eye, shows an apparently diploid karyotype, and the cells demonstrate DNA content per nucleus of about 0.4 pg/haploid genome size, ~15% of that in human cells. This property is characteristic of pufferfish and is unlike most orders of fish, which in rare instances can show genome sizes several hundred times higher.

Cultures also have been derived from marine elasmobranchs (cartilaginous fish): the spiny dogfish shark (*Squalus acanthias*) and little skate (*Raja erinacea*) (Forest *et al.*, 2007; Mattingly *et al.*, 2004; Parton *et al.*, 2007) (Fig. 2). In general, for these species LDF medium is supplemented with the following:

EGF (50 ng/ml),

Fetal bovine serum (heat inactivated, 2%),

FGF (10 ng/ml),

L-glutamine (25 mM),

Insulin (1 μg/ml),

2-Mercaptoethanol (55 μM),

Nonessential amino acids (concentrations equal to that used for mammalian cell culture),

Selenous acid (100 ng/ml).

The medium is also supplemented with a chemically defined lipid solution formulated for mammalian cell culture medium supplementation. These lipids, although helpful, may not represent the ideal lipid mix for cells of cold water animals like the dogfish shark or little skate. This material also can oxidize easily, creating potentially toxic by-products, and must be stored with care. In early cultures of cold water cartilaginous fish, this lipid formulation was replaced with an extract from shark egg yolk, but this is not feasible for most laboratories. Exotic shark-specific medium supplements and elasmobranch blood components, such as urea and trimethylamine-oxide, are not necessary for the embryo-derived lines, none of the supplements are fish-specific, and all are commercially available. Recent progress in cloning growth factors from fish (e.g., Biga *et al.*, 2005) may provide even better medium formulations and application to more species. Dogfish sharks use glutamine as a major source of energy, through enzymatic transamination to alpha keto glutamate and oxidation of this metabolite in the citric acid cycle.

The cells also are plated on a collagen gel matrix. A commercially available human collagen preparation is commonly used for this purpose, and the use of elasmobranch-derived extracellular matrix also is being explored. The first continuous, multipassage cell line from cartilaginous fish was the SAE embryo-derived line from the dogfish shark, and represents a mesenchymal stem cell type (Forest *et al.*, 2007; Parton *et al.*, 2007). Details of culture and cell properties are described in later sections of this chapter.

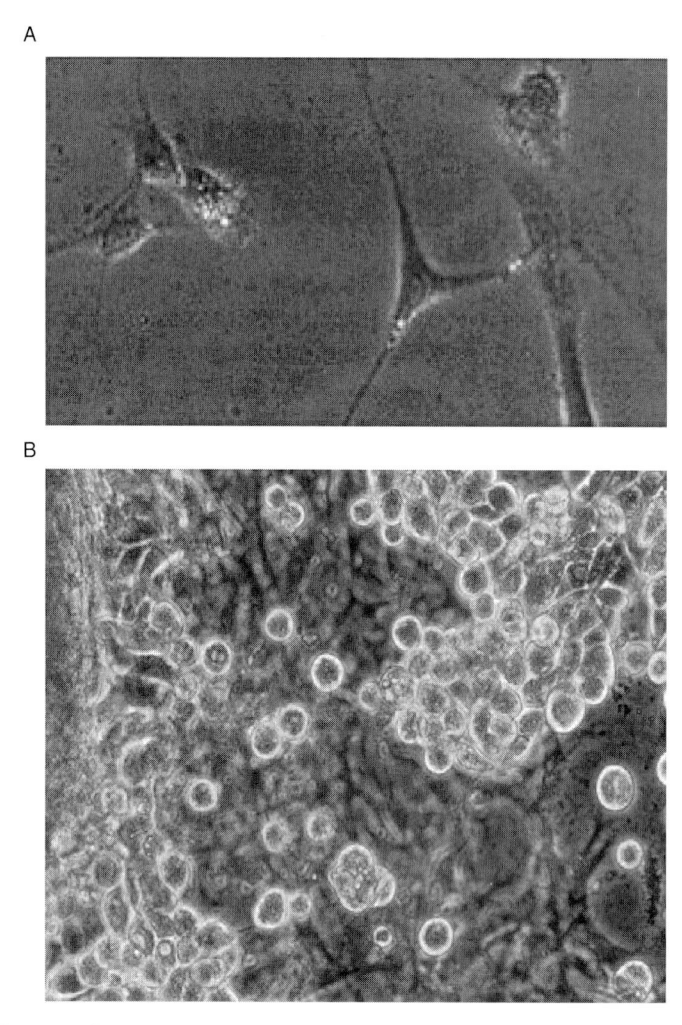

Fig. 2 Cell cultures from the cartilaginous fish *Raja erinacea* (little skate). (A) Skate embryo culture initiated from developmental stage 22. (B) Skate spleen-derived leukocyte culture initiated after concentration by density gradient centrifugation.

II. Embryonal Stem Cells

Mouse embryonal stem (ES) cell cultures that maintain the capacity to contribute to the germ cell lineage of a host embryo were derived several decades ago, and remain a critical tool for the introduction of targeted mutations and production of conditional transgenic animals (Gertsenstein et al., 2007). ES cells carrying targeted mutations and introduced into host embryos participate in development and contribute to the germ cell lineage. Breeding of germ line chimeric animals then can be pursued to produce heterogeneous or homogeneous gene knockout lines. The technique can be used for both generation of knockout mice for studies of gene function and for site-specific insertion of a transgene to achieve optimal or tissue-specific expression. This approach improves on the traditional transgenic method involving pronuclear injection and random insertion (Gertsenstein *et al.*, 2007). In an even more sophisticated approach, site-specific recombinases can be used to generate conditional transgenic or conditional knockout models in which expression or loss of function is manifested in a tissue-specific or temporal manner (Dymecki *et al.*, 2004).

A. *In Vitro* Methods

To date the full potential of approaches with ES cells has not been realized with any fish species, but ES-like cell lines that are suitable for gene transfer studies are available. ES-like cell lines have been derived from zebrafish, medaka, seabream, red seabream, sea perch, and rainbow trout. These lines exhibit various degrees of *in vitro* characteristics of pluripotency exhibited by mouse ES cells, including high levels of alkaline phosphatase activity, expression of pluripotency markers, embryoid body formation, and the ability to differentiate into multiple cell types in culture (Bejar *et al.*, 2002; Bols *et al.*, 2004; Chen *et al.*, 2003a,b; Hong *et al.*, 1998; Ma *et al.*, 2001). Lines from zebrafish (ZEB, ZEG), medaka (MES1), and seabream (SaBE-1c) can participate in normal development following introduction into a host embryo, leading to viable, chimeric fish.(Bejar *et al.*, 2002; Hong *et al.*, 1998). ZEB, ZEG, and MES1 cells contribute extensively to multiple tissues of the host. The degree of MES1 chimerism was enhanced when the cells were selected for expression of a pluripotency marker by introduction of a plasmid encoding a neomycin-resistance (neo)-lacZ fusion protein under the control of the mouse Oct4 promoter (Hong *et al.*, 2004c). Colonies of pluripotent cells were selected in G418. Markers for ES cells or primordial germ cells (PGCs) in zebrafish and medaka have been identified or are under study (Amsterdam *et al.*, 2004; Bejar *et al.*, 2003; Burgess *et al.*, 2002; Scholz *et al.*, 2004).

Marking of potential ES cells with a reporter gene under control of an ES-specific promoter is an extremely useful technological approach. Successful generation of ES-based targeted gene knockouts and conditional transgenics requires that the cells maintain the capacity to contribute to the germ line following introduction into a host embryo, and maintain this capacity for a sufficient length of time in culture to allow for gene targeting by homologous recombination followed by *in vitro* selection of homologous recombinants. Toward this goal, culture conditions recently were developed that successfully preserve the germ line competency of zebrafish ES cells *in vitro* (Fan *et al.*, 2006; Ma *et al.*, 2001). A key component of the cell culture system is the use of a feeder layer derived from the rainbow trout spleen cell line, RTS34st (Ganassin and Bols, 1999) These cells provide factors essential to maintain pluripotency and germ line competency (Ma *et al.*, 2001) (Fig. 3). Two zebrafish ES cell lines (ZEB and ZEG) have been derived

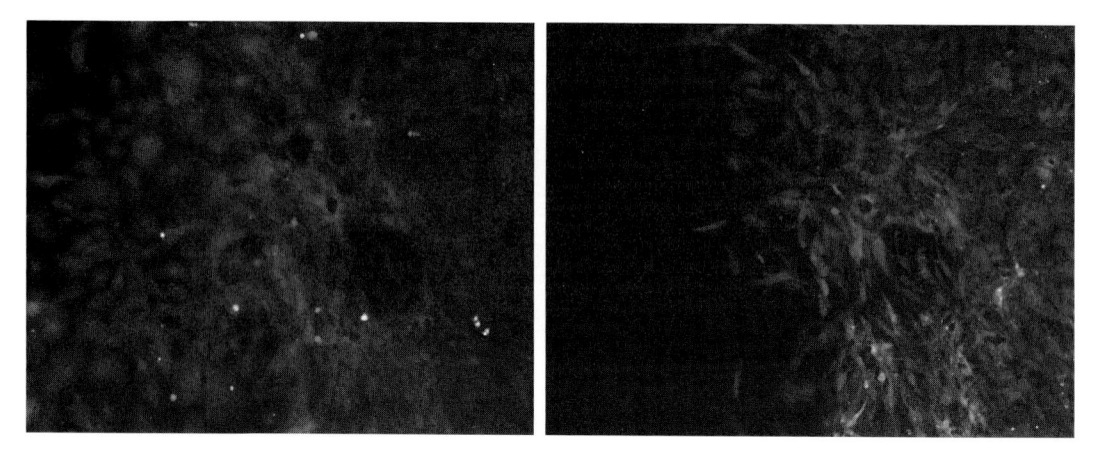

Fig. 3 Homologous recombination in zebrafish embryonic stem cell cultures. Two G418-resistant zebrafish ES cell colonies that were generated by introducing a plasmid designed to insert into the *no tail* gene (*ntl*). The plasmid contained *neo* located within the *ntl* homologous arms and red fluorescent protein (RFP) outside of the homologous region. The colony that incorporated the plasmid in a targeted fashion by homologous recombination is identified by the loss of RFP expression (arrow). Both colonies express GFP under the control of the *β-actin* promoter. (See Plate no. 24 in the Color Plate Section.)

using this system. These cells remain germ line competent for multiple passages in culture. Founder chimeras were bred in which the F1 individuals inherited the marker gene and pigmentation pattern donated by the cultured ES cells. However, because of the low efficiency of germ line transmission by ES cells transplanted into host embryos, production of germ line chimeric animals capable of passing on the appropriate genetics to the next generation remains problematic.

In mammals, this problem has been circumvented in a few situations by the use of nuclear transfer techniques to generate knockouts without the need to produce germ line chimeras (Denning *et al.*, 2001; Kolber-Simonds *et al.*, 2004). Nuclei from cells with the targeted mutation are transferred to enucleated oocytes, generating animals possessing the targeted mutation (clones). Several recent reports describing zebrafish and medaka cell cultures of gonadal stem cells capable of differentiating into functional sperm point toward a second potential means of obviating the need to generate ES cells capable of contributing to germ line (Hong *et al.*, 2004b). However, success in this technology, like that of ES technology, depends on maintaining normalcy of the cultures through the period of transfection and selection necessary to generate the targeted mutant. Although nuclear transfer and sperm-generation techniques have been applied successfully to zebrafish, medaka, and carp, the technical difficulty of these procedures and the low frequency of success make it desirable to also develop an ES cell-based approach to gene targeting (Lee *et al.*, 2002a; Shaoyi *et al.*, 1990).

B. Concepts and Procedures for Transgene Integration, Expression, and Homologous Recombination

Established, non-ES fish cell lines such as RTG-2 from rainbow trout gonad, CHSE-214, and some zebrafish cell lines have been employed for development of gene transfer technology through the introduction of reporter genes or viral gene delivery vectors (Carvan *et al.*, 2000, 2001; Overturf *et al.*, 2003; Sharps *et al.*, 1992). Application of this technology may be limited in some applications because these fish cell lines may not express the appropriate and critical tissue-specific transcription factors needed to ask specific molecular questions. However, this approach has been quite useful for some applications. For instance, some fish cell lines have been genetically engineered to make them more sensitive indicators of toxin exposure.

As an example, the luciferase reporter gene under the control of dioxin-responsive enhancers was introduced into RTH-149 cells to generate the recombinant cell line designated remodulated lightning trout or RTL 2.0 (Richter *et al.*, 1997). This cell line can be used in a high-throughput *in vitro* assay system for evaluating dioxin-like potency of halogenated aromatic hydrocarbons (Richter *et al.*, 1997; Villeneuve *et al.*, 2001). A gene transfer approach using rainbow trout gonad (RTG)-2 cells also has been developed to create an *in vitro* assay system for the detection of estrogenic compounds in environmental samples (Ackerman *et al.*, 2002: Rutishauser *et al.*, 2004). Similarly, the zebrafish embryo cell line, ZEM2s (zebrafish embryo 2-serum adapted), was transfected with plasmids containing a reporter gene under the control of either aromatic hydrocarbon, heavy metal, or electrophile response elements (Carvan *et al.*, 2000), so that expression from these transgenes indicated induction upon exposure to compounds belonging to each class of inducer. In another application of this technology, plasmid-based metallothionein expression was investigated in transfected CHSE-214 cells (Kling and Olsson, 2000), identifying this system in protecting the cells against oxidative stress.

In addition to these applied emphases, fish cell lines have been used to express exogenous DNA for a variety of more basic objectives. This includes identification of enhancer or promoter elements upstream from specific genes, and evaluations of the relative strength of the activities of these noncoding regions (Bearzotti *et al.*, 1992; Chan and Devlin, 1993; Friedenreich and Schartl, 1990; Fu and Aoki, 1991; Moav *et al.*, 1992). Promoters derived from several sources have been shown to function in fish cell lines in experiments emphasizing transient, stable, or inducible expression. These include the sockeye salmon metallothionein-B, histone H3 promoters, and the abalone actin promoter. Viral promoters derived from human cytomegalovirus, simian virus 40, and Rous sarcoma virus are also active on zebrafish and other fish species (Bearzotti *et al.*, 1999; Driever and Rangini, 1993; Gomez-Chiarri *et al.*, 1999; Sharps *et al.*, 1992). Recombinant retrovirus and baculovirus vectors have been used for transgene expression in fish cell lines, as has a pseudotyped rhabdovirus/retrovirus that can integrate into the germ line of zebrafish embryos and generate transgenic fish (Gaiano *et al.*, 1996). The recombinant retrovirus also generates proviral insertions at a sufficient frequency to make it the basis of insertional mutagenesis experiments (Burgess *et al.*, 2002; Gaiano *et al.*, 1996; Leisy *et al.*, 2003).

To be useful for biological experiments using the targeted gene approach, fish ES cells must not only remain germ line competent in culture but also stably incorporate vector DNA in a targeted fashion. This is accomplished by introducing foreign DNA into the coding region of a specific gene by homologous recombination, followed by *in vitro* selection and expansion of the successfully targeted colonies. In mouse ES cell cultures, colonies of homologous recombinants are commonly isolated using a duel drug selection procedure, and some colonies that survive long-term drug selection remain pluripotent in that they contribute to the formation of chimeras and the differentiation to multiple cell types *in vitro*. Gene targeting has been achieved in zebrafish ES cell cultures and colonies of homologous recombinants isolated using a visual screening method (Fig. 4).

The screening strategy involved the use of a targeting vector that contained the *neo* marker gene located within the region of the vector that was homologous to the gene being targeted and the red fluorescent protein (RFP) gene located outside of the homologous region. Using this approach, the cell colonies that incorporated the vector randomly were RFP positive, and those that had undergone homologous recombination and targeted insertion of the vector were RFP negative. The vector was introduced into the ES cells by electroporation and following G418 selection to eliminate nontransfected cells, the colonies that completely lacked

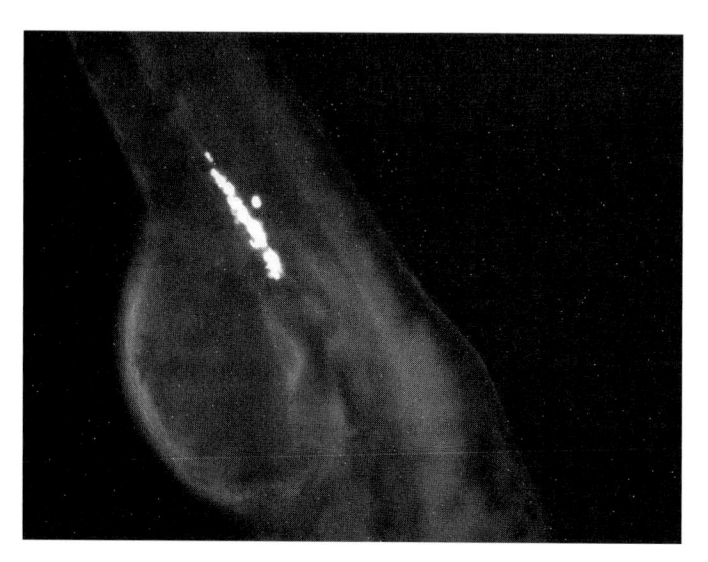

Fig. 4 Transgenic zebrafish larva expressing red fluorescent protein (RFP) in the primordial germ cells. RFP expression is driven by the *vasa* promoter. (See Plate no. 25 in the Color Plate Section.)

RFP expression were manually selected from the dish and expanded. In addition to the use of a visual selection method to identify homologous recombinants, a positive/negative selection approach was also used to isolate colonies of zebrafish ES cells that had undergone targeted plasmid insertion (Fan *et al.*, 2006). This approach involved the use of neo to select against nontransfected cells together with the diphtheria toxin gene to eliminate cells that incorporated the plasmid by random insertion.

Both the visual and positive/negative selection methods have been used to successfully isolate several colonies of zebrafish ES cells that carry targeted plasmid insertions (Fan *et al.*, 2006). However, application of this strategy toward the production of knockout lines of fish will require further work to optimize the ES cell culture system, identify markers of pluripotency, and improve the efficiency of germ line chimera production. Recent production of transgenic fish that express fluorescent marker genes under the control of PGC-specific promoters has been reported, and may provide a valuable complement to the ES cell-based work (Krovel and Olsen, 2002) (see also Fig. 3).

III. Tissue-Specific Stem Cells

The term "stem cell" often is applied to either tissue-specific stem cells such as hematopoietic cells or to embryonic stem cells capable of differentiating to any tissue. Studies of a variety of tissue-specific stem cells appear in the literature. For instance, in mammals a multipotent neural precursor cell type exists, capable of differentiating into neuronal and glial cell types, as well as oligodendrocytes (Murayama *et al.*, 2001). Furthermore, these cells may also differentiate into muscle or blood. Evidence of tissue-specific stem cells in many organs requires a reevaluation of the nature of the cell types derived in culture from some tissues. This may be of particular importance for fish cell cultures, because fish exhibit regenerative potential considerably beyond that of mammals. Fish models and cell cultures showing differentiation potential from regenerating organs, like liver, kidney, heart, retina, or cells of the immune system, may represent good models of tissue-specific stem cells.

A. Stem Cell Lines and the Methodological Understanding of Cell Culture Development

In addition to providing specialized media formulations as described above for the initiation and early passage of cell lines of a specific nature, the processing of tissue and primary plating can be of critical importance. An example is the derivation of zebrafish brain and ovary-derived cell cultures. In this procedure, brains are pooled from juvenile zebrafish, and ovaries pooled from reproductively active females. Tissues are washed twice with LDF medium supplemented with 2% heat-inactivated fetal bovine serum, 0.25% trout serum, trout embryo extract

(50 µg/ml), 2-mercaptoethanol (55 µM), chemically defined lipids (1:1000), peptide growth factors, and nutritional components. These cells are cultured at 27.5 °C.

To initiate primary cultures, washed tissues are minced, dispersed by pipetting, pelleted at 1500 rpm, and resuspended in LDF twice before plating in 24-well plates (1 ml medium per well at $\sim 10^5$ cells per well). Wells also are precoated with fibronectin (10 µg/ml). Individual cells attach and spread by 48 h after plating, and the medium changed at this time to remove unattached cell aggregates. Cultures reach confluence after about 10 days. To produce secondary cultures, cells in confluent wells are trypsinized at room temperature, pelleted, washed, and plated at a split ratio of 1:2. Cultures have been maintained for multiple passages in this manner. The brain cells exhibit an astrocytic morphology when plated at low cell densities, but develop a more epithelial morphology as the cultures approach monolayer density. Ovary cells retain a primarily bipolar, fibroblast-like morphology.

A likely example of a cell line containing tissue-specific stem cells is the XM line developed from *Xiphophorus* (Barnes *et al.*, 2006) (Fig. 1D). Procedures similar to those used to establish zebrafish brain and ovary cultures were employed, but cells not adhering to the culture plates in the initial plating were transferred several times to other plates in the following days after plating. This procedure is useful in cases in which cultivation temperature is low and a large amount of noncellular material is present in the initial plating preparations. For XM cells the source of material was an external melanoma. This tissue contains a large amount of extracellular matrix as well as dead cells, both malignant and normal skin cells and melanophores possibly derived from partial or complete differentiation of initially malignant cells. In addition, as an external tissue, it was potentially heavily microbially contaminated. For this reason, minced tissue was washed and initially processed in LDF basal nutrient medium containing a high-concentration antibiotic solution (1.25 units penicillin G, 1.25 g streptomycin sulfate, 1.25 g neomycin sulfate, and 12,500 units bacitracin per milliliter).

The cells were disaggregated in medium containing 0.01 mg/ml dispase II for 30–45 min, minced, and again washed in medium. Tissue clumps were dispensed into bovine fibronectin-coated wells of a 48-well plate. Growth medium for the initial derivation of the cell line was LDF basal nutrient medium supplemented with insulin, selenious acid, holo-transferrin, chemically defined lipids, recombinant human EGF, recombinant human fibroblastic growth factor, recombinant human interleukin-6, recombinant human ciliary neurotropic factor (CNTF), non-essential amino acids, β-mercaptoethanol (55 µM), trout embryo extract (50 µg/m), and L-glutamine, at concentrations previously published (Barnes *et al.*, 2006).

This large number of supplements was used because it was unclear which components might be stimulatory for the cells. Each component had been found previously to be stimulatory for some type of fish cell in culture, and none had been found to be inhibitory on any fish culture. Subsequently each component was tested for effects on the cells, and those having no effect were eliminated from the medium as soon as possible. Antibiotics for the plating were reduced to a concentration 10-fold lower than the concentrations used for the initial washing. Cells were incubated at 24 °C.

The next day, medium with floating tissue from the initial plating was transferred into new fibronectin-coated wells, and fresh medium supplemented as described above was added to the initially plated wells with fetal bovine serum (2%) and trout serum (0.25%) as additional components. It has been previously observed that plating fish cells directly into serum-containing medium can prevent attachment. Two days later, the floating cells and tissue clumps were transferred to a third set of collagen-coated wells. Cells attached and healthy cells began to proliferate in wells from all three platings of the initial material. After 2–4 weeks, the cultures were passed at a splitting density of 1:2.

Beyond three passages, several basal nutrient media were evaluated for cell proliferation, and the medium was changed to M154, originally formulated for human keritinocytes, supplemented as stated above with antibiotics: penicillin (200 international units/ml), streptomycin sulfate (200 μg/ml), and ampicillin (25 μg/ml). For passage, cultures at confluency were dissociated with 0.2% trypsin, 1 mM EDTA in HEPES-buffered saline. After examination of growth factor requirements over ∼10 months of culture, L-glutamine was raised to 2.0 mM and CNTF, interleukin-6, and trout extract were removed from culture. Platelet-derived growth factor (PDGF) was also found to be stimulatory at this point, and subsequently was routinely added to the medium. The large majority of cells initially in culture were melanotic, but continuously proliferating cells were not, a phenomenon also seen with mammalian melanomas and possibly indicative of selection for the stem cells.

Flow cytometric analysis (see below) suggested that XM cells were diploid or near-diploid, and the cells were demonstrated to act as melanoma and melanocyte stem cells by injection into zebrafish embryos. To identify injected XM cells with certainty in the embryos, the cells were previously transfected with the pDsRed-Express-N1 vector containing the RFP gene under control of the human cytomegalovirus early promoter. For drug-resistant selection, the neomycin-resistance gene under control of the SV40 promoter also was present on the plasmid. Transfection was carried out with Cellfectin reagent by procedures described by the manufacturer. G418 disulfate at concentrations up to 1.0 mg/ml was used as the drug selection. Under these conditions frequency of stable tranfectants was ∼1 per 1000 cells. Drug selection was maintained for 5 months before injection into zebrafish embryos.

Zebrafish embryos at early gastrula, a stage before the development of immune-related cells and about 20 h before the development of pigmented cells, were dechorionated with 0.5 mg/ml pronase in Holtfreter's solution and overlayed into agar depressions in petri dishes. The reporter gene-expressing XM cells were removed from culture dishes by treatment with trypsin/ ethylenediaminetetraacetic acid (EDTA) in a HEPES-buffered NaCl solution. Phosphate-buffered saline routinely is not used for these or other cells in manipulations of live cells. The trypsin was inhibited with fetal calf serum in M154, washed, and resuspended in a small volume. Cells were loaded into a pulled Pasteur pipette and injected into embryos (Collodi et al., 1992). Postinjection embryos were incubated at 28 °C in 6 cm diameter dishes, in sterile tank water. XM cells were capable of producing tumors when injected in these embryos, with pigmented, melanoma-like cells appearing at about 24 h postfertilization. Some of these cells also migrated and appeared to differentiate consistent with the behavior of normal melanocytes.

Initiation and passage of the cartilaginous fish SAE cell line was pursued by a yet different initial approach (Parton et al., 2007). Separate cultures were initiated from a mixture of small embryos ranging from embryonic stages 20–23 and pooled cells from two embryos at stages 24–26 and stages 27–29. These developmental stages represent early points in development, no later than the appearance of gill filaments or formation of a capillary yolk network and before eye pigmentation. Cells were cultured in the cartilaginous fish medium described earlier. Fetal bovine serum at 2% is critical to the growth of cells. A small growth effect of FGF is apparent, and transforming growth factor-beta (TGFb) also increases cell number, but was not included in the medium in which the line was originally derived. TGFb was found to affect the cells in experiments in which components of the medium were evaluated for stimulation in order to simplify the medium as described above for XM cells. The basal nutrient medium is LDF, and cells are maintained at 18 °C or less in ambient CO_2. By the seventh passage, the cells in all cultures appeared morphologically identical with equivalent growth rates, and were pooled. At this point the minimal population doubling time was ~10 days. This long population doubling time is consistent with the slow growth of these cold water animals and relatively low temperature at which the cells are cultured. SAE cells can be cryopreserved in medium with 10% glycerol; recovery of frozen cells is ~50%.

Identification of the SAE cell type was determined by examining the sequences of expressed sequence tags (ESTs) from a normalized SAE cDNA library. Five thousand ESTs were sequenced from the 5' end, assembled as contigs and subjected to BLAST analysis. ESTs showing expected (e) values of high confidence that the appropriate protein had been identified (e^{-20} minimum) were further examined. This cautionary approach identified only shark proteins strongly related to homologues of higher vertebrates, and ESTs representing other less homologous shark proteins that may be of equal significance were not considered in the analysis. ESTs meeting the required criteria were scrutinized more closely for tissue-specific markers. A number of mesenchymal stem cell markers were identified, including connective tissue growth factor (CTGF) and mesenchymal stem cell protein. A variety of extracellular matrix molecules were detected, some commonly associated with mesenchymal stem cells, including collagens and proteoglycans. A variety of proteins influencing proliferation, differentiation, and development also were seen, including bone morphogenetic proteins (BMPs 1 and 4, TGFb 2, TGFb-1-binding protein, and FGF receptor; Parton et al., 2007). These exhibit activities related to mesenchymal stem cell differentiation, but are also more widespread in function.

Other markers and signaling molecules were identified (see Parton *et al.*, 2007), and also overlap in relevance to mesenchymal stem cells. These signaling markers affect differentiation into related tissues, including effects on muscloskeletal and neural development. Perhaps not surprisingly, osteoblast markers (bone sialoprotein, osteopontin, the matrix protein osteonectin, and the mineral-binding protein osteocalcin) were not detected. In addition, no markers for adipocytes (peroxide proliferation activation receptor gamma, adipocyte fatty acid transporter, or lipid-binding protein) were detected. Sharks primarily store lipid in liver with little or no fat accumulating elsewhere, and we have found no reports of adipocytes in elasmobranchs in the literature or evidence of adipocytes in examination of the dogfish shark.

Cartilage-specific markers in SAE cells are prevalent, including chondrocyte-specific protein, cartilage-associated protein, chondroitin polymerizing factor, ch-runt/runx, and chondrocyte-specific collagens (Parton *et al.*, 2007). Some of these are associated with early stages of osteogenesis in other vertebrates. Cartilage formation is a prerequisite to bone formation of higher organisms, and these proteins in elasmobranchs likely function in cartilaginous tissue regulation and formation. Neural markers also were found. This is reasonable, since evidence exists that a common mesenchymal-derived type of stem cell may give rise to both neural and muscle cells in mammals, and it has been suggested that muscle and neural markers present in mesenchymal stem cell cultures may be expressed at low levels in the entire culture population, or, alternatively, in a subpopulation of more differentiated cells. Taken as a whole, the results suggest that SAE cells represent embryonic mesenchymal stem cells.

Analysis of ESTs sequenced from the 3′ end of SAE cells also identified a number of highly evolutionarily conserved, potentially regulatory, untranslated regions (UTRs) (Forest *et al.*, 2007). All eight of the genes identified in this way are involved in cell growth and development Phylogenetic footprints as long as 203 nucleotides in length were identified. In at least one gene, a similar homology was found to exist in the chimera (Holoencephili) *Callorhinchus milii*, a more primitive cartilaginous fish than sharks (Forest *et al.*, 2007). Repetitive elements specific to cartilaginous fish also were identified by this analysis of ESTs from SAE cells (Parton *et al.*, 2007).

The complicated medium formulation of the more difficult cell types makes culture expensive and tedious. However, it has been possible in several instances to derive cell lines by highly complicated means and then simplify the medium formulation and procedures so that the system was more practical for other laboratories. Success with this approach has been achieved previously with both pufferfish and zebrafish embryo-derived cell lines (Barnes and Collodi, 2005; Bradford *et al.*, 1997). For instance, the ZEM2S zebrafish embryo-derived cell line, which is used for a variety of purposes in a number of laboratories, is the result of adaptation to a medium supplemented only with fetal calf serum (see below). The risk in this approach, of course, is that selection has occurred in culture for a cell type that is less representative of the original cultures or tissue of origin. Recently attempts have been made to simplify the highly specialized elasmobranch culture medium toward the goal of making cell lines from these species a tractable *in vitro* system available in laboratories that do not have access to sharks or low-temperature incubators. Adaptation of elasmobranch cell lines is being accomplished through a combination of cellular selection and minor adjustments to the basal nutrient medium, such as increased fetal calf serum and calcium levels. The necessity for a collagenous matrix substratum also has been eliminated, and temperature tolerance increased. Continued adaptation may further simplify the medium formulation, as well as improve growth rate, but may also introduce alterations in the phenotype and genotype of the cell.

It is often observed with both mammalian and fish cell lines that the cells once established in culture show an abnormally hyperploid karyotype, often grossly so. This phenomenon is seen, in particular, when cell lines are derived under conventional conditions. One example is the SHF-1 fin-derived line from sheepshead. Karyotyping showed the cell line to be aneuploid with a large marker metacentric chromosome (Gregory *et al.*, 1980). Other examples are a pufferfish cell line derived from F. niphobles (Bradford *et al.*, 1997) and the ZEM2S zebrafish cell line. Fish cell cultures with slightly hypoploid karyotype also have been reported (Clem *et al.*, 1961). It is a clear advantage for the interpretation of any experiment for the culture to remain genotypically and phenotypically as normal as possible. Stem cell cultures may fulfill these criteria if carefully maintained, but must be monitored to determine if either karyotypic or phenotypic drift is occurring.

Examples of diploid or near-diploid teleost fish cell lines predominate in cases in which the lines are established and maintained under species-specific conditions in which a variety of supplements are present in the medium in addition to serum. Examples are the previously discussed lines from *Fugu rubripes* and *Xiphophorus*, and some *Danio* lines (Barnes and Collodi, 2005; Barnes *et al.*, 2006; Bradford *et al.*, 1997). In the case of SAE, minimal information is available on the karyotype of *S. acanthias*, including chromosomal morphology, dye-banding, or gene location. However, flow cytometry as an indication of DNA content per nucleus may be used to estimate karyotype and this approach indicates that SAE cell DNA content is consistent with a diploid or slightly hypoploid karyotype (Parton *et al.*, 2007). Translocations or deletions existing in individual chromosomes would not be detected by this method.

Flow cytometry also has been used for analysis of leucocyte and other immune cell cultures from the little skate, *R. erinacea* (Fig. 2B). In elasmobranchs lymphocytes exist in blood and several types of immune organs (e.g., spleen). To create an enrichment of lymphocytes, red blood cells may be removed from the skate peripheral blood or single-cell suspensions prepared from the spleen by density gradient centrifugation using RediGrad reagent (Amersham Biosciences). A population that is primarily lymphocytes is trapped on surface of the gradient (density of 1.095 g/ml), and thus separated from both red blood cells and granulocytes that segregate to the pellet. These results were confirmed by May-Grunwald

Giemsa staining (Walsh and Luer, 2004) and by flow cytometric analysis with comparison to control samples taken before the centrifugation.

Flow cytometric analysis for fish is based on previously developed procedures. After fixing in 70% ethanol overnight, the resulting nuclei are resuspended in a staining solution containing 50 µg/ml propidium iodide, 100 units/ml RNaseA, and phosphate-buffered saline without calcium or magnesium. Nuclei are suspended at 3×10^6/ml and stained for 30 min. Immediately before measurement the samples are filtered through a 40 µm mesh. For an accurate statistical analysis, a minimum of 20,000 events per sample should be accumulated. This procedure showed that the lymphocyte population in the spleen was greatly enriched after the centrifugation (93%) compared with controls (62%). Similar results were obtained with peripheral blood cells. Skate lymphocyte cultures can be maintained for extended periods at very low temperatures (e.g., 4 °C) and remain active in several tests of immune function.

A number of other tissue-specific cell lines have been developed from teleosts. These include lines developed from normal liver and immune-mediating cells of several species, suggesting that tissue-specific stem cells may be a common phenomenon in fish (Barnes and Collodi, 2005). Tissue-specific stem cells might be expected to exist in normal tissues linked with the immune system of fish, and it may not be surprising that cell lines have been derived from these tissues. Pioneers in this work include Drs. W. Clem and G. Litman (Barnes and Collodi, 2005). The laboratory of Dr. Niels Bols has developed trout liver-derived lines, as well as a wealth of other differentiated fish cell lines, and demonstrated that several are transfectable (Lee *et al.*, 1993; Romoren *et al.*, 2004; Tom *et al.*, 2001). The ZFL zebrafish cell line, established from normal *Danio rerio* liver, has been used by several laboratories for studies of xenobiotic metabolism as well as a test cell line to evaluate effectiveness of plasmid constructions before use in generation of transgenic fish (Carvan *et al.*, 2001; Collodi *et al.*, 1992, 1994; Miranda *et al.*, 1993).

Stem cell lines also have been developed from goldfish, derived from tumors of integumental erythrophores. These have the capability to express differentiated products of other tissues, such as the crystallins of lens, and to undergo melanogenesis, formation of platelets, formation of teeth and fin rays, and expression of neuronal markers (Akiyama *et al.*, 1986; Matsumoto *et al.*, 1989). Goldfish and zebrafish also are common models for neural stem cell proliferation and differentiation in regenerating retinal tissue. Cell lines also have been developed from grouper brain and trout pituitary (Chi *et al.*, 1999; Tom *et al.*, 2001). These may represent tissue-specific stem cells, or cultures of mixed phenotype resulting from *in vitro* differentiation from a subpopulation of stem cells.

B. Applications to Biology and Medicine

Historically, fish cell lines have provided extremely useful tools for identification of the sources of fish diseases significant in commercial settings, studies of mechanisms of microbial infection, and the associated pathology (Bearzotti *et al.*, 1999;

Hogan et al., 1999; Hong and Wu, 2002; Hong et al., 1999a; Imajoh et al., 2003; Joseph et al., 2003; Lee et al., 2002b; Liu and Collodi, 2002). Applications to biotechnology related to veterinary medicine and aquaculture, as well as human biomedicine from a comparative point of view, are increasing. For instance, grass carp cells have been propagated in microcarrier suspension using polyacrylamide, polystyrene, and DEAE-dextran beads, allowing scale-up of channel catfish virus production and creating the potential for high-level manufacturing capacity of attenuated vaccine (Chen et al., 1992; Hogan et al., 1999; Leong and Fryer, 1993).

Fish cells as models are applicable to biomedicine in several aspects, such as the identification of new therapeutic compounds and the potential to serve as substrates for the safer production of recombinant molecules through minimization of contamination with fortuitous infective agents potentially dangerous to humans (Laville et al., 2004). The use of fish cells for toxicology (Bols et al., 2001, 2003; Ganassin et al., 2000) has been mentioned briefly above. Contributions of fish cell cultures to this discipline at a biochemical level include studies of xenobiotics and identification of the metabolites that are produced through CYP (P450)-mediated oxidation (epoxide formation and breakdown), and gluthione or glucuronide conjugation (Bols et al., 1999; Collodi et al., 1994; Jung et al., 2001; Tom et al., 2001; Zabel et al., 1996). The aryl-hydrocarbon receptor (AHR) has been studied in a number of fish cell types and experimental situations (Choi and Oris, 2003; Hestermann et al., 2002a,b; Huuskonen et al., 1998). In addition, cellular and molecular aspects of fish immunology critically relevant to comparative biomedicine and human health have been investigated using both primary cell cultures and immortalized cell lines (Bols et al., 2001, 2003; Clem et al., 1996).

Sharks have received recent publicity for providing insights into cancer biology. Although public media reports claiming that sharks do not get cancer largely are scientifically unsubstantiated (Ostrander et al., 2004), the spiny dogfish shark and other elasmobranchs are sources of antibiotics and angiogenesis inhibitors that may be promising cancer treatments (Kang et al., 2003b). Some of the public claims of shark compounds affecting malignancy have related to cartilage-derived compounds. Primary culture of skate cartilage cells has been reported, and the SAE cell line produces a number of cartilage-related molecules as described above (Fan et al., 2003; Parton et al., 2007). Should reports of biomedical effects of shark cartilage-derived compounds be substantiated, these cell models may allow the identification of potential therapeutic molecules, examination of synthetic pathways, and possible mechanisms of pharmaceutical action.

Fish clearly have interesting properties regarding aging, cancer, proto-oncogene and telomerase expression, stem cells, and immortalization (Barnes and Collodi, 2005; Barnes et al., 2005; Bradford et al., 1997; Forest et al., 2007). Many species continually increase in body size if food and space are unrestricted, and organs similarly increase in size. Several lines of evidence suggest that stem cells of considerable potential exist well into adulthood (Barnes and Collodi, 2005; Barnes et al., 2005; Mattingly et al., 2004). The use of fish for regenerative studies of some tissues, such as fin and retina, is well established, and other uses are appearing. For instance, a region of continual renal regeneration has been identified in the little skate, with new tubules forming continually through adulthood (Elger *et al.*, 2003), and cardiac regeneration in zebrafish has been demonstrated (Keating, 2004).

IV. Outlook and Future Contributions

As genomic sequence data accumulate for many fish species and genetic tools and reagents expand, fish models will continue to increase in importance for biological and biomedical studies. Large-scale genomic sequence data are now available for pufferfish, medaka, and zebrafish. Projects are underway to extend this knowledge to other fish species, including the following:

Salmonids (Altantic salmon, trout, and several species of Pacific salmon)
Ciclid species (Cichlasoma)
Cartilagenous fish (e.g., *Raja erinacea*)
Xiphophorus species. (e.g., *X. helleri, maculatus*)
Stickleback species (Gasterosteus)

With expanding applications of comparative genomic approaches to fish models, cell lines derived from fish embryos and differentiated tissues will increase in technological applications such as gene trap vectors and the production of gene trap libraries, functional characterization and use of RNAi and understanding of microRNAs, and proteomic profiling by microarray. Appropriate cell culture models for these applications will provide innovative approaches. Identification and use of purified recombinant fish growth factors and cytokines specifically from fish is in its infancy, and increased knowledge in that area almost surely will make it possible to derive and maintain new cell lines. Reliable markers of pluripotency and differentiation also will be necessary to characterize the cultures, design experiments, and interpret data. As progress continues in this area, fish cell lines will increasingly complement the numerous advantages of whole animal models for biological and biomedical research. Application of gene transfer methods leading to genetically engineered fish expressing genes under the control of tissue-specific or inducible promoters are producing valuable resources. Continual advances in identification of promoters and other regulatory elements in the genomes of model fish species will accelerate this work. Taken together, these approaches allow fish models to continue to contribute to understanding stem cells and aging in humans through comparative genomic and physiological studies.

Acknowledgments

Portions of this work were supported by NIH Grants RR-19732, RR017336, RR016463, and GM69384. Core support was provided through NIH-ES03828. The scientific expertise and participation by Dr. Steven Kazainis, The Wistar Institute, was critical in parts of the work described in this chapter. The work of Dr. Christopher J. Bayne was similarly critical. The authors also acknowledge the pervasive contributions of Dr. Niels Bols to the fields reviewed in this chapter. This chapter is dedicated to Dr. Bill Clem, former Chairman of the Department of Microbiology, University of Mississippi Medical School. Bill was an innovator. He took risks. More importantly, he was a kind, unassuming, and giving individual. DWB thanks Joe Ely, Gil Evans, and Amber Miller.

References

Ackerman, G. E., Brombacher, E., and Fent, K. (2002). Development of a fish reporter gene system for the assessment of estrogenic compounds and sewage treatment plant effluents. *Environ. Toxicol. Chem.* **21,** 1864–1875.

Akiyama, T., Matsumoto, J., Ishikawa, T., and Eguchi, G. (1986). Production of crystallins and lens-like structures in differentiation-induced neoplastic pigment cells (goldfish erythrophoroma cells) in vitro. *Differentiation* **33,** 34–44.

Amsterdam, A., Nissen, R. M., Sun, Z., Swindell, E. C., Farrington, S., and Hopkins, N. (2004). Identification of 315 genes essential for early zebrafish development. *Proc. Natl. Acad. Sci. USA* **35,** 12792–12797.

Babich, H., Goldstein, S. H., and Borenfreund, E. (1990). In vitro cyto- and genotoxicity of organo-mercurials to cells in culture. *Toxicol. Lett.* **50,** 143–149.

Barnes, D., and Collodi, P. (2005). In "Fish Cell Lines and Stem Cells in the Physiology of Fishes" (D. Evans, and J. B. Claiborne, eds.), pp. 553–575. CRC Press, Boca Rotan.

Barnes, D., Dowell, L., Forest, D., Parton, A., Pavicevic, P., and Kazianis, S. (2006). Characterization of XM, a novel Xiphophorus melanoma-derived cell line. *Zebrafish* **3,** 371–381.

Barnes, D., and Sato, G. H. (2000). In "Cell Culture Systems in Tissue Engineering" (P. Lazarrini, ed.), pp. 111–118.

Barnes, D. W., Mattingly, C. J., Parton, A., Dowell, L. M., Bayne, C. J., and Forrest, J. N. (2005). Marine organism cell biology and regulatory sequence discovery in comparative functional genomics. *Cytotechnology* **43,** 123–137.

Bearzotti, M., Perrot, E., Michard-Vanhee, C., Jolivet, G., Attal, J., Theron, M. C., Piossant, C., Dreano, M., Kopchick, J. L., Powell, R., et al. (1992). Gene expression following transfection of fish cells. *J. Biotechnol.* **26,** 315–325.

Bearzotti, M., Delmas, B., Lamoureux, A., Loustau, A. M., Chilmonczyk, S., and Bremont, M. (1999). Fish rhabdovirus cell entry is mediated by fibronectin. *J. Virol.* **73,** 7703–7709.

Bejar, J., Hong, Y., and Alvarez, M. C. (2002). An ES-like cell line from the marine fish Sparus aurata: Characterization and chimaera production. *Transgenic Res.* **11,** 279–289.

Bejar, J., Hong, Y. H., and Schartl, M. (2003). Mitf expression is sufficient to direct differentiation of medaka blastula derived stem cells to melanocytes. *Development* **130,** 6545.

Bernard, J., and Bremont, M. (1995). Molecular biology of fish viruses: A review. *Vet. Res.* **26,** 341–351.

Biga, P. R., Roberts, S. B., Iliev, D. B., McCaouey, L. A., Moon, J. S., Collodi, P., and Goetz, F. W. (2005). The isolation, characterization and expression of a novel dgf11 gene and a second myostatin form in zebrafish danio rerio. *Comp. Biochem. Physiol. B Biochem. Mol. Biol.* **141,** 218–230.

Bols, N. C., Schirmer, K., Joyce, E. M., Dixon, D. G., Greenberg, B. M., and Whyte, J. J. (1999). Ability of polycyclic aromatic hydrocarbons to induce 7-ethyoxyresorufin-o-deethylase activity in a trout liver cell line. *Ecotoxicol. Environ. Saf.* **44,** 118–128.

Bols, N. C., Brubacher, J. L., Ganassin, R. C., and Lee, L. E. J. (2001). Ecotoxicology and innate immunity in fish. *Dev. Comp. Immunol.* **25,** 853–873.

Bols, N. C. (2003). Cell culture approaches in aquatic immunotoxicology. In "Vitro Methods in Aquatic Toxicology" (C. Mothersill, and B. Austin, eds.), pp. 399–420. Springer, London.

Bols, N. C. (2004). Development and characterization of a cell line from a blastula stage rainbow trout embryo [abstract]. *In Vitro Cell. Dev. Biol.* **40,** 80A.

Borenfreund, E., and Puerner, J. A. (1984). A simple quantitative procedure using monolayer cultures for cytotoxicity assays. *J. Tissue Cult. Methods* **9,** 7–12.

Bradford, C., Nishiyama, K., Shirahata, S., and Barnes, D. (1997). Characterization of cell cultures derived from Fugu, the Japanese pufferfish. *Mol. Mar. Biol. Biotechnol.* **6,** 270–288.

Buck, C., Helmrich, A., Mericko, P., Toumadje, A., Kusumoto, K., Walsh, C., Davis, C., and Barnes, D. (2001). Cell culture and retrovirus expression from a tumor of a moray eel. *Mar. Biotechnol.* **3,** 193–202.

Burgess, S., Reim, G., Chen, W., Hopkins, N., and Brand, M. (2002). The zebrafish spiel-ohne-grenzen (spg) gene encodes the POU domain protein Pou2 related to mammalian Oct4 and is essential for formation of the midbrain and hindbrain, and for pre-gastrula morphogenesis. *Development* **129,** 905–916.

Caipang, C. M., Hirono, I., and Aoki, T. (2003). In vitro inhibition of fish rhabdoviruses by Japanese flounder, Paralichthys olivaceus Mx. *Virology* **317,** 373–382.

Carvan, M. J., Solis, W. A., Gedamu, L., and Nebert, D. W. (2000). Activation of transcription factors in zebrafish cell cultures by environmental pollutants. *Arch. Biochem. Biophy.* **376,** 320–327.

Carvan, M. J., 3rd, Sonntag, D. M., Cmar, C. B., Cook, R. S., Curran, M. A., and Miller, G. L. (2001). Oxidative stress in zebrafish cells: Potential utility of transgenic zebrafish as a deployable sentinel for site hazard ranking. *Sci. Total Environ.* **274,** 183–196.

Chan, W. K., and Devlin, R. H. (1993). Polymerase chain reaction amplification and functional characterization of sockeye salmon histone H3, metallothionein-B and protamine promoters. *Mol. Mar. Biol. Biotechnol.* **2,** 308–318.

Chen, Z., Chen, Y., Shi, Y., Yie, X., and Yang, G. (1992). Microcarrier culture of fish cells and viruses in cell culture bioreactor. *Can. J. Microbiol.* **38,** 222–225.

Chen, S. L., Ye, H. Q., Sha, Z. X., and Hong, Y. (2003a). Derivation of a pluripotent embryonic cell line from red sea bream blastulas. *J. Fish Biol.* **63,** 1–11.

Chen, S. L., Sha, Z. X., and Ye, H. Q. (2003b). Establishment of pluripotent embryonic cell line from sea perch (Lateolabrax japonicus) embryos. *Aqualculture* **218,** 141–151.

Chi, S. C., Lin, S. C., Su, H. M., and Hu, W. W. (1999). Temperature effect on nervous necrosis virus infection in grouper cell line and in grouper larvae. *Virus Res.* **63,** 107–114.

Choi, J., and Oris, J. T. (2003). Assessment of the toxicity of anthracene photo-modification products using the topminnow (Poeciliopsis lucida) hepatoma cell line (PLHC-1). *Aquat. Toxicol.* **65,** 243–251.

Clem, L. W., Bly, J. E., Wilson, M., Chinchar, V. G., Stuge, T., Barker, K., Luft, C., Rycyzyn, M., Hogan, R. J., van Lopik, T., and Miller, N. W. (1996). Fish immunology: The utility of immortalized lymphoid cells—A mini review. *Vet. Immunol. Immunopathol.* **54,** 137–144.

Clem, L. W., Moewus, L., and Sigel, M. M. (1961). Studies with cells from marine fish in tissue culture. *Proc. Soc. Exp. Biol. Med.* **108,** 762–766.

Collodi, P., and Barnes, D. (1990). Mitogenic activity from trout embryos. *Proc. Natl. Acad. Sci.* **87,** 3498–3502.

Collodi, P., Kame, Y., Ernst, T., Miranda, C., Buhler, D. R., and Barnes, D. W. (1992). Culture of cells from zebrafish (Brachydanio rerio) embryo and adult tissues. *Cell Biol. Toxicol.* **8,** 43–61.

Collodi, P., Miranda, C. L., Zhao, X., Buhler, D. R., and Barnes, D. W. (1994). Induction of zebrafish (Brachydanio rerio) P450 *in vivo* and in cell culture. *Xenobiotica* **24,** 487–493.

Denning, C., Burl, S., Ainslie, A., Bracken, J., Dinnyes, A., Fletcher, J., King, T., Ritchie, M., Ritchie, W. A., Rollo, M., de Sousa, P., and Travers, A. A. J. (2001). Deletion of the alpha(1,3) galactosyl transferase (GGTA1) gene and the prion protein (PrP) gene in sheep. *Nat. Biotechnol.* **19,** 559–562.

Driever, W., and Rangini, Z. (1993). Characterization of a cell line derived from zebrafish (Brachydanio rerio). *In Vitro Cell Dev. Biol.* **29A,** 749–754.

Dymecki, S. M. (2004). Switching on lineage tracers using site-specific recombination. In "Methods in Molecular Biology" (K. Turksen, ed.), Vol. 185, pp. 309–334.

Elger, M., Hentschel, H., Litteral, J., Wellner, M., Kirsch, T., Luft, F. C., and Haller, H. (2003). Nephrogenesis is induced by partial nephrectomy in the elasmobranch leucoraja erinacea. *J. Am. Soc. Nephrol.* **56,** 1506–1518.

Fan, I., Moon, I., Crodian, J., and Collodi, P. (2006). Homologous recombination in zebrafish ES cells. *Transgenic Res.* **1**, 21–30.

Fan, T. J., Jin, L. Y., and Wang, X. F. (2003). Initiation of cartilage cell culture from skate (Raja porasa Gunther). *Mar. Biotechnol. (NY)* **5**(1), 64–69.

Forest, D., Nishikawa, R., Kobayashi, H., Parton, A., Bayne, C. J., and Barnes, D. W. (2007). RNA expression in a cartilaginous fish cell line reveals ancient 3′ noncoding regions highly conserved in vertebrates. *Proc. Natl. Acad. Sci. USA* **104**, 1224–1229.

Friedenreich, H., and Schartl, M. (1990). Transient expression directed by homologous and heterologous promoter and enhancer sequences in fish cells. *Nucleic Acids Res.* **18**, 3299–3305.

Fu, L., and Aoki, T. (1991). Expression of CAT gene directed by the 5′-upstream region from yellowtail alpha-globin gene in tissue cultured fish cells. *Nippon Suisan Gakkaishi* **57**, 1689–1696.

Ganassin, R. C., and Bols, N. C. (1999). A stromal cell from rainbow trout spleen, RTS34ST, that supports the growth of rainbow trout macrophages and produces conditioned medium with mitogenic effects on leukocytes. *In Vitro Cell. Dev. Biol.* **35**, 80–86.

Ganassin, R. C., Schirmer, K., and Bols, N. C. (2000). Methods for the use of fish cell and tissue cultures as model systems in basic and toxicology research. *In* "The Laboratory Fish" (G. K. Ostrander, ed.), pp. 631–651. Academic Press, San Diego, CA.

Gaiano, N., Allende, M., Amsterdam, A., Kawakami, K., and Hopkins, N. (1996). Highly efficient germ line transmission of proviral insertions in zebrafish. *Proc. Natl. Acad. Sci. USA* **93**, 7777–7782.

Gertsenstein, M., Lobe, C., and Nagy, A. (2007). ES cell-mediated conditional transgenesis. *In* "Methods in Molecular Biology" (K. Turksen, ed.), Vol. 185, pp. 285–307.

Gomez-Chiarri, M., Kirby, V. L., and Powers, D. A. (1999). Isolation and characterization of an actin promoter from the red abalone (Haliotis rufescens). *Mar. Biotechnol.* **1**, 269–278.

Gregory, P. E., Howard-Peebles, P. N., Ellender, R. D., and Martin, B. J. (1980). Analysis of a marine fish cell line from a male sheepshead. *J. Hered.* **71**, 209–211.

Grutzner, F., Lutjens, G., Rovira, C., Barnes, D. W., Ropers, H. H., and Haaf, T. (1999). Classical and molecular cytogenetics of the pufferfish Tetraodon nigroviridis. *Chromosome Res.* **7**, 655–662.

Hart, D., Frerichs, G. N., Rambaut, A., and Onions, D. E. (1996). Complete nucleotide sequence and transcriptional analysis of snakehead fish retrovirus. *J. Virol.* **70**, 3606–3616.

Hasegawa, S., Nakayasu, C., Okamoto, N., Nakanishi, T., and Ikeda, Y. (1997). Fin cell line from isogeneic ginbuna crucian carp. *In Vitro Cell. Dev. Biol. Anim.* **33**, 232–233.

Hestermann, E. V., Stegeman, J. J., and Hahn, M. E. (2002a). Relationships among the cell cycle, cell proliferation, and aryl hydrocarbon receptor expression in PLHC-1 cells. *Aquat. Toxicol.* **58**, 201–213.

Hestermann, E. V., Stegeman, J. J., and Hahn, M. E. (2002b). Serum withdrawal leads to reduced aryl hydrocarbon receptor expression and loss of cytochrome P4501A inducibility in PLHC-1 cells. *Biochem. Pharmacol.* **63**, 1405–1414.

Hogan, R. J., Taylor, W. R., Cuchens, M. A., Naftel, J. P., Clem, L. W., Miller, N. W., and Chinchar, V. G. (1999). Induction of target cell apoptosis by channel catfish cytotoxic cells. *Cell. Immunol.* **195**, 110–118.

Hong, J. R., and Wu, J. L. (2002). Induction of apoptotic death in cells via Bad gene expression by infectious pancreatic necrosis virus infection. *Cell Death Differ.* **9**, 113–124.

Hong, J. R., Lin, T. L., Yang, J. Y., Hsu, Y. L., and Wu, J. L. (1999a). Dynamics of nontypical apoptotic morphological changes visualized by green fluorescent protein in living cells with infectious pancreatic necrosis virus infection. *J. Virol.* **73**, 5056–5063.

Hong, Y., Chen, S., Gui, J., and Schartl, M. (2004a). Retention of the developmental pluripotency in medaka embryonic stem cells after gene transfer and long-term drug selection for gene targeting in fish. *Transgenic Res.* **13**, 41–50.

Hong, Y., Liu, T., Zhao, H., Xu, H., Wang, W., Liu, R., Chen, T., Deng, J., and Gui, J. (2004b). Establishment of a normal medakafish spermatogonial cell line capable of sperm production *in vitro*. *Proc. Natl. Acad. Sci. USA* **101**, 8011–8016.

Hong, Y., Winkler, C., Liu, T., Chai, G., and Schartl, M. (2004c). Activation of the mouse Oct4 promoter in medaka embryonic stem cells and its use for ablation of spontaneous differentiation. *Mech. Dev.* **121,** 933–943.

Hong, Y., Winkler, C., and Schartl, M. (1998). Production of medaka fish chimeras from a stable embryonic stem cell line. *Proc. Natl. Acad. Sci. USA* **95,** 3679–3684.

Huuskonen, S. E., Hahn, M. E., and Lindstrom-Seppa, P. (1998). A fish hepatoma cell line (PLHC-1) as a tool to study cytotoxicity and CYP1A induction properties of cellulose and wood chip extracts. *Chemosphere* **36,** 2921–2932.

Imajoh, M., Yagyu, K., and Oshima, S. (2003). Early interactions of marine birnavirus infection in several fish cell lines. *J. Gen. Virol.* **84,** 1809–1816.

Joseph, T., Kibenge, M. T., and Kibenge, F. S. (2003). Antibody-mediated growth of infectious salmon anaemia virus in macrophage-like fish cell lines. *J. Gen. Virol.* **84,** 1701–1710.

Jung, D. K., Klaus, T., and Fent, K. (2001). Cytochrome P450 induction by nitrated polycyclic aromatic hydrocarbons, azaarenes, and binary mixtures in fish hepatoma cell line PLHC-1. *Environ. Toxicol. Chem.* **20,** 149–159.

Kamer, I., Douek, J., Tom, M., and Rinkevich, B. (2003). Metallothionein induction in the RTH-149 cell line as an indicator for heavy metal bioavailability in a brackish environment: Assessment by RT-competitive PCR. *Arch. Environ. Contam. Toxicol.* **45,** 86–91.

Kang, J., Kim, J., Song, H., Bae, M., Yi, E., Kim, K., and Kim, Y. (2003b). Anti-angiogenic and antitumor invasive activities of of tissue inhibitor of metalloproteinase-3 from shark, *Scyliorhinus torazame*. *Biochim. Biophys. Acta.* **1620,** 59–64.

Keating, M. T. (2004). Genetic approaches to disease and regeneration. *Philos. Trans. R. Soc. Lond. B Biol. Sci.* **359,** 795–798.

Kling, P. G., and Olsson, P. E. (2000). Involvement of differential metallothionein expression in free radical sensitivity of RTG-2 and CHSE-214 cells. *Free Radic. Biol. Med.* **28,** 1628–1637.

Kolber-Simonds, D., Lai, L., Watt, S. R., Denaro, M., Arn, S., Augenstein, M. L., Betthauser, J., Carter, D. B., Greenstein, J. L., Hao, Y., Im, G. S., Liu, Z., *et al.* (2004). Production of alpha 1,3-galactosyltransferase null pigs by means of nuclear transfer with fibroblasts bearing loss of heterozygosity mutations. *Proc. Natl. Acad. Sci. USA* **101,** 7335–7340.

Krovel, A., and Olsen, L. C. (2002). Expression of a *vasEGFP* transgene in primordial germ cells of the zebrafish. *Mech. Dev.* **116,** 141–150.

Laville, N., Ait-Aissa, S., Gomez, E., Casellas, C., and Porcher, J. M. (2004). Effects of human pharmaceuticals on cytotoxicity, EROD activity and ROS production in fish hepatocytes. *Toxicology* **196,** 41–55.

Lee, K.-Y., Huang, H., Ju, B., Yang, Z., and Lin, S. (2002a). Cloned zebrafish by nuclear transfer from long-term-cultured cells. *Nat. Biotechnol.* **20,** 795–799.

Lee, K., Chi, S., and Chen, T. (2002b). Interference of the life cycle of fish nodavirus with fish retrovirus. *J. Gen. Virol.* **83,** 2469–2474.

Lee, L. E., Clemons, J. H., Bechtel, D. G., Caldwell, S. J., Han, K. B., Pasitschniak-Arts, M., Mosser, D. D., and Bols, N. C. (1993). Development and characterization of a rainbow trout liver cell line expressing cytochrome P450-dependent monooxygenase activity. *Cell Biol. Toxicol.* **9,** 279–294.

Lee, L. E. J., Vijayan, M. M., and Dixon, B. (in press). Toxicogenomic technologies for *in vitro* aquatic toxicology. *In* "Vitro Aquatic Toxicology" (B. Austin, and C. Mothersill, eds.), Springer, London.

Leisy, D. J., Lewis, T. D., Leong, J. A., and Rohrmann, G. F. (2003). Transduction of cultured fish cells with recombinant baculoviruses. *J. Gen. Virol.* **84,** 1173–1178.

Leong, J. C., and Fryer, J. L. (1993). Viral vaccines for aquaculture. *Annu. Rev. Fish Dis.* **3,** 225–240.

Liu, X., and Collodi, P. (2002). Novel form of fibronectin from zebrafish mediates infectious hematopoietic necrosis virus infection. *J. Virol.* **76,** 492–498.

Lopez-Doriga, M. V., Smail, D. A., Smith, R. J., Doménech, A., Castric, J., Smith, P. D., and Ellis, A. E. (2001). Isolation of salmon pancreas disease virus (SPDV) in cell culture and its ability to protect against infection by the 'wild-type' agent. *Fish Shellfish Immunol.* **11,** 505–522.

Lorenzen, E., Carstensen, B., and Olesen, N. J. (1999). Inter-laboratory comparison of cell lines for susceptibility to three viruses: VHSV, IHNV, and IPNV. *Dis. Aquat. Organ.* **37,** 81–88.

Lu, Y. A., Lannan, C. N., Rohovec, J. S., and Fryer, J. L. (1990). Fish cell lines: Establishment and characterization of three new cell lines from grass carp (*Ctenopharyngodon idella*). *In Vitro Cell. Dev. Biol.* **26,** 275–279.

Ma, C., Fan, L., Ganassin, R., Bols, N., and Collodi, P. (2001). Production of zebrafish germ-line chimeras from embryo cell cultures. *Proc. Natl. Acad. Sci. USA* **98,** 2461–2466.

Martin-Alguacil, N., Babich, H., Rosenberg, D. W., and Borenfreund, E. (1991). *In vitro* response of the brown bullhead catfish cell line, BB, to aquatic pollutants. *Arch. Environ. Contam. Toxicol.* **20,** 113–117.

Matsumoto, J., Wada, K., and Akiyama, T. (1989). Neural crest cell differentiation and carcinogenesis: Capability of goldfish erythrophoroma cells for multiple differentiation and clonal polymorphism in their melanogenic variants. *J. Invest. Dermatol.* **92,** 255S–260S.

Mattingly, C., Parton, A., Dowell, L., Rafferty, J., and Barnes, D. (2004). Cell and molecular biology of marine elasmobranchs: Squalus acanthias and Raja erinacea. *Zebrafish* **1,** 111–120.

Mcallister, P. E. (1997). Susceptibility of 12 lineages of Chinook salmon embryo cells (CHSE-214) to four viruses from salmonid fish implications for clinical assay sensitivity. *J. Aquat. Anim. Health* **9,** 291–294.

Miranda, C. L., Collodi, P., Zhao, X., Barnes, D. W., and Buhler, D. R. (1993). Regulation of cytochrome P450 expression in a novel liver cell line from zebrafish (Brachydanio rerio). *Arch. Biochem. Biophys.* **305,** 320–327.

Moav, B., Liu, Z. J., and Hackett, P. (1992). Selection of promoters for gene transfer into fish. *Mol. Mar. Biol. Biotechnol.* **1,** 338–345.

Moredock, S., Nairn, R. S., Johnston, D. A., Byrom, M., Heaton, G., Lowery, M., and Mitchell, D. L. (2003). Mechanisms underlying DNA damage resistance in a Xiphophorus melanoma cell line. *Carcinogenesis* **24,** 1967–1975.

Mork, C., Hershberger, P., Kocan, R., Batts, W., and Winton, J. (2004). Isolation and characterization of a rhabdovirus from starry flounder (Platichthys stellatus) collected from the northern portion of Puget Sound, Washington, USA. *J. Gen. Virol.* **85,** 495–505.

Murayama, K., Singh, N. N., Helmrich, A., and Barnes, D. W. (2001). Neural Cell Lines. *In* "Protocols for Neural Cell Cultures" (S. Federhoff, and A. Richards, eds.), pp. 219–228. Humana Press, Totowa, NJ.

Ostrander, G. K., Cheng, K. C., Wolf, J. C., and Wolfe, M. J. (2004). Shark cartilage, cancer and the growing threat of pseudoscience. *Cancer Res.* **64,** 8485–8491.

Overturf, K., LaPatra, S., and Reynolds, P. N. (2003). The effectiveness of adenoviral vectors to deliver and express genes in rainbow trout, Oncorhynchus mykiss (Walbaum). *J. Fish Dis.* **26,** 91–101.

Parton, A., Forest, D., Kobayashi, H., Dowell, L., Bayne, C., and Barnes, D. (2007). Cell and molecular biology of SAE, a cell line from the spiny dogfish shark, Squalus acanthias. *Comp. Biochem. Physiol.* **145,** 111–119.

Perez-Prieto, S. I., Rodriguez-Saint-Jean, S., Garcia-Rosado, E., Castro, D., Alvarez, M. C., and Borrego, J. J. (1999). Virus susceptibility of the fish cell line SAF-1 derived from gilt-head seabream. *Dis. Aquat. Organ.* **35,** 149–153.

Richter, C. A., Tieber, V. L., Denison, M. S., and Giesy, J. P. (1997). An *in vitro* rainbow trout cell bioassay for aryl hydrocarbon receptor-mediated toxins. *Environ. Toxicol. Chem.* **16,** 543–550.

Romoren, K., Thu, B. J., Bols, N. C., and Evensen, O. (2004). Transfection efficiency and cytotoxicity of cationic liposomes in salmonid cell lines of hepatocyte and macrophage origin. *Biochim. Biophys. Acta* **1663,** 127–134.

Ruiz-Leal, M., and George, S. (2004). An *in vitro* procedure for evaluation of early stage oxidative stress in an established fish cell line applied to investigation of PHAH and pesticide toxicity. *Mar. Environ. Res.* **58,** 631–635.

Rutishauser, B. V., Pesonen, M., Escher, B. I., Ackermann, G. E., Aerni, H. R., Suter, M. J., and Eggen, R. I. (2004). Comparative analysis of estrogenic activity in sewage treatment plant effluents involving three *in vitro* assays and chemical analysis of steroids. *Environ. Toxicol. Chem.* **23,** 857–864.

Scholz, S., Domaschke, H., Kanamori, A., Ostermann, K., Rodel, G., and Gutzeit, H. (2004). Germ cell-less expression in medaka. *Mol. Reprod. Dev.* **67,** 15–18.

Shaoyi, Y. (1990). Developmental incompatibility between cell nucleus and cytoplasm as revealed by nuclear transplantation experiments in teleost of different families and orders. *Int. J. Dev. Biol.* **34,** 255–265.

Sharps, A., Nishiyama, K., Collodi, P., and Barnes, D. (1992). Comparison of activities of mammalian viral promoters directing gene expression *in vitro* in zebrafish and other fish cell lines. *Mol. Mar. Biol. Biotechnol.* **1,** 426–431.

Shimizu, C., Shike, H., Malicki, D. M., Breisch, E., Westerman, M., Buchanan, J., Ligman, H. R., Phillips, R. B., Carlberg, J. M., Van Olst, J., and Burns, J. C. (2003). Characterization of a white bass (*Morone chrysops*) embryonic cell line with epithelial features. *In Vitro Cell. Dev. Biol. Anim.* **39,** 29–35.

Smith, S. F., Snell, P., Gruetzner, F., Bench, A. J., Haaf, T., Metcalfe, J. A., Green, A. R., and Elgar, G. (2002). Analyses of the extent of shared synteny and conserved gene orders between the genome of Fugu rubripes and human 20q. *Genome Res.* **12,** 776–784.

Sun, L., Bradford, C. S., and Barnes, D. W. (1995). Feeder cell cultures for zebrafish embryonal cells *in vitro*. *Mol. Mar. Biol. Biotechnol.* **4,** 43–50.

Syed, M., Vestrheim, O., Mikkelsen, B., and Lundin, M. (2003). Isolation of the promoters of Atlantic salmon MHCII genes. *Mar. Biotechnol.* **5,** 253–260.

Tom, D. J., Lee, L. E., Lew, J., and Bols, N. C. (2001). Induction of 7-ethoxyresorufin-O-deethylase activity by planar chlorinated hydrocarbons and polycyclic aromatic hydrocarbons in cell lines from the rainbow trout pituitary. *Comp. Biochem. Physiol. A Mol. Integr. Physiol.* **128,** 185–198.

Villeneuve, D. L., Khim, J. S., Kannan, K., and Giesy, J. P. (2001). *In vitro* response of fish and mammalian cells to complex mixtures of polychlorinated naphthalenes, polychlorinated biphenyls, and polycyclic aromatic hydrocarbons. *Aquat. Toxicol.* **54,** 124–141.

Wakamatsu, Y., Ozato, K., and Sasado, T. (1994). Establishment of a pluripotent cell line derived from a medaka (*Oryzias latipes*) blastula embryo. *Mol. Mar. Biol. Biotechnol.* **3,** 185–191.

Walsh, C. J., and Luer, C. J. (2004). Elasmobranch husbandry manual. *In* "Elasmobranch Hematology: D. Identification of Cell Types and Practical Applications" (M. Smith, D. Warmolts, D. Thoney, and R. Hueter, eds.), pp. 307–323. Ohio Biological Survey, Inc., Columbus, Ohio.

Wergeland, H. I., and Jakobsen, R. A. (2001). A salmonid cell line (TO) for production of infectious salmon anaemia virus (ISAV). *Dis. Aquat. Organ.* **44,** 183–190.

Wolf, K. (1988). Fish viruses and fish diseases. *In* "Comstock Publishing Associates." Cornell University Press, Ithaca, NY.

Zabel, E. W., Pollenz, R., and Pederson, R. E. (1996). Relative potencies of individual dibenzofuran, and biphenyl congeners and congener mixtures based on induction of cytochrome P4501A mRNA in a rainbow trout gonadal cell line (RTG-2). *Environ. Toxicol. Chem.* **15,** 2310–2318.

Zhang, Q. Y., Ruan, H. M., Li, Z. Q., Yuan, X. P., and Gui, J. F. (2003). Infection and propagation of lymphocystis virus isolated from the cultured flounder Paralichthys olivaceus in grass carp cell lines. *Dis. Aquat. Organ.* **57,** 27–34.

INDEX

A

Accutase™ enzyme, 168–169
Acute tubular necrosis (ATN), 97
Adhesion receptor expression, 289–291; *See also* Mesenchymal progenitor cells (MPCs)
  activation of MAP kinase pathway, 290
  adhesion receptor subunit CD49e of, 289, 291
  β1-dependent adhesion and integrin-mediated interactions, 291
Adipocytes, in elasmobranchs, 357
Adipogenesis assays for MSCs differentiation, 127
Adipose-derived adult stromal/stem (ADAS) cells, 258
Adipose-derived MSCs, 102
Adipose tissue, hMSCs isolation from, 263–265
Adult diploid and polyploid cells in liver tissue, 143
Alpha-modified minimum essential medium, 261
Amniotic fluid, 86
  cell population within, 87–88
  isolation of stem cells from, 89–91
    characterization, 88, 90
Antibiotics, in kidney epithelial cultures, 252–253
Assisted hatching; *See also* Blastocyst preparation
  acidic Tyrode's solution by, 6
  by laser, 6
  partial zona dissection, 7

B

Basal media, biological fluid for cell culture, 176–179
Basic Fibroblast growth factor (bFGF), 9, 17, 173, 228, 236, 282, 316, 346–347, 356
Biopore PTFE membranes, 34
Blastocyst preparation, 4
  immunosurgery, 7–8
  sequential culture media, 5
  thawing cryopreserved embryos, 5
  zona pellucida, removal of, 6–7
Bluegill BG/F-2, cell lines, 344
Bone marrow-derived MSCs (BM-MSCs)
  clinical trials, 102
  proliferative capacity, 129
  Wnt signaling pathway, 128–129
Bone marrow-derived MSCs, clinical trials, 102
Bovine serum albumin (BSA), 69
BrdU staining procedure, 193
Brown bullhead catfish (BB), cell line, 344
Buffer preparation, 207–208

C

E-Cadherin, 76
Calcium content, in I/3F medium, 244–245
California Institute of Regenerative Medicine (CIRM), 2
Cancer Pathway Finder gene array, 129
Cancer Stem Cell Line, 330
Cancer stem cells (CSCs), 312, 326–327; *See also* Tumor cell lines, isolation and establishment
  medium selection for growth of, 332–333
  phenotypes growth conditions, importance of, 337–338
  serum effect on, 333–335
Carp epithelial cell (EPC), 344
Cartilage-associated protein, 357
Cartilage-specific markers, in SAE cells, 357
CD133 and CD44 glycoprotein, 90, 197–198, 248, 253, 305, 313–314, 318–321, 337
CD45⁻Suspension cell fraction, 284–286; *See also* Mesenchymal progenitor cells (MPCs)
Cell adhesion, *See* Adhesion receptor expression
Cellular subpopulation
  epithelia and mesenchymal cells, lineage stages of, 206
  maturational stages of, 140
Cerebellar granule cell phenotype, in v-myc transfected mouse, 228–229
CFU-osteoblast (O) assays, for osteogenic capacity, 286
Chelation buffer, preparation of, 207
Chondrogenesis assays for MSCs differentiation, 126
Chondrogenic differentiation, of hMSCs, 272–273, 275–276
Chondroitin polymerizing factor, 357

CHSE-214, cell lines, 344
c-kit^{pos} cells, 90
c-kit positive population, 86–87
CNS neural cells, markers for immunochemical characterization, 236
Collagenase enzyme, 150
   buffer, preparation of, 208
   and dispase, preparation of, 253
Colon cancer stem cells, 312–314
   identification of, 314–315
   isolation and propagation of
      culture of, 316–319
      FACS technology, 320–322
      MACS technology, 319–320
      tissue digestion, 315–316
Colon, hematoxylin-eosin analysis, 312–313
Colony forming unit-fibroblast (CFU-F), 282
   mesenchymal progenitors evaluation, 288
Committed progenitors, culture conditions for, 143
Conditioned medium (CM), 16
   in kidney epithelial cultures, 253
Connective tissue growth factor (CTGF), 356
Culture media
   requirements of
      basal media, 176–177
      hormones and growth factors, 178
      serum and lipids, 177
Cypriniformes fish, 344

D

Digestion buffer, 210
Dimethyl ether (DME), 10
Donor-derived spermatogenesis, in infertile recipient mouse testes, 61
Drosophila, Msi-1 in, 312
Dulbecco's Modified Eagle's Medium (DMEM), 251, 275

E

EGF-Dependent mouse SFME, 228
Elasmobranchs
   adipocytes in, 357
   antibiotics and angiogenesis inhibitors, sources of, 360
   cultures, derived from, 347
   lymphocytes and immune organs, 358
Elastase enzyme for liver digestion, 150

Electron microscopy, sample preparation for, 53–54
Embryoid bodies (EBs), 31; *See also* Membrane-based cell culture
   formation of, 10
   hanging drop method, formation of, 48–49
   immunochemical analysis of cardiac marker expression in, 49–50
Embryonic germ (EG) cells, 66
Embryonic stem cells (ESCs), 32, 66, 86, 142, 297
   expansion and passage of, 32
   membrane-based differentiation of, 46
      membrane-based cardiomyocyte, 46–50
   membrane-based expansion, of pluripotent colonies, 38
   multiwell expansion of murine, 39–42
EpCAM cells, 156
   antibody of, 76
   enrichment of, adult human liver, 198
   from mature parenchymal cells, 199
Epidermal growth factor (EGF), 316
Epithelial cells
   co-culturing of, 141
   relations with mesenchymal cells, 139–140
Epithelial-mesenchymal relationship, 139–141
Escherichia coli LacZ gene, 61
Expressed sequence tags (ESTs), 356
Extracellular calcium, role of, 244
Extracellular matrix (ECM), 31, 34
   cell culture protocols for, 175
   chemistry and physical features of, 174
   components, 174, 176
   within spheroids, 186

F

FACS tubes, 110–111
Feedback loop, stem cells, 182, 201
Feeder cells
   hUVECs of, 173
   paracrine signals, 173–174
   parenchyma and, 170
   preparation from
      fetal livers, 170–171
      STO cells, 172–173
Feeder preparation
   mitotically inactivating, 3–4
   observe clean technique, 2–3
   observe sterile technique, 3
Fetal bovine serum (FBS), 68, 265, 316
Fetal kidney epithelial progenitor cell, cultures development of, 243–245

Index

historical perspectives of, 242–243
materials in, 250–253
methods for
cryopreservation and reculturing, 249–250
isolation and primary culture, 245–247
subculturing, 247–249
Fetal liver cells suspensions, fractionation of, 167–168
Fetal MSCs
life of, 102
and UCMSCs, 103
Fetal Schwann cell cultures, 228
Fibroblasts, 65, 67, 74, 78
Fibronectin, 229
Fish as models for stem cell studies, 344
biomedicine and, 360
embryonal stem cells, conditional transgenic or knockout models, 348
transgene integration, expression, and homologous recombination, 351–353
in vitro methods, 349–350
fish cell and tissue culture systems
concepts and approaches, to derivation of, 345
teleost and chondricthyian cells *in vitro*, 345–348
tissue-specific stem cells, 353
applications, 359–361
stem cell lines and cell culture development, 353–359
Flow cytometry analysis
HUCPVCs for, 128
human UCMSCs for, 110–111, 115–116
Oct-4, SSEA-4, and Tra-1-60 in hES-3 cells, 23–24
Fluorescence-activated cell sorting (FACS) technology, 67, 320–322
Freshwater fish, 344
Fructose-based culture medium, *See* I/3F medium
Fugu rubripes, 358

G

β-Galactosidase, 61
Germ cell nuclear antigen 1 (GCNA1), 74
Germ cells, 60
GF, marine fish cell line, 344
Glial cell line derived neurotrophic factor (GDNF), 62, 96
Glial precursor cell line, 228
Green fluorescence protein (GFP), 61, 90

transduction with lentivirus codifying for, 94–95
Growth factors and substrate, in kidney epithelial cultures, 251–252

H

Haemulon sciurus, 344
Hank's balanced salt solution (HBSS), 67
Hematopoetic progenitor cells, 87
Hematopoietic stem cells (HSCs), 280, 291
Hepatic progenitors
human, 168
antigenic profiles of pluripotent, 197
rodent, 170–171
Hepatic stem cells (hHpSCs)
antigenic profile, 142
coculture with natural partners, 141
markers for, 146–148
monolayer culture, in differentiated state, 184
monolayer culture, in growth state
isolation of, 181–182
passaging of, 182–183
Hepatoblasts (HBs), 143
Hepatocyte Medium (HM), 188
Hepatocytes, *See* Liver cells
hESC-derived mesenchymal stem cell (MSC) line, 17
Hormonally defined medium
serum-free
epithelial cells, influence on, 178
heparins, 179–180
hormones and growth factors for, 177–178
tissue-specific gene expression, 178–179
Human amniotic fluid stem cells (hAFSC), 87
Human bone marrow-derived cells (hBMDC), 281
Human embryonic stem cells (hESC)
characterization of, 19–21
flow cytometry analysis, 19–20
immunocytochemistry, 20
karyotyping, 21
in vivo SCID mouse models, 20
cultures, 16–17
HuES9.E1 CM for feeder-free cultures, 19
HuES9.E1 for feeder cocultures, plating of, 18
inactivated feeders, preparation of, 18
Matrigel plates for feeder-free cultures, 19
passaging of hESC, 17–18
derivation of
blastocyst preparation, 4–8

Human embryonic stem cells (hESC) (*cont.*)
  characterization, 10–12
  cryopreservation, 10
  feeder preparation, 1–4
  passage and expansion, 8–9
  growth and morphology of, 21–22
  membrane-based expansion, indirect coculture, 42–46 (*see also* Membrane-based cell culture)
    gelatin coating and MEF seeding, 44
    passage and expansion of hESC colonies, 45–46
    passage of hESCs, 44–45
    thawing of cryopreserved hESCs, 43
Human hepatic stem cells, 139
  colonies from
    adult liver, 203
    human fetal liver, 202
  in hyaluronan hydrogel, 204
  vs. hepatoblasts, 200
Human liver tissue
  cell isolation from
    gall bladder, 161
    portal vein and hepatic artery, 162
  two-step perfusion methods for, 157
    blood vessels, 158–159
    buffers/reagents/supplies, 158
    cell pellet, 161
    centrifugation, Percoll suspension, 160
    collagenase buffer, 159
    plating medium, 159
Human MSCs (hMSCs)
  adipogenic differentiation of, 270–272, 274–275
  from adipose tissue, 263–265
  cryopreservation, of, 265–267
  culture in osteogenic, adipogenic, and chondrogenic media, 270–276
  growth and cryopreservation of, 265–267
  surface markers for, 267–270
  from trabecular bone, 259–262
Human UCMSCs
  arteries and vein, 105
  blood vessels, 106, 108
  cryopreservation of, 110
  culture characteristics of, 114–116
  isolation of, 103–107
  passing, 107–109
  phase contrast micrograph of, 118
  RNA yield from, 114
Human umbilical cord perivascular cells (HUCPVC)
  anatomy of, 123
  clonal analysis of, 127

cryopreservation of, 126
culture of, 125–126
differentiation of
  adipogenesis, 127
  osteogenesis and condrogenesis, 126
dissection table, 125
enzymatic digestion of, 125–126
fluorescent micrograph of, 133
isolation of, 124
mixed lymphocyte culture (MLC), 132–133
phenotypic characterization of
  cancer gene array, 129–130
  flow cytometric measurements, 128
  immunosuppression, 130, 132
  telomerase activity, 129
  transduction, 132–133
  in vitro alloreactivity, 130
  Wnt signaling pathway, 128–129
Hyaluronan hydrogels
  BrdU staining procedure, 193
  cultures in
    cell recovery solution, 192
    HpSCs and HBs, 190
  preparation of
    crosslinking, 190
    mechanical properties, 191
  sterilization of, 191

I

IBMX, *See* Isobutylmethylxanthine
Immunofluorescent staining for Tra-1-60, 25
Immunoselection
  hHpSCs of, 167
  red blood cells elimination for, 165
  rodent HB purification, 165–166
Immunosuppression, 130, 132
Inner cell mass (ICM), 1
Institutional Review Board (IRB), 2, 259
Intracellular markers, 128
In vitro alloreactivity, 130
Irradiated human foreskin fibroblasts (ATCC), 2
Isobutylmethylxanthine, 270

K

Kubota's Medium (KM), 209

L

Laminin protein, 229
Leffert's buffer, preparation of, 207

Index 373

Lentiovirus, transduction, with, 94
Leydig cells, 65–66
Lineage biology
 cellular subpopulation, 140
 ex vivo growth potential for, 181
 liver
  epithelial-mesenchymal partners in, 141
  mouse, rat and human, 145
  schematic of, 197
Live cell imaging (LCI), 304
Liver acinus
 histology of, 195
 regenerative capacity of, 204
 schematic of, 196
 zone of, 195
  periportal, 175
Liver cells
 cryopreservation of, adult, 149
 fractionation of
  antigenic properties, 165–170
  cell density, 164
  ploidy analysis, 162–164
  size analysis, 164
 isolation (*see* Liver perfusion process)
 lineage stages
  requirements for culture of, 200
  schematic of, 197
 maturational stages of
  culture conditions for, 140
 pediatric and adult
  mature parenchymal cell death, 156
  procurement, 155
 plating medium and culture
  medium for, 208
 spheroid cultures of, 185
  cell viability and quality, 188–189
  differentiated functions over, 186
  shape and size, 187
 stem cells in, 142
Liver perfusion process
 enzymes used in, 150
 fetal livers, 153
  culture, 155
  hemopoietic cells separation, 154
 in mice and rats, 150
  cell isolation, 152–153
  culture, 153
  supplies and perfusion set up for, 151
  surgery and perfusion, 152
 neonatal, pediatric, and adult
  processing of, 157–162
  sourcing of, 155–157

M

MACS Thy-1 cells on STO feeders, 74
Madin-Darby canine kidney epithelial cell line,
 culture of, 242
Magnetic cell sorting (MACS)
 technology, 319–320
Mantle cell lymphoma, 337
Markers, for hepatic stem/progenitors, 146–148
Marrow stromal cells (MSCs), 258
Matrix chemistry, *See* Extracellular matrix
 (ECM)
Maturational process, liver cells
 cellular subpopulation, 140–143
 from zygote to mature liver cells
  committed progenitors and diploid cells, 143
  ES cells and hepatoblasts, 142
  matrix chemistry, 201
  polyploid cells, 143–144
α-MEM, *See* Alpha-modified minimum essential
 medium
Membrane-based cell culture, 30–32
 ES cells with MEFs for pluripotent expansion,
  coculture of, 39–46
 general considerations of, 32–34
 membrane-Based Differentiation, of ES
  Cells, 46–50
 membranes, preparation of, 34–38
 microscopic and immunochemical
  analysis, 50–54
Membrane-based epithelial cell systems, 31
Membranes preparation, for tissue culture;
 See also Membrane-based cell culture
 coating procedures, for ECM proteins, 34
 extracellular matrix coatings
  collagen Type 1 coating, 35
  fibronectin, 35–36
  laminin coating, 36–37
  matrigel™ coating, 37
Mesenchymal feeder cells, *See* Feeder cells
Mesenchymal progenitor cells (MPCs), 87, 280
 suspension culture of
  bioreactors setting and sources, 281
  cell growth, in suspension cultures, 284
  cytokines, choice of, 281–282
  expansion in suspension, mechanism
   of, 288–291
  growth media and passaging of cells, 283
  phenotypic analysis, 284–288
  total viable cells, 283–284
Mesenchymal stem cells (MSCs), 16, 258,
 280, 297
 adipose tissue from, 263–265

Mesenchymal stem cells (MSCs), (cont.)
    bone marrow derived, 122
    clinical utility of, 122
    differentiation of, 102
        into mesenchymal phenotypes, 126–127
    immunocharacterization of, 130
    isolation from trabecular bone, 259–262
Microporous membranes, 30–32
Millicell, media volumes used for, 34
Modified mouse SFM, 70
Monolayer cultures
    in differentiated state
        hepatic stem cells, 184
        normal mature cells, 185
    in growth state
        under completely defined
            conditions, 180–181
        mature parenchymal cells, normal, 183–184
    hepatic stem cells
        isolation of, 181–182
        passaging of, 182–183
Mouse embryonic fibroblast (MEF), 16, 66
Mouse Sertoli cell line (MSC-1), 73
Mouse serum-free medium, 70
Mouse vasa homologue (MVH), 74
Multilineage differentiation; *See also*
        Mesenchymal progenitor cells (MPCs)
    adipogenesis and CFU-M assay, 288
    osteogenesis and chondrogenesis, 287–288
    suspension-grown cells, potential of, 287
Multipotent adult progenitor cells (MAPC), 297
Murine embryonic stem cells, multiwell
        expansion, 39–42; *See also* Embryonic stem
        cells (ESCs)
    ESC/MEF indirect coculture, 41
    fibronectin coating and MEF seeding, 40–41
    passage of ESCs, 41–42
Murine melanoma cell line, 327
Murine neural progenitor cell, 228
Musashi 1 (Msi-1) protein, in colon stem
        cells, 312–313
Muscle-derived stem cells (MDSC), 297
Myoendothelial cells, 297–298
    cell analysis and purification, by flow
            cytometry
        cell sorting, 300–301
        characterization, in fetal and adult skeletal
                muscle, 299
        tissue procurement, 298–299
    clonal growth, 302–303
    long-term cultured cells, genotypic and
            phenotypic analyses of, 304

    flow cytometry analysis, 305
    malignant transformation analysis, 306
    proliferation rates of, 303
    time-lapsed microscopy, use of, 304

N

National Academy of Sciences (NAS), 2
Neonatal livers
    cell isolation procedure, 161–162
    sources of, 157
NEP cell line, *See* Neuroepithelial precursor cell
        line
Neural plate
    E9 dissection of, 229–231
    primary cocultures and ESC, 232
Neural progenitor cells, 228
Neuroepithelial precursor cell line
    isolation and establishment of, 229
        characterization of NEP, 235
        differentiation of NEP, 235–237
        E9 neural plate, dissection of, 229–232
        Esc coculture and conditioned medium, 231
        long-term culture, 232–235
        morphology of, 233
        NEP cell line, serial passage and, 235
Notch signaling pathway, 313

O

OCT-4 and SSEA-4, marker for pluripotential
        capability, 90
Osteoblast markers, 357
Osteogenic differentiation, of hMSCs, 270–272,
        274

P

Paracrine signals, 141
pDsRed-Express-N1 vector, 355
Percoll® buffer, preparation of, 208
Pericytes cells, 297–298
    analysis and purification, by flow cytometry
        cell sorting, 300–301
        characterization, from fetal and adult
                tissues, 299
        tissue procurement, 298–299
    cell culture, 301–302
    clonal growth of, 302–303
    long-term cultured cells, genotypic and
            phenotypic analyses of, 304

Index

flow cytometry analysis, 305
immunocytochemistry on, 305
malignant transformation analysis, 306
proliferation rates of, 304
Peroxisome proliferator-activated receptor-gamma, 270
p-HEMA coated 6-well plates, 187
Phenotypic markers, 128
Phosphate buffered saline (PBS), 128, 251, 261
Phylogenetic footprints, 357
Platelet-derived growth factor (PDGF), 355
Ploidy analysis
 diploid cells, 162
 on nuclei, 163
 on viable cells, 163–164
Pluripotent cells, 16, 86
Pluripotent markers, characterization, 21–22
Polyethylene terephthalate (PET), 32–33
Polyploid cells
 in liver, 143
 parenchymal cells, age effects on, 143–144
Porcine pituitary extract, in kidney epithelial cultures, 252
Porous membrane-based stem cell culture, 32
Porous substrates vs. typical 2-D plastic culture, 31
POU5F1 expression, 78
PPAR γ, *See* Peroxisome proliferator-activated receptor-gamma
Primary antibodies, sources of, 212
Primary embryonic fibroblast, 2
Primordial germ cells (PGCs), 349
Progenitor cell therapies, 149
Promyelocytic leukaemia zinc finger (PLZF), 76
Proteases for liver digestion, 150

R

Recombinant human stem cell factor (rhSCF), 283
Red blood cells (RBCs), 281
Renilla luciferase, 129
Reporter luciferase activity, 129
Reverse transcriptase–polymerase chain reaction (RT–PCR), 10
RPMI 1640 cell lines, 178
RTH-149, cell lines, 344

S

Salmoniformes fish, 344
Schwann cell, 228, 232, 238
Secondary antibodies, sources of, 212
Seminiferous tubules, 63
Sertoli cells, 65–66
Serum-free culture, 66
 hormonally defined medium, 65
 mammalian cells for, 228
Serum-Free Medium (SFM), 62, 64, 66, 69–70, 327
Serum-supplemented media, 177
Severe combined immunodeficiency (SCID), 16–17, 330, 336, 338
Shaking and static methods, 186
SHF-1 fin-derived cell line, 358
Siluriformes fish, 344
Single-cell seeded clones, culture of, 127; *See also* Human umbilical cord perivascular cells (HUCPVC)
Soluble signals
 from feeders, 174–175
 in hormonally defined medium, 177–178
Soybean trypsin inhibitor, in kidney epithelial cultures, 253
Spermatogonial stem cells (SSCs), 60
 components, culture system, 64–67
  feeder cells, 66–67
  serum-free defined medium, 65–66
 culture concept, 63–64
 mouse SSC culture
  culture of Thy-1$^+$ germ cells, 73
  feeder cell, preparation, 71
  phenotypic characteristics of, 74–76
  prepare mouse SFM, 69–70
  testis cell suspension and enrichment of SSCs, 67–69
 rat SSC culture, 76–78
  characteristics, 78
  enrichment, 76–77
Spheroid cultures
 formation and morphology, factors affecting
  cell viability and quality, 188–189
  choice of, 186–188
  composition of, 188
 liver cells, 185–186
 p-HEMA coated 6-well plates, 187
 preparation protocols, 189
 shaking and static methods, 186
SSM, *See* Serum-supplemented media
Stem cell biology, 32
Stem cells, 60; *See also* Amniotic fluid
 ex vivo and *in vivo* application of, 94
  adult mouse kidney, integration into, 96–97
  mouse embryonic kidneys, integration into, 94–96

Stem cells, 60; *See also* Amniotic fluid *(cont.)*
  fish and fish cell cultures for study of, 344
  therapeutic strategies for utilization, 280
  in vitro potential of differentiation, 91
    adipogenic and myogenic, 91
    endothelial and hepatocytes, 93
    neurogenic differentiation, 93–94
    osteogenic differentiation, 92
Stem cells, embryonic
  cryopreservation, 144, 149
  feedback loop, 201
  maturational stages of, 140
  problems with determined, 149
  purification of, 140
STO cells
  subclone
    feeder layers, preparation of, 173
    feeder stocks, preparation of, 172
Stock preparation, 209–210
SV40 large T-antigen transfection, 228

T

Telomerase activity in HUCPVCs, 129
Teratoma
  CD24-negative cell, reducing contamination, 27
  formation of, 11–12, 17
  hAFSC, implanted *in vivo* environment., 98
  hESC differentiated to form, SCID mice, 16, 20–21
  usefulness clinically, 86
Three-dimensional culture systems
  hyaluronan hydrogels, 189
    cells, fixation and sectioning in, 193–194
    for culturing HpSCs and HBs, 190
    preparation of, 190–191
    sterilization of, 191
  spheroid culture
    factors affecting, 186–189
    liver cells, 185–186
Three-dimensional (3-D) growth substrate, 30
Thy-1 antibody, 76
Thy-1$^+$ germ cells, 73
Tissue progenitor cell lines
  media requirements, for tumor stem cell isolation, 333
Tissue-specific gene expression
  requirements for, 181
  in serum-free media, 178–180
TOP-FLASH luciferase plasmid, 129

Trabecular Bone, hMSCs isolation from, 259–262
Transforming Growth Factor-beta (TGFb), 356
Transforming growth factor β3, 258, 275
Transplantation assay, for SSCs, 60
TRAP assay, *See* Telomerase activity
Trypsin/EDTA, in kidney epithelial cultures, 253
Tumor cell lines, isolation and establishment, 327
  basal cell carcinoma BCCA1, isolation of, 330
  cells from xenograft, 330–331
  cell surface markers, expression of, 337
  cell surface phenotype, 331–332
  dissociation/matrix, and adherence properties, parameters for, 329
  effect of serum on, 333–335
    prostate tumor cells, primary outgrowth of, 334
  growth conditions, for CSCs phenotype, 337–338
    growth factors and animal passage, effect of, 339
    hundred cells, from RECA1208 cell line, 338
    media exhaustion in CRCA1115, 337
  immunohistochemistry, 335
  isolation and growth, media components used in, 335
  medium selection, for CSCs, 332–333
  renal capsule grafting, 335–336
  tissue treatment, 328–330
  tumor-forming capability
    cell lines tested for mice, 336
Two-dimensional (2-D) plastic substrates, 30
Tyrode's solution, 6

U

Umbilical cord matrix cells (UCMSCs)
  human
    arteries and vein, 105
    blood vessels, 106, 108
    cryopreservation of, 110
    culture characteristics of, 114–116
    isolation of, 103–107
    passing, 107–109
    RNA yield from, 114
  properties of, 103
  viability of, 115
Umbilical cord tissue, cells of perivascular region, 122

Index

Umbilical vessels
  cryopreservation of, 126
  enzymatic digestion, 124
Untranslated regions (UTRs), 357
Ureteric bud (UB), 96

W

Wharton's Jelly
  cells of, 123–124
  stromal cells from, 124
Wharton's Jelly-derived mesenchymal cells
  isolation protocols, Human
    UCMSCs, 103
    blood vessels removal, 104
    enzymatic digestion, 105–106
  enzyme solutions for, 111
  freezing, 110, 114
  growth medium, 111–112
  trypsinization reaction, 107

X

Xiphophorus species, 344, 354, 358

Z

ZEM2S zebrafish embryo-derived
  cell line, 357
Zona occludin-1 (ZO-1), 96
Zona pellucida, 6–7

VOLUMES IN SERIES

Founding Series Editor
DAVID M. PRESCOTT

Volume 1 (1964)
Methods in Cell Physiology
Edited by David M. Prescott

Volume 2 (1966)
Methods in Cell Physiology
Edited by David M. Prescott

Volume 3 (1968)
Methods in Cell Physiology
Edited by David M. Prescott

Volume 4 (1970)
Methods in Cell Physiology
Edited by David M. Prescott

Volume 5 (1972)
Methods in Cell Physiology
Edited by David M. Prescott

Volume 6 (1973)
Methods in Cell Physiology
Edited by David M. Prescott

Volume 7 (1973)
Methods in Cell Biology
Edited by David M. Prescott

Volume 8 (1974)
Methods in Cell Biology
Edited by David M. Prescott

Volume 9 (1975)
Methods in Cell Biology
Edited by David M. Prescott

Volume 10 (1975)
Methods in Cell Biology
Edited by David M. Prescott

Volume 11 (1975)
Yeast Cells
Edited by David M. Prescott

Volume 12 (1975)
Yeast Cells
Edited by David M. Prescott

Volume 13 (1976)
Methods in Cell Biology
Edited by David M. Prescott

Volume 14 (1976)
Methods in Cell Biology
Edited by David M. Prescott

Volume 15 (1977)
Methods in Cell Biology
Edited by David M. Prescott

Volume 16 (1977)
Chromatin and Chromosomal Protein Research I
Edited by Gary Stein, Janet Stein, and Lewis J. Kleinsmith

Volume 17 (1978)
Chromatin and Chromosomal Protein Research II
Edited by Gary Stein, Janet Stein, and Lewis J. Kleinsmith

Volume 18 (1978)
Chromatin and Chromosomal Protein Research III
Edited by Gary Stein, Janet Stein, and Lewis J. Kleinsmith

Volume 19 (1978)
Chromatin and Chromosomal Protein Research IV
Edited by Gary Stein, Janet Stein, and Lewis J. Kleinsmith

Volume 20 (1978)
Methods in Cell Biology
Edited by David M. Prescott

Advisory Board Chairman
KEITH R. PORTER

Volume 21A (1980)
Normal Human Tissue and Cell Culture, Part A: Respiratory, Cardiovascular, and Integumentary Systems
Edited by Curtis C. Harris, Benjamin F. Trump, and Gary D. Stoner

Volume 21B (1980)
Normal Human Tissue and Cell Culture, Part B: Endocrine, Urogenital, and Gastrointestinal Systems
Edited by Curtis C. Harris, Benjamin F. Trump, and Gray D. Stoner

Volume 22 (1981)
Three-Dimensional Ultrastructure in Biology
Edited by James N. Turner

Volume 23 (1981)
Basic Mechanisms of Cellular Secretion
Edited by Arthur R. Hand and Constance Oliver

Volume 24 (1982)
The Cytoskeleton, Part A: Cytoskeletal Proteins, Isolation and Characterization
Edited by Leslie Wilson

Volume 25 (1982)
The Cytoskeleton, Part B: Biological Systems and *In Vitro* Models
Edited by Leslie Wilson

Volume 26 (1982)
Prenatal Diagnosis: Cell Biological Approaches
Edited by Samuel A. Latt and Gretchen J. Darlington

Series Editor
LESLIE WILSON

Volume 27 (1986)
Echinoderm Gametes and Embryos
Edited by Thomas E. Schroeder

Volume 28 (1987)
***Dictyostelium discoideum*: Molecular Approaches to Cell Biology**
Edited by James A. Spudich

Volume 29 (1989)
Fluorescence Microscopy of Living Cells in Culture, Part A: Fluorescent Analogs, Labeling Cells, and Basic Microscopy
Edited by Yu-Li Wang and D. Lansing Taylor

Volume 30 (1989)
Fluorescence Microscopy of Living Cells in Culture, Part B: Quantitative Fluorescence Microscopy—Imaging and Spectroscopy
Edited by D. Lansing Taylor and Yu-Li Wang

Volume 31 (1989)
Vesicular Transport, Part A
Edited by Alan M. Tartakoff

Volume 32 (1989)
Vesicular Transport, Part B
Edited by Alan M. Tartakoff

Volume 33 (1990)
Flow Cytometry
Edited by Zbigniew Darzynkiewicz and Harry A. Crissman

Volume 34 (1991)
Vectorial Transport of Proteins into and across Membranes
Edited by Alan M. Tartakoff

Selected from Volumes 31, 32, and 34 (1991)
Laboratory Methods for Vesicular and Vectorial Transport
Edited by Alan M. Tartakoff

Volume 35 (1991)
Functional Organization of the Nucleus: A Laboratory Guide
Edited by Barbara A. Hamkalo and Sarah C. R. Elgin

Volume 36 (1991)
Xenopus laevis: Practical Uses in Cell and Molecular Biology
Edited by Brian K. Kay and H. Benjamin Peng

Series Editors
LESLIE WILSON AND PAUL MATSUDAIRA

Volume 37 (1993)
Antibodies in Cell Biology
Edited by David J. Asai

Volume 38 (1993)
Cell Biological Applications of Confocal Microscopy
Edited by Brian Matsumoto

Volume 39 (1993)
Motility Assays for Motor Proteins
Edited by Jonathan M. Scholey

Volume 40 (1994)
A Practical Guide to the Study of Calcium in Living Cells
Edited by Richard Nuccitelli

Volume 41 (1994)
Flow Cytometry, Second Edition, Part A
Edited by Zbigniew Darzynkiewicz, J. Paul Robinson, and Harry A. Crissman

Volume 42 (1994)
Flow Cytometry, Second Edition, Part B
Edited by Zbigniew Darzynkiewicz, J. Paul Robinson, and Harry A. Crissman

Volume 43 (1994)
Protein Expression in Animal Cells
Edited by Michael G. Roth

Volume 44 (1994)
***Drosophila melanogaster:* Practical Uses in Cell and Molecular Biology**
Edited by Lawrence S. B. Goldstein and Eric A. Fyrberg

Volume 45 (1994)
Microbes as Tools for Cell Biology
Edited by David G. Russell

Volume 46 (1995)
Cell Death
Edited by Lawrence M. Schwartz and Barbara A. Osborne

Volume 47 (1995)
Cilia and Flagella
Edited by William Dentler and George Witman

Volume 48 (1995)
***Caenorhabditis elegans:* Modern Biological Analysis of an Organism**
Edited by Henry F. Epstein and Diane C. Shakes

Volume 49 (1995)
Methods in Plant Cell Biology, Part A
Edited by David W. Galbraith, Hans J. Bohnert, and Don P. Bourque

Volume 50 (1995)
Methods in Plant Cell Biology, Part B
Edited by David W. Galbraith, Don P. Bourque, and Hans J. Bohnert

Volume 51 (1996)
Methods in Avian Embryology
Edited by Marianne Bronner-Fraser

Volume 52 (1997)
Methods in Muscle Biology
Edited by Charles P. Emerson, Jr. and H. Lee Sweeney

Volume 53 (1997)
Nuclear Structure and Function
Edited by Miguel Berrios

Volume 54 (1997)
Cumulative Index

Volume 55 (1997)
Laser Tweezers in Cell Biology
Edited by Michael P. Sheetz

Volume 56 (1998)
Video Microscopy
Edited by Greenfield Sluder and David E. Wolf

Volume 57 (1998)
Animal Cell Culture Methods
Edited by Jennie P. Mather and David Barnes

Volume 58 (1998)
Green Fluorescent Protein
Edited by Kevin F. Sullivan and Steve A. Kay

Volume 59 (1998)
The Zebrafish: Biology
Edited by H. William Detrich III, Monte Westerfield, and Leonard I. Zon

Volume 60 (1998)
The Zebrafish: Genetics and Genomics
Edited by H. William Detrich III, Monte Westerfield, and Leonard I. Zon

Volume 61 (1998)
Mitosis and Meiosis
Edited by Conly L. Rieder

Volume 62 (1999)
Tetrahymena thermophila
Edited by David J. Asai and James D. Forney

Volume 63 (2000)
Cytometry, Third Edition, Part A
Edited by Zbigniew Darzynkiewicz, J. Paul Robinson, and Harry Crissman

Volume 64 (2000)
Cytometry, Third Edition, Part B
Edited by Zbigniew Darzynkiewicz, J. Paul Robinson, and Harry Crissman

Volume 65 (2001)
Mitochondria
Edited by Liza A. Pon and Eric A. Schon

Volume 66 (2001)
Apoptosis
Edited by Lawrence M. Schwartz and Jonathan D. Ashwell

Volume 67 (2001)
Centrosomes and Spindle Pole Bodies
Edited by Robert E. Palazzo and Trisha N. Davis

Volume 68 (2002)
Atomic Force Microscopy in Cell Biology
Edited by Bhanu P. Jena and J. K. Heinrich Hörber

Volume 69 (2002)
Methods in Cell–Matrix Adhesion
Edited by Josephine C. Adams

Volume 70 (2002)
Cell Biological Applications of Confocal Microscopy
Edited by Brian Matsumoto

Volume 71 (2003)
Neurons: Methods and Applications for Cell Biologist
Edited by Peter J. Hollenbeck and James R. Bamburg

Volume 72 (2003)
Digital Microscopy: A Second Edition of Video Microscopy
Edited by Greenfield Sluder and David E. Wolf

Volume 73 (2003)
Cumulative Index

Volume 74 (2004)
Development of Sea Urchins, Ascidians, and Other Invertebrate Deuterostomes: Experimental Approaches
Edited by Charles A. Ettensohn, Gary M. Wessel, and Gregory A. Wray

Volume 75 (2004)
Cytometry, 4th Edition: New Developments
Edited by Zbigniew Darzynkiewicz, Mario Roederer, and Hans Tanke

Volume 76 (2004)
The Zebrafish: Cellular and Developmental Biology
Edited by H. William Detrich, III, Monte Westerfield, and Leonard I. Zon

Volume 77 (2004)
The Zebrafish: Genetics, Genomics, and Informatics
Edited by William H. Detrich, III, Monte Westerfield, and Leonard I. Zon

Volume 78 (2004)
Intermediate Filament Cytoskeleton
Edited by M. Bishr Omary and Pierre A. Coulombe

Volume 79 (2007)
Cellular Electron Microscopy
Edited by J. Richard McIntosh

Volume 80 (2007)
Mitochondria, 2nd Edition
Edited by Liza A. Pon and Eric A. Schon

Volume 81 (2007)
Digital Microscopy, 3rd Edition
Edited by Greenfield Sluder and David E. Wolf

Volume 82 (2007)
Laser Manipulation of Cells and Tissues
Edited by Michael W. Berns and Karl Otto Greulich

Volume 83 (2007)
Cell Mechanics
Edited by Yu-Li Wang and Dennis E. Discher

Volume 84 (2007)
Biophysical Tools for Biologists, Volume One: *In Vitro* Techniques
Edited by John J. Correia and H. William Detrich, III

Volume 85 (2008)
Fluorescent Proteins
Edited by Kevin F. Sullivan

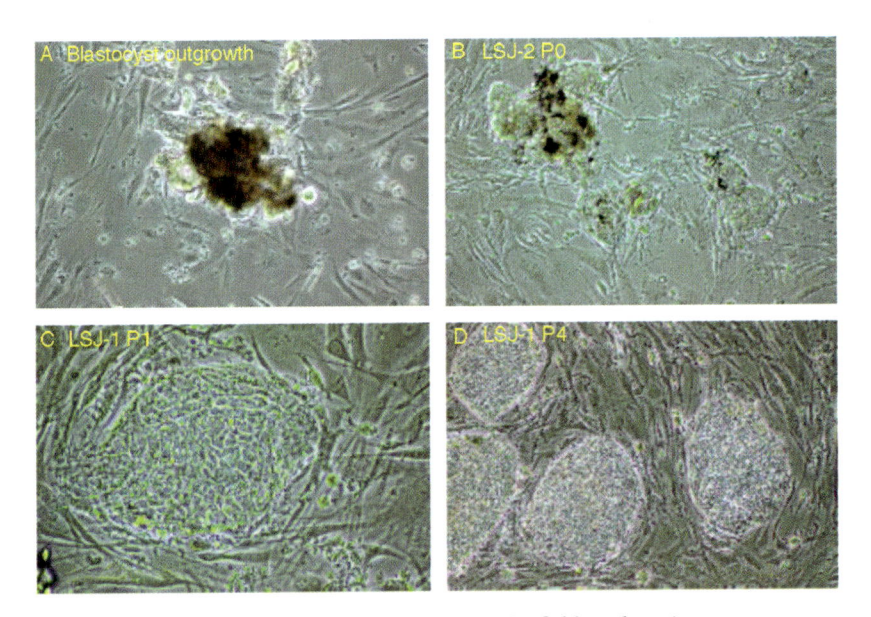

Plate 1 (Figure 1.1 on page 8 of this volume)

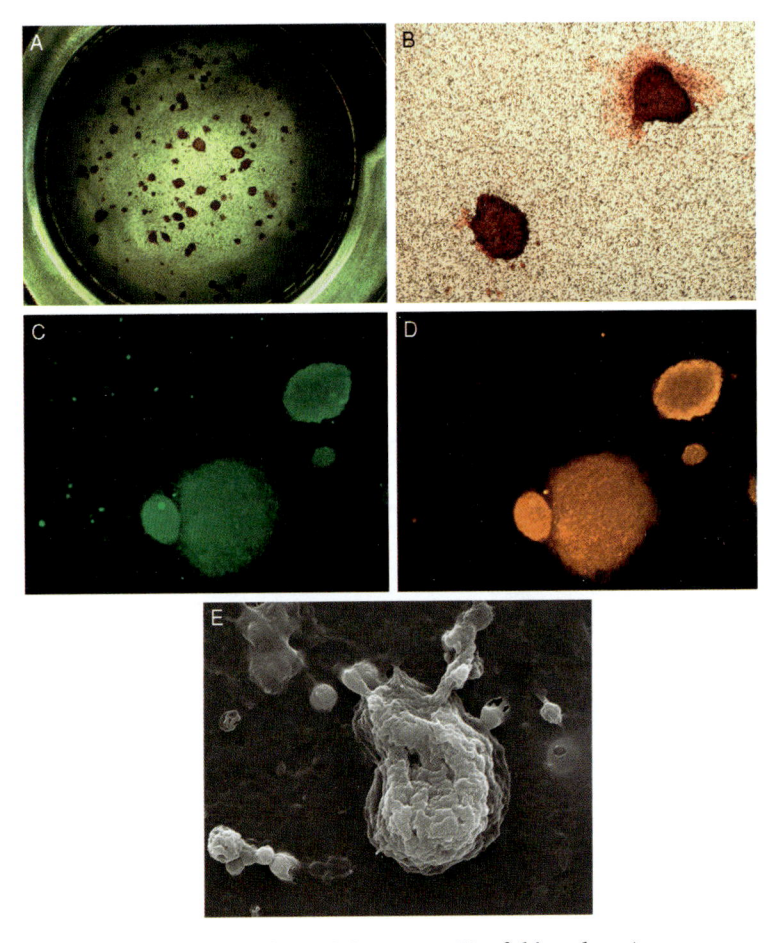

Plate 2 (Figure 3.2 on page 38 of this volume)

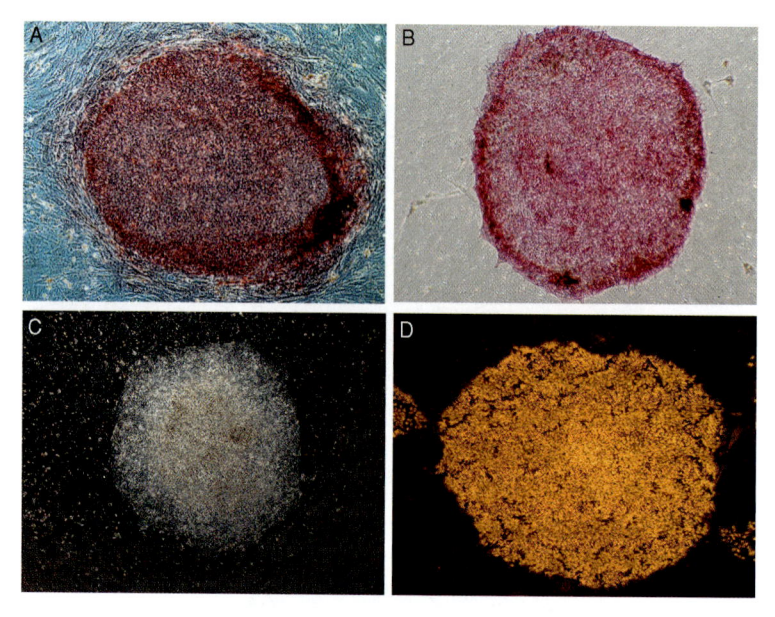

Plate 3 (Figure 3.3 on page 43 of this volume)

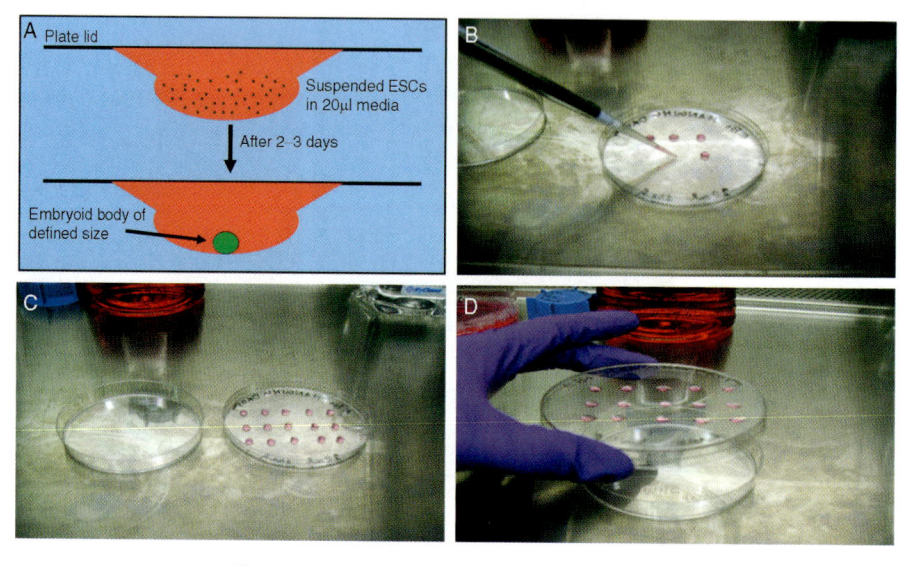

Plate 4 (Figure 3.4 on page 47 of this volume)

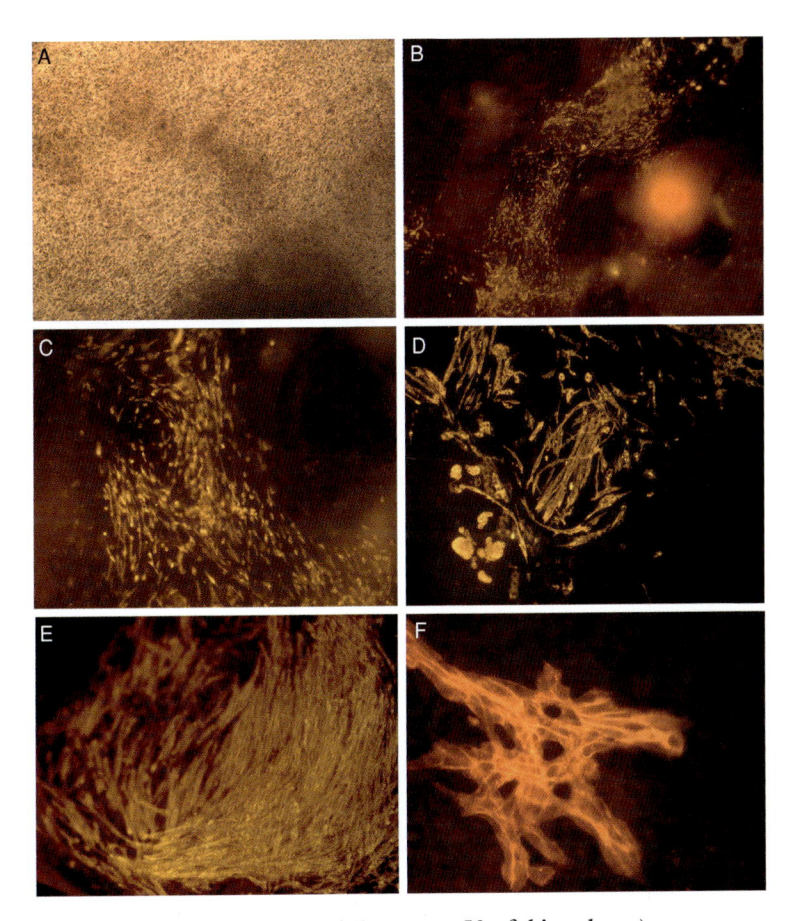

Plate 5 (Figure 3.5 on page 50 of this volume)

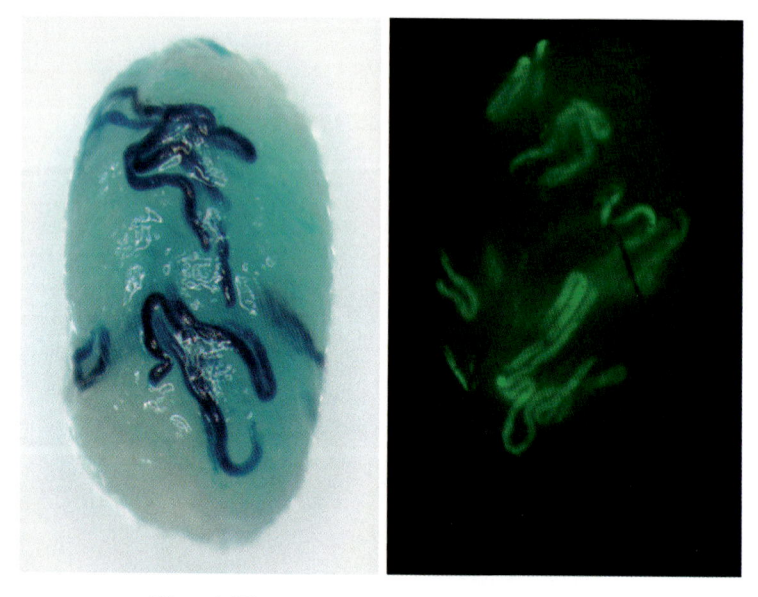

Plate 6 (Figure 4.1 on page 61 of this volume)

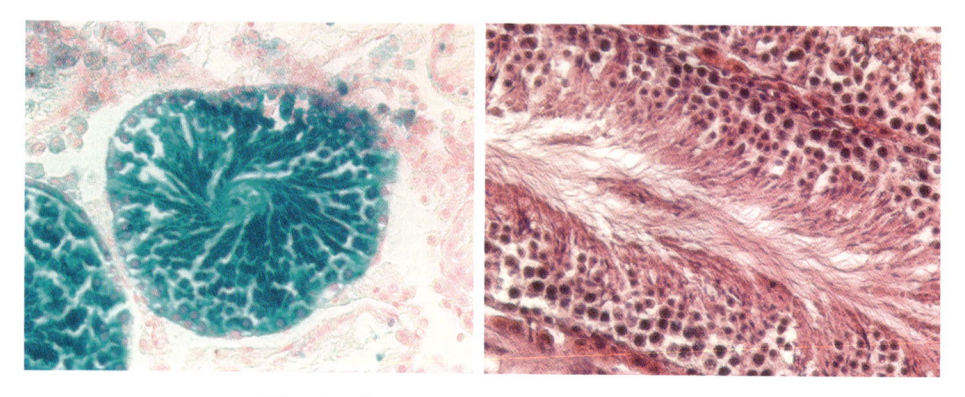

Plate 7 (Figure 4.2 on page 63 of this volume)

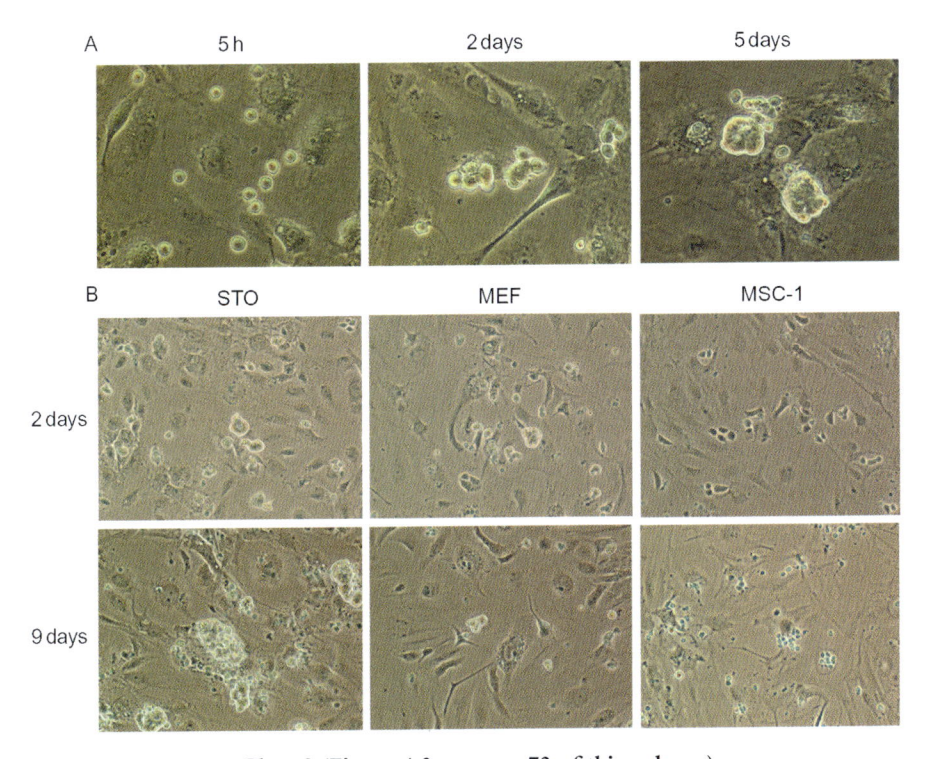

Plate 8 (Figure 4.3 on page 72 of this volume)

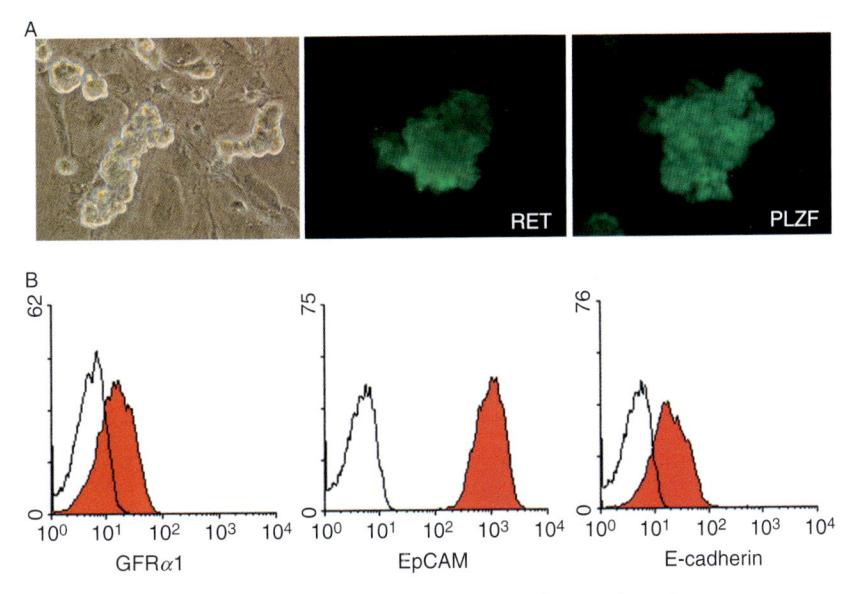

Plate 9 (Figure 4.4 on page 75 of this volume)

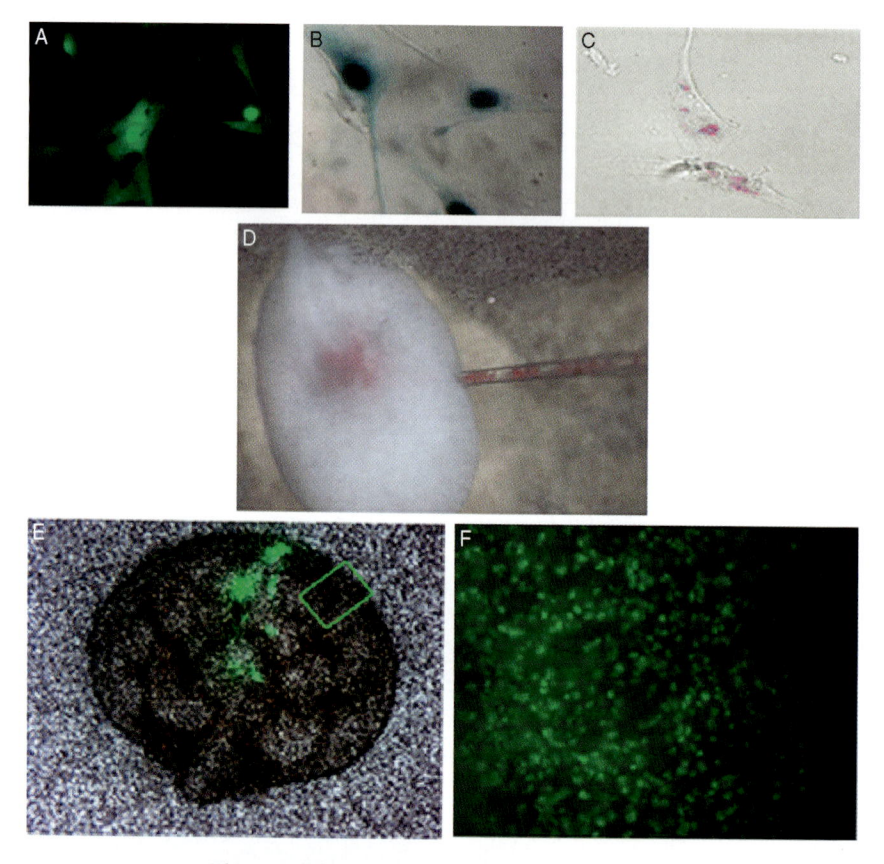

Plate 10 (Figure 5.5 on page 95 of this volume)

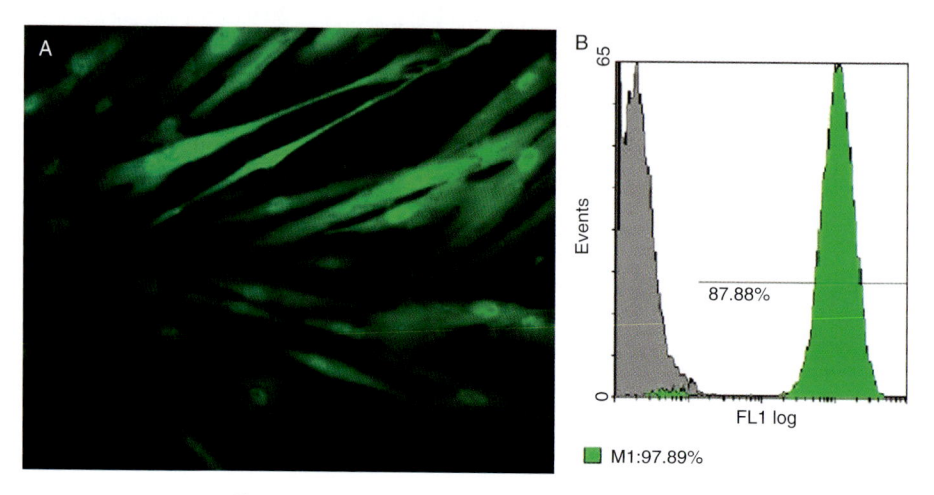

Plate 11 (Figure 7.6 on page 133 of this volume)

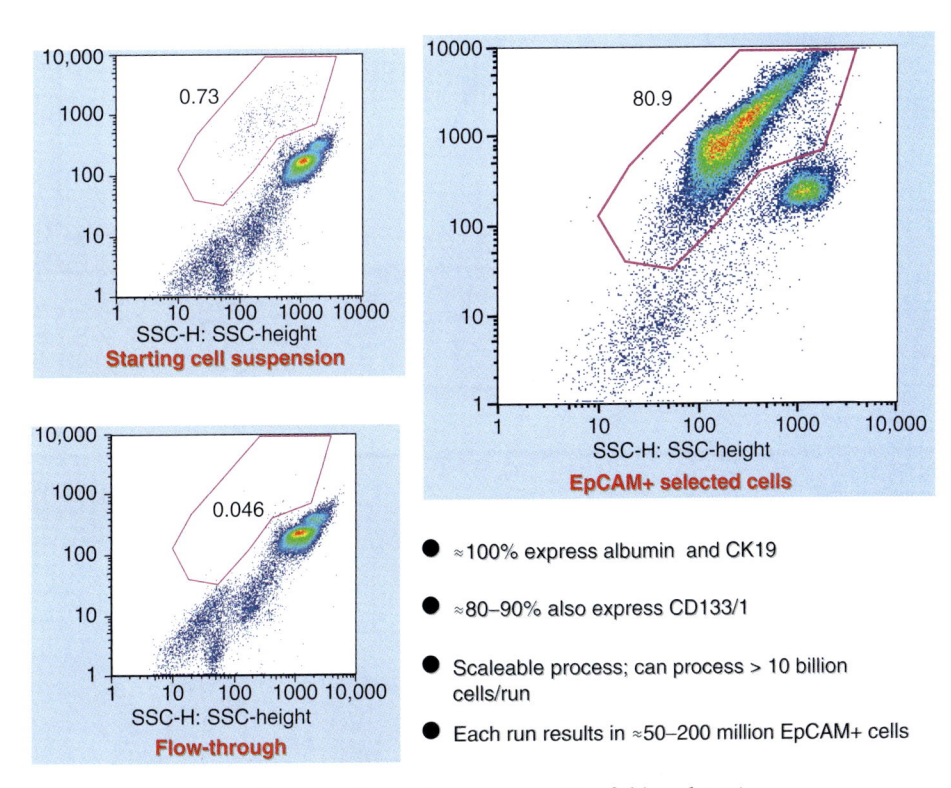

Plate 12 (Figure 8.5 on page 198 of this volume)

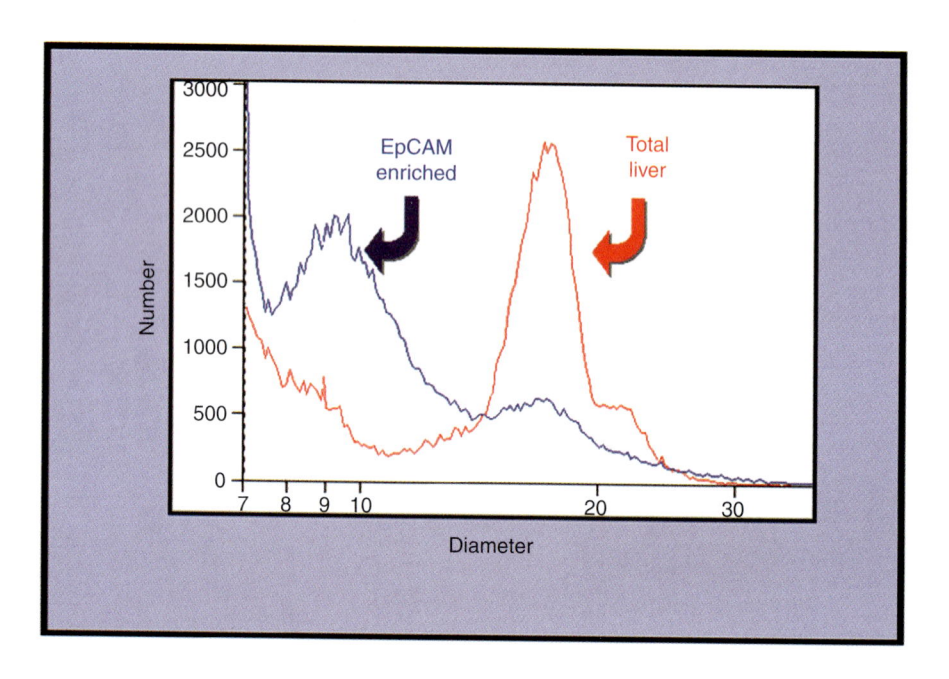

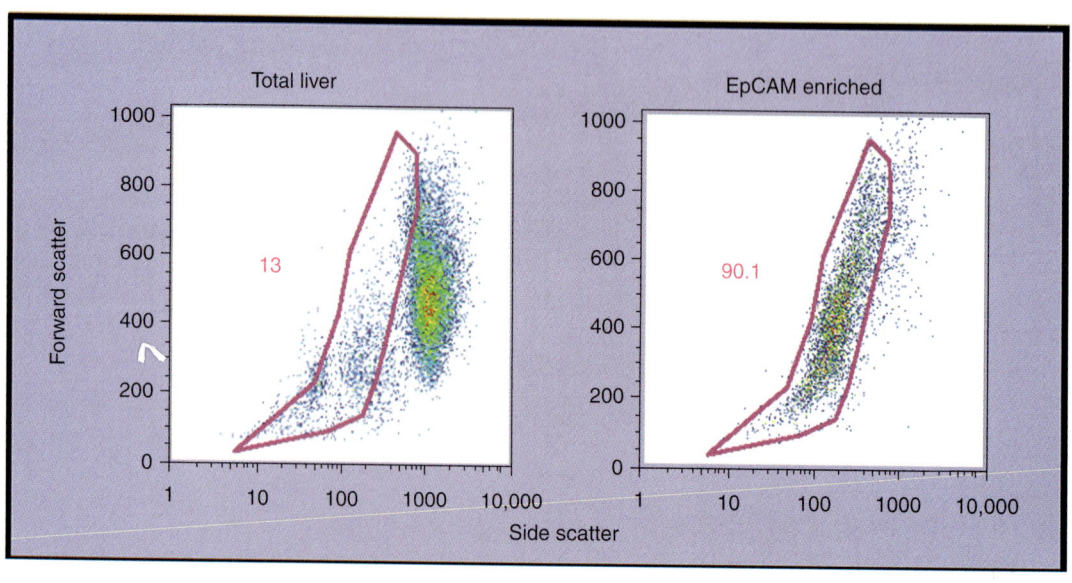

Plate 13 (Figure 8.6 on page 199 of this volume)

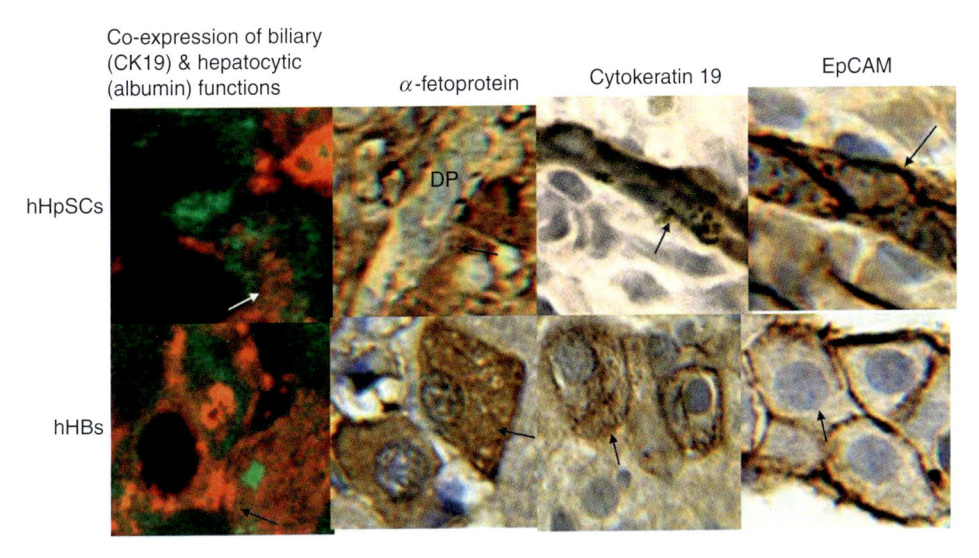

Plate 14 (Figure 8.7 on page 200 of this volume)

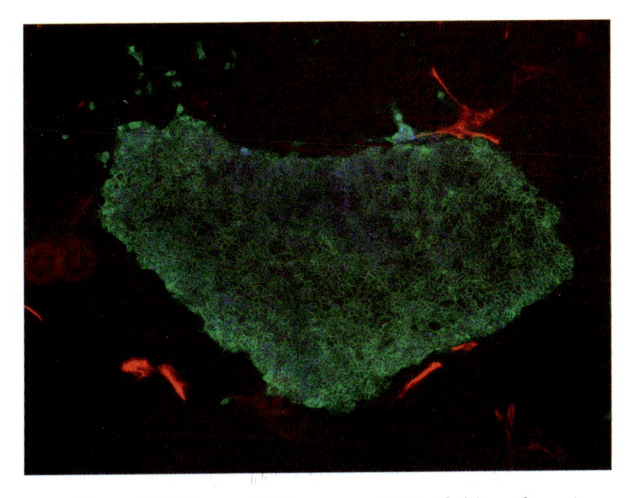

Plate 15 (Figure 8.11 on page 202 of this volume)

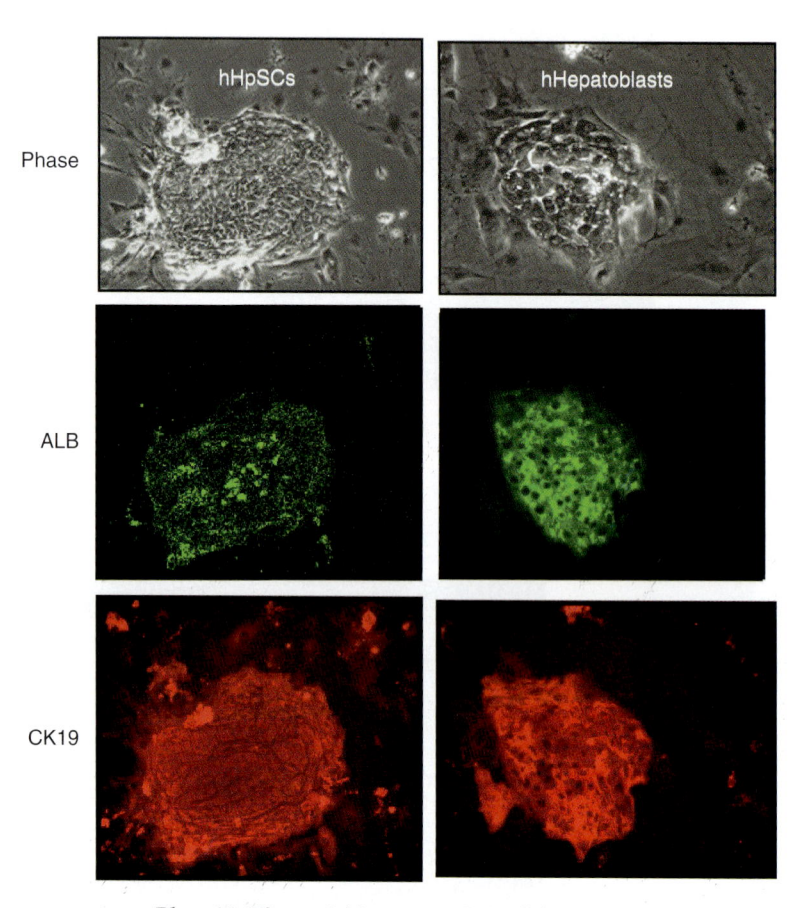

Plate 16 (Figure 8.12 on page 203 of this volume)

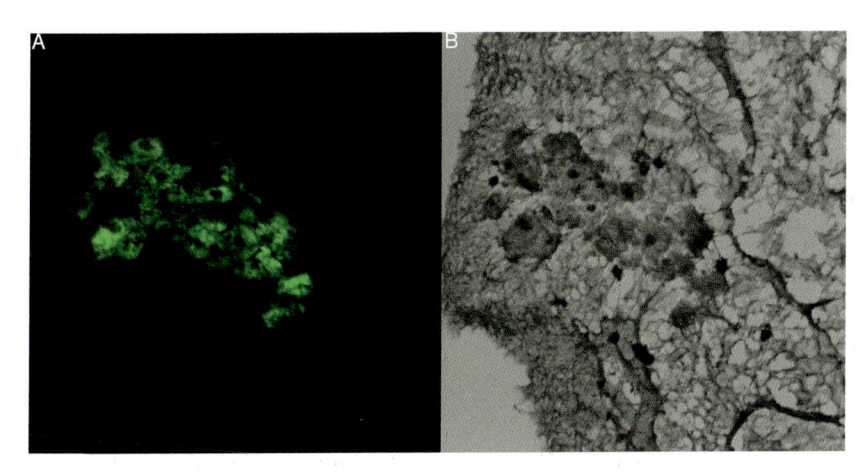

Plate 17 (Figure 8.13 on page 204 of this volume)

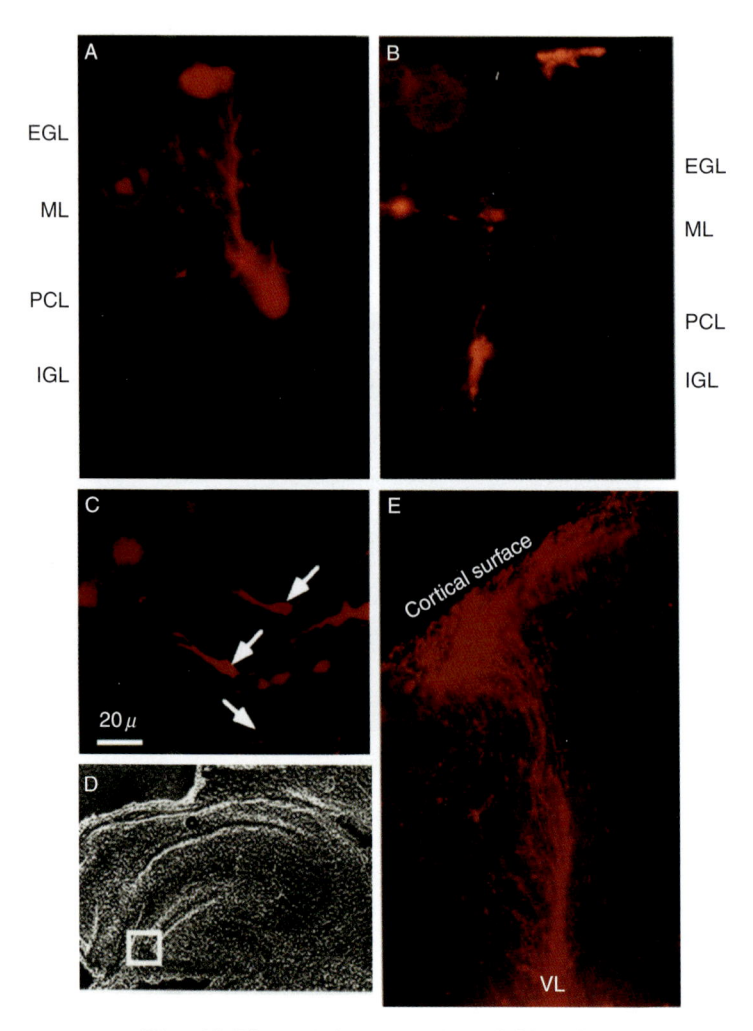

Plate 18 (Figure 9.4 on page 234 of this volume)

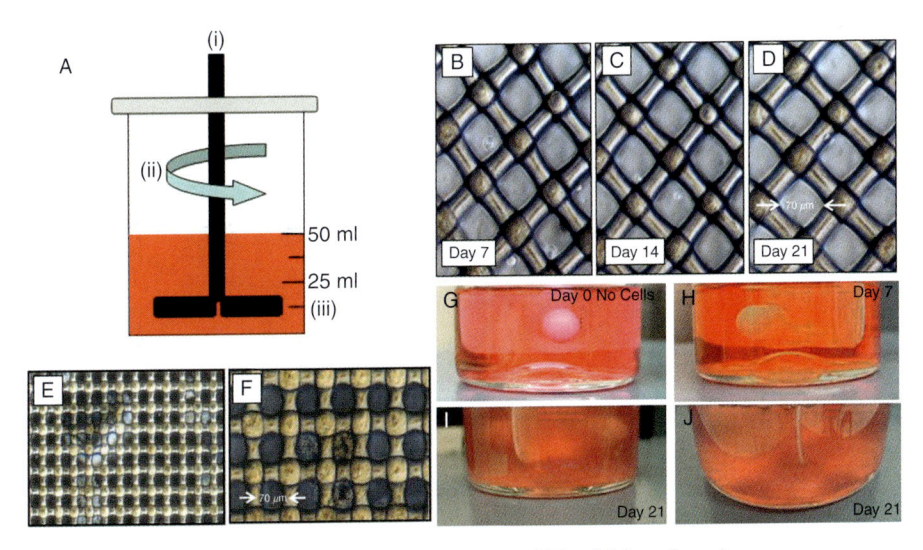

Plate 19 (Figure 12.1 on page 282 of this volume)

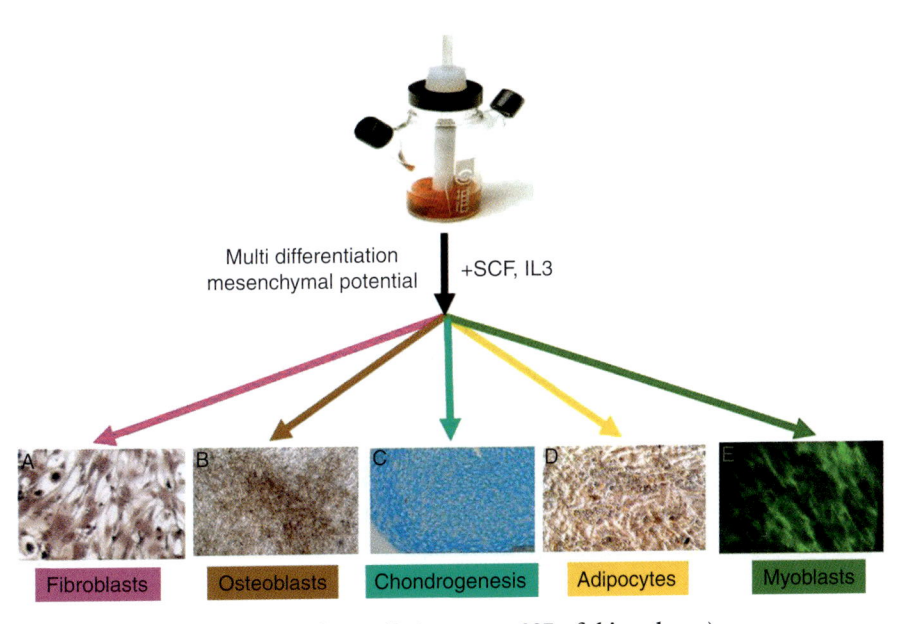

Plate 20 (Figure 12.4 on page 287 of this volume)

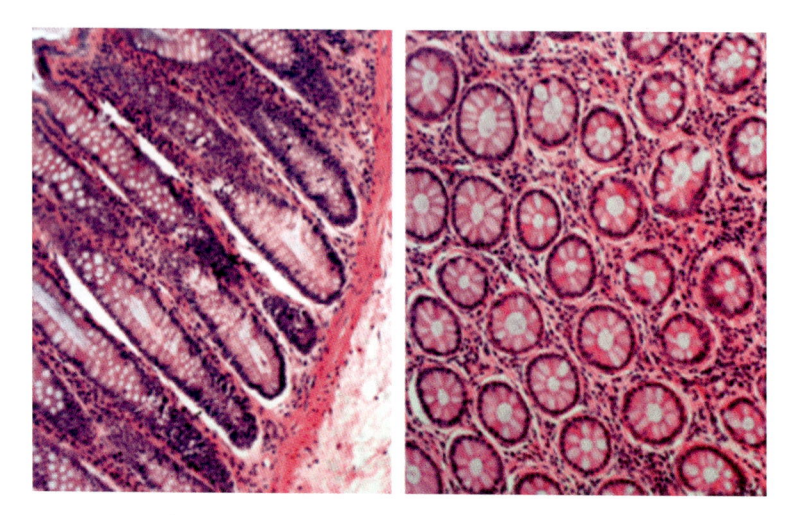

Plate 21 (Figure 14.1 on page 313 of this volume)

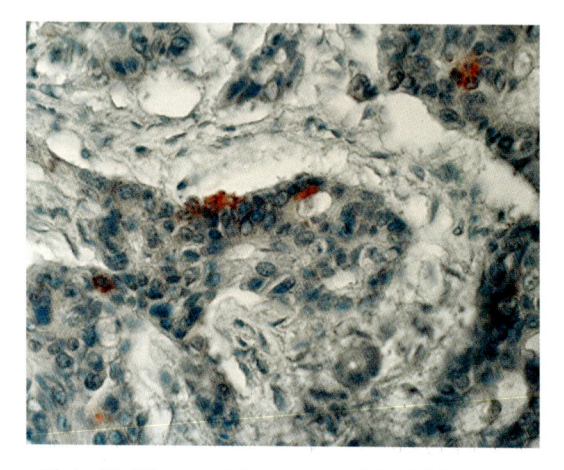

Plate 22 (Figure 14.2 on page 314 of this volume)

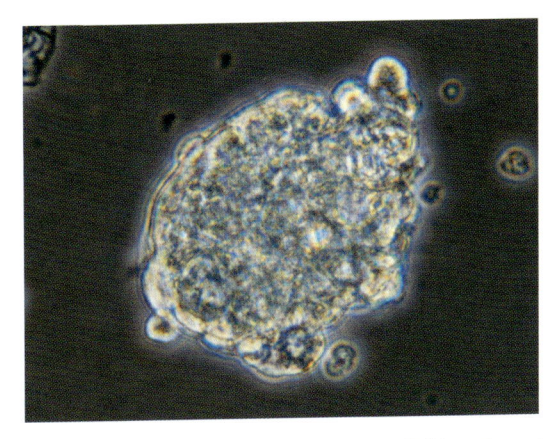

Plate 23 (Figure 14.3 on page 316 of this volume)

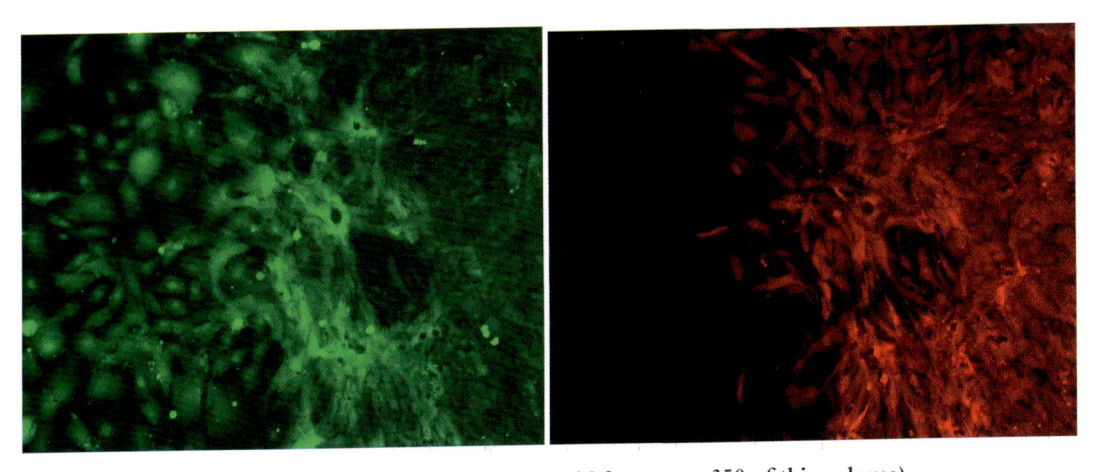

Plate 24 (Figure 16.3 on page 350 of this volume)

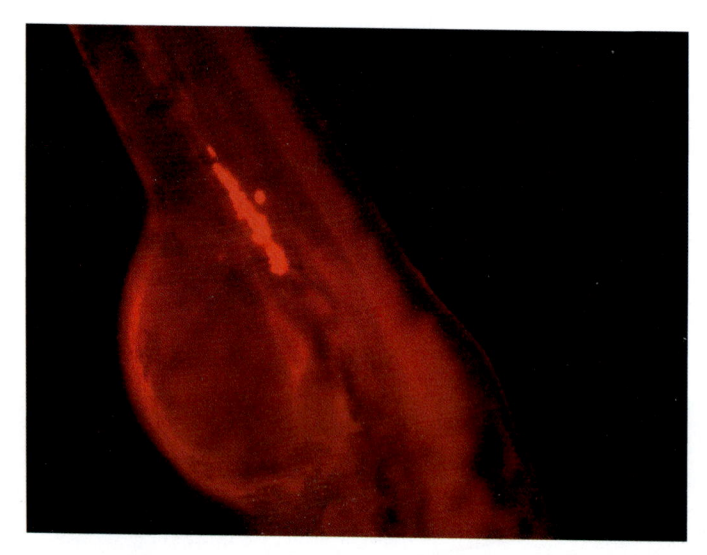

Plate 25 (Figure 16.4 on page 352 of this volume)

Natural History Museum

My Nature Collection

IN THE RAINFOREST

CAMERON MENZIES and MARC PATTENDEN

WAYLAND

First published in Great Britain in 2025 by Wayland
Copyright © Hodder and Stoughton, 2025
All rights reserved

Commissioning editor: Grace Glendinning
Editorial consultants: Sarah Ridley and Miranda Moore
Designer: Lisa Peacock

HB ISBN: 978 1 5263 2438 2
PB ISBN: 978 1 5263 2441 2
EB ISBN: 978 1 5263 2760 4

Printed and bound in China

Wayland, an imprint of
Hachette Children's Group
Part of Hodder and Stoughton
Carmelite House
50 Victoria Embankment
London EC4Y 0DZ
An Hachette UK Company

www.hachette.co.uk
www.hachettechildrens.co.uk

MIX
Paper | Supporting responsible forestry
FSC® C104740

The website addresses (URLs) included in this book were valid at the time of going to press. However, it is possible that contents or addresses may have changed since the publication of this book. No responsibility for any such changes can be accepted by either the author or the Publisher.

CONTENTS

LAYERS OF LIFE IN THE RAINFOREST	4
WAY UP HIGH	6
FILLING THE CANOPY	8
THE UNIQUE UNDERSTOREY	10
FILLING THE FOREST FLOOR	12
GARDENING WITH LEAF-CUTTER ANTS	14
ON THE RAINFOREST RIVERBANK	16
TEENY-TINY POOLS OF LIFE	18
THE HEART OF THE CONGO	20
SPECIAL ISLAND RAINFOREST	22
NIGHT IN THE AMAZON	24
IN AND AROUND AN ORANG-UTAN NEST	26
ON AND UNDER A COCOA TREE	28
ANSWERS	30
GLOSSARY	32

LAYERS OF LIFE IN THE RAINFOREST

Earth's tropical rainforests grow where it's hot and humid all year round. Combined with lots of rain, this means lots of life – from the tip-tops of the trees to the underground layers of the rainforest.

Rain

There are four layers to every tropical rainforest.

Emergent layer

Rain clouds

Canopy

Understorey

Brazil nut tree

Woolly monkey

Hoatzin

Shrubs

Liana vine

Forest floor

CAN YOU FIND THE TRICKIEST TREASURES?

This book contains scenes paired with featured treasures to spot. Some of these will be trickier to find – a real treasure hunt!

For example, on this page, the *fruit bat* is hard to see, as it hangs off a branch, asleep while most of the rainforest is buzzing with life. Can you spot it?

Two camouflaged frogs are hiding in the scene below, too. Where can they be?

Check your answers on page 30.

Fruit bat

Tree frog

Leaf frog

Royal Flycatcher

Scarce bamboo page butterfly

Green anaconda

Rainforests cover less than five per cent of our planet, but they hold **half** of all the plant and animal species on Earth. It's the perfect place to build a nature collection ...

WAY UP HIGH

The tallest trees in the rainforest tower above the rest as the emergent layer, touching the clouds. Here, birds soar above the treetops, and the sun and rain beat down.

Brazil nut tree

These trees can live for hundreds of years!

Harpy eagle

Harpy eagle chick

This eagle is a top predator of the top layer of the rainforest, hunting monkeys and sloths from high above.

Scarlet macaw

Kapok tree

Brazil nut pods

Super-hard seed protectors for a super-long fall to the ground.

Kapok blossom

Fruit bat

Some fruit bats eat the nectar of the kapok blossom and help to pollinate the tree as they gather their food.

Capuchin monkey

These monkeys tend to travel in family groups and almost never descend to the forest floor.

Water vapour

Parakeet

The trees of the emergent layer create clouds. They soak up so much rain that they have to release some of it as water vapour, which collects as clouds hovering above the rainforest.

The two brightly coloured bird species in this scene, the macaw and the parakeet, are part of the same order of birds. Do you know what this group is called? Check page 30 for the answer!

FILLING THE CANOPY

We've moved down one layer to the canopy. This canopy is in South America. It is so dense with leaves and branches that it forms a roof over the rest of the rainforest below and is teeming with life.

Sleeping sloth

Three-toed sloths' usually take short naps throughout the day *and* the night.

Fruiting fig tree

Canopy beetle

Flowering matayba tree

Stingless bee

Canopy beetles, like stingless bees, are expert pollinators of the Amazon rainforest. But some beetle species stick strictly to pollinating small white blossoms.

Many of these monkey species travel around the canopy all year to feed, as rainforest trees produce fruit at different times.

Emerald tree boa

An expert climber!

Howler monkey

Squirrel monkey

Emperor tamarin

Spider monkey

A spider monkey's strong tail and long limbs can grab on to branches as it swings through the canopy.

Toucan

Mealy Amazon parrot

Green iguana

It's mainly a herbivore but does sometimes eats bugs if necessary.

Macaw nest

Many rainforest birds build their nests and raise their young in the protected and food-filled canopy.

Screech owl

There are two types of reptile in this scene. Do you know which animals they are? Turn to page 30 to check your answer!

9

THE UNIQUE UNDERSTOREY

We're creeping down into the warm, damp and dimly lit understorey. Much less sunlight reaches here, as it's shaded by the canopy's leaves and branches.

Tree frog calling for a mate

Jaguars can climb trees. Many hide in the understorey and wait for prey to pass beneath before they pounce.

Jaguar

Heliconia plant

Large leaves

Acai palm

Orchids

Orchids here have adapted to grow on tree trunks instead of in the soil.

The bigger the leaf, the more chance of it catching some sunlight in these shady conditions.

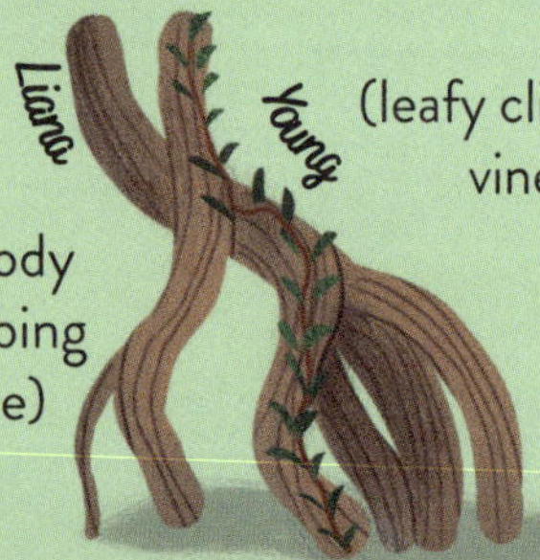

Hummingbird

Fern

Liana (woody climbing vine)

Yaung (leafy climbing vine)

Gecko

Shrubs and small trees provide excellent homes for small creatures.

Passionflower vine

Hercules beetle

Blue morpho caterpillar

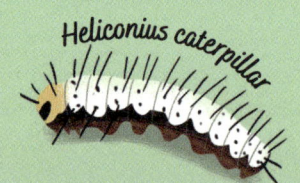

Heliconius caterpillar

10

Heliconia plants thrive in the conditions of the understorey, and rainforest hummingbirds love the nectar they make. As they drink, they pollinate the heliconia.

Can you guess which butterfly in the scene above is the adult version of each caterpillar on page 10? Check page 30 for the answers!

11

FILLING THE FOREST FLOOR

Rotting leaves

Fallen leaves carpet the floor and are decomposed by minibeasts, bacteria, fungi and slime moulds. This process puts nutrients into the soil and so feeds the plants of the rainforest.

It's dark and wet on the rainforest floor. But these are perfect conditions for many living things to forage for food, hide from predators or do the work of decomposition.

Millipede

Dung beetle

Both creatures are super-helpful decomposers!

Buttress roots

Wide, strong roots to support extra-tall trees.

Tapirs defend themselves from predators with the tough, thick hide on their backs and a strong, snappy bite.

Agouti

Paca

Tapir

Agoutis have strong teeth to gnaw open tough brazil nuts!

Rotting fallen fruit provides energy and nutrients for hungry pacas and butterflies.

Fungi fruiting bodies

Slime mould

Hercules beetle larva (eats rotting wood)

Blue morpho butterfly

Bird-eating spider (eats insects, frogs, lizards, snakes, mice, bats and birds!)

Can you spy the legless creature on the forest floor? Can you guess what it is? Turn to page 30 for the answer.

The fruiting bodies of various fungi are seen on the surface, but most of a fungus is hidden underground in a huge network of thin threads.

13

GARDENING WITH LEAF-CUTTER ANTS

Living together in a colony works well for leaf-cutter ants! They create an astonishingly complex habitat, keeping the rainforest healthy and full of life.

Ant cutting a leaf

Guard ant

Ant carrying

Ant cutting a fallen fig

These bigger soldier ants attack at the first sign of a threat!

Queen ant

Gardener ant

These smaller ants chew up the leaves and feed the pulp to the fungus.

There's only one queen for the whole colony, and she might lay millions of eggs in her lifetime.

Two-way system

Both directions follow a scent trail to and from a food source.

Fungus garden

The fungus breaks down bits of the leaves that the ants can't digest. Then the ants eat the nutritionally rich fungus they have 'farmed'.

Rich soil

Hungry armadillo

They can eat up to 40,000 ants in one sitting!

Ant egg

Ant larva

An ant colony like this is called a *superorganism*. All the ants have a part to play in keeping the system running. The whole process keeps the surrounding soil healthy by making spaces for water and oxygen to reach plant roots, and by helping to add nutrients into the soil.

Can you guess what the two bottom chambers are filled with? Hint: leaf-cutter ants like to keep their gardens very tidy ...

Do you have an idea? Check page 30 for the answer!

Sometimes fruit or flowers are on the menu for the fungus garden – it's not just leaves.

ON THE RAINFOREST RIVERBANK

Rainforest rivers, such as the Amazon River, provide a unique habitat for many fascinating living things. The Amazon is 6,400 km long and contains the most water of any river in the world!

Acai palms

Butterfly mud-puddling

Both butterflies and capybara come to the river's edge for a drink.

Ocelot

This small cat will dive into the river to hunt fish and crabs.

Capybara

Giant otter

River turtle

Caiman

These build burrows next to the river, with an underground entrance straight into the water.

Amazon manatee

Pink river dolphin

The largest of all freshwater dolphins.

The smallest of all manatee species.

Archer fish

These spit water to hit beetles on nearby plants. The beetles fall into the water and become the archer fish's lunch!

Anaconda

Piranha

Infamous for its sharp teeth.

Acai palms thrive in wet habitats. They produce a lot of fruit on each tree, so they are a vital food source for many.

Some *fish* will even nibble away at an acai berry that lands and floats on the water's surface!

The three reptiles in this scene – the caiman, turtle and anaconda – are all great swimmers, but only one is an omnivore (eats plants **and** animals). Can you guess which? Check your answer on page 30.

17

TEENY-TINY POOLS OF LIFE

Some tough plants in the rainforest never touch the forest floor. One special species that does this is the bromeliad, which also captures water and creates entire mini habitats to support life.

Pollinating bee

Sturdy, mossy branches

The bromeliads cling to high-up branches, but don't take any nutrients from them.

Bromeliad pools

Bromeliad roots

'Tank bromeliads' can hold pools of up to 20 litres of water!

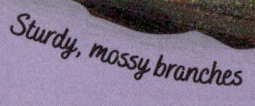

Mosquito and larva

Diving beetle — Eats mosquito larvae!

Coati

Poison dart frog and tadpole

Bromeliad crab

Some unique crabs live their whole lives in trees – born as eggs in bromeliad pools.

Snail — Snails are a source of food hunted by mother bromeliad crabs for her young.

Thirsty spider monkey

Paradise tanager

Birds of the rainforest will take a bath, a drink and have a snack of insects in the high-up pools of water.

Water this high above the forest floor is a safe haven, a nursery and a great source of hydration for living things, big and small. The moisture and flowers also attract lots of insects, which attract lots of small, hungry predators.

Can you spot a tiny, camouflaged amphibian hunting among the bromeliads? Check page 31 for the answer!

Poison dart frogs come in different bright colours to warn predators they are toxic. They lay their eggs on the ground, but raise their tadpoles in the safe, food-rich bromeliad pools.

The tadpoles hitch a ride on their mother's back for the long climb up into the trees.

THE HEART OF THE CONGO

Filling the centre of the African continent is the second biggest rainforest in the world: the Congo Basin. And it is home to some extraordinary natural treasures, including the impressive mountain gorilla.

Africa's longest venomous snake!

Black mamba

Kola nut

Mountain gorilla

Both of these huge mammals travel in family groups.

Baby forest elephant

Famous cola soft drinks are flavoured to taste like the kola nut.

Young plants

Bongo (a type of antelope)

Forest buffalo

Hornbill

Forest buffalo prefer the clearings in the Congo, where they can graze on grass and young plants.

Mahogany tree

Termites and mound

Golden-bellied mangabey monkey

Agama lizard

20

The Congo Basin is a perfect habitat for hungry monkeys. They love to eat everything from fruits to nuts, insects to leaves.

Tall, strong mahogany trees provide shade and cover for animals, plus sturdy trunks for climbing and sometimes for hosting termite mounds.

Amazingly, the termites know how to build overlapping brackets to direct all the rainwater away from the nest.

Can you find two birds in this scene that are not featured on page 20? Check your answer and learn their names on page 31.

SPECIAL ISLAND RAINFOREST

Millions of years ago, the islands of the Australasian area were connected. That's why, today, the islands' rainforests share some amazing features, while also having their own unique characteristics. The rainforests of New Guinea are particularly well known for their astounding biodiversity.

Flying fox

Buff-breasted paradise kingfisher

Agile wallaby

Tree kangaroo

Bird of paradise

Wallabies and tree kangaroos are both marsupials — they carry and feed their young in a pouch.

Long-beaked echidna

Echindas are the oldest mammal species alive on Earth today.

Cassowary

Rail bird

Both are flightless birds.

World's smallest frog

It's only as big as a housefly!

(actual size)

New Guinea ground boa

New Guinea is the most diverse island in the world in terms of plant variety.

Glossy laurel

Freycinetia

Orchids

Terminalia

22

Flying foxes only eat flowers, nectar and fruit, while the kingfishers are carnivores and only eat animals!

There are **three** species of bird of paradise shown in this scene, in male and female pairs. Can you spot them all? Check page 31 for the answer.

There are about 45 species of bird of paradise, and most varieties live only on New Guinea.

23

NIGHT IN THE AMAZON

At night in the Amazon rainforest, many groups of living things wake up to eat in the coolness and safety of the cover of darkness.

Bullet ant

Vampire bat

Pacarana (rodent)

Rat (rodent)

Mouse opossum (not a rodent – a tiny marsupial)

Nine-banded armadillo

The female armadillo always gives birth to identical quadruplets.

Kinkajou

Also called a honey bear because it steals honey from bees' nests!

Firefly

Moonflower

This cactus flowers at night with a bright white bloom for attracting pollinators.

Margay

Sloth – awake!

Velvet worm

Cockroach

The velvet worm squirts quick-hardening slime onto its prey like a super-strong net!

Potoos are nocturnal and have huge eyes for seeing at night.

Potoo bird

Many of the rainforest's vulnerable amphibians also come out at night when there are fewer predators and many more bugs to eat.

Various frogs and geckos with sticky feet climb around in the moonlight, hunting insects and earthworms.

Can you spot the frog with a special skill? Turn to page 31 for details on this gliding wonder ...

IN AND AROUND AN ORANG-UTAN NEST

The islands of Borneo and Sumatra in South East Asia have rainforests packed with unique animals, including the beautiful orang-utan.

Lychee tree

Mango tree

Mango, lychee and fig trees bear lots of fruit in the rainforest. They provide food for many, including orang-utans in the trees and tapirs on the forest floor.

Forest elephant

Sleeping baby orang-utan
Can you spot the hidden baby orang-utan? Check page 31 for the answer.

Sumatran tiger

Malayan tapir

Amazing gliders

The Draco lizard, the paradise tree snake, the colugo and some tree frogs have amazing body shapes to allow them to glide long distances. Can you spot them all? See page 31 for the answers.

Orang-utan

Proboscis monkey

Civet

Oil palm

The pitcher plant has sticky nectar on its 'lid'. Tree shrews perch on the plant, lick the nectar and poo into the pitcher, feeding the plant!

Wild boar eat the seeds of the *dipterocarp tree* and spread the seeds far and wide when they poo!

Sumatran rhino

Tree shrew

Pitcher plant

Clouded leopard

Pangolins are the only mammal covered in scales. They are mostly nocturnal, and sometimes climb trees in search of ants' nests.

Wild boar

Sunda pangolin

ON AND UNDER A COCOA TREE

Wild cocoa trees grow in the shade of other rainforest trees. They thrive where the sun is not as strong and offer a delicious and safe shelter for many.

Forcipomyia midge

Fruit and flowers
The cocoa tree can have fruit and flowers growing on it at the same time – and both grow straight out of the trunk of the tree.

Peccary (wild pig)

Midges and mosquitos
Tiny insects pollinate the flowers. Some of the flowers then grow into fruit called cocoa pods.

Sweet pulp and bitter seeds
The pods have sweet, juicy pulp inside, which many animals of the rainforest love to eat! They gnaw into or crack open the pods to get at the pulp, but often spit out the bitter seeds. This is a key way in which the seeds of new cocoa trees are spread across the forest floor.

Toucan

Cocoa seedling
Can you spot a new baby cocoa tree, grown from a seed, in the scene below? Check page 31 to see if you're right.

Vanilla bean vine

Across the world
Originally from South America, the cocoa plant was later brought to Africa, where most cocoa beans are grown today. Do you know what food we make out of cocoa beans? Check page 31 for the answer!

Rat

29

ANSWERS

Page 5: The fruit bat is hiding here:

And the tree frogs are here:

Page 7: Macaws and parakeets are both from the parrot family.

Page 9: Emerald tree boas and green iguanas are both reptiles. Check the glossary for a definition of reptile!

Page 11: The blue morpho butterfly is the adult stage of the blue morpho caterpillar:

The heliconius butterfly is the adult stage of the heliconius caterpillar:

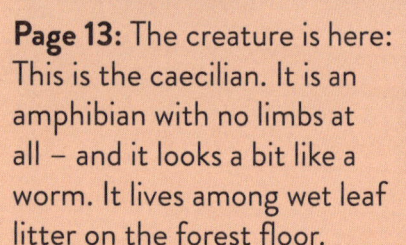

Page 13: The creature is here: This is the caecilian. It is an amphibian with no limbs at all – and it looks a bit like a worm. It lives among wet leaf litter on the forest floor.

Page 15: The bottom chambers in the leaf-cutter ant colony are for collecting rubbish! The ants remove anything dead, used-up or unsanitary from the nest and place it down here, where it can decompose away from the queen and the young.

Page 17: The river turtle is the omnivore of the three reptiles. The anaconda and the caiman are both carnivores.

Page 19: The hiding amphibian is here: This is a tree salamander!

Page 21: The two extra birds are here: They are the African grey parrot and the Congo peafowl.

The grey parrot lives a long time in the wild, usually 60 but up to 80 years! It is one of the cleverest of all animal species on Earth.

The peafowl is the national bird of the Republic of Congo. It is the only species of true pheasant native to Africa.

Page 23: The three pairs of bird of paradise are here:

Superb bird of paradise
Male – black and blue with blue around the eyes.

Six-plumed bird of paradise
Male – black with an iridescent throat.

Greater bird of paradise
Male – brown, gold and white.

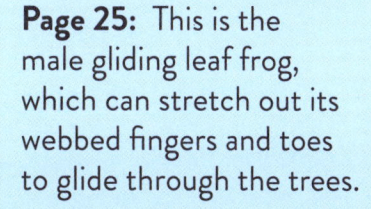

Page 25: This is the male gliding leaf frog, which can stretch out its webbed fingers and toes to glide through the trees.

Page 26: The baby orang-utan is sleeping in its nest here:

Page 27: The four gliders are here:

Page 29: The cocoa seedling is here:

We use cocoa beans to make chocolate!

31

GLOSSARY

amphibian an animal that usually lives in the water when young and later on land, such as a frog or newt

bacteria tiny single-celled living things
biodiversity the variety of plant and animal life in a habitat

camouflaged coloured or patterned to blend in with the surroundings
carnivore a meat-eating animal
colony a community of animals or plants of one kind living together

decompose to decay or rot

flightless describes birds that can't fly
freshwater found in fresh water, not the sea
fungi organisms such as mushrooms or moulds that produce spores and get nutrients from plant or animal matter

habitat the usual home of an animal, plant or other organism
haven a safe place
herbivore an animal that eats plants
humid damp air holding lots of water vapour
hydration water intake to avoid drying out

larva the young of insects

mammal a warm-blooded animal with a backbone; the mother produces milk to feed her young

nursery a safe place for young animals to grow up
nutrients substances that living things use to grow and remain healthy

omnivore an animal that eats plants and animals
order a rank used to describe types of animal, plant or other organisms. For example, reindeers share an order with other hoofed mammals such as cows, pigs, antelope and giraffes

pollinate to deposit pollen on a flower or plant so that seeds can be produced
pulp the soft, wet part of a fruit

quadruplets four babies born in the same birth

reptile an animal (such as a snake, lizard, crocodile or turtle) with dry, scaly skin and a backbone, that lays soft-shelled eggs

rodent a gnawing mammal with strong, sharp teeth, such as a rat, mouse, squirrel or porcupine

slime mould single-celled organisms that clump together and reproduce to form a slimy mass
species a kind of living thing
superorganism a community of many organisms that live together, each performing a job that benefits them all